Other Related Pergamon Titles of Interest

Books

ARIS & VARMA: The Mathematical Understanding of Chemical Engineering Systems (Selected Papers of Neal R. Amundson)

COULSON & RICHARDSON: Chemical Engineering (6 Vols all in SI units)

Volume 1:	Fluid Flow, Heat Transfer and Mass Transfer, 3rd edition
Volume 2:	Unit Operations, 3rd edition
Volume 3:	Chemical Reaction Design, Biochemical Reaction Engineering including Computational Techniques and Control, 2nd edition
Volume 4:	Solutions to the Problems in Volume 1
Volume 5:	Solutions to the Problems in Volume 2
Volume 6:*	Introduction to Chemical Engineering Design

Journals

Chemical Engineering Science
Computers & Chemical Engineering
International Journal of Heat and Mass Transfer
Letters in Heat and Mass Transfer

**In preparation*

Full details of all titles/free specimen copy of any Pergamon journal available on request from your nearest Pergamon office.

INSIGHTS INTO CHEMICAL ENGINEERING

(Selected Papers of P. V. Danckwerts)

by

P. V. DANCKWERTS

F.R.S., F.Eng., Foreign Ass. N.A.E.,

Emeritus Professor, University of Cambridge

PERGAMON PRESS

OXFORD · NEW YORK · TORONTO · SYDNEY · PARIS · FRANKFURT

U.K.	Pergamon Press Ltd., Headington Hill Hall, Oxford OX3 0BW, England
U.S.A.	Pergamon Press Inc., Maxwell House, Fairview Park, Elmsford, New York 10523, U.S.A.
CANADA	Pergamon Press Canada Ltd., Suite 104, 150 Consumers Rd., Willowdale, Ontario M2J 1P9, Canada
AUSTRALIA	Pergamon Press (Aust.) Pty. Ltd., P.O. Box 544, Potts Point, N.S.W. 2011, Australia
FRANCE	Pergamon Press SARL, 24 rue des Ecoles, 75240 Paris, Cedex 05, France
FEDERAL REPUBLIC OF GERMANY	Pergamon Press GmbH, 6242 Kronberg-Taunus, Hammerweg 6, Federal Republic of Germany

First edition 1981

British Library Cataloguing in Publication Data

Danckwerts, P. V.
Insights into chemical engineering.
1. Chemical engineering
I. Title
660.2 TP155 80-42316

ISBN 0-08-026250-3

In order to make this volume available as economically and as rapidly as possible the author's typescript has been reproduced in its original form. This method unfortunately has its typographical limitations but it is hoped that they in no way distract the reader.

Printed in Great Britain by A. Wheaton & Co. Ltd., Exeter

Foreword

It gives me great pleasure to be associated with this volume of some of the publications which Professor Danckwerts has made in chemical engineering science. His output over the past 30 years has been rich and varied and I was fortunate, as an undergraduate at Cambridge, to be one of those on whom some of his early ideas were tried out.

Many will associate him with the theory of mass transfer operations, particularly in the field of gas absorption, and this work represents a very significant contribution to chemical engineering. It is so well known and has been used so extensively in plant design that it needs no introduction from me.

Perhaps fewer will be familiar with his early work on mixing but this was an important attempt to quantify a very practical process of great industrial significance. The deceptively simple and elegant concepts of the scale and intensity of segregation to describe the degree of mixing were much debated at the time but were not taken up as quickly as they should have been. However, it is noteworthy that they are now being increasingly applied to great effect in widely ranging fields, polymer processing being a typical example, in which temperature or colour uniformity can be of great importance.

The concept of residence time distribution in continuous processes is another topic which Danckwerts pioneered and this has proved to be of great benefit in many areas, including reactor design, blending operations, and many more.

Whereas chemical engineers in industry owe much to his work, those in education have also a lot to be thankful for. Many of the concepts he developed appear to have been designed to stretch the minds of undergraduates and perhaps they were! Certainly many of the examination papers with which they inevitably became confronted contain questions which unmistakably stem from his work.

It is perhaps a little sad to reflect that the gulf between chemical engineering science as developed in the universities and chemical engineering practice is still so wide. Even the best-intentioned engineers in industry find great difficulty in using all but the very simple and approximate design procedures and they have to rely heavily on expensive and time-consuming pilot plant work or the inclusion of extravagant safety factors. We should strive to do better and there are signs that we are now succeeding for, whereas over-exposure to computers at an early age tends to dull the brain, their widespread use in industry is clearly proving to be of great benefit insofar as they allow more realistic design procedures to be used. However, a necessary pre-requisite is a thorough understanding of the basic physical and chemical concepts of the system and it is to this that Danckwerts has made a lasting contribution.

His work has been basic but, more importantly, it has been objective basic in that it has had as its main purpose the development of a better understanding of some of the chemical engineering phenomena which provide a sound basis for plant design.

I feel sure that many will enjoy reading these original papers and come to reflect that they are much easier to understand than some of the mutilated 'interpretations' of his work which appear in many textbooks.

W L WILKINSON

President, Institution of Chemical Engineers

24 November 1980

Contents

Autobiographical Note

DANCKWERTS, Prof. Peter Victor, b 14 Oct. 1916, GC 1940; MBE 1943; FRS 1969; FI ChemE; Shell Professor of Chemical Engineering, Cambridge University, 1959-77. Educ. Winchester Coll.; Balliol Coll., Oxford; Massachusetts Inst. of Technology, BA(chemistry)Oxon, 1938; SM (Chemical Engineering Practice),MIT, 1948; MA Cantab 1948; Hon.D.Sc. Bradford, 1978. RNVR, 1940-46. Commonwealth Fund Fellow, MIT 1946-48; Demonstrator and Lecturer, Dept. of Chemical Engineering, Cambridge Univ., 1948-54; Deputy Director of Research and Development, Industrial Group, UK Atomic Energy Authority, 1954-56; Professor of Chemical Engineering Science, Imperial College of Science and Technology, 1956-59. MIChemE 1955 (President, 1965-66). For. Hon. Mem., Amer. Acad. of Arts and Scis, 1964; For. Associate, Nat. Acad. of Engineering,USA. 1978. E.V. Murphree Award, ACS. 1973. Address: The Abbey House, Abbey Road, Cambridge CB5 8HQ.

Born on 14 Oct. 1916, the son of a naval officer whose family were brought up in a state of genteel poverty, I attended various dame-schools and then went to a prep-school (= U.S. primary boarding school), a singular establishment in some ways reminiscent of Evelyn Waugh's Llanabbas Towers. Between the ages of 10 and 14 I wasted many dreary hours on Latin and even Greek, not to mention the dates of the kings of England. However, among the masters there were a few to whom I owe a debt for a thorough grounding in the writing of English and a basic appreciation of the principles of mathematics. The academic standards were not high and it was regarded as little short of a miracle when I got a place at

Winchester College (founded by William of Wykeham in 1382). This was a public school (= U.S. prep-school) which was unique in valuing intellectual at least as highly as athletic attainment. In spite of its notable classical tradition, it provided an excellent scientific education. Despite the long hours devoted to the ineffectual study of Latin and Greek, I got a remarkably good grounding in physics and chemistry, but for some reason made no progress in mathematics. However, my training in English was reinforced. Churchill used to say that he was regarded as a stupid boy at Harrow and was thus allowed to learn English instead of Latin. I was not so lucky, but at least I managed to avoid tackling Homer.

There was a scholarship, perhaps more closed than any other, for a Winchester scientist to study chemistry at Balliol College, Oxford (founded about 1260 by John de Balliol and Devorgilla, his wife). There were only two candidates in my year and I was the rank outsider. Neverless I won it, to universal dismay, and became a Scholar of Balliol, a notable title. I must say that I have been somewhat cynical about honours and awards ever since. Happy to escape from school I spent a year in Austria, that charmed country, living in Vienna and Salzburg, learning a fair amount of German and conceiving a passion for the Baroque and for the mountains.

Balliol was a college with a formidable tradition of raising the rulers (both political and executive) of Britain and other countries, but I had no contacts in such circles; I spoke once to Edward Heath but never to Denis Healey. The course in chemistry had a curious structure. It lasted four years but the all-important final examinations, which determined the pecking-order, took place in the summer of the third year. The fourth was spent on a light-hearted piece of research which meant that in one's final summer one was free to punt on the river, go to balls and otherwise lead the *dolce vita*. The year was 1939.

I cannot claim that I was particularly diligent during my first two years, either, so that in my third year it was necessary to do some serious work. I acquired some excellent text-books which I read from cover to cover, I studied the predilections of the examiners and tested myself on previous examination-papers. As a result I did rather well in my examinations, a result which I believe gave my tutor the idea that something was wrong with the system. It is true that I could have learned a great deal more than I did. Chemistry was an essentially literary subject at Oxford at that time. I was taught no physics and no mathematics. It was said that if an Oxford chemist encountered a differential coefficient in a book he turned the page; if he saw an integral sign he shut the book. Perhaps the most valuable parts of my intellectual heritage from Balliol were a sense of scientific rigour combined with scepticism - the latter summed-up in the imputations of the word "bogus".

In 1939 I joined a small chemical company, doing a job which - as I now recognise - should have been done by a chemical engineer (an almost unknown species in England in those days). By the summer of 1940 I had had enough; in order to have a go at the Germans who were supposed to be about to invade the country it was clearly necessary to get into the armed forces. I was interviewed at the Admiralty on a Wednesday and on the following Sunday, in the uniform of a Sub-Lieutenant R.N.V.R., I was on my way North to train at the R.A.F. Bomb Disposal School. No one ever taught me the things a young naval officer should know, such as who and how to salute, and that the Loyal Toast was drunk sitting down. Moreover, my role throughout the war was non-combatant. Such an inconsequential chain of events is typical of war; however, my whole career has run on similar lines of chance and hazard.

I then went to the R.N. Torpedo School, H.M.S. Vernon, at Portsmouth. During a lecture by one of those rock-faced Chief Petty Officers about how to render safe enemy mines there was a sharp explosion. "Ah" said our instructor "they're trying-out one of those double depth-charge throwers". As it happened, two experts had been dismantling a recovered German magnetic mine, which had been booby-trapped. For some reason (probably bureaucracy in the Kriegsmarine) the Germans had not put the 1 kg booby-trap charge in a position where it would detonate the main 750 kg charge of explosive. Otherwise our course would have come to an abrupt end. As it was, the episode served as a salutary practical lesson.

I was posted to the Port of London in time for the Blitz, which began on 7 Sept. 1940. I became for a time the incumbent disposer of German parachute mines. These were magnetic mines and would have been infinitely more cost-effective if used as such to close British harbours to shipping. However, they were the biggest bombs the Germans had and in the pursuit of Schrecklicheit were dropped on land. As it turned out, the mechanism which was designed to detonate them if they dropped on land,rather than into water, was grossly inefficient and after each raid a number were found standing on the top floors of houses, hanging from trees and bridges, etc. There was a sporting chance of actuating the "bomb fuse" while trying to remove it, but usually it buzzed for 10-15 s before the explosion. Those were stirring times.

After many adventures abroad and a great deal of boredom I fetched up early in 1944 in Combined Operations Headquarters in Whitehall, where I was an Experimental Staff Officer. COHQ had been set up by Lord Louis Mountbatten in his usual princely style. It had assembled the prettiest girls from all three services and had a car-pool, a private cinema, the only all-ranks bar I encountered during the war and a canteen, run by society ladies, which served the best food in Whitehall. Many of the famous and the notorious passed through the HQ. I remember particularly our genius-in-residence, Geoffrey Pyke who invented Habbakuk, the giant iceberg ship, and the concept of Power-Driven Rivers; and the brilliant Marxist physicist J.D.Bernal, with whom I shared an office. Whatever may have transpired later about leftward security leaks he seemed to be genuinely surprised when the Hiroshima bomb was dropped.

So passed an agreeable year, devising and attending trials often involving large quantities of explosives, which caused either apathy among the assembled brass when nothing went off at all or indignation when the intensity of blast and missiles among the spectators was too great.

In 1946 I was "on the beach" and wondering about a job. At the same time a friend advised me to apply for a Commonwealth Fund Fellowship (financed by the Commonwealth Fund of New York, based on the Harkness-Standard Oil fortune). I met an Oxford don who had spent the war in Washington and had heard of a profession called "chemical engineering". This happy combination of suggestions led me to a Fellowship for the study of Chemical Engineering at MIT (founded 1861 by William Barton Rogers). The Commonwealth Fund Fellowships were conceived on a generous scale. One was expected to own a car and to spend at least six months of the Fellowship travelling about the U.S.A.

My introduction to MIT was a considerable cultural shock. At Oxford I had been a scholar and a gentleman, in the Navy an officer and a gentleman. At MIT I was one of a multitude, mostly enjoying the GI Bill of Rights, who were being processed by a vast and somewhat impersonal machine working double shifts. I was bewildered by everything - "material balance" meant nothing to me, I didn't know what a differential equation was and (at the age of 30) I had never used a slide-rule. I scored zero on my first quizz and 100% on my last one. The principles were not so difficult; the main things I had to learn were how to apply them and put numbers into them, and how to adopt a commonsense and pragmatic attitude towards problems rather than a purely intellectual one. I believe these were the most important attributes of the MIT course and should properly be the basis of any engineering course in its extended sense. Of course, there are subjects such as mathematics, fluid mechanics and chemical thermodynamics which should be taught at a university if they are ever to be learned; but never should it be implied that they constitute the whole of an engineer's training.

The MIT Chemical Engineering Practice School was then at its prime and I spent six months divided between different locations from Bangor, Maine to Buffalo, working on the plants. We were a group of six, of whom I was the only non-U.S. citizen and the only owner of a car. It was a unique opportunity to become acclimatised to a country and its people, and I have felt at home in the U.S.A. ever since. The course was brilliantly conceived and executed. At each plant, apart from carrying out measurements and experiments we learned about the history and the state of the art of the process and about scientific topics related to it. In the Lackawanna Steel works, for instance, I learned more about the theory and the practical aspects of radiation than I could have done in any other way. Even Boltzmann, I like to think, could not have had a better intuitive appreciation of the meaning of "black-body radiation" than I acquired by peering into furnaces. One member of the group was Ray Baddour, who subsequently became head of the Chemical Engineering Department at MIT.

Returning to England in 1948 with a Master's Degree in Chemical Engineering Practice, I began to look around for a job. I found that invariably I had to explain to potential employers what a chemical engineer was - they had always got along perfectly well with chemists and mechanical engineers (in which they followed the highly-successful German model). Not feeling that I had the fortitude to revolutionise the British chemical industry, I was grateful for a chance to join the staff of the newly-formed Chemical Engineering Department at Cambridge University. This had been foisted on the University by an irresistable cash benefaction from the

Shell group of companies and had been judged unsuitable to join the department which housed the more traditional branches of engineering. It became, therefore, an independent and anomalous entity, not attached to any Faculty - a situation which continues to this day and carries many advantages as regards autonomy. The fact is that engineers don't understand chemistry and chemists don't understand engineering, and it is not uncommon for university departments of chemical engineering, both in the U.K. and in the U.S.A., to be happily independent of both.

The first Shell professor of chemical engineering was Terence Fox, a Cambridge mechanical engineer with a singularly rigorous cast of mind - just what was needed, in fact, to make the subject respectable in Cambridge eyes. He soon licked us all, staff and students, into shape. He was a brilliant teacher but never launched into original research.

During the years 1948-54, although there was plenty of teaching to be done, I sat on no committees and supervised only one graduate student. I look back on this as my period of "academic indolence", during which I had time to shove my feet up on the desk and actually think. Such "insights" as I have experienced originated mainly in this period, stimulated by summer visits to chemical plants. Alas, indolence is no longer in fashion in universities where, as in industry, one feels that the cult of busyness is antipathetic to thinking. University people sit on committees, travel the world, generate reams of administrative paper, go to conventions; all too seldom does one hear the cry "Eureka !". It was also true that "chemical engineering science" was in its infancy in those days; relatively few people had applied the methods of science to distinctively chemical engineering problems. The scene was very different from that of today when (dare I say it?) too many academics are pursuing too many non-problems. We had an almost virgin field to plough and very few competitors in launching some important new ideas.

In 1954 I joined the U.K.Atomic Energy Authority in a job which was essentially administrative and for which I was not qualified. It happened that the U.K., having launched the atomic energy project ("Tube Alloys") in 1940, having sold the idea to the Americans and having provided much of the expertise involved in the Manhattan Project, was abruptly cut-off from all communication with the U.S.A. by the McMahon Act of 1946. This was too much even for Clement Attlee's labour government who decided in 1947 (in an inner cabinet committee of four or five persons) to thumb their noses at the Americans and go it alone. We had, of course, a good number of people who had been among the mainstays of the Manhattan Project and knew how to go about the two routes for making a bomb - plutonium or enriched uranium. John Cockroft (later Lord) was put in charge of the research establishment at Harwell; William Penny (later Lord) in charge of the Atomic Weapons Research Establishment at Aldermaston. Christopher Hinton (later Lord) was my boss and with very close personal attention to detail built and commissioned the Springfields plant for the manufacture of extremely pure uranium; the Capenhurst plant for the enrichment of U_{235} by gaseous diffusion; and the Windscale plant which consisted of two air-cooled graphite-moderated reactors and the associated chemical facilities for the extraction of plutonium and the separation of fission products and unchanged uranium. Hinton is, I believe, one of the great British engineers. Having had to learn the science and technique from scratch he succeeded by sheer brilliance and unremitting driving force to get his plants designed, built and commissioned on time. It was a reign of terror and the degree of hyperactivity was comparable with that associated with the Manhattan Project. Harwell soon became indistinguishable from a university campus and the hard facts required for the Industrial Group to carry out its job led to the need for a completely separate and dedicated R. and D. organisation to be set up. Such an experience, I imagine, is not unique to the Atomic Energy Authority. I was Deputy Director of this R. and D. organisation for two years which were fascinating, adventurous and precarious. It was in this job that I learnt the meaning of the phrase "empire building" - not without its analogues in academic life. I was mercifully released in 1959 by the offer of a chair of Chemical Engineering Science at Imperial College, London. Not long after I left our No.1 reactor went up in flames because of the uncontrolled release of Wigner energy from the graphite. One can imagine the kind of hullabaloo which would have ensued if this had happened in 1980; fortunately in those days the anti-nuclear hysteria had not yet incubated and the worst that happened was that a few hundred gallons of milk had to be thrown away.

My three years at Imperial College were indeed strenuous, but since I was a bachelor able to live five minutes from my work and with all the delights of London at my disposal they were not exiguous. My hero at this time, equal in stature to Hinton although very different in personality, was Prof. D.M. Newitt. Soldier, sailor, saboteur and a chemical engineer before the profession had become fashionable, he was humane and deeply sympathetic to his many third-world students. He could cap any story: e.g. "When I was sent to arrest Lawrence of Arabia ... ". Imperial College is one of the richer anomalies of the English university system. Although for practical purposes autonomous it is officially one of the colleges of London University, which means a double ration of Faculty Boards, committees and so on. I have always favoured a unilateral declaration of independence to set up a University of South Kensington - Imperial College, the Royal College of Music, the Royal College of Art, the Royal College of Organists, the Royal School of Needlework, the Victoria and Albert Museum, the Science Museum, the Natural History Museum, the Geological Museum, the Royal Marsden Hospital - even perhaps the Albert Hall; most of these institutions having been founded on the profits (unique) of the Great Exhibition of 1851 and owing their foundation to Prince Albert.

In 1959 Terence Fox resigned from the Shell chair at Cambridge (he died the following year) and I felt (perhaps not for the first time) the firm guiding hand of Sir Harold Hartley, evergreen Balliol man, who was, among many other appointments, Brigadier General and controller of chemical warfare in the first world war, Vice-President of the L.M.S. Railway Co., Chairman of British European Airways and a Director of the "Times"; speech-writer to the Duke of Edinburgh, physical chemist extraordinary, energetic promoter of chemical engineering and éminence grise in general regarding the scientific and technological establishment. So I was impelled to Cambridge again, cheerfully leaving behind me a lot of unsolved questions about the succession at Imperial College.

In 1960-74 I held the chair and was head of the department at Cambridge. On the whole I think these were successful years, although the number of our students fluctuated wildly according to fashion and job opportunities. I cannot say that the reputation of the department's graduates (its primary product) or its research has suffered as a result.

Ill-health removed me from the scene for a large part of 1974. In 1975-6 I spent a year as Visiting Professor at North Carolina State University which was highly enjoyable but more of a holiday than a study-leave, I fear.

The Headship of the department had passed to John Davidson in 1975 and in 1977, when I decided to retire, he was elected to the Shell chair. I hope he will prove to have more staying-power than his two predecessors.

At the moment I continue to pursue my researches on gas-absorption in a modest way, act as Executive Editor of "Chemical Engineering Science", and am about to publish an account of the owners of the land on which my house stands - namely the site of Barnwell Priory (founded 1112, dissolved 1538, demolished 1578), who include the Royal Physician to Henry VIII, Edward VI, Mary and Elizabeth I, and also the winner of the Derby in 1786.

The general squalor of travelling conditions is such that I seldom leave Cambridge and never attend committees unless I am chairman (preferably in my own home). I leave it to others to decide whether it is irresponsible or wise to "opt out" before one is forced to retire.

General Introduction

The title of this book is somewhat pretentious, but a title was needed and I think it does in fact contain, here and there, certain ideas which were novel and thought-provoking at the time of their publication. Some of these have inspired a considerable literature and, I hope, proved useful to practical chemical engineers; others seem to have passed unnoticed and I am glad of an opportunity to give them a second airing here.

The papers which follow are selected from research work conducted at Cambridge and Imperial College, mainly during the years 1950-1954 and 1957-1973 (after which ill-health put an end to research for some time).

The bulk of the material is contained in SECTION B - "The Design of Gas Absorbers" - but I would like to think that I have provided some growing-points in other fields which I have not been able to follow-up personally.

In some cases papers of interest have had to be represented only by their Abstracts, in order to keep the number of pages within limits. Each section is provided with a commentary and I hope that these may encourage interested readers to refer to the originals of the missing papers. These commentaries are in no sense review articles because the proliferation of the literature subsequent to (and sometimes based on) the papers reproduced here has been enormous.

One particular paper, which I believe has been widely used in industry, but which has not been reproduced here because of its length and because it is essentially a review, is "The Absorption of Carbon Dioxide into Solutions of Alkalies and Amines" (The Chem. Engr Oct. 1966, pp. 245-280) which was written in collaboration with M.M.Sharma. Not all the papers published under my name appear here, nor is there a complete bibliography. I am afraid that reproduction of photographic plates in some of the papers may be unsatisfactory but the information they were intended to convey is duplicated to a large extent in the text.

My heartfelt thanks go to my co-authors (whose permission I have sought wherever possible). They did all the experimental work and much of the thinking. I must also thank publishers of the various journals for permission to reproduce papers.

SECTION A

Mathematics of Diffusion with and without Chemical Reaction

This section includes my earliest work, done at a time when I had no experimental facilities and was much taken up with the transient processes which occur when a gas comes into contact with a semi-infinite liquid (particularly one containing a reagent) and was also somewhat fascinated by the possibilities of mathematics (to which I had only recently been introduced after a lapse of many years). It was admittedly unsophisticated mathematics but it set the scene for the more realistic matters discussed in SECTION B. I did not indulge much in this sort of mathematical exercise after 1952.

Paper A1, dealing with "surface resistance" to gas absorption is based on Higbie's experimental results, which are certainly inaccurate; this is not surprising considering the complexity of his apparatus. A number of subsequent attempts have been made to detect surface resistance using, for instance, the laminar-jet technique. The results are inconclusive but recently C.A.Ward (J. Chem. Phys. 1977 67 229) has suggested that the rate of attainment of equilibrium at the interface is not governed by the simple linear equation which I and others had assumed and that, moreover, calculation based on physical principles agrees with published data in suggesting that the time required for oxygen to equilibrate with a water surface is much greater than that ($\sim$20 ms) characteristic of a laminar-jet experiment and that consequently such an experiment gives an ostensible value of the diffusivity which is too low. Dr. Ward has privately communicated his calculations for carbon dioxide which show that this gas equilibrates much more rapidly with water and that a laminar-jet experiment should yield the value of the diffusivity with little error. (The integral in Paper A1 is evaluated analytically in Paper B2 eqn. 2 (SECTION B)).

Paper A2 was my first attempt to tackle the matter of gas absorption with chemical reaction - the latter being first-order or pseudo-first order with respect to the dissolved gas (a state of affairs which is seldom encountered in practice as foreshadowed in Part 3 of this paper). Mathematically this is an adaptation of the case of heat-conduction in a wire as presented by Carslaw and Jaeger ("Conduction of Heat in Solids" O.U.P., 1959).

Paper A3 is a generalisation of the mathematics describing the process of gas absorption with "instantaneous" reaction in a semi-infinite medium, which was touched on in Paper A1. It can be used to deal with various other aspects of gas absorption, tarnishing reactions, freezing of liquids, etc. It is rather removed from the main theme of my work but once having realised the key to such problems - namely that the position of the boundary moves as the square-root of the time - it seemed a pity not to exploit the discovery as far as it would go.

Paper A4 is also something of the nature of the by-product of a mathematical trick. The transformation by which the equations for diffusion alone are changed to those for diffusion accompanied by first-order reaction can be applied by analogy to the many situations for which the heat-conduction equations have been solved. The extension to falling liquid drops is perhaps of particular interest to chemical engineers, despite the constraints upon the applicability of the solution.

Papers A5 and A6 belong together. Paper A5 sets out to determine the rise in temperature at the surface of a liquid into which a gas is being absorbed while undergoing a first-order reaction. It was inspired by the need to determine, in the framework of a rational analysis of gas absorption, the conditions under which a temperature rise would be negligible. The solution was given in the form of an infinite series with coefficients laboriously determined by computation. In Paper A6 the relevant Laplace transform has been inverted and the solution given in terms of Bessel functions. Also, the treatment is extended to provide a test for the degree of depletion of a reactant (or augmentation of product) when the gas undergoes a pseudo-first order reaction with a dissolved reagent.

Paper A7 gives the rather surprising result that when bubbles of a mixture of soluble and insoluble gases are dispersed in a liquid the rate of absorption is proportional to the partial pressure of the soluble gas in the bubbles originally introduced into the liquid, in spite of the fact that the change in partial pressure in the immersed bubble may vary from 100% to 0%. The result is subject to severe restraints but subsequent workers have shown that in many cases these can be relaxed.

A1

Kinetics of the Absorption of Carbon Dioxide in Water

WHEN a gas is in contact with a liquid, molecules of the former strike the interface at the rate :

$$G' = \frac{p}{\sqrt{2\pi RTM}} \text{ mols per unit area per unit time} \qquad \dots (1)$$

(where p = pressure of gas, R = gas constant, T = absolute temperature, M = molecular weight of gas). If a fraction $(1 - K)$ of the incident molecules is reflected from the surface, then gas molecules enter the liquid at the rate $G = KG'$. (The *net* rate of transference is generally not equal to G, however, because there is simultaneous passage of molecules in the opposite sense.)

While K has been determined for the condensation of the vapours of several substances into the corresponding liquids[1], there do not appear to have been any genuine determinations for the absorption of soluble gases into liquids[2, 3]. However, some recorded measurements by HIGBIE[4] of the rate of absorption of carbon dioxide into water enable K to be calculated for this case. Higbie was able to measure the rate of absorption during the first few hundredths of a second of contact. For practical purposes the water could be regarded as stagnant and infinitely deep, and under these circumstances the following relation should hold if the surface of the solution is constantly saturated :

$$Q = 2c^* \sqrt{\frac{Dt}{\pi}} \qquad \dots (2)$$

where Q is the quantity of gas absorbed in time t by unit area of surface, c^* being the saturated concentration and D the diffusion coefficient in the liquid. Fick's first law of diffusion is assumed to apply.

However, the quantity absorbed was consistently less than this, and as Higbie pointed out, was compatible with the assumption that the rate of absorption was proportional to the difference between c^* and c_0, the actual concentration at the surface at any time.

If Henry's law applies to the system, and K is independent of c_0, it can easily be shown that :

$$\frac{dQ}{dt} = G\left(1 - \frac{c_0}{c^*}\right) \qquad \dots (3)$$

and provided Fick's law holds in the liquid, the variation of c_0 with t can be obtained from the solution of the analogous problem for heat-conduction[5]:

$$1 - \frac{c_0}{c^*} = \exp\left[\frac{G^2 t}{(c^*)^2 D}\right] \cdot \operatorname{erfc}\left[\frac{G}{c^*}\sqrt{\frac{t}{D}}\right] \qquad \dots (4)$$

Since from equation 3 :

$$\frac{QG}{(c^*)^2 D} = \int_{t=0}^{t} \left(1 - \frac{c_0}{c^*}\right) d\left(\frac{G^2 t}{(c^*)^2 D}\right) \qquad \dots (5)$$

the relation between the dimensionless quantities $\frac{QG}{(c^*)^2 D}$ and $\frac{tG^2}{(c^*)^2 D}$ can be obtained by integration of equation 4. This has been performed numerically, and the resulting function used to find the value of K which best fits Higbie's results for atm., 10°C (those for 25°C, which were obtained by a less accurate technique, are too scattered to allow K to be estimated).

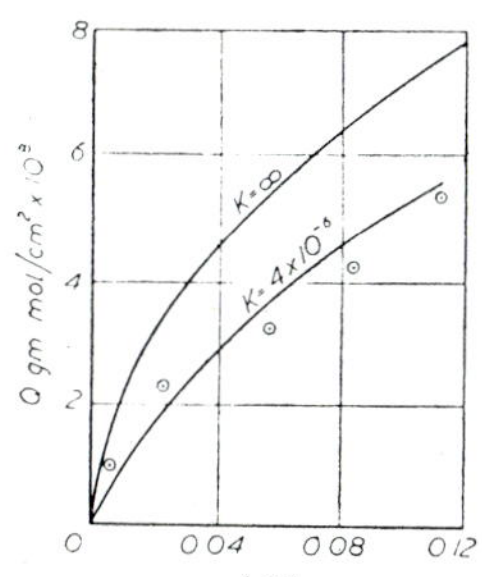

Figure 1. Absorption of carbon dioxide by water at 10°C

Putting $c^* = 5{\cdot}3 \times 10^{-5}$ gm mol/cc, $D = 1{\cdot}41 \times 10^{-5}$ cm²/sec, the best value of K is found to be 4×10^{-6}. *Figure 1* shows the experimental points and the calculated curve. The curve for equation 2, corresponding to $K = \infty$, is also shown ; this is virtually indistinguishable, on this scale, from that for $K = 1$.

It can be shown that neither the probable error in the value of D, nor the presence of insoluble impurities (about 0·05 per cent) in the gas, nor the heat of solution of carbon dioxide would account for the observed results if K were actually in the neighbourhood of unity.

It is hoped to make further determinations of K for this system, using a falling film of liquid.

P. V. DANCKWERTS

Department of Chemical Engineering
Tennis Court Road
Cambridge

(Received September 1949)

REFERENCES

[1] WYLLIE, G. *Proc. roy. Soc.* *A* 197 (1949) 383
[2] BOHR, C. *Ann. der Physik* 3 (1899) 500 ; 4 (1900) 244
[3] BELL, R. P. *J. Phys. Chem.* 33 (1929) 99
[4] HIGBIE, R. *Trans. Amer. Inst. chem. Engrs* 31 (1935) 365
[5] CARSLAW, H. S. and JAEGER, J. C. *The Conduction of Heat in Solids* Oxford, 1947, p. 53

ABSORPTION BY SIMULTANEOUS DIFFUSION AND CHEMICAL REACTION

By P. V. Danckwerts

Received 21st November, 1949

Equations are derived for the rate of absorption and the concentration distribution when a solute of limited solubility diffuses into a semi-infinite medium with which it undergoes a first-order or pseudo first-order reaction. Consideration is given to the case where the medium is a mixture, of which only one component reacts. A method is given for determining whether the above-mentioned equations can be applied in spite of local depletion of this component.

An equation is presented for the rate of absorption when the absorbed substance reacts instantaneously with a component of the medium which is present in limited concentration.

---

1. Introduction.—The problem here considered is the calculation of the rate of absorption of a substance (which will be called the "solute") which diffuses into a stagnant medium in which it has a limited solubility and with which it undergoes an irreversible first-order or pseudo first-order reaction. The medium has a plane surface and is of infinite depth (or, for practical purposes, is of such a depth that the concentration does not change appreciably at the bottom during the period of time considered). The surface of the medium is continuously saturated with the solute. As examples may be cited: the absorption of a gas into a liquid or solid, the dissolution of a solid into a supernatant liquid, or the absorption of one solid or liquid material into another. In each case the solute is assumed to be destroyed by a process which proceeds at a rate solely dependent on, and proportional to, its concentration in the medium.

If the medium is a fluid, the expressions developed here will only be valid if the density of the fluid always increases consistently in a downward direction (or, of course, if it remains uniform); otherwise convection currents will be set up and will increase the rate of absorption (as, for instance, when CO_2 is absorbed into a horizontal water surface). Even in the latter case, however, the fluid may be regarded as stagnant for a short initial period.

2. Exact and Approximate Solutions.—It is assumed that the ideal diffusion law expressed by (2.1) below holds for the diffusion of unreacted solute through the medium, and also that the mass of solvent per unit volume of the medium is virtually constant throughout, so that there are no volume changes in the medium when solute is absorbed. These conditions are the more likely to be fulfilled, the lower the solubility of the solute.

Let c^* be the saturated concentration of the solute reigning at the surface ($x = 0$), with c the concentration at a distance x below the surface, t the time, and k the velocity constant of the reaction between solute and medium. The initial concentration of solute in the medium is uniformly equal to zero,

Now the rate at which solute crosses unit area of any plane of constant x in the direction of increasing x is

$$\left(\frac{\mathrm{d}Q}{\mathrm{d}t}\right)_x = - D\frac{\partial c}{\partial x}, \quad . \quad . \quad . \quad . \quad (2.1)$$

(D being the diffusion coefficient of the solute in the medium) and the rate at which chemical reaction destroys the solute (per unit volume) is kc.

Consider an element of volume of unit cross-sectional area between planes x and $x + \mathrm{d}x$. The following changes in its content of solute occur in time $\mathrm{d}t$:—

Diffusing in : $\quad -D\frac{\partial c}{\partial x}\mathrm{d}t.$

Diffusing out : $\quad -\left[- D\mathrm{d}t\left(\frac{\partial c}{\partial x} + \frac{\partial^2 c}{\partial x^2}\mathrm{d}x\right)\right].$

Reacting : $\quad - kc\mathrm{d}t \,.\, \mathrm{d}x.$

Net increase : $\quad \left(D\frac{\partial^2 c}{\partial x^2} - kc\right)\mathrm{d}t \,.\, \mathrm{d}x.$

If the resulting increase in concentration is $\mathrm{d}c$, the net increase above can be equated to $\mathrm{d}c \,.\, \mathrm{d}x$, giving

$$\frac{\partial c}{\partial t} = D\frac{\partial_2 c}{\partial x^2} - kc. \quad . \quad . \quad . \quad . \quad (2.2)$$

With the boundary conditions,

$$\begin{aligned} c &= c^*, & x &= 0, & t &> 0 \\ c &= 0, & x &> 0, & t &= 0 \\ c &= 0, & x &= \infty, & t &> 0, \end{aligned}$$

the solution to eqn. (2.2) is

$$\frac{c}{c^*} = \frac{1}{2}\exp\left(- x\sqrt{\frac{k}{D}}\right) . \operatorname{erfc}\left[\frac{x}{2\sqrt{tD}} - \sqrt{kt}\right]$$
$$+ \frac{1}{2}\exp\left(x\sqrt{\frac{k}{D}}\right) . \operatorname{erfc}\left[\frac{x}{2\sqrt{tD}} + \sqrt{kt}\right], \quad (2.3)$$

where

$$\operatorname{erfc} z = 1 - \operatorname{erf} z = 1 - \frac{2}{\sqrt{\pi}}\int_0^z e^{-y^2}\,dy.$$

(The mathematical problem and its solution are identical with those for the conduction of heat along a thin rod which loses heat from its periphery at a rate proportional to its temperature.[1])

Differentiating (2.3) with respect to x and setting $x = 0$, we find

$$\left(\frac{\partial c}{\partial x}\right)_{x=0} = - c^* \sqrt{\frac{k}{D}}\left[\operatorname{erf}\sqrt{kt} + \frac{e^{-kt}}{\sqrt{\pi kt}}\right].$$

The rate of absorption dQ/dt per unit area of surface is given by

$$\frac{dQ}{dt} = - D\left(\frac{\partial c}{\partial x}\right)_{x=0} = c^* \sqrt{Dk}\left[\operatorname{erf}\sqrt{kt} + \frac{e^{-kt}}{\sqrt{\pi kt}}\right], \quad . \quad (2.4)$$

whence the quantity Q absorbed in time t is

$$Q = c^* \sqrt{\frac{D}{k}}\left[(kt + \tfrac{1}{2})\operatorname{erf}\sqrt{kt} + \sqrt{\frac{kt}{\pi}}\,e^{-kt}\right] \quad . \quad . \quad (2.5)$$

The dimensionless quantities $\frac{Q}{c^*}\cdot\sqrt{\frac{k}{D}}$ and kt are plotted in Fig. 1, which enables the quantity absorbed at any time to be readily found.

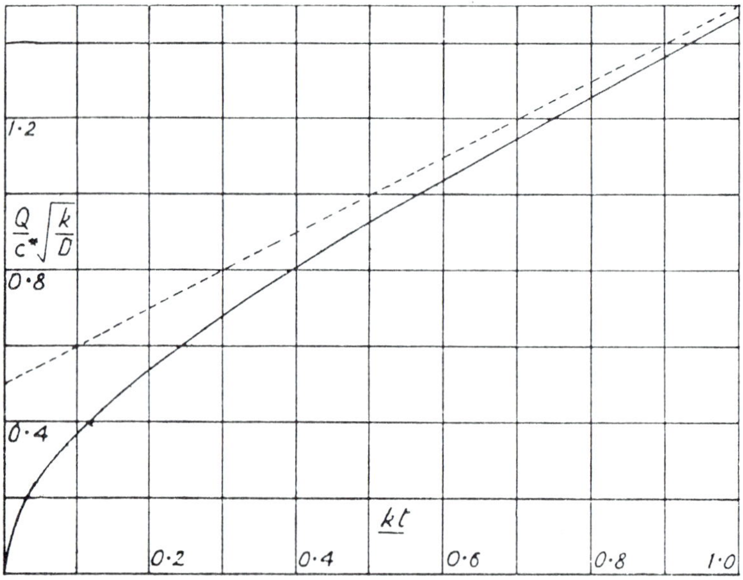

FIG. 1.

[1] Carslaw and Jaeger, *Conduction of Heat in Solids* (Oxford University Press, Oxford, 1947), p. 111.

The function $\frac{Q}{c^*}\sqrt{\frac{k}{D}} = \left[(kt + \tfrac{1}{2}) \text{ erf } \sqrt{kt} + \sqrt{\frac{kt}{\pi}}\, e^{-kt}\right]$ and asymptote

$$\frac{Q}{c^*}\sqrt{\frac{k}{D}} = (kt + \tfrac{1}{2}).$$

When kt is sufficiently large erf $\sqrt{kt} \approx 1$, and eqn. (2.3), (2.4) and (2.5) become

$$\frac{c}{c^*} \approx \exp\left(-x\sqrt{\frac{k}{D}}\right), \quad \ldots \quad (2.6)$$

$$\frac{dQ}{dt} \approx c^* \sqrt{Dk}, \quad \ldots \quad (2.7)$$

$$Q \approx c^* \sqrt{Dk}\left(t + \frac{1}{2k}\right). \quad \ldots \quad (2.8)$$

(The error in (2.7) and (2.8) is less than 2 % if $kt > 4$.) In other words, the rate of absorption and the concentration at any point tend to a steady value.

When kt is very small (a condition which is of interest in connection with the theory of absorption into agitated liquids) we find, on expanding erf $\sqrt{kt}$ and e^{-kt} and neglecting powers of kt higher than the first, that (2.4) and (2.5) become

$$\frac{dQ}{dt} \approx c^* \sqrt{\frac{D}{\pi t}}\,(1 + kt) \quad \ldots \quad (2.9)$$

$$Q \approx 2c^* \sqrt{\frac{Dt}{\pi}}\left(1 + \frac{kt}{2}\right) \quad \ldots \quad (2.10)$$

which reduce, when $kt = 0$, to the well-known values for diffusion without reaction.

3. Depletion of Second Reactant.—We may wish to apply the results to the case where the medium is a mixture, of which only one component, B, reacts with the absorbed solute, A, the reaction being first order with respect to both reactants. It is therefore of interest to determine the extent to which depletion of reactant B is likely to affect the rate of absorption.

Suppose the initial concentration of B in the medium is uniformly equivalent to c'_0. The concentrations, c and c', of A and B respectively at x at time t would be given by the solution, subject to the appropriate boundary conditions, of

$$\left.\begin{aligned} \frac{\partial c}{\partial t} &= D\frac{\partial_2 c}{\partial x^2} - k'cc' \\ \frac{\partial c'}{\partial t} &= D'\frac{\partial_2 c'}{\partial x^2} - k'cc' \end{aligned}\right\}, \quad \ldots \quad (3.1)$$

where D' is the diffusion coefficient of component B in the medium, k' the second-order velocity constant ($k'c'$ is equivalent to k in eqn. (2.2) to (2.10).)

Putting

$$\partial c/\partial t = \partial c'/\partial t = 0$$

in (3.1), we find

$$D\frac{\partial_2 c}{\partial x^2} = D'\frac{\partial_2 c'}{\partial x^2},$$

whence

$$D\frac{\partial c}{\partial x} = D'\frac{\partial c'}{\partial x} + A, \quad . \quad . \quad . \quad (3.2)$$

where A is a constant. However, this is incompatible with the necessity that $\partial c'/\partial x = 0$ when $x = 0$, $\partial c'/\partial x = \partial c/\partial x = 0$ when $x = \infty$. This system, therefore, does not tend to a steady state, and we may suppose that the longer absorption proceeds the more widely does the rate of absorption depart from that given in (2.4). However, it is clear that if at a given time the quantity of B which has reacted per unit volume of the medium is at all points an insignificant fraction of its initial concentration c_0', then the quantity of A absorbed, and its concentration at any point, will not differ appreciably from that predicted by eqn. (2.3) and (2.5). Since the concentration of A, and hence the rate of reaction, is always greater at the surface ($x = 0$) than elsewhere, we may take the quantity of reaction occurring here as the criterion. In time t the quantity of B which has reacted per unit volume of the medium in the neighbourhood of the surface is Q_0, where $Q_0 < k'c_0'c^*t$, since the surface concentration of B will always be less than c_0'. A sufficient condition that the variation of c' may be ignored is, as we have seen, that $Q_0/c_0' \ll 1$. Substituting for Q_0 we have

$$k'c^*t \ll 1 \quad . \quad . \quad . \quad . \quad . \quad (3.3)$$

as a sufficient condition for the applicability of eqn. (2.3), (2.4) and (2.5). This condition may be unnecessarily stringent, particularly if $D' \gg D$, but should be a useful guide.

If the reaction between A and B is instantaneous, A and B cannot both be present in the same part of the medium at the same time; there will, in fact, be a zone near the surface containing A alone. It can by shown [2] that under these circumstances the rate of absorption is given be

$$\frac{dQ}{dt} = \frac{c^*}{\operatorname{erf}\,[\beta/\sqrt{D}]}\sqrt{\frac{D}{\pi t}}, \quad . \quad . \quad . \quad (3.4)$$

β being defined (so long as $\beta \neq 0$) by the relation

$$\left(\frac{n'c^*}{\sqrt{D'}}\right)\exp\left[\frac{\beta^2}{D'}\right]\operatorname{erfc}\left[\frac{\beta}{\sqrt{D'}}\right] - \left(\frac{nc_0'}{\sqrt{D}}\right)\exp\frac{\beta^2}{D}\Bigg]\operatorname{erf}\left[\frac{\beta}{\sqrt{D}}\right] = 0, \quad (3.5)$$

where n moles of A react with n' moles of B, and the concentrations are expressed in moles per unit volume.

To provide a clearer idea of the significance of (3.4) and (3.5) the coefficients of diffusion of A and B may be set equal (this will approximately represent the facts in many cases). Putting $D = D'$, eqn. (3.5) becomes

$$\operatorname{erf}\left[\frac{\beta}{\sqrt{D}}\right] = \frac{n'c^*}{n'c^* + nc'_0} \qquad . \quad . \quad . \quad (3.6)$$

so that (3.4) becomes

$$\frac{dQ}{dt} = c^*\left(1 + \frac{nc'_0}{n'c^*}\right)\sqrt{\frac{D}{\pi t}}. \qquad . \quad . \quad . \quad (3.7)$$

If the initial concentration of B in the medium is equivalent to the saturated concentration of A at the surface, the rate of absorption will be twice as great as if there were no reaction taking place.

Department of Chemical Engineering,
Tennis Court Road,
Cambridge.

[2] Unpublished work by author.

A3

UNSTEADY-STATE DIFFUSION OR HEAT CONDUCTION WITH MOVING BOUNDARY

BY P. V. DANCKWERTS

Received 20th January, 1950

A general method of solution is described for a class of problems in unsteady-state linear heat conduction or diffusion, which involve two phases or regions separated by a moving plane interface.

The following examples are used to illustrate the method : the absorption by a liquid of a single component from a mixture of gases ; tarnishing reactions ; condensation of a vapour at the surface of a cold liquid or on a cooled surface ; gas reacting at solid surface to form gaseous product ; solution of a gas in a liquid, followed by reaction with solute ; and progressive freezing of a liquid.

---

1. **Introduction.**—There is a category of problems in unsteady-state linear heat conduction or diffusion, involving contiguous phases or regions with a moving interface, for which special solutions have been produced from time to time. Among these have been (*a*) the progressive freezing of a liquid,[1] (*b*) the absorption by a liquid of a single component from a mixture of gases,[2] and (*c*) tarnishing reactions.[3] These are variants of a single mathematical problem, but have all been solved independently. The more general method of solution presented here may be of service in enabling solutions of new variants to be obtained with the minimum of fresh mathematical work. In addition, the method can be used to give a more general form of the solution to problem (*a*) above, and a more rigorous and highly-developed solution to (*b*), than appear to have been available hitherto.

In all the cases considered the two regions are separated by a plane surface, and diffusion or heat conduction takes place only along an axis perpendicular to this plane. The concentration or temperature is initially uniform in each region. The process of heat conduction or diffusion may cause changes which bring about the disappearance or appearance of matter at the interface in one or both of the regions, and a resulting bodily movement of the matter in one or both regions relative to the interface. As a simple example consider the melting of ice in contact with water; as heat flows from water to ice, ice disappears and water appears at the interface, so that both the ice and the water are in bodily movement with respect to the interface. In all the cases considered the rates of bodily motion of the matter in the two regions with respect to the interface are related by constant proportionality (e.g. the volume of ice melted is proportional to the volume of water formed). Variations in density (or partial density) of either medium due to variations in temperature or concentration are ignored.

In the next section the conditions of the problem will be stated in a form sufficiently general to include all likely applications of the solution. The extent to which various types of real system obey these conditions will be discussed in connection with the examples given later. The practical value of the results is limited by the fact that they cannot be applied to systems in which convection currents occur, or in which gradients of both temperature and concentration exist.

2. Conditions of the Problem.—We consider two *media* which are stagnant in the sense that they are free from convection currents, but which may be in relative bodily motion along the x-axis (which is perpendicular to the interface). Location in medium 1 will be referred to co-ordinates of the x_1 system, which are stationary with respect to medium 1; location in medium 2 is referred to x_2 co-ordinates, which are stationary with respect to medium 2 (Fig. 1). The media are separated at time t

[1] Carslaw and Jaeger, *Conduction of Heat in Solids* (Oxford University Press, 1947), pp. 71, 227.

[2] Arnold, *Trans. Amer. Inst. Chem. Eng.*, 1944, **40**, 361.

[3] Booth, *Trans. Faraday Soc.*, 1948, **44**, 796.

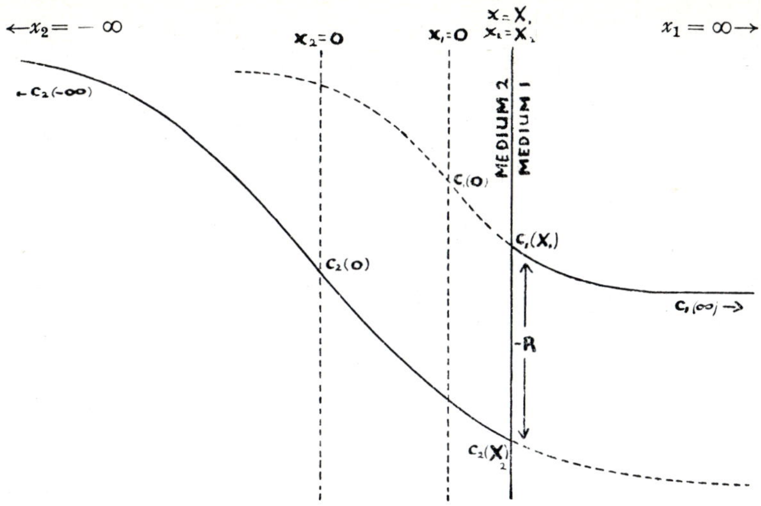

FIG. 1

by the plane $x_1 = X_1$, $x_2 = X_2$, which is initially at $x_1 = x_2 = 0$. Medium 1 occupies all or part of the space $X_1 < x_1 < \infty$, medium 2 all or part of the space $-\infty < x_2 < X_2$.

Present in both media is a *diffusant* (matter or heat); diffusional processes cause transport of the diffusant relative to the x_1 and x_2 coordinates, and transfer from one medium to the other. The concentration of the diffusant at time t is denoted by c_1 at x_1 and c_2 at x_2. Initially c_1 has the value $c_1(0)$ at $x_1 = 0$ and $c_1(\infty)$ for all $x_1 > 0$, c_2 the value $c_2(0)$ at $x_2 = 0$ and $c_2(-\infty)$ for all $x_2 < 0$. The following diffusion equations are obeyed in the two media:

$$\frac{\partial c_1}{\partial t} = D_1 \frac{\partial_2 c_1}{\partial x_1{}^2}, \quad . \quad . \quad . \quad . \quad (2.1)$$

$$\frac{\partial c_2}{\partial t} = D_2 \frac{\partial_2 c_2}{\partial x_2{}^2}, \quad . \quad . \quad . \quad . \quad (2.2)$$

D_1 and D_2 being the respective diffusivities, which are assumed to be independent of c_1 and c_2 respectively. (In diffusion in liquids, this assumption will be justified only if variations in the concentration of the diffusant are small.)

At any time the concentrations $c_1(X_1)$ and $c_2(X_2)$, at either side of the interface, are related by an equilibrium expression:

$$c_2(X_2) = Qc_1(X_1) + R, \quad . \quad . \quad . \quad (2.3)$$

where Q and R are physical constants (e.g. for absorption of a gas obeying Henry's law, Q = solubility of gas (in suitable units), $R = 0$).

The diffusant is conserved at the interface, as elsewhere, so that

$$D_1\left(\frac{\partial c_1}{\partial x_1}\right)_{x_1=X_1} - D_2\left(\frac{\partial c_2}{\partial x_2}\right)_{x_2=X_2} + c_1(X_1)\frac{dX_1}{dt} - c_2(X_2)\frac{dX_2}{dt} = 0. \quad (2.4)$$

Eqn. (2.4) merely expresses the fact that the quantity of the diffusant disappearing from medium 1 is equal to the quantity appearing in medium 2.

There is constant proportionality between X_1 and X_2 (since there is constant proportionality between the rates of movement of the two media relative to the interface—although one or both of these rates may be zero). Hence

$$X_2 = PX_1 \quad . \quad . \quad . \quad . \quad (2.5)$$

where P is determined by the conditions of the problem.

At least two of the quantities $c_1(\infty)$, $c_1(0)$, $c_1(X_1)$, $c_2(-\infty)$, $c_2(0)$ must be specified in the data of the problem. In addition, in order to define the problem completely, it is necessary either that a third of these quantities should also be specified (solutions of Class B, see § 6) or that there should be constant proportionality of specified magnitude between the quantity of diffusant crossing the interface and the values of X_1 and X_2 (solutions of Class A, see § 5).

3. Solutions of Diffusion Equations for Infinite Media.—Consider an *infinite* medium in which eqn. (2.1) holds, together with the following boundary conditions:

$$c_1 = c_1(\infty),\ t = 0,\ \text{all } x_1 > 0, \qquad c_1 = c_1(0),\ \text{all } t,\ x_1 = 0,$$
$$c_1 = c_1(-\infty),\ t = 0,\ \text{all } x_1 < 0, \qquad c_1 = c_1(-\infty),\ \text{all } t,\ x_1 = -\infty,$$
$$2c_1(0) = c_1(\infty) + c_1(-\infty), \qquad c_1 = c_1(\infty),\ \text{all } t,\ x = \infty,$$

where $c_1(\infty)$, $c_1(-\infty)$, $c_1(0)$ are all constant.

Employing the well-known solutions for the conduction of heat,[4] we find

$$\frac{c_1(\infty) - c_1}{c_1(\infty) - c_1(0)} = 1 - \operatorname{erf}\left[\frac{x_1}{2\sqrt{D_1 t}}\right] \quad . \quad . \quad (3.1)$$

as the solution of (2.1), c_1 being the concentration of the migrant at x_1, t. Similarly the solution of (2.2) under analogous boundary conditions is

$$\frac{c_2(\infty) - c_2}{c_2(\infty) - c_2(0)} = 1 - \operatorname{erf}\left[\frac{x_2}{2\sqrt{D_2 t}}\right], \quad . \quad . \quad (3.2)$$

which may be re-written as

$$\frac{c_2(-\infty) - c_2}{c_2(-\infty) - c_2(0)} = 1 + \operatorname{erf}\left[\frac{x_2}{2\sqrt{D_2 t}}\right]. \quad . \quad . \quad (3.3)$$

We shall also have occasion to use the following results, applicable under the same sets of boundary conditions:

$$V_1 = 2[c_1(\infty) - c_1(0)]\sqrt{\frac{D_1 t}{\pi}}, \quad . \quad . \quad . \quad . \quad (3.4)$$

$$V_2 = 2[(c_2(0) - c_2(-\infty))]\sqrt{\frac{D_2 t}{\pi}}, \quad . \quad . \quad . \quad (3.5)$$

[4] Ref. (1), p. 43.

where V_1, V_2 are the quantities of diffusant crossing the planes $x_1 = 0$, $x_2 = 0$ respectively in direction of decreasing x in time t.

4. Application to Media bounded by Plane Surfaces.—Eqn. (3.1) and (3.3) refer to an infinite medium; however, each applies equally well to a region bounded by one or two x-planes, either stationary or moving, provided that (*a*) the *initial* concentration at every point in this restricted region is the same as for the same value of x_1 or x_2 in the infinite medium; and (*b*) the concentration at the boundary plane (or planes) is *at all times* the same as for the same value of x_1 or x_2 in the infinite medium.

The problems here discussed have the common feature that, although they concern media bounded by one or two planes, they fulfil the above conditions; the equations for infinite media can therefore be used. In other words eqn. (3.1) and (3.3) are, as will be shown, compatible with (*a*) eqn. (2.3), (2.4) and (2.5), and (*b*) the conditions determined by the data of the problem, which may be of Class A or Class B, as described below.

It may be noted that values of c_1 outside medium 1 and of c_2 outside medium 2 will have no physical significance.

5. Problems of Class A.—For this class of problem the conditions mentioned under (*b*) above are that two of the quantities $c_1(\infty)$, $c_1(0)$, $c_1(X_1)$, $c_2(-\infty)$, $c_2(0)$ should have specified values and that the magnitudes of X_1 and X_2 should at all times be proportional to the quantity of diffusant which has crossed the interface ($x_1 = X_1$, $x_2 = X_2$). Hence we may put X_1/S = (quantity of diffusant crossing interface in direction of decreasing x), whence

$$\frac{dX_1}{dt} = S\left[D_1\left(\frac{\partial c_1}{\partial x_1}\right)_{x_1=X_1} + c_1(X_1)\,\frac{dX_1}{dt}\right], \qquad (5.1)$$

where S is a proportionality constant characteristic of the system. Combining this with (2.4) and (2.5) we have

$$\frac{dX_2}{dt} = PS\left[D_2\left(\frac{\partial c_2}{\partial x_2}\right)_{x_2=X_2} + c_2(X_2)\,\frac{dX_2}{dt}\right]. \qquad (5.2)$$

Substituting (2.3), (2.5), (3.1) and (3.3) in (5.1) and (5.2)

$$\frac{dX_1}{dt}\left[\frac{1}{S} - c_1(X_1)\right] = [c_1(\infty) - c_1(0)]\sqrt{\frac{D_1}{\pi t}}\exp\left[\frac{-X_1^2}{4D_1t}\right], \qquad (5.3)$$

$$\frac{dX_1}{dt}\left[\frac{1}{S} - PR - PQc_1(X_1)\right] = [c_2(0) - c_2(-\infty)]\sqrt{\frac{D_2}{\pi t}}\exp\left[\frac{-P^2X_1^2}{4D_2t}\right], \qquad (5.4)$$

while, from (3.1), putting $x_1 = X_1$, $c_1 = c_1(X_1)$,

$$c_1(X_1) = c_1(0) + [c_1(\infty) - c_1(0)]\,\mathrm{erf}\left[\frac{X_1}{2\sqrt{D_1t}}\right]. \qquad (5.5)$$

It is clear that (5.3), (5.4) and (5.5) can be simultaneously true for any value of t if, and only if, $X_1/\sqrt{t}$ is constant. Put

$$X_1 = 2\alpha\sqrt{D_1 t}, \quad . \quad . \quad . \quad . \quad (5.6)$$

(5.3), (5.4) and (5.5) become respectively :

$$\sqrt{\pi}\,\alpha \exp[\alpha^2] \,.\, \left[\frac{1}{S} - c_1(X_1)\right] = c_1(\infty) - c_1(0), \quad . \quad (5.7)$$

$$\sqrt{\pi}\,.\,\alpha\sqrt{\frac{D_1}{D_2}} \exp\left[\frac{\alpha^2 P^2 D_1}{D_2}\right] . \left[\frac{1}{S} - PR - PQc_1(X_1)\right] = c_2(0) - c_2(-\infty), \quad (5.8)$$

$$c_1(X_1) = c_1(0) + [c_1(\infty) - c_1(0)]\,\mathrm{erf}\,[\alpha]\,; \quad . \quad . \quad (5.9)$$

while from (3.3),

$$\frac{Qc_1(X_1) + R - c_2(-\infty)}{c_2(0) - c_2(-\infty)} = 1 + \mathrm{erf}\left[P\alpha\sqrt{\frac{D_1}{D_2}}\right] \quad . \quad (5.10)$$

Eqn. (5.7), (5.8), (5.9), (5.10) are independent and contain, besides physical constants and the parameters P and S, the six quantities, $c_1(0)$, $c_1(\infty)$, $c_1(X_1)$, $c_2(0)$, $c_2(-\infty)$, α. Hence if two of the concentrations are given the equations can be solved for the other three and for α. Those constant concentrations appearing in eqn. (3.1) and (3.3) which were not given explicitly in the data of the problem can now be substituted in terms of known quantities; the resulting equations for c_1 in terms of x_1 and t, and for c_2 in terms of x_2 and t, fulfil both the boundary conditions of the problem and the general diffusion equations, and hence constitute the required solution. Substitution for α in (5.6) gives X_1 (and hence X_2) in terms of t and known quantities.

It will be noted that the interfacial concentrations $c_1(X_1)$ and $c_2(X_2)$ are necessarily constant.

6. Problems of Class B.—Here the condition expressed by eqn. (5.1) and (5.2) no longer holds. Instead, the data of the problem require that three of the five concentrations $c_1(0)$, $c_1(\infty)$, $c_1(X_1)$, $c_2(0)$, $c_2(-\infty)$ should have specified values.

Substituting (2.3), (2.5), (3.1), and (3.3) into (2.4) :

$$[c_1(\infty) - c_1(0)]\sqrt{\frac{D_1}{\pi t}} \exp\left[\frac{-X_1^2}{4D_1 t}\right] - [c_2(0) - c_2(-\infty)]\sqrt{\frac{D_2}{\pi t}} \exp\left[\frac{-P^2 X_1^2}{4D_2 t}\right]$$
$$+ \frac{dX_1}{dt}[(1 - PQ)c_1(X_1) - PR] = 0 \quad (6.1)$$

Taken with (5.5) this can be true for all values of t if, and only if, $X_1/\sqrt{t}$ is constant. Putting

$$X_1 = 2\beta\sqrt{t} \quad . \quad . \quad . \quad . \quad (6.2)$$

(6.1) becomes

$$\sqrt{\frac{D_1}{\pi}}[c_1(\infty) - c_1(0)] \exp\left[\frac{-\beta^2}{D_1}\right] - \sqrt{\frac{D_2}{\pi}}[c_2(0) - c_2(-\infty)] \exp\left[\frac{-P^2\beta^2}{D_2}\right]$$
$$+ \beta[(1 - PQ)c_1(X_1) - PR)] = 0. \quad (6.3)$$

From (5.5)

$$c_1(X_1) = c_1(0) + [c_1(\infty) - c_1(0)] \operatorname{erf}\left[\frac{\beta}{\sqrt{D_1}}\right], \quad . \quad . \quad (6.4)$$

while from (3.3)

$$c_2(X_2) = Qc_1(X_1) + R = c_2(0) + [c_2(0) - c_2(-\infty)] \operatorname{erf}\left[\frac{P\beta}{\sqrt{D_2}}\right]. \quad (6.5)$$

Eqn. (6.3), (6.4) and (6.5) are independent; since three of the five quantities $c_1(\infty)$, $c_1(X_1)$, $c_1(0)$, $c_2(0)$, $c_2(-\infty)$ are known, the equations determine the values of the other two and of β. Hence X_1 can be found as a function of t from eqn. (6.2), and c_1 or c_2 as functions of t and x_1 or x_2 from (3.1) and (3.3).

7. Examples: Class A.—(*a*) ABSORPTION BY A LIQUID OF A SINGLE COMPONENT FROM A MIXTURE OF GASES.—An ideal mixture of a soluble gas A and an insoluble gas is in contact with a liquid. Initially the mole fraction of A in the gas is $c_1(\infty)$ and the concentration (vol. gas/vol. liquid) of dissolved A in the liquid is uniformly equal to $c_2(-\infty)$; $c_1(\infty)$ and $c_2(-\infty)$ are given. There is constant equilibrium between gas and liquid at the interface, where Henry's law is obeyed. There is no appreciable change in the partial volume or the temperature of the liquid when A is absorbed. Let the liquid be medium 2 and the gas medium 1. Provided the difference between $c_2(-\infty)$ and $c_2(X_2)$ (see below) is not too great, the law of diffusion expressed by (2.2) is obeyed in the liquid, dissolved A being the diffusant (Fick's law). If the x_1-origin moves so that there is no net flow of gas molecules across any plane of constant x_1, then equation (2.1) is obeyed in the gas (Stefan-Maxwell law). For this to be so, X_1 must be equal at all times to the volume of A absorbed by unit area of the liquid surface, i.e. $X_1 = V$. It can be seen that the conditions of the problem correspond to those specified for Class A, with $S = 1$ in eqn. (5.1) and (5.2).

The x_2 co-ordinate of the liquid surface does not change, hence $X_2 = 0$, $c_2(X_2) = c_2(0)$ for all t; in eqn. (2.5), $P = 0$. Eqn. (2.3) be-becomes

$$R = 0, \; c_2(X_2) = c_2(0) = Q \,.\, c_1(X_1),$$

where Q is the solubility of A in the liquid, expressed as volumes of gaseous A per unit volume of liquid per unit mole fraction of A in the gas.

Making the appropriate substitutions and eliminating the unknowns except for α, eqn. (5.7) to (5.9) become

$$\frac{\alpha\sqrt{\dfrac{\pi D_1}{D_2}} - Qc_1(\infty) + c_2(-\infty)}{\alpha\sqrt{\dfrac{\pi D_1}{D_2}} - Q + c_2(-\infty)} = \sqrt{\pi} \,.\, \alpha \exp[\alpha^2] \operatorname{erfc}[\alpha] = f(\alpha), \quad (7.1)$$

where $$V = 2\alpha\sqrt{D_1 t} \quad . \quad . \quad . \quad . \quad (7.2)$$

Fig. 2 and 3 show f as a function of α, which can thus readily be found by trial, and used to find V as a function of t by means of (7.2).

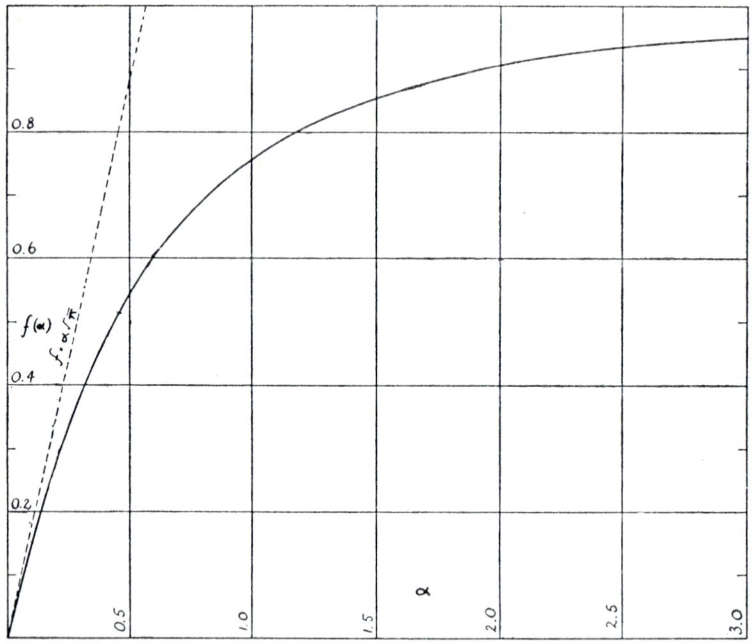

FIG. 2.—$f(\alpha) = \sqrt{\pi} \,.\, \alpha \,.\, e^{\alpha^2} \,.\, \mathrm{erfc}\,(\alpha)$ for positive values of α.

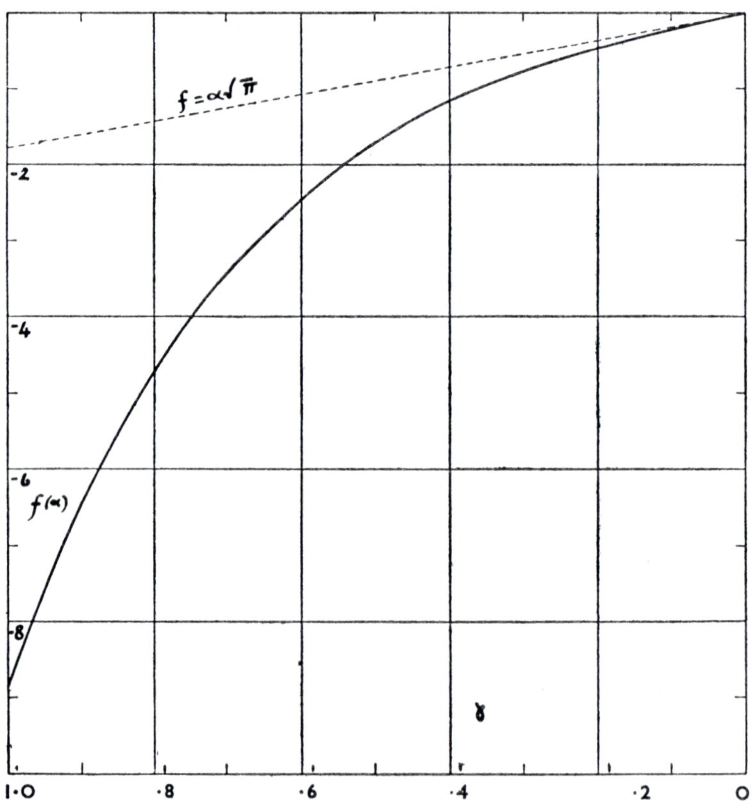

FIG. 3.—$f(\alpha) = \sqrt{\pi} \,.\, \alpha \,.\, e^{\alpha^2} \,.\, \mathrm{erfc}\,(\alpha)$ for negative values of α.

It is further easily shown that

$$c_1(X_1) = \frac{c_1(\infty) - f}{1 - f} = \frac{c_2(-\infty) + \alpha\sqrt{\dfrac{\pi D_1}{D_2}}}{Q} \quad . \quad . \quad (7.3)$$

$$c_1 = c_1(\infty) + [c_1(X_1) - c_1(\infty)] \frac{\operatorname{erfc}\left[\dfrac{X_1}{2\sqrt{D_1 t}}\right]}{\operatorname{erfc}[\alpha]}, \quad . \quad . \quad (7.4)$$

$$c_2 = Qc_1(X_1) + [Qc_1(X_1) - c_2(-\infty)] \operatorname{erf}\left[\frac{X_2}{2\sqrt{D_2 t}}\right]. \quad . \quad (7.5)$$

The same equations may be used, *mutatis mutandis*, for the escape of dissolved A from solution. In this case α will be found to be negative, and the quantity of A leaving the solution in time t is $-V$.

The solution to this problem obtained by Arnold [2] is in its essentials identical with that given above. However, Arnold's derivation is not mathematically rigorous, nor is the application to the general case, where $c_2(0) \neq c_2(-\infty)$, very clearly demonstrated.

It should be noted that the applicability of these equations to real systems depends on the absence of convection currents (this applies to all the systems discussed in this paper which contain fluids). For instance, if CO_2 is absorbed from air at a horizontal water surface the rate of absorption is very much higher than that given,[5] because changes in density cause convection in both gas and liquid (although, as will be shown, only the latter is of importance in this case). On the other hand, the escape of CO_2 from water into air, or the absorption of ammonia from air into water would produce systems in which the density consistently increased in a downward direction, and which would therefore remain stagnant.

It must also be assumed that the heat of solution and rate of absorption of A are not such as to cause serious departure from isothermal conditions, and that variations of the diffusion coefficient with concentration in the liquid will not be great enough to prevent the application of the diffusion equation (2.2). These assumptions are the more likely to be justified the smaller the value of $c_2(X_2) - c_2(-\infty)$ (i.e. when A is not very soluble, or the liquid is initially nearly saturated, or the mole fraction of A in the gas is small), but their validity should be considered in each case.

In cases where Henry's law does not apply, eqn. (2.3) may be assigned values of Q and R which will make it a good approximation to the facts in the neighbourhood of the expected value of $c_1(X_1)$. Eqn. (7.1) will then no longer apply, and eqn. (5.7) to (5.10) must be solved anew.

A special case arises when $c_1(X_1)$ (the interfacial concentration) is determined by some factor other than diffusion in a liquid. Examples are the isothermal evaporation of a liquid into still air ($c_1(X_1)$ determined

[5] Becker, *Ind. Eng. Chem.*, 1924, **16**, 1220.

by vapour pressure) or absorption of a gas into a liquid which is rapidly stirred without deforming the surface ($Q \, . \, c_1(X_1) = c_2(-\infty)$). For such cases, as is easily shown, (5.7) and (5.9) become

$$f(\alpha) = \frac{c_1(X_1) - c_1(\infty)}{c_1(X_1) - 1}, \quad . \quad . \quad . \quad (7.6)$$

where $f(\alpha)$ and α have the same significance as before.

(*b*) Effect of Presence of Insoluble Gas on the Rate of Absorption of a Soluble Gas.—This is an application of the previous solution. Considering the absorption of A into pure liquid ($c_2(-\infty) = 0$), we see from (3.5) that the quantity absorbed at time t will be

$$V = 2Qc_1(X_1)\sqrt{\frac{D_2 t}{\pi}}. \quad . \quad . \quad . \quad . \quad (7.7)$$

It is of interest to see how widely, under typical conditions, the mole fraction of A at the surface departs from that in the body of the gas, that is, how closely the rate of absorption can be approximated by putting $c_1(\infty)$ for $c_1(X_1)$ in eqn. (7.7), and also the extent to which the presence of a small quantity of an insoluble impurity will slow down the absorption of a soluble gas.

The values of D_1 and D_2 will, in the case of most gases and liquids at ordinary temperatures, be in the neighbourhood of 0·1 and 2×10^{-5} cm.^{2}/sec. respectively, making $\sqrt{D_1/D_2} \approx 70$. Now suppose the value of Q is unity (a solubility comparable to that of CO_2 in water at ordinary temperatures); trial in eqn. (7.1) shows that α must be very small. Expanding exp $[\alpha^2]$ and erfc $[\alpha]$ we find that when $\alpha \ll 1$, $f \approx \alpha\sqrt{\pi}$. Hence from (7.1)

$$f \approx \frac{70f - c_1(\infty)}{70f - 1}, \quad . \quad . \quad . \quad . \quad (7.8)$$

whence $f \approx c_1(\infty)/70$. Thus, from (7.3), $c_1(X_1)/c_1(\infty) \approx 0{\cdot}986$, so the error involved in putting $c_1(\infty)$ for $c_1(X_1)$ in (7.7) will be less than 2 %.

In contrast to this, consider the case where $Q = 500$ (a value comparable with the solubility of ammonia in water at ordinary temperatures). If the bulk mole fraction, $c_1(\infty)$, of A is 0·95, $c_1(X_1)$ is found to be about 0·50.

This illustrates the well-known fact that diffusion in the gas does not play an important part in determining the rate of absorption unless the soluble gas has a fairly high solubility; with substances of low or moderate solubility, the concentration-gradient in the gas is small. Similarly, the "blanketing" effect of a small quantity of an insoluble gas will be appreciable only if the solubility of the soluble gas is fairly high.

(*c*) Tarnishing Reactions (Gas Diffusing).—A film of tarnish is formed on the surface of a metal by reaction with a gas. The reaction proceeds by diffusion of dissolved gas through the film to the surface of the metal, where its concentration $c_1(X_1)$ is assumed to be zero (i.e. the reaction is assumed to be so rapid that the rate of tarnishing is controlled entirely by the diffusion process). The outer surface of the film is constantly saturated with the gas.

Let the film be medium 1. Since $c_1(X_1)$ is not determined by diffusion in another medium, eqn. (5.8) and (5.10) are not required.

Let W be the mass fraction of the gaseous component in the compound which it forms with the metal, ρ the density of this compound (which is assumed independent of the concentration of the dissolved gas) and c_1 the concentration of dissolved gas (mass per unit volume of the film) at a distance x_1 beneath the surface of the film. The outer surface of the film is at $x_1 = 0$, the metal surface at $x_1 = X_1$, the film-thickness being X_1.

From eqn. (5.1)

$$S = -1/W\rho.$$

$c_1(0)$ is the saturated concentration of gas at the outer surface of the film; $c_1(X_1) = 0$. Then from (5.7) and (5.9)

$$\frac{c_1(0)}{W\rho} = \sqrt{\pi}\,.\,\alpha \exp[\alpha^2] \operatorname{erf}[\alpha] = g(\alpha) \qquad \cdot \qquad \cdot \qquad (7.9)$$

while

$$X_1 = 2\alpha\sqrt{D_1 t} \qquad \cdot \qquad \cdot \qquad \cdot \qquad \cdot \qquad (5.6)$$

where D_1 is the diffusivity of dissolved gas in the film. Fig. 4 gives g as a function of α, so that values of α corresponding to given values of $c_1(0)/W\rho$ can be found directly and used to calculate film thickness as a function of time from (5.6).

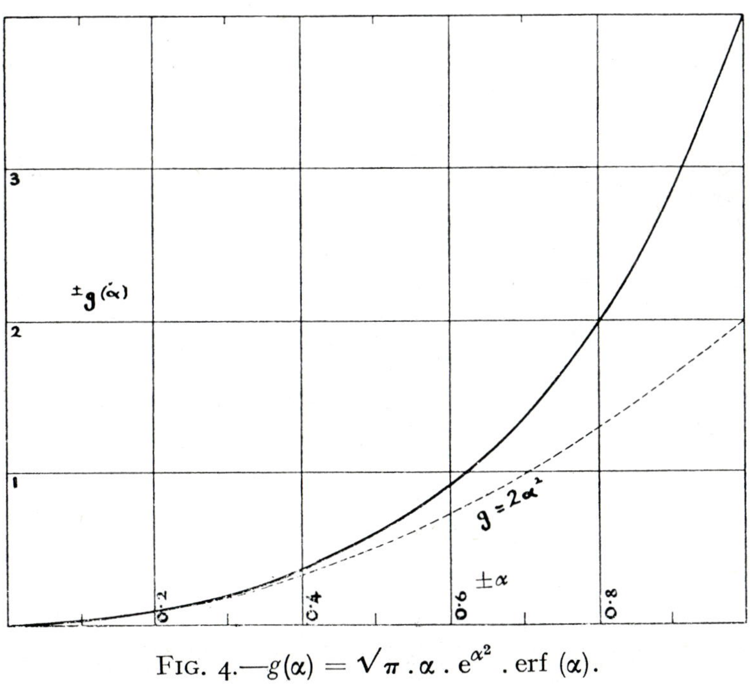

FIG. 4.—$g(\alpha) = \sqrt{\pi}\,.\,\alpha\,.\,e^{\alpha^2}\,.\operatorname{erf}(\alpha)$.

If $c_1(0) \ll W\rho$, expansion of exp $[\alpha^2]$ and erf $[\alpha]$ shows that $g \approx 2\alpha^2$, whence

$$X_1 = \sqrt{\frac{2D_1c_1(0)t}{W\rho}}. \qquad (7.10)$$

In other words, if the solubility of the gas in the film material is sufficiently small, the thickness may be calculated without appreciable error from the conventional "parabolic" formula which assumes that the concentration gradient in the film is uniform.

Eqn. (7.9) and (5.6) are equivalent to the solution to this problem obtained by Booth,[3] who points out also that the approximate form will be very close to the exact one for the type of system normally encountered.

(*d*) TARNISHING REACTIONS (METAL DIFFUSING).—Eqn. (7.9), (5.6) and (7.10) will apply in this case also if $c_1(0)$ = saturated solubility of metal in film (mass/unit vol.), D_1 = diffusivity of metal dissolved in film, ρ = density of film, W = mass-fraction of metal in film-compound.

It may be noted that the methods of this paper could be used to deal with the case where both metal and gas diffuse through the film in opposite directions, provided that the reaction between them is fast enough to ensure that the concentration of one diffusant is virtually zero wherever the concentration of the other is appreciable.

(*e*) CONDENSATION OF A VAPOUR AT THE SURFACE OF A COLD LIQUID OF INFINITE DEPTH.—Here heat is the diffusant, and the liquid is taken to be medium 1. The vapour (which contains no uncondensable gases) is at its saturation temperature T_X, and is at equilibrium with the surface of the liquid at the same temperature. The heat content of the vapour is higher than that of the liquid at the surface by an amount equal to the latent heat of vaporization L (per unit mass). The liquid is initially at a uniform temperature T_∞.

The surface of the liquid is originally at $x_1 = 0$ and is at $x_1 = X_1$ at time t; X_1 is always a negative value of x_1, so that the depth of liquid which has condensed up to time t is $-X_1$, and the mass M condensed per unit area is $-X_1\rho$.

Taking the heat content of the liquid at the surface to be zero, we have

$$c_1(X_1) = 0, \quad c_1 = \rho\sigma(T - T_X),$$

where ρ, σ are the density and specific heat of the liquid (both taken to be independent of the temperature), and T is the temperature of the liquid at a depth x_1 below the surface. From (5.1) we see that

$$S = 1/L\rho.$$

Eqn. (5.7) and (5.9) become

$$\frac{\sigma(T_\infty - T_X)}{L} = f(\alpha), \qquad (7.11)$$

where $f(\alpha)$ is the function of α given in (7.1) and plotted in Fig. 2 and 3. From (5.6)

$$M = -X_1\rho = -2\alpha\rho\sqrt{D_1 t}, \quad . \quad . \quad . \quad (7.12)$$

where D_1 is the thermal diffusivity of the liquid, being equal to $\kappa/\rho\sigma$, where κ is the thermal conductivity. Hence (7.12) can be written

$$M = -X_1\rho = -2\alpha\sqrt{\frac{\kappa t\rho}{\sigma}}. \quad . \quad . \quad . \quad (7.13)$$

For the temperature distribution in the liquid we have from (3.1)

$$T = T_X - (T_X - T_\infty)\left(1 - \operatorname{erfc}\left[\frac{X_1}{2}\sqrt{\rho\sigma/\kappa t}\right]\Big/\operatorname{erfc}[\alpha]\right) \quad (7.14)$$

When $\sigma(T_X - T_\infty) \ll L$, f is small, hence $f \approx \alpha\sqrt{\pi}$, and (7.12) becomes

$$M \approx \frac{2}{L}(T_X - T_\infty)\sqrt{\frac{\kappa t\rho\sigma}{\pi}}. \quad . \quad . \quad . \quad (7.15)$$

The same equations may be used for evaporation, in circumstances such that no convection currents arise.

(*f*) Vapour Condensing on a Cooled Surface.—The cooled surface ($x_1 = 0$) is assumed to be maintained at a constant temperature T_0. The depth of liquid which has condensed on the surface at time t is $-X_1$, and $-X_1\rho$ the mass condensed per unit area. Using the symbols of (*e*) above, we find

$$\frac{\sigma(T_X - T_0)}{L} = g(\alpha), \quad . \quad . \quad . \quad . \quad (7.16)$$

where $g(\alpha)$ is the function of α given in (7.9) and plotted in Fig. 4. Here α will give the value of M or X_1 when substituted in eqn. (7.12).

(*g*) Reaction of a Gas at a Solid Surface.—A pure gas A reacts with a solid to form a gaseous compound B. The reaction is so rapid that the concentration $c_1(X_1)$ of A at the interface is negligible. n_1 moles of A react to give n_2 of B. The gas is taken to be medium 1, and occupies the region $0 < x_1 < \infty$. The plane $x_1 = 0$ moves relative to the solid surface in such a way that there is no net flow of gas molecules across any plane of constant x_1. Hence

$$X_1 = V(1 - n_2/n_1),$$

where V is the volume of A reacted in time t, and X_1 is the x_1-co-ordinate of the surface of the solid. From (5.1) we see that

$$S = (1 - n_2/n_1).$$

c_1 is the mole fraction of A, $(1 - c_1)$ the mole-fraction of B, at x_1; the initial mole fraction of A in the gas $c_1(\infty)$ will be taken to be unity. Eqn. (5.7) and (5.9) now become

$$1 - n_2/n_1 = f(\alpha), \quad . \quad . \quad . \quad . \quad (7.17)$$

where $f(\alpha)$ is the function of α specified in (7.1) and plotted in Fig. 2 and 3.

Now if $n_1 \neq n_2$, from (5.6)

$$V = \frac{X_1}{(1 - n_2/n_1)} = \frac{2\alpha\sqrt{D_1 t}}{(1 - n_2/n_1)} \quad . \quad . \quad (7.18)$$

while if $n_1 = n_2$, then $c_1(X_1) = c_1(0) = 0$, and hence from (3.4)

$$V = 2\sqrt{\frac{D_1 t}{\pi}} \quad . \quad . \quad . \quad . \quad . \quad . \quad (7.19)$$

8. Examples : Class B.—(*a*) SOLUTION OF GAS IN A LIQUID, FOLLOWED BY REACTION WITH SOLUTE.—The surface of the liquid is constantly saturated with the gas B ; it is assumed either that the latter is undiluted with inert gases, or that the solubility of B is sufficiently small to permit the assumption that the composition of the phase is uniform (see example, § 7(*b*)). The case where this restriction is not imposed is discussed later.

The liquid contains a solute A which reacts virtually instantaneously and irreversibly with the dissolved gas. Laws of the form (2.1) hold for the diffusion of A and B in the liquid. Any part of the liquid will contain A or B but not both. There will be a plane, $x_1 = X_1$, $x_2 = X_2$, at which the concentrations of both will be zero. The part of the liquid containing A is taken to be medium 1, that containing B is medium 2. n_1 moles of A react with n_2 moles of B. The surface of the liquid is permanently at $x_1 = x_2 = 0$, so that $X_1 = X_2$, $P = 1$ (eqn. (2.5)). Also

$$c_1(X_1) = c_2(X_2) = 0,$$

hence $R = 0$ (eqn. (2.3)). In order to fulfil the conservation condition (2.4), put $c_1 = -m_1 n_2/n_1$, where m_1 is the concentration of A at x_1 (moles per unit volume). c_2 is the concentration of B at x_2 in the same units. $c_2(0)$ is the saturated concentration of B at the surface of the liquid. $m_1(\infty)$ is the initial concentration of A in the liquid.

Eqn. (6.3) and (6.5) then become

$$n_1 c_2(0) f\left[\frac{\beta}{\sqrt{D_1}}\right] - n_2 m_1(\infty) g\left[\frac{\beta}{\sqrt{D_2}}\right] = 0, \quad . \quad . \quad (8.1)$$

where f, g are the functions given (with α as variable) in eqn. (7.1) and (7·9) respectively, and plotted in Fig. 2, 3 and 4. Hence from (3.5), (6.5) we find the volume of gaseous B absorbed in time t to be

$$V = \frac{2c_2(0)}{\operatorname{erf}\left[\frac{\beta}{\sqrt{D_2}}\right]}\sqrt{\frac{D_2 t}{\pi}}. \quad . \quad . \quad . \quad (8.2)$$

The value of β can be found from (8.1) by trial, using Fig. 2, 3 and 4, and substituted in (8.2) to find V.

If the gas is diluted by an insoluble component, the surface concentration ($c_2(0)$) of B in the liquid will be determined by diffusion in the gas. In this case, to obtain an exact solution, (7.2) and (7.6) must be taken in conjunction with (8.1) and (8.2) ; by trial a value of the concentration at the surface of the liquid must be found which will make the rate of absorption the same when calculated from either set of equations. The quantities D_1, $c_1(\infty)$ and $c_1(X_1)$ in (7.2) and (7.6) will refer respectively to the diffusion coefficient in the gas and to the mole fraction of B in the bulk of the gas and at the interface

(*b*) PROGRESSIVE FREEZING OF A LIQUID.—Let the liquid be medium 1 and the solid medium 2. The temperature of the exposed surface of the solid is maintained constant at T_0; since no material crosses this boundary $x_2 = 0$ for all t at the surface. If there is a change in volume on freezing, there will be relative movement of the planes $x_1 = 0$, $x_2 = 0$.

The diffusant in this case is heat. L is the latent heat of fusion (per unit mass); ρ_1, ρ_2 are the densities, σ_1, σ_2 the specific heats and D_1, D_2 the thermal diffusivities of liquid and solid respectively. The variation of these quantities with temperature will be ignored. If κ_1, κ_2 are the thermal conductivities of liquid and solid respectively, then

$$D_1 = \kappa_1/\rho_1\sigma_1, \quad D_2 = \kappa_2/\rho_2\sigma_2.$$

The initial temperature of the liquid is uniformly equal to T_∞, and the solid-liquid interface ($x_1 = X_1$, $x_2 = X_2$) is always at the melting point T_X. The heat content of the solid at T_X is taken to be zero; hence that of the liquid at T_X is L. From (2.3), since $c_2(X_2) = 0$,

$$c_1(X_1) = -R/Q = L\rho_1.$$

Since $X_1/X_2 = \rho_2/\rho_1$, from (2.5)

$$P = \rho_1/\rho_2.$$

We also have

$$c_1 = \rho_1\sigma_1(T_1 - T_X + L/\sigma_1)$$

and

$$c_2 = \rho_2\sigma_2(T_2 - T_X),$$

where T_1 is the temperature at x_1 (in the liquid) and T_2 the temperature at x_2 (in the solid).

Eliminating $c_1(0)$, $c_2(-\infty)$ from eqn. (6.3) to (6.5), and substituting as above, we find

$$\frac{\sigma_1(T_\infty - T_X)}{f\left[\dfrac{\beta}{\sqrt{D_1}}\right]} + \frac{\sigma_2(T_0 - T_X)}{g\left[\dfrac{\rho_1\beta}{\rho_2\sqrt{D_2}}\right]} + L = 0, \qquad . \quad . \quad (8.3)$$

where f and g are the functions given (with α as variable) in (7.1) and (7.9) respectively, and plotted in Fig. 2, 3 and 4.

The thickness of the solid at time t is (from (6.2))

$$X_2 = X_1\rho_1/\rho_2 = 2\rho_1\beta\sqrt{t}/\rho_2 \qquad . \quad . \quad . \quad (8.4)$$

The temperature T_2 at any point in the solid is

$$T_2 = T_X - (T_X - T_0)\left(1 - \frac{\operatorname{erf}[x_2/2\sqrt{D_2 t}]}{\operatorname{erf}[\rho_1\beta/\rho_2\sqrt{D_2}]}\right), \qquad . \quad (8.5)$$

while the temperature in the liquid is

$$T_1 = T_X - (T_X - T_\infty)\left(1 - \frac{\operatorname{erfc}[x_1/2\sqrt{D_1 t}]}{\operatorname{erfc}[\beta/\sqrt{D_1}]}\right). \qquad . \quad (8.6)$$

Hence (8.3) may be solved by trial, with the aid of Fig. 2, 3 and 4, and the resulting value of β used to calculate the thickness of the solid and the temperature at any point in the system as functions of time.

If we put $\rho_1 = \rho_2$ in this solution (i.e. no change of volume on freezing) we obtain the form found in the text-books ; [1] this form applies exactly to such phenomena as the freezing of damp soil, where there is in fact no appreciable change in volume on freezing.

It should be noted that a real system of liquid and solid will only behave in the way described by the formulae given above if the density increases consistently in a downward direction (or is uniform), as otherwise convection currents will arise. Thus the formulae could be used for the freezing from the top of water initially below 4° C, or for the freezing from the bottom of a liquid whose density increases consistently with decreasing temperature.

The above results apply, *mutatis mutandis*, to the melting of a solid which is initially at a uniform temperature.

Department of Chemical Engineering,
Tennis Court Road,
Cambridge.

A4

ABSORPTION BY SIMULTANEOUS DIFFUSION AND CHEMICAL REACTION INTO PARTICLES OF VARIOUS SHAPES AND INTO FALLING DROPS

BY P. V. DANCKWERTS

Received 17th January, 1951

A substance diffuses into an absorbent in which it is destroyed by a first-order process. A method is given by which expressions for the transient concentration-distribution and rate of absorption can be obtained by transformation of expressions for diffusion without reaction. For absorbent particles of simple shapes the latter correspond to solutions of well-known problems in heat-conduction and can be found in text-books. The method is illustrated by application to the sphere, parallelepiped, cylinder and semi-infinite absorbent, and also to a liquid drop moving through another liquid. Steady-state solutions are also discussed.

---

1. Introduction.—The process considered is the absorption of some substance, called the diffusant, by another which will be called the absorbent. The absorbent may be in the form of a particle of finite size, or may be a thread, rod or sheet of infinite extent. It is immersed in fluid atmosphere or solution of the diffusant. As the latter diffuses into the absorbent, it is destroyed by a process which is first-order with respect to the diffusant.

The expressions obtained by the methods of this paper may prove useful in connection with unsteady state diffusion into porous adsorbents and catalysts, liquid drops in two-phase reaction systems, or living cells, including the micro-organism employed in the fermentation and other biochemical industries. Expressions for the absorption of oxygen into red blood-cells have already been obtained by Roughton [1] and by Roughton and Nicolson.[3]

In practice, the reaction by which the diffusant is destroyed will not generally be first order. Nevertheless, a first-order rate equation may often serve as an approximation for sufficiently short times of contact. The conditions under which a second-order reaction between diffusant and absorbent can be represented by a first-order equation have been examined elsewhere.[2] Roughton and Nicolson [3] have obtained numerical solutions for a second-order process for one shape of absorbent.

2. Conditions of the Problem.—It is assumed that there are no convection currents in the absorbent (except in the case of the falling drop, § 6) that it does not swell or shrink during the process of absorption, and that the diffusivity of the diffusant in the absorbent is uniform and constant. Under these conditions, as shown elsewhere,[2] the equation representing the concentration c of the diffusant in the absorbent is

$$\frac{\partial c}{\partial t} = D\frac{\partial^2 c}{\partial x^2} - kc, \qquad (2.1)$$

where D = diffusivity, k = first-order reaction-velocity constant, when diffusion in the x-direction only is considered. This is also the equation representing the conduction of heat along a wire which loses heat from its surface at a rate proportional to its temperature. Some solutions may be found in the text-books.[4] The general 3-dimensional equation, as can easily be shown, is

$$\left.\begin{aligned}\frac{\partial c}{\partial t} &= D\left(\frac{\partial^2 c}{\partial x^2} + \frac{\partial^2 c}{\partial y^2} + \frac{\partial^2 c}{\partial z^2}\right) - kc,\\ &= D\nabla^2 c - kc.\end{aligned}\right\} \qquad (2.2)$$

[1] Roughton, *Proc. Roy. Soc. B*, 1932, **111**, 1.

[2] Danckwerts, *Trans. Faraday Soc.*, 1950, **46**, 300.

[3] Roughton and Nicolson, *Proc. Roy. Soc. B* (in press).

[4] Carslaw and Jaeger, *Conduction of Heat in Solids* (Oxford University Press, 1947), Chap. 4.

In the problems considered here the boundary conditions at the surface of the absorbent will be one of the following:

(*a*) SURFACE-SATURATION CONDITION.—The surface is at equilibrium at all times with the surrounding atmosphere or solution which is assumed to be of constant composition, so that the concentration in the absorbent at the interface has a constant value c^*.

(*b*) SURFACE-RESISTANCE CONDITION.—The rate of absorption per unit area of surface is proportional at any time to the difference between the saturated concentration c^* and the actual concentration in the absorbent at the interface. If the concentration-gradient in the absorbent is $\partial c/\partial N$ at any point on the surface (measured in an outward direction along the normal to the surface), the local rate of absorption per unit area is $D\frac{\partial c}{\partial N}$. Hence the surface-resistance condition may be written

$$D\frac{\partial c}{\partial N} = k_s(c^* - c_s), \qquad (2.3)$$

where c_s is the surface-concentration at the point to which $\partial c/\partial N$ refers, and k_s is a constant. A condition of this form will arise if the absorbent is covered by a thin membrane of low permeability, as will be the case with many micro-organisms and cells. In this case,

$$k_s = D'/\delta, \qquad (2.4)$$

where D' is the diffusivity in the membrane and δ its thickness. The condition may also be used as a working approximation if diffusion of the diffusant up to the outer surface of the medium, through the surrounding solution or atmosphere, is of importance in determining the rate of absorption. The assumption that condition (2.3) is obeyed is commonly made when considering such mass-transfer processes, and often represents the facts quite well. The mass-transfer coefficient k_s will in this case depend on the degree of agitation of the fluid. There is also the possibility that the passage of diffusant molecules across the interface is intrinsically a slow process. In this case condition (2.3) will apply if the process is kinetically first-order in both directions.

Condition (2.3) is equivalent to the " radiation " boundary-condition in heat-transfer, and in order to conform to the nomenclature of Carslaw and Jaeger [5] will be re-written

$$\partial c/\partial N = h(c^* - c_s), \qquad (2.5)$$

where

$$h = k_s/D.$$

3. General Mathematical Method.—The method consists in the transformation by a standard process of the solution of the corresponding problem in diffusion *without* reaction. That is, the solution c of eqn. (2.2), can easily be obtained if the solution c' of

$$\left.\begin{aligned}\frac{\partial c'}{\partial t} &= D\left(\frac{\partial^2 c'}{\partial x^2} + \frac{\partial^2 c'}{\partial y^2} + \frac{\partial^2 c'}{\partial z^2}\right)\\ &= D\,.\,\nabla^2 c'\end{aligned}\right\} \qquad (3.1)$$

[5] *Idem, ibid.*, p. 13.

is known, c and c' having the same boundary conditions. Appropriate solutions to eqn. (3.1) have been obtained for the conduction of heat in various simple shapes, and can be found in the text-books of the subject.

In the problems under consideration the initial concentration in the absorbent is zero, so that the boundary conditions are :

$$c = 0,\ t = 0, \text{ all points in absorbent,} \qquad (3.2)$$

$$c = c^*,\ t > 0, \text{ all points at surface of absorbent,} \qquad (3.3)$$

$$\partial c/\partial N = h(c^* - c),\ t > 0, \text{ all points at surface of absorbent} \qquad (3.3a)$$

(either (3.3) or (3.3*a*) to be used according to circumstances). Let c' be the solution of eqn. (3.1), the boundary conditions on c' being the same as those on c. The required solution to eqn. (2.2) is then

$$c = k\int_0^t e^{-kt} \,.\, c' \,.\, dt + c' \,.\, e^{-kt} \qquad (3.4)$$

This is easily shown as follows. Differentiating (3.4) :

$$\frac{\partial c}{\partial t} = kc' e^{-kt} - kc' e^{-kt} + \frac{\partial c'}{\partial t} \,.\, e^{-kt} = \frac{\partial c'}{\partial t} \,.\, e^{-kt} \qquad (3.5)$$

and

$$\nabla^2 c = k\int_0^t \nabla^2 c' \,.\, e^{-kt} dt + \nabla^2 c' \,.\, e^{-kt}.$$

Substituting from eqn. (3.1) into the above

$$D\nabla^2 c = k\int_0^t \frac{\partial c'}{\partial t} e^{-kt} dt + e^{-kt}\frac{\partial c'}{\partial t}.$$

Substituting from eqn. (3.5),

$$D\nabla^2 c = kc + \frac{\partial c}{\partial t},$$

so that (3.4) is a solution of eqn. (2.2). When $t = 0$ we find immediately from eqn. (3.4) that $c = c'$, so that c obeys the required initial conditions. For points at which $c' = c^*$, independent of t,

$$c = kc^*\int_0^t e^{-kt}\, dt + c^* e^{-kt} = c^*.$$

For points at which

$$\partial c'/\partial N = h(c^* - c'),$$

h and c^* being independent of time,

$$\begin{aligned}\frac{\partial c}{\partial N} &= k\int_0^t \frac{\partial c'}{\partial N} e^{-kt}\, dt + \frac{\partial c'}{\partial N} e^{-kt}\\ &= hk\int_0^t (c^* - c')e^{-kt}\, dt + (c^* - c')h\, e^{-kt}\\ &= h(c^* - c).\end{aligned}$$

The solution given in eqn. (3.4) therefore fulfils the required boundary conditions.

The quantity Q of diffusant absorbed in time t is usually of more interest than c. From eqn. (3.4)

$$R = \frac{dQ}{dt} = \iint D \frac{\partial c}{\partial N} ds = k \int_0^t R' . e^{-kt} . dt + R' . e^{-kt}. \quad . \quad (3.6)$$

The area integral is taken over the whole surface of the absorbent, and R' is the corresponding rate of absorption when no chemical reaction takes place; i.e.

$$R' = \frac{dQ'}{dt} = \iint D \frac{\partial c'}{\partial N} ds = \frac{d}{dt} \iiint c' dx dy dz. \quad . \quad . \quad (3.7)$$

The volume integral is taken over the whole volume of the absorbent.

Many of the solutions for diffusion *without* reaction are available in the form of infinite series which can be written

$$c'/c^* = 1 - \sum f\{x, y, z\} e^{-\nu t} \quad . \quad . \quad . \quad . \quad (3.8)$$

f and ν are different for each term in the series, but are not functions of t; ν is not a function of x, y, z. Applying (3.4) to this, we find

$$\frac{c}{c^*} = 1 - \sum f\{x, y, z\} \left(\frac{k + \nu . \exp[-t(k + \nu)]}{k + \nu} \right). \quad . \quad (3.9)$$

Similarly, since in such cases R' will be of the form

$$R'/c^* = \sum g . e^{-\nu t}, \quad . \quad . \quad . \quad . \quad (3.10)$$

where g is different for each term in the series and is not a function of x, y, z, t, we obtain from (3.6),

$$\frac{R}{c^*} = \sum g \left(\frac{k + \nu . \exp[-t(k + \nu)]}{k + \nu} \right), \quad . \quad . \quad (3.11)$$

and $$Q = \int_0^t R dt = c^* \sum g \left(\frac{kt(k + \nu) - \nu \exp[-t(k + \nu)] + \nu}{(k + \nu)^2} \right). \quad (3.12)$$

4. Examples.—The method can be applied, by transformation of well-known expressions for c', to the following shapes of absorbent: semi-infinite with plane surface, sheet or slab, rectangular edge, rectangular corner, rectangular rod, rectangular block or parallelepiped, sphere, cylindrical rod or thread, cylinder (ends and side). Expressions for the semi-infinite absorbent with surface saturation [2] and for the sheet with and without surface resistance [1, 3] are already available. The method will be illustrated by its application to three cases.

(*a*) SPHERE WITH SURFACE RESISTANCE ($0 < r < a$; radius $= a$).—The solution for diffusion without reaction is [6]

$$\frac{c'}{c^*} = 1 - \frac{2h}{r} \sum_{n=1}^{\infty} \frac{e^{-D\alpha_n^2 t}}{\alpha_n^2} \left[\frac{a^2\alpha_n^2 + (ah - 1)^2}{a^2\alpha_n^2 + ah(ah - 1)} \right] \sin a\alpha_n . \sin r\alpha_n, \quad (4.1)$$

[6] Carslaw and Jaeger, *Conduction of Heat in Solids* (Oxford University Press, 1947), p. 203.

where α_n are the successive roots of

$$a\alpha \, . \cot a\alpha + ah - 1 = 0, \quad . \quad . \quad . \quad (4.2)$$

which have been tabulated.[7] Hence

$$R' = 4\pi a^2 D \left(\frac{\partial c'}{\partial r}\right)_{r=a} = 8\pi h^2 c^* Da^2 \sum_{n=1}^{\infty} \frac{e^{-D\alpha_n^2 t}}{a\alpha_n^2 + h(ah - 1)}, \quad . \quad (4.3)$$

and from (3.11)

$$R = 8\pi h^2 c^* Da^2 \sum_{n=1}^{\infty} \frac{k + D\alpha_n^2 \, . \exp\,[-\, t(k + D\alpha_n^2)]}{(k + D\alpha_n^2)[a\alpha_n^2 + h(ah - 1)]} \quad . \quad (4.4)$$

and from (3.12)

$$Q = 8\pi h^2 c^* Da^2 \sum_{n=1}^{\infty} \frac{kt(k + D\alpha_n^2) - D\alpha_n^2\,(\exp\,[-\, t(k + D\alpha_n^2)] - 1)}{(k + D\alpha_n^2)^2[a\alpha_n^2 + h(ah - 1)]} \quad (4.5)$$

(*b*) Rectangular Block or Parallelepiped with Surface Saturation.—($-\, a < x < a, \quad -\, b < y < b, \quad -\, c < z < c$; block has edges of length $2a$, $2b$, $2c$).—The solution for diffusion without reaction is [8]

$$\frac{c'}{c^*} = 1 - \frac{64}{\pi^3} \sum_{l=0}^{\infty} \sum_{m=0}^{\infty} \sum_{n=0}^{\infty} \frac{(-\, 1)^{l+m+n}}{(2l + 1)(2m + 1)(2n + 1)} \, . \cos \frac{(2l + 1)\pi x}{2a}$$

$$. \cos \frac{(2m + 1)\pi y}{2b} \, . \cos \frac{(2n + 1)\pi z}{2c} \, . \, e^{-\alpha_{l,m,n} t} \quad (4.6)$$

where

$$\alpha_{l,m,n} = \frac{\pi^2 D}{4}\left[\frac{(2l + 1)^2}{a^2} + \frac{(2m + 1)^2}{b^2} + \frac{(2n + 1)^2}{c^2}\right]. \quad . \quad (4.7)$$

Hence

$$R' = \frac{d}{dt} \int_{-a}^{a} \int_{-b}^{b} \int_{-c}^{c} c' \, . \, dz dy dx$$

$$= \frac{(64)^2}{\pi^6} c^* \, . \, abc \sum_{l=0}^{\infty} \sum_{m=0}^{\infty} \sum_{n=0}^{\infty} \frac{\alpha_{l,m,n} e^{-\alpha_{l,m,n} t}}{(2l + 1)^2(2m + 1)^2(2n + 1)^2}, \quad (4.8)$$

and from (3.11)

$$R = \frac{(64)^2}{\pi^6} c^* \, . \, abc \sum_{l=0}^{\infty} \sum_{m=0}^{\infty} \sum_{n=0}^{\infty} \frac{\alpha_{l,m,n}\,(k + \alpha_{l,m,n} \exp\,[-t(k + \alpha_{l,m,n})])}{(k + \alpha_{l,m,n})(2l + 1)^2(2m + 1)^2(2n + 1)^2} \quad (4.9)$$

Q is easily obtained from (3.12).

(*c*) Infinite Cylinder with Surface Resistance ($0 \leqslant r \leqslant a$; radius $= a$).—We take the curved surface of unit length of the cylinder to be the absorbing surface. The expression for diffusion without reaction is found to be [9]

[7] *Idem, ibid.*, p. 378.

[8] *Idem, ibid.*, p. 163.

[9] Carslaw and Jaeger, *Conduction of Heat in Solids* (Oxford University Press, 1947), p. 177.

$$\frac{c'}{c^*} = 1 - \frac{2h}{a} \sum_{n=1}^{\infty} \frac{J_0(r\alpha_n) e^{-D\alpha_n^2 t}}{J_0(a\alpha_n) \cdot (h^2 + \alpha_n^2)}, \quad . \quad . \quad (4.10)$$

where α_n are the roots of

$$\alpha J_1(a\alpha) = h J_0(a\alpha), \quad . \quad . \quad . \quad . \quad (4.11)$$

and J_0, J_1 are Bessel functions of the first kind, zero and first order respectively. (Values of α have been tabulated.) [10] Hence

$$R' = \frac{d}{dt} \int_0^a 2\pi r c' \cdot dr = 4\pi h^2 c^* D \sum_{n=1}^{\infty} \frac{e^{-D\alpha_n^2 t}}{(h^2 + \alpha_n^2)}, \quad . \quad (4.12)$$

(the method of integration is given by Carslaw and Jaeger [11]), and from (3.11)

$$R = 4\pi h^2 c^* D \sum_{n=1}^{\infty} \frac{k + D\alpha_n^2 \exp[-t(k + D\alpha_n^2)]}{(k + D\alpha_n^2)(h^2 + \alpha_n^2)}. \quad . \quad (4.13)$$

Q is easily found from (3.12).

Other solutions for R will be given without details of their derivation.

(*d*) Sphere with Surface Saturation (radius $= a$).—

$$R = 8\pi a D c^* \sum_{n=1}^{\infty} \frac{ka^2 + Dn^2\pi^2 \cdot \exp\left[-t\left(k + \frac{Dn^2\pi^2}{a^2}\right)\right]}{ka^2 + Dn^2\pi^2} \quad . \quad (4.14)$$

(*e*) Infinite Cylinder with Surface Saturation (radius $= a$; absorption into unit length of curved surface).—

$$R = 4\pi D c^* \sum_{n=1}^{\infty} \frac{k + D\alpha_n^2 \cdot \exp[-t(k + D\alpha_n^2)]}{k + D\alpha_n^2}, \quad . \quad (4.15)$$

where α_n are the roots of

$$J_0(a\alpha) = 0, \quad . \quad . \quad . \quad . \quad . \quad (4.16)$$

which have been tabulated.[10]

(*f*) Finite Cylinder with Surface Saturation (radius $= a$; length $= 2l$; absorption into ends and curved surface).—

$$R = \frac{64Dlc^*}{\pi} \sum_{n=0}^{\infty} \sum_{m=1}^{\infty} \left[\frac{\beta_{m,n}}{\alpha_m^2 (2n+1)^2}\right] \left[\frac{k + D\beta_{m,n} \exp\{-t(k + D\beta_{m,n})\}}{k + D\beta_{m,n}}\right], \quad (4.17)$$

where α_m are the roots of eqn. (4.16) and

$$\beta_{m,n} = \alpha_m^2 + \frac{(2n+1)^2 \pi^2}{4l^2}.$$

(*g*) Semi-infinite Solid with Plane Surface and Surface Resistance (absorption into unit area).—The solution is derived by the

[10] *Idem, ibid.*, p. 379 (the values in Table III for $C = \infty$ correspond to the roots of $J_0(a\alpha) = 0$).

[11] *Idem, ibid.*, p. 173.

application of eqn. (3.6) to the corresponding solution for diffusion without reaction.[12]

$$R=\frac{hDc^*}{h^2D-k}\left[h\sqrt{Dk}\ .\ \mathrm{erf}\ \sqrt{kt}+h^2D\ .\ \mathrm{erfc}\ h\sqrt{Dt}\ .\ \exp\{t(h^2D-k)\}-k\right] \quad (4.18)$$

5 Steady-state Solutions.—As a result of the chemical reaction, in all systems of the type under consideration the rate of absorption and the concentration of diffusant at any point in the absorbent tend to a steady finite value as the time of contact increases. The steady-state solutions may be obtained by setting $\partial c/\partial t$ equal to zero in the appropriate form of eqn. (2.2), which may then be solved to give the steady-state solution directly. When the expression for c or R takes the form of eqn. (3.9) or (3.11), the steady-state solution may be obtained merely by omitting the time-dependent term, which tends to zero as t tends to infinity. Solutions of this type in series form may be less convenient in use than those obtainable by other methods in closed form.

In certain cases the following method may be found convenient. Putting $t = \infty$ in eqn. (3.4) and (3.6) it is found that

$$\frac{c_\infty}{k} = \int_0^\infty c'\ .\ e^{-kt}\ .\ dt, \quad (5.1)$$

$$\frac{R_\infty}{k} = \int_0^\infty R'\ .\ e^{-kt}\ .\ dt. \quad (5.1)$$

The expressions on the right-hand side of these equations are the Laplace transforms[13] of c' and R' respectively, the parameter (usually written as p or s) taking the value k. For example, for the semi-infinite solid with surface resistance (eqn. (4.18)) we have[12]

$$R' = hD(c^* - c_s') = hDc^*\, e^{h^2Dt}\ .\ \mathrm{erfc}\ (h\sqrt{Dt}). \quad (5.3)$$

The Laplace transform of the right-hand expression is

$$hDc^*/(k + h\sqrt{Dk}),$$

whence

$$R_\infty = hDc^*\Big/\left(1 + h\sqrt{\frac{D}{k}}\right). \quad (5.4)$$

Some steady-state solutions (which can be obtained by the first method mentioned) are given below, preceded by references to the corresponding unsteady-state expressions.

[12] Carslaw and Jaeger, *Conduction of Heat in Solids* (Oxford University Press, 1947), p. 53.
[13] For tables of Laplace transforms, see ref (4), p. 380; Churchill, *Modern Operational Mathematics in Engineering* (McGraw-Hill, 1944), p. 295.

SPHERE WITH SURFACE RESISTANCE (eqn. (4.4)).—

$$\frac{R_\infty}{c^*} = 4\pi a^2 hD\left[\frac{a\sqrt{\frac{k}{D}}\coth\left(a\sqrt{\frac{k}{D}}\right) - 1}{ah + a\sqrt{\frac{k}{D}}\coth\left(a\sqrt{\frac{k}{D}}\right) - 1}\right] \quad . \quad . \quad . \quad (5.5)$$

INFINITE CYLINDER WITH SURFACE RESISTANCE (eqn. (4.13)).—

$$\frac{R_\infty}{c^*} = 2\pi ahD\left[\frac{\sqrt{\frac{k}{D}}I_1\left(a\sqrt{\frac{k}{D}}\right)}{\sqrt{\frac{k}{D}}I_1\left(a\sqrt{\frac{k}{D}}\right) + hI_0\left(a\sqrt{\frac{k}{D}}\right)}\right] \quad . \quad . \quad . \quad (5.6)$$

(I_0, I_1, are modified Bessel functions of the first kind, zero and first order respectively).

SPHERE WITH SURFACE SATURATION (eqn. (4.14)).—

$$\frac{R_\infty}{c^*} = 4\pi aD\left[a\sqrt{\frac{k}{D}}\coth\left(a\sqrt{\frac{k}{D}}\right) - 1\right] \quad . \quad . \quad . \quad (5.7)$$

INFINITE CYLINDER WITH SURFACE SATURATION (eqn. (4.15)).—

$$\frac{R_\infty}{c^*} = 2\pi a\sqrt{kD}\,\frac{I_1(a\sqrt{k/D})}{I_0(a\sqrt{k/D})} \quad . \quad . \quad . \quad . \quad (5.8)$$

6. Absorption into Falling Drop with Surface Saturation.—Kronig and Brink[14] have carried out a mathematical investigation into the problem of extraction of a dissolved substance from a drop of liquid which is falling or rising through another liquid. The effect of the relative motion of the drop and the surrounding liquid is to cause a circulatory motion within the drop, and the transport of material by this means increases the rate of extraction as compared with that from a stationary drop.

The treatment is restricted to cases in which certain terms in the equations of motion can be neglected. As far as the motion of the outside liquid is concerned the condition will probably be virtually the same as that for the applicability of Stokes' law to a solid sphere,[15] namely, the Reynolds number must be less than about 1·0,

$$2au\rho_2/\eta_2 < 1,$$

where a = radius of drop ; u = velocity ; ρ_2 = density and η_2 = viscosity of outside liquid. It will be assumed that in these circumstances the motion of the liquid inside the drop also fulfils the necessary conditions, although there does not seem to be any experimental evidence available

[14] Kronig and Brink, *Appl. Sci. Res. A*, 1950, **2**, 142.
[15] Allen, *Phil. Mag.*, 1900, **50** (5), 323. Arnold, *Phil. Mag.*, 1911, **22** (6), 755.

on this point. The motion of both liquids will be streamline under the conditions in which the treatment is valid. In practice the limitation on the Reynolds number usually implies either extremely small drops or an extremely viscous outside liquid.

In certain circumstances, which will be discussed later, the rate of circulation within the drop is so rapid that the time taken for an element of liquid to make a complete circuit of its streamline path is very much less than the time required for the mean concentration within the drop to undergo a change of observable magnitude. (This cannot be true for paths which pass very close to the surface; but in the case considered here it is assumed that the dimensions of the region in which the condition is not fulfilled are vanishingly small.) The concentration will then be the same at every point along a given streamline, and under these conditions the differential equation for diffusion inside the drop takes the form

$$\frac{\partial}{\partial \xi}\left[p(\xi)\frac{\partial c'}{\partial \xi}\right] = \frac{a^2}{16D}\, q(\xi)\,\frac{\partial c'}{\partial t}, \qquad (6.1)$$

where ξ, $p(\xi)$, $q(\xi)$ are functions of the space co-ordinates only. Kronig and Brink solve this for the case where the concentration in the drop is initially uniform, and the concentration at the surface is zero at all times. From this solution (modified to fit the case where the concentration inside the drop is initially zero, and that at the surface has a constant value c^*) we obtain

$$R' = 8\pi a c^* D \sum_{n=1}^{\infty} A_n^2 \mu_n \exp\left(-\mu_n \cdot \frac{16Dt}{a^2}\right) \qquad (6.2)$$

for the rate of absorption without reaction. The values of the coefficients A_n and μ_n for the first two terms are

$$A_1 = 1{\cdot}32, \qquad \mu_1 = 1{\cdot}678,$$
$$A_2 = 0{\cdot}73, \qquad \mu_2 = 9{\cdot}83.$$

If the absorbed substance undergoes a first-order reaction in solution, eqn. (6.1) must be replaced by

$$\frac{\partial}{\partial \xi}\left[p(\xi)\frac{\partial c}{\partial \xi}\right] = \frac{a^2}{16D}\, p(\xi)\left[\frac{\partial c}{\partial t} + kc\right]. \qquad (6.3)$$

It is easily shown that the solution of eqn. (6.3) with boundary conditions (3.2) and (3.3), and the solution of eqn. (6.1) with the same boundary conditions, are related by eqn. (3.4).

When no reaction takes place

$$R' = D\iint \left(\frac{\partial c'}{\partial \xi}\right)_{r=0}\left(\frac{\partial \xi}{\partial r}\right)_{r=0} ds, \qquad (6.4)$$

(the integral being taken over the whole surface of the drop) whereas when reaction occurs

$$R = D\iint \left(\frac{\partial c}{\partial \xi}\right)_{r=0}\left(\frac{\partial \xi}{\partial r}\right)_{r=0} ds. \quad . \quad . \quad . \quad (6.5)$$

It can be seen from eqn. (6.4), (6.5) and (3.4) that R and R' are related by eqn. (3.6). Hence from eqn. (6.2)

$$R = 8\pi ac^*D \sum_{n=1}^{\infty} A_n{}^2\mu_n \left[\frac{k + \frac{16D\mu_n}{a^2}\exp\left(-t\left[k + \frac{16D\mu_n}{a^2}\right]\right)}{k + 16D\mu_n/a^2}\right]. \quad (6.6)$$

(Fig. 1 shows comparative curves of $R/8\pi ac^*D$ against $16Dt/a^2$ for eqn. (6.6), (6.2), (4.14) and for a sphere with neither chemical reaction nor internal circulation taking place (A, C, B, D respectively). k has arbitrarily been made equal to $16D/a^2$.)

As mentioned above, these results will only be valid if the rate of circulation in the drop is sufficiently rapid. This condition may be stated quantitatively as follows:

$$\frac{t_d a\mathbf{g}\,|\,\rho_1 - \rho_2\,|}{6(3\eta_1 + 2\eta_2)} \gg 1, \quad . \quad . \quad . \quad . \quad (6.7)$$

where $\mathbf{g}$ = acceleration due to gravity, ρ_1, ρ_2 are the densities of the liquids inside and outside the drop respectively, and η_1, η_2 their viscosities. t_d is a " diffusion time " characteristic of the system. For diffusion without reaction Kronig and Brink arbitrarily identify t_d with the time taken for the total amount of solute in a stationary drop to fall to $1/e$ its original value. Then

$$t_d = 0{\cdot}056\, a^2/D. \quad . \quad . \quad . \quad . \quad (6.8)$$

Substitution of this in eqn. (6.7) gives the order of magnitude of the minimum drop-radius for which the treatment is valid. However, the radius must not be too large, or the Reynolds number will become greater than 1·0. If we put the Reynolds number equal to 1·0, which we have assumed to be its maximum permissible value, and substitute the velocity given by Stokes' law as modified for liquid drops,[16] we have:

$$\frac{(\eta_1 + \eta_2)4\rho_2 a^3\mathbf{g}\,|\,\rho_1 - \rho_2\,|}{3\eta_2{}^2(3\eta_1 + 2\eta_2)} = 1. \quad . \quad . \quad . \quad (6.9)$$

Substituting (6.8) and (6.9) in (6.7)

$$\frac{7 \times 10^{-3}\eta_2}{D\rho_2(1 + \eta_1/\eta_2)} \gg 1. \quad . \quad . \quad . \quad . \quad (6.10)$$

[16] Lamb, *Hydrodynamics* (Cambridge University Press, 1932), p. 600. Bond, *Phil. Mag.*, 1927, **4** (7), 889.

Taking representative values of the physical quantities (e.g. in c.g.s. units: $\eta_1 = \eta_2 = 10^{-2}$, $\rho_2 = 1$, $D = 10^{-5}$) we find the left-hand side of (6.10) to have a value of 3·5. Eqn. (6.2) may therefore only be applicable as an approximation, and to drops whose radius approaches the value given by eqn. (6.9).

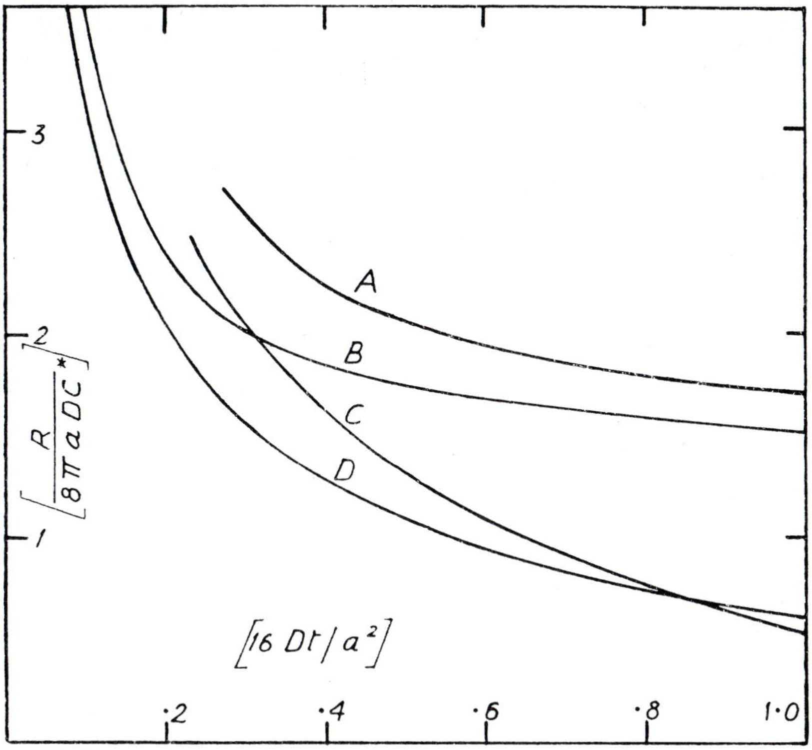

FIG. 1.

When the solute undergoes a first-order reaction the diffusion-time may conveniently be given the form

$$t_d = 16 \cdot \frac{0{\cdot}056}{k + 16D/a^2}, \quad . \quad . \quad . \quad . \quad (6.11)$$

which varies from 0·056 a^2/D when k is very small to 0·9/k when k is very large; in both cases, therefore, it represents approximately the time taken for the amount of solute in a stationary drop to fall to 1/e its original value. In general the value of a required to satisfy (6.7) will be greater than when no reaction occurs, and the range of drop-sizes to which the treatment can be applied with a given degree of exactitude will be reduced by an extent which depends on the relative values of k and D/a^2, and may vanish completely.

It is clear that caution is necessary when using eqn. (6.2) in a case where no reaction occurs, and is even more necessary when applying eqn. (6.6) to a system where reaction does occur. However, the range of applicability of eqn. (6.2) and (6.6) may be greater than condition (6.10) implies, since it is possible that under certain circumstances the internal motion of the drop may not depart widely from the postulated behaviour even when the Reynolds number exceeds 1·0.

A further difficulty may arise from the fact that in some cases, particularly when the interfacial tension is great and the drop is very small, the expected circulation of liquid within the drop apparently does not occur.[15, 17] Eqn. (4.14) would then apply.

Department of Chemical Engineering,
Tennis Court Road,
Cambridge.

[17] Bond and Newton, *Phil. Mag.*, 1928, **5** (7), 794.

A5

TEMPERATURE EFFECTS ACCOMPANYING THE ABSORPTION OF GASES IN LIQUIDS

by P. V. DANCKWERTS

Department of Chemical Engineering, Cambridge

§ 1. *Introduction.* It has been suggested [1]) that the rates of absorption of gases into stirred and flowing liquids in industrial equipment may be related, by a simple transformation, to the rates of absorption into stagnant liquids. It is therefore potentially important to be able to determine whether the latter will be appreciably affected by the rise in temperature which usually accompanies the absorption. The temperature effect can be of controlling importance when ammonia, for instance, is absorbed by water. In other cases, even when the mean rise in temperature is small, there may be an appreciable rise at the surface, which will have a marked influence on the rate of absorption.

In the following treatment the temperature rise at the surface of the liquid is calculated on the assumption that it will be too small to affect the solubility or diffusivity of the solute gas, or the reaction-velocity constant. If the calculated temperature rise contradicts this assumption, the expressions for the rate of absorption under isothermal conditions cannot be used.

The following assumptions will be made. The gas and liquid are at all times at equilibrium at the interface. The liquid is stagnant, and of effectively infinite depth; it is initially at uniform temperature. The loss of heat from the liquid to the gas is negligible. The dissolved gas undergoes a first-order reaction in solution. The diffusivity D, thermal diffusivity a and reaction-velocity constant k are uniform and constant. The concentration C (at x, t) and rate of absorption per unit area R are then given by the expressions [2])

$$\frac{C}{C^*} = \tfrac{1}{2}\exp\left[-x\sqrt{\frac{k}{D}}\right]\operatorname{erfc}\left[\frac{x}{2\sqrt{tD}} - \sqrt{kt}\right] + $$

$$+ \tfrac{1}{2}\exp\left[x\sqrt{\frac{k}{D}}\right]\operatorname{erfc}\left[\frac{x}{2\sqrt{tD}} + \sqrt{kt}\right], \quad (1)$$

$$R/C^*\sqrt{Dk} = \operatorname{erf}\sqrt{kt} + \frac{e^{-kt}}{\sqrt{\pi kt}}, \quad (2)$$

C^* being the saturated concentration of unreacted gas at the interface and t the time of contact between gas and liquid.

The total temperature rise is the sum of two effects: firstly, the heat of solution, which produces a flux of heat across the surface proportional to R; and secondly the heat of reaction, which causes production of heat at a rate (per unit volume) proportional to the local value of C. The two effects are additive, and will be calculated separately.

§ 2. *The heat of solution.* The flux of heat across unit area of the surface ($x = 0$) at time t is $F(t)$, where

$$F(t) = RH_1, \quad (4)$$

H_1 being the heat of solution (positive for an exothermic process). The resulting rise in temperature at the surface v_0 is given by the expression [3])

$$v_0 = \frac{1}{\varrho\sigma\sqrt{\pi\alpha}} \int_0^t \frac{F(t-\theta)}{\sqrt{\theta}} \mathrm{d}\theta, \tag{5}$$

where ϱ and σ are the density and specific heat of the liquid. The integration can be carried out by using the convolution theorem [4])

$$L\{v_0\} = \frac{1}{\varrho\sigma\sqrt{\pi\alpha}} L\{F(t)\}\, L\left\{\frac{1}{\sqrt{t}}\right\}, \tag{6}$$

where $L\{v_0\}$, $L\{F(t)\}$ and $L\{1/\sqrt{t}\}$ are the Laplace transforms of v_0, RH_1 and $1/\sqrt{t}$ respectively. Hence

$$\frac{\varrho\sigma}{H_1C^*}\sqrt{\frac{\alpha}{D}}\, L\{v_0\} = \frac{\sqrt{1+k/p}}{p}, \tag{7}$$

p being the parameter of the Laplace transform. Using the binomial expansion

$$\frac{\varrho\sigma}{H_1C^*}\sqrt{\frac{\alpha}{D}}\, L\{v_0\} = \frac{1}{p} + \frac{1}{p}\sum_{n=1}^{\infty} \frac{(\frac{1}{2})\ldots(\frac{3}{2}-n)}{n!}\left(\frac{k}{p}\right)^n. \tag{8}$$

Inverting equation (8)

$$\frac{\varrho\sigma}{H_1C^*}\sqrt{\frac{\alpha}{D}}\, v_0 = 1 + \sum_{n=1}^{\infty} \frac{(\frac{1}{2})\ldots(\frac{3}{2}-n)}{n!}\,\frac{(kt)^n}{n!}. \tag{9}$$

The behaviour of v_0 for small and large values of kt may also be of interest:

$$\frac{\varrho\sigma}{H_1C^*}\sqrt{\frac{\alpha}{D}}\, v_0 \approx 1 + \frac{kt}{2}, \quad kt \ll 1, \tag{10}$$

$$\frac{\varrho\sigma}{H_1C^*}\sqrt{\frac{\alpha}{D}}\, v_0 \approx 2\sqrt{\frac{kt}{\pi}}, \quad kt \gg 1. \tag{11}$$

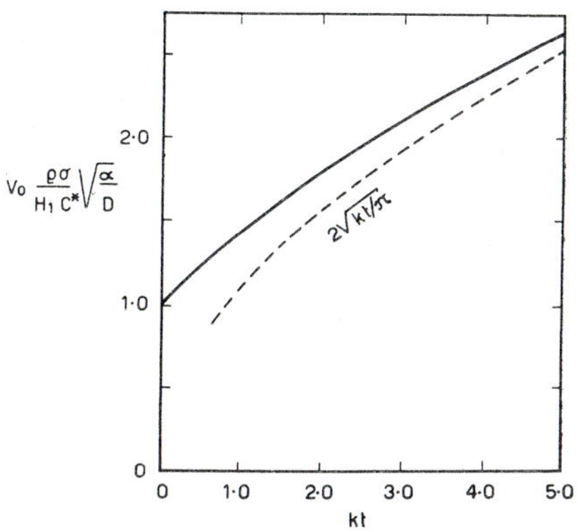

Fig. 1. Surface temperature as a function of time as a result of heat of solution.

A plot of

$$\frac{\varrho\sigma}{H_1C^*}\sqrt{\frac{\alpha}{D}}\, v_0$$

against kt is shown in fig. 1.

If no reaction takes place, the temperature of the surface is constant at the value

$$v_0 = \frac{H_1C^*}{\varrho\sigma}\sqrt{\frac{D}{\alpha}}.$$

§ 3. *The heat of reaction.* If H_2 is the heat of reaction of unit quantity of the dissolved gas, the equation governing the temperature rise v at (x, t) due to the reaction is

$$\frac{\partial v}{\partial t} = \alpha\frac{\partial^2 v}{\partial x^2} + \frac{kH_2C}{\varrho\sigma}; \quad \begin{cases} v = 0,\ t = 0,\ x > 0. \\ v = 0,\ t > 0,\ x = \infty. \\ \dfrac{\partial v}{\partial x} = 0,\ t > 0,\ x = 0. \end{cases} \tag{12}$$

Substituting (1) and applying the Laplace transformation

$$\frac{\mathrm{d}^2L\{v\}}{\mathrm{d}x^2} = \frac{p}{\alpha}L\{v\} - \frac{kH_2C^*}{\varrho\sigma\alpha p}e^{-x\sqrt{(p+k)/D}}; \quad \left\{\begin{array}{l} L\{v\} = 0,\ x = \infty. \\ \dfrac{\mathrm{d}L\{v\}}{\mathrm{d}x} = 0,\ x = 0. \end{array}\right\} \tag{13}$$

Hence

$$\frac{\varrho\sigma(\alpha-D)}{kH_2C^*D}L\{v_0\} = \frac{\sqrt{\frac{\alpha}{D}}\sqrt{1+\frac{k}{p}} - 1}{p\{p + k\alpha/(\alpha-D)\}} \tag{14}$$

Expanding $\sqrt{1+k/p}$ and inverting, we have

$$\frac{\varrho\sigma}{H_2C^*}\sqrt{\frac{D}{\alpha}}\left(\frac{\alpha-D}{D}\right)v_0 = \frac{(\alpha-D)}{\alpha}\left(1-\sqrt{\frac{D}{\alpha}}\right)(1-e^{-\beta t}) +$$

$$+ e^{-\beta t}\sum_{n=1}^{\infty}\frac{(\frac{1}{2})\ldots(\frac{3}{2}-n)}{n!}k^n\int_0^t t^n e^{\beta t}\mathrm{d}t, \tag{15}$$

where $\beta = k\alpha/(\alpha - D)$. Expanding $e^{\beta t}$ and integrating term by term, we have

$$\frac{\varrho\sigma}{H_2C^*}\sqrt{\frac{D}{\alpha}}\left(\frac{\alpha-D}{D}\right)v_0 = \left(\frac{\alpha-D}{\alpha}\right)\left(1-\sqrt{\frac{D}{\alpha}}\right)(1-e^{-\beta t}) +$$

$$+ ke^{-\beta t}\sum_{n=1}^{\infty}\frac{(\frac{1}{2})\ldots(\frac{3}{2}-n)}{(n!)^2}k^n\left[\frac{t^{n+1}}{n+1} + \sum_{m=1}^{\infty}\frac{\beta^m t^{n+m+1}}{m!(n+m+1)}\right]. \tag{16}$$

Collecting all terms in t^{n+1}

$$\frac{\varrho\sigma}{H_2C^*}\sqrt{\frac{D}{\alpha}}\,\frac{(\alpha-D)}{D}\,v_0 = \left(\frac{\alpha-D}{\alpha}\right)\left(1-\sqrt{\frac{D}{\alpha}}\right)(1-e^{-\beta t}) +$$

$$+ e^{-\beta t}\sum_{n=1}^{\infty}\frac{(kt)^{n+1}}{n+1}\left[\frac{(\frac{1}{2})\ldots(\frac{3}{2}-n)}{(n!)^2} + \sum_{m=1}^{n-1}\frac{(\frac{1}{2})\ldots(\frac{3}{2}-n+m)}{[(n-m)!]^2 m!}\left(\frac{\beta}{k}\right)^m\right]. \tag{17}$$

For many solutes in water, $\alpha \sim 100\,D$, as a first approximation we can put $\beta \approx k$, and the series can be written approximately

$$\frac{\varrho\sigma}{H_2C^*}\sqrt{\frac{\alpha}{D}}\,v_0 \approx 1 - e^{-kt} + e^{-kt}\sum_{n=1}^{\infty}\mu_n(kt)^{n+1}. \tag{18}$$

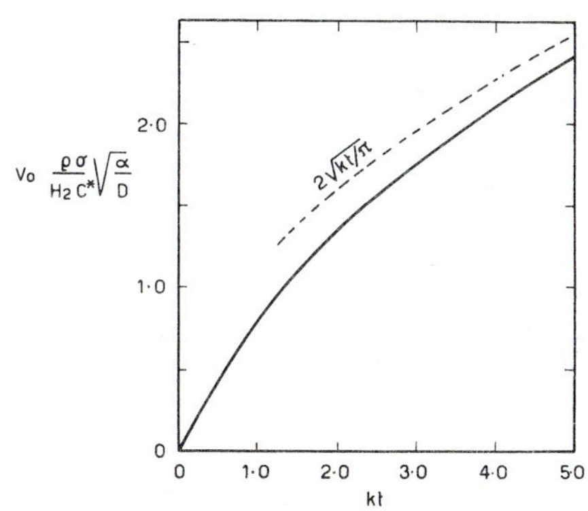

Fig. 2. Surface temperature as a function of time as a result of heat of reaction.

The first seven coefficients of (kt^{n+1}) are:

n =	1	2	3	4	5	6	7
μ_n =	0.250	0.1457	0.0495	0.01218	0.00237	0.000384	0.0000532.

A plot of

$$\frac{\varrho\sigma}{H_2C^*}\sqrt{\frac{\alpha}{D}}\,v_0.$$

(as calculated from (18)) against kt is shown in fig. 2. The function (17) tends to $2\sqrt{kt/\pi}$ as kt tends to infinity.

§ 3. *Absorption of CO_2 in a buffer solution.* As an example of the use of these expressions, they will be applied to the temperature at the surface of a carbonate-bicarbonate buffer solution, with an initial temperature of 25°C, in which CO_2 dissolves and reacts with OH' to form HCO_3', the reaction being pseudo-first-order with a velocity constant 10 s^{-1}. The net reaction is:

$$CO_2 + CO_3'' + H_2O \rightarrow 2HCO_3'.$$

The values of C^*, α, D, ϱ and σ are assumed to be the same as for pure water, namely

$C^* = 3.28 \times 10^{-5}$ g mole/cm^3, $a = 1.46 \times 10^{-3}$ cm^2/s,
$D = 2 \times 10^{-5}$ cm^2/s, $\varrho\sigma = 1$ cal/°C cm^3.

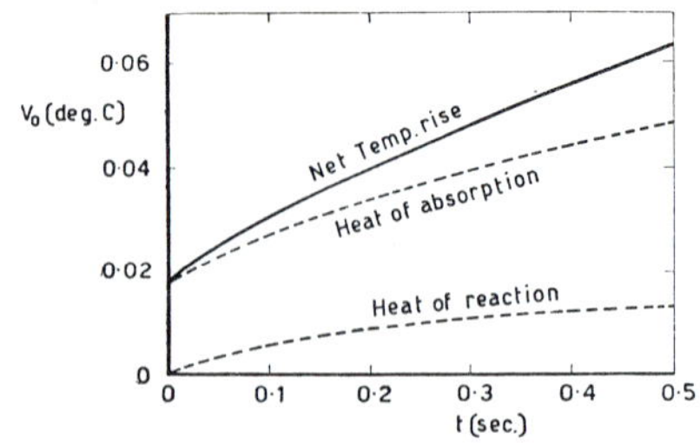

Fig. 3. Surface temperature as a function of time for the absorption of CO_2 in a buffer solution.

The heats of solution and reaction are approximately [5]:

$H_1 = 4760$ cal/g mole, $H_2 = 1{,}500$ cal/g mole.

The rise in surface temperature is shown as a function of time in fig. 3. It will be seen that for periods of contact of the order of $\frac{1}{2}$ s the rise in temperature is too small to affect the rate of absorption appreciably.

The author wishes to thank Mr A. M. Kennedy for his help in checking the mathematics and arithmetic.

Received 19th July, 1952.

REFERENCES

1) Danckwerts, P. V., Industr. Engng. Chem. **43** (1951) 1460.

2) Danckwerts, P. V., Trans. Faraday Soc., **46** (1950) 300.

3) Carslaw, H. S. and Jaeger, J. C., "Conduction of Heat in Solids", p. 57, Oxford University Press (1947).

4) Carslaw, H. S., and Jaeger, J. C., "Conduction of Heat in Solids", p. 243, Oxford University Press (1947).

5) Bichowsky, F. R. and Rossini, F. D., "The Thermochemistry of the Chemical Substances", Reinhold (1936).

A6

Gas-absorption accompanied by first-order reaction: concentration of product, temperature-rise and depletion of reactant

WHEN a gas is absorbed into a quiescent, semi-infinite liquid with which it undergoes an irreversible first-order or pseudo-first order reaction, the distribution of unreacted dissolved gas in the liquid, as a function of time and of distance from the surface, is given by a well-known expression [1]. The following quantities may also be of interest:

(a) Change in temperature at the surface resulting from the heat of absorption and the heat of reaction.
(b) Concentration at the surface of the product of the reaction.
(c) Depletion at the surface of reactant dissolved in liquid (in case of pseudo-first order reaction).

Temperature

The method of calculation has been given by DANCKWERTS [2], but the solutions were given in the form of infinite series. This is not necessary, as the Laplace transforms involved can be inverted to give the following expressions involving Bessel functions:

$$v_A \frac{\rho\sigma}{H_A C^*} \sqrt{\frac{\alpha}{D}} = e^{-kt/2}[(1+kt)I_0(kt/2) + ktI_1(kt/2)] = V_A \qquad (1)$$

$$v_R \frac{\rho\sigma}{H_R C^*} \sqrt{\frac{\alpha}{D}} = kte^{-kt/2}[I_0(kt/2) + I_1(kt/2)] = V_R \qquad (2)$$

Here v_A, v_R are the temperature rises at the surface due to the absorption and the subsequent reaction of the gas, respectively; H_A and H_R the heats of absorption and of reaction (positive if exothermic); C^* the constant concentration of dissolved gas at the surface; ρ, σ the density and specific heat of the liquid; a its thermal diffusivity ($K/\rho\sigma$, where K is the thermal conductivity); D the diffusivity of the dissolved gas in solution; k the reaction-rate constant; and t the time after first exposure of liquid to gas. These results are valid when $a >> D$, as is the case in practice. The total temperature rise is $(v_A + v_R)$. When kt is large, both V_A and V_R tend to $2\sqrt{kt/\pi}$. The functions V_A and V_R are shown in Fig. 1.

Product

The basic equation for product concentration, a, is the same as that for temperature-change due to reaction, with (a/y) replacing $(v_R\rho\sigma/H_R)$, where y is the number of moles of product arising from the reaction of one mole of dis-

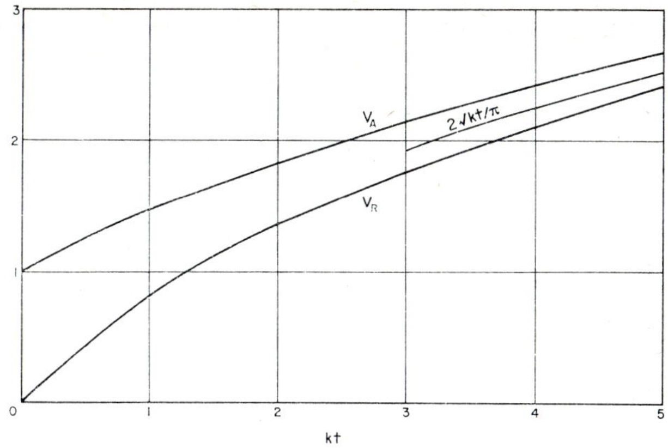

FIG. 1

solved gas; and D', the diffusivity of the product, replacing a. The diffusivities of product and reactant will often be nearly equal. If they are put equal, the Laplace transform of (a/yC^*) (Ref. [2], Eq. (14)) becomes identical to that of (V_A-1) (Ref. [2], Eq. (7)), so that we have

$$\frac{a}{yC^*}=V_A-1 \tag{3}$$

where V_A is defined as a function of kt by Eq. (1). This result is valid if the two diffusivities are equal, and thus will in most cases be approximately true.

Reactant

A second-order reaction between dissolved gas and a reactant in solution behaves as a pseudo-first order reaction so long as the concentration of reactant at the surface is not significantly depleted by the reaction. The depletion can be calculated, on the assumption that it is small enough for the reaction to be pseudo-first order, and the result used to check the assumption. DANCKWERTS and KENNEDY [3] have shown that if the diffusivities of solute gas and of reactant are assumed equal,

$$\frac{b_0-b}{zC^*}=V_A-1 \tag{4}$$

where b_0 is the initial uniform concentration of reactant, b its concentration at the surface at time t, and z the number of moles of reactant reacting with one mole of gas. In evaluating V_A, k is put equal to k_2b_0, where k_2 is the second-order rate constant.

The reactant may be a buffer-solution. When CO_2 is absorbed by carbonate/bicarbonate solutions, for instance, the local rate of reaction is $k_2[CO_2][OH^-]$. The reaction can be regarded as pseudo-first order if the concentration $[OH^-]$ is not significantly depleted at the surface. Since $[OH^-]$ is proportional to the ratio $[CO_3^=]/[HCO_3^-]$, it is the value of the latter quantity at the surface which must be investigated. If one mole of gas reacts with z moles of one component of the buffer to give y moles of the other, and if the concentrations of the two components are c_0, d_0 initially and c, d at the surface at time t, then as shown by KENNEDY and DANCKWERTS [3]

$$1-\frac{c_0d}{cd_0}=1-\frac{d_0}{c_0}\left[\frac{c_0-zC^*V_A}{d_0+yC^*V_A}\right] \tag{5}$$

This quantity is the fractional depletion at the surface of the buffer-ratio c_0/d_0, and hence the fractional depletion of the pseudo-first order rate constant at the surface. In evaluating V_A, k is put equal to the pseudo-first order rate-constant in the undepleted buffer solution.

P. V. DANCKWERTS

Dept. of Chemical Engineering
Pembroke Street
Cambridge

REFERENCES

[1] DANCKWERTS P. V., *Trans. Faraday Soc.* 1950 **46** 300.
[2] DANCKWERTS P. V., *Appl. Scient. Res.* 1952 **A3** 385.
[3] DANCKWERTS P. V. and KENNEDY A. M., *Chem. Engng Sci.* 1958 **8** 201.

A7

Absorption from bubbles of dilute gas

Two types of process for absorbing a gas in a liquid are considered here:

(a) The gas is introduced into the liquid in the form of discrete bubbles—for instance through a porous plate or perforated tube.

(b) The gas is entrained into the liquid from its free surface by an agitator, all or most of the absorption taking place from the entrained bubbles.

If the soluble gas is not diluted with an insoluble gas, the mole-fraction of soluble gas in a bubble does not alter as absorption proceeds. The bubble may disappear before reaching the free surface. If an insoluble gas is present, the mole-fraction of soluble gas in the bubble is initially equal to that in the inflowing stream (in case a) or to that in the bulk of the gas in the space above the liquid (in case b), but it falls progressively as absorption proceeds. In the limit the gas remaining in the bubble will consist only of insoluble gas and the vapour of the liquid.

At first sight one might not, therefore, expect the rate of absorption of a given gas in such a system to be simply proportional to its mole-fraction in the bulk or inflowing stream (other things being equal). The driving force for absorption from a bubble of diluted soluble gas falls progressively during the sojourn of the bubble in the liquid, its average value depending on how much of the gas is absorbed before the bubble leaves the liquid. As against this, however, the bubbles do not shrink so much as bubbles of pure gas, and at any time after immersion there is a larger surface area from which absorption can take place.

In the following analysis certain simplifying assumptions are made. There is assumed to be no resistance to mass transfer on the gas side of the interface, and the resistance on the liquid side is assumed to be independent of the size or situation of the bubble. The gas in the bubble is always saturated with solvent vapour. The pressure is assumed to be constant. The concentration of dissolved gas in the bulk of the liquid is effectively zero. The bubbles considered are all the same shape.

A bubble is formed at $t = 0$, when:

Volume of bubble $= v_0$

Partial volume of inerts $= Mv_0(1 - s)$

Partial volume of solvent vapour $= v_0 s$

Partial volume of soluble gas $= v_0(1 - M)(1 - s)$

where M is the initial mole-fraction of insoluble gas, calculated on a solvent-free basis.

After time t, the situation is:

Volume of bubble $= v$

Partial volume of inerts $= Mv_0(1 - s)$

Partial volume of solvent vapour $= vs$

Partial volume of soluble gas $= (v - Mv_0)(1 - s)$

Mole-fraction of soluble gas $= (1 - Mv_0/v)(1 - s)$.

Now the rate of absorption is proportional to the mole-fraction of the soluble gas and to the surface-area of the bubble: i.e.

$$-(1 - s)\frac{dv}{dt} = K\,v^{2/3}\left(1 - M\frac{v_0}{v}\right)(1 - s)$$

or

$$Kt = \int_v^{v_0} \frac{dv}{v^{2/3}\,[1 - (Mv_0/v)]} \tag{1}$$

By putting $M = 0$, we find that for a bubble containing no insoluble gas

$$Kt = 3(v_0^{1/3} - v^{1/3}) \tag{2}$$

and that the time, τ, for such a bubble to disappear completely is $3v_0^{1/3}/K$. Equation (1) can now be re-written, with change of variable, $x = (v/v_0M)^{1/3}$:

$$t/\tau = M^{1/3}\int_{x=(v/vM)^{1/3}}^{(1/M)^{1/3}} \frac{x^3dx}{(x^3 - 1)} \tag{3}$$

The indefinite integral has the form

$$I(x) = \int \frac{x^3dx}{x^3 - 1} = x + \frac{1}{6}\ln\frac{(x-1)^2}{(x^2+x+1)} - \frac{1}{\sqrt{3}}\arctan\left[\frac{2x+1}{\sqrt{3}}\right] \tag{4}$$

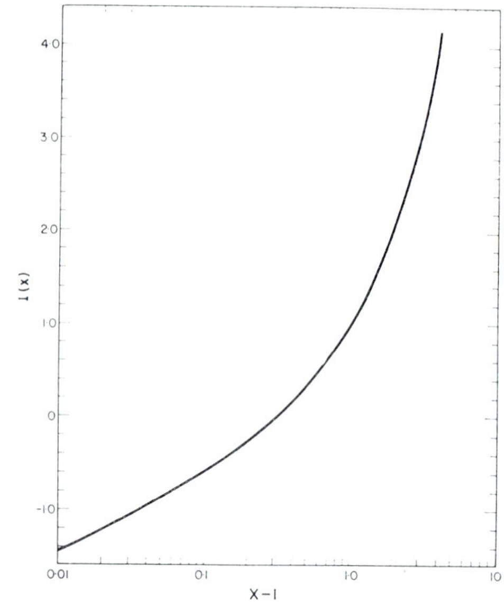

FIG. 1

and equation (3) becomes

$$\frac{t}{\tau} = M^{1/3}\left[I\left(\frac{1}{M}\right)^{1/3} - I\left(\frac{v}{v_0M}\right)^{1/3}\right] \tag{5}$$

Values of I(x) are shown in Fig. 1, plotted against (x − 1).

It is thus possible to calculate, for any given value of M, the fraction y of the soluble gas absorbed in a given time, t, and to compare it with the fraction $y' = 1 - (1 - t/\tau)^3$ absorbed from a pure bubble of the same initial volume in the same time.

Simple expressions can be obtained for the behaviour under limiting conditions. Thus if the bubbles are mostly immersed for a time much less than τ, the *fraction* of the soluble gas absorbed is almost exactly the same for pure and dilute bubbles, and the rate of absorption is thus approximately proportional, other things being equal, to the mole-fraction, $(1 - M)$, of soluble gas in the bulk or inflowing gas. If conditions are such that if the gas were undiluted most of the bubbles would disappear before re-emerging, then when the gas is diluted most of the bubbles will spend a time greater than τ in the liquid, which implies that at least 95 per cent of the soluble gas will be absorbed from them (see Fig 2); in this case, too, the rate of absorption is approximately proportional to $(1 - M)$.

For very dilute bubbles $(1 - M \ll 1)$

$$\frac{t}{\tau} \approx \frac{1}{3}\ln\left[(1 - M)\Big/\left(\frac{v_0}{v} - M\right)\right] \tag{6}$$

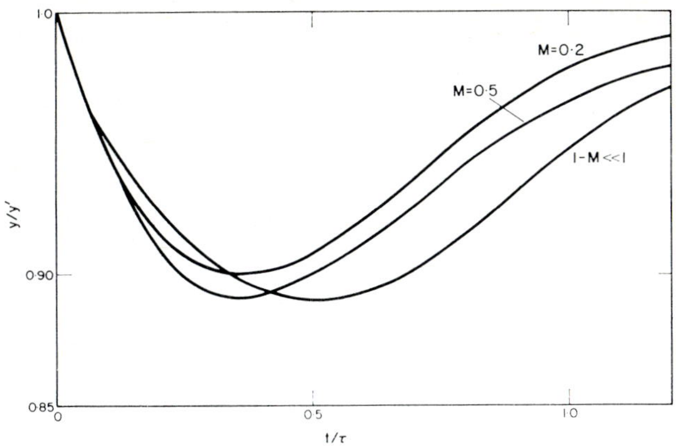

FIG. 2

whence

$$\frac{y}{y'} \approx \frac{1 - \exp(-3\,t/\tau)}{1 - (1 - t/\tau)^3}, \frac{t}{\tau} < 1$$

$$\frac{y}{y'} \approx 1 - \exp\left(-\frac{3t}{\tau}\right), \quad \frac{t}{\tau} > 1 \qquad (7)$$

From these expressions it is calculated that y/y' is never less than about 0·89 in very dilute bubbles.

For nearly pure bubbles ($M \ll 1 - y$)

$$\frac{t}{\tau} \approx 1 - (1 - y)^{1/3} - \frac{M}{2}[1 - (1 - y)^{2/3}] \qquad (8)$$

whence

$$\frac{y}{y'} \approx 1 + M - \frac{3M}{2}\left[\frac{1 - (1 - v)^{2/3}}{y}\right] \qquad (9)$$

Thus the effect of small traces of diluent is correspondingly small.

Fig. 2 shows some values of the ratio y/y' for different values of M and of t/τ. The ratio is never less than about 0·89, whatever the degree of dilution or the fraction of gas absorbed. Thus, in dynamically comparable situations, the rate of absorption of a diluted gas from a swarm of bubbles is equal, with an error of less than 11 per cent, to the rate of absorption of the pure gas multiplied by $(1 - M)$, where M is the mole-fraction of diluent gas in the bulk of the gas or in the inflowing gas mixture (calculated on a solvent vapour free basis); in other words, the rate of absorption is proportional to the partial pressure of soluble gas initially present in the bubbles.

However, this conclusion may not be valid if the dynamical behaviour of the dilute bubbles is very different from that of the pure bubbles. Consider, for instance, conditions such that when a pure gas is used most of the bubbles disappear before leaving the liquid; if a dilute gas were used, and the dilute bubbles followed the same trajectories as the pure bubbles, most of the bubbles would spend a time greater than τ in the liquid and hence would lose almost all their soluble gas by absorption, and the use of the initial partial pressure as driving force would give the right answer. But the dilute bubble is at all times after formation larger than a pure bubble of the same initial size; this may cause the dilute bubbles to spend a time substantially less than τ in the liquid, and thus reduce the absorption rate.

In the case of very small bubbles which move so slowly relative to the liquid that transfer of dissolved gas away from the bubble is governed entirely by diffusion, to the exclusion of convection, the mass-transfer coefficient varies inversely as the diameter of the bubble. The preceding analysis does not apply if a substantial part of the absorption occurs from bubbles as small as this. It will not apply, either, if there is a substantial chance that a bubble coalesces with another before leaving the liquid, or if the depth of liquid is such that there is a substantial change in the pressure on a bubble during its immersion.

The conditions in which the analysis given above is exactly applicable are somewhat circumscribed, therefore; but the general conclusion may nevertheless be valid in many circumstances of practical interest.

Department of Chemical Engineering, P. V. DANCKWERTS
Pembroke Street,
Cambridge.

NOTATION

$I(x)$	Integral given in equation (4)
K	Mass-transfer factor
M	Initial mole-fraction of inert gas, solvent-vapour free basis
s	Mole-fraction of solvent vapour
t	Time since immersion of bubble
v	Volume of bubble at time t
v_0	Initial volume of bubble
y	Fraction of soluble gas absorbed from dilute bubble in time t
y'	Fraction of soluble gas absorbed from pure bubble in time t.
x	$(v/Mv_0)^{1/3}$
τ	time taken for pure bubble to disappear.

SECTION B

Gas Absorption and the Design of Absorbers

B7	The design of gas absorbers - I. Methods for predicting rates of absorption with chemical reaction in packed columns, and tests with 1½ in. Raschig rings P.V. Danckwerts & A.J. Gillham Trans. Instn. Chem. Engrs 1966 44 T42.	122
B8	The design of gas absorbers. Part II: Effective interfacial areas for several types of packing P.V. Danckwerts & S.F. Rizvi Trans. Instn. Chem. Engrs. 1971 49 124.	135
B9	Design of gas absorbers: Part III: Laboratory "point" model of a packed column absorber P.V. Danckwerts & E. Alper Trans. Instn. Chem. Engrs. 1975 53 34.	139
B10	Laboratory scale model of a complete packed column absorber E. Alper & P.V. Danckwerts Chem. Engg. Sci. 1976 31 599.	145
B11	Surface instability during the absorption of CO_2 by monoethanolamine solutions P.V. Danckwerts & A. Tavares da Silva Chem. Engng. Sci. 1967 22 1513.	156
B12	Gas absorption with instantaneous reaction P.V. Danckwerts Chem. Engng. Sci. 1968 23 1045.	157
B13	The effects of diffusivity in the liquid on rates of gas absorption A. Tavares da Silva & P.V. Danckwerts I. Chem. E. Symposium Series No.28 (1968: Instn. Chem. Engrs., London).	163
B14	Chemical reaction and effective interfacial areas in gas absorption G.E.H. Joosten & P.V. Danckwerts Chem. Engng. Sci. 1973 28 453.	171
B15	The effects of catalysis on rates of absorption of CO_2 into aqueous amine-potash solutions P.V. Danckwerts & K.M. McNeil (abstract only) Chem. Engng. Sci. 1967 22 925.	179
B16	The absorption of carbon dioxide into aqueous amine solutions and the effects of catalysis (abstract only) P.V. Danckwerts & K.M. McNeil Trans. Instn. Chem. Engrs. 1967 45 T32.	180
B17	Desorption of carbon dioxide from aqueous potash solutions with and without the addition of arsenite as a catalyst C.N.S. McLachlan & P.V. Danckwerts Trans. Instn. Chem. Engrs. 1972 50 300.	180

This series of papers deals in the main with attempts to rationalise the design of gas absorbers by the use of laboratory and pilot-scale apparatus. The objective was to produce a model of the packed column containing two parameters (e.g. surface renewal rate and effective interfacial area) which would enable these parameters to be measured once and for all for different types of packing and different liquid-flow rates, and then to use either available physico-chemical data, or data provided by laboratory-scale models, which could be combined with these parameters to predict local absorption rates (per unit packed volume) when particular chemically-reactant gases and absorbents were used. These objectives, pursued sporadically over a good many years, were not attained completely. However, I think that the work represented here throws a great deal of light on the subject - sufficient, at least, to discredit the older and less sophisticated design procedures. It suffers from several limitations; first of all, much of the work on packed columns was done on the laboratory bench with small packings and small columns. Experiments on packings and columns of a scale approaching those used in industry generally require to be made on industrial premises. The limited number of experimental campaigns I was able to mount on this scale (0.5 m columns, 25 mm packings) indicated that the whole methodology could be applied more successfully than on the laboratory scale. No doubt there has been much work done in industry, but not published, which would throw further light on the matter. I conclude that the body of the work described in this section goes a long way towards providing rational methods of design.

The work needs extension to higher temperatures, reversible reactions and to desorption. The only work on desorption at boiling-point reported here (Paper B17) unfortunately does not lead to any firm conclusions.

Paper B1 has been widely regarded as providing a true "insight". The film model of gas absorption was a convenient fiction, but could not be regarded as providing a realistic picture of events occurring at the mobile surface of a turbulent liquid. The idea of "surface renewal" had been conceived by Higbie (Ref. 7) but it was clearly improbable that each element of liquid brought to the surface would be exposed to the gas for the same length of time. It occurred to me that if surface renewal did in fact occur (e.g. in turbulent flow or in film flow over a randomly arranged packing) then any part of the surface was equally likely to be replaced, regardless of the length of time for which it had already been exposed. The rest of the argument follows and produces expressions for the rate of absorption when linear processes (simple diffusion, first-order reaction and instantaneous reaction) are involved. An important feature of the surface-renewal models is that in principle small-scale measurements of transient absorption rates into stagnant liquids, without any prior knowledge of the physico-chemical factors involved (e.g. solubility kinetics and equilibria), can be used to predict absorption rates in large-scale columns. Any surface-renewal model leads to the prediction that the rate of absorption depends on the $\frac{1}{2}$ power of the diffusivity of the dissolved gas as opposed to the first power in the case of the film model. Such experiments as have been done support the former and Paper B13 provides a different test which supports the surface-renewal hypothesis. Many other models have been proposed from time to time which predict the possibility of the absorption rate depending on any power of the diffusivity between 0 and 1, but in general they require two parameters to define the hydrodynamic conditions. A good model has as few parameters as are compatible with observed facts and required for design purposes. The film, Higbie and Danckwerts models each require only two (effective interfacial area, together with film thickness, exposure time or surface-renewal rate respectively) and the precision of experimental measurements and of the physico-chemical factors involved in experiment and design do not seem to justify further refinement.

The part of Paper B1 dealing with second-order reaction between dissolved gas and solute is less than satisfactory. Not until 1974 was an approximate solution proposed (W.J. de Coursey, Chem. Engg. Sci. 1974 29 1867) and in 1978 it was shown by computation to be surprisingly accurate (R.M. Wellek, R.J. Brunson and F.H. Law, Can. J. Chem. Engg 1978 56 181). Paper B12 showed how instantaneous reversible reactions can be dealt with in the framework of the surface renewal models.

Paper B2 Part I sprang from the sudden realization that the three models which I have been considering gave rise to predictions which were remarkably similar considering the dissimilar appearance of the algebraic expressions which they generate. However, the comment that the film model and the surface-renewal models give nearly the same results for instantaneous irreversible reaction (p.S51) is incorrect unless the diffusivities of dissolved gas and dissolved reactant are nearly equal. The same is true for the second-order reaction of finite speed. In this case computation has given numerical solutions for the film (D.W. van Krelen and P.J. Hoftijzer, Rec. Trav. Chim. 1948 67 563), Higbie (P.L.T.Brian, J.F. Hurley and E.H. Hasseltine , A.I.Ch.E.J. 1961 7 226) and Danckwerts (Wellek et.al., op. cit.) models. The results for the two surface-renewal models are numerically very close and there is a close analogy with those for the film model provided that the values of the two diffusivities are introduced in the appropriate way. The expression of de Coursey is to be preferred as, unlike the others, it is an explicit formula for the absorption rate and based on the most realistic (i.e. Danckwerts) model.

Paper B2 Part II was, I believe, the first publication to deal with transient rates of absorption since Higbie, whose very complicated technique gave unreliable results (see Paper A1). However, the rotating-drum method described suffered from the formation of a "stagnant film" of surface-active material which builds up at the exit of the liquid and reduces the rate of absorption (see Fig. 4). The same effect was observed and eliminated in the falling-film absorber (Paper B4) which, with the laminar-jet absorber, came to serve the purpose more conveniently. The paper was concerned to reconcile transient rates of absorption into inert and reactive liquids with (a) known or estimated physico-chemical quantities and (b) rates of absorption in a small packed column. Modest success was achieved as regards (a), but objective (b) was not attained as it proved impossible to assign values of the surface-renewal rate and the effective interfacial area of the packing which would reconcile the transient absorption rates on the drum with absorption rates in the column. This problem is one of the threads which runs through SECTION B and which was not squarely faced, and its likely answer formulated, until nearly 20 years later (Paper B14). Rational design of absorbers requires a knowledge of a (the effective interfacial area per unit packed volume) and s (the surface-renewal rate) separately; the conclusion of Paper B14 is that the value of a (other things being equal) depends on the ratio of the two effects of a dissolved reactant, namely (increase in capacity/increase in absorption rate). Chemical methods of measuring a (see, for example Papers B6,7,8) make use of reactions which maximise a. In the case of "instantaneous" reactions or non-reactive absorbents a reaches its minimum value. In the case of the small packings used in many laboratory experiments the maximum value of a may only be about 30% of the geometrical area. The explanation given in Paper B14, namely that there are regions where the liquid layer is thin or slowly replenished and thus becomes saturated seems quite reasonable, but it is discouraging from the point of view of the designer as it introduces a further parameter which is determined both by the hydrodynamic factors and the kinetics of the reaction. However, the observation in Papers B7 and B8 that in a larger column with larger packings the effective interfacial area may equal or even slightly exceed the geometrical area suggests that on the industrial scale the problem of the "effective" interfacial area may not arise. (I have avoided using the phrase "wetted area" because the naked eye shows that the whole surface is wet in the case of ceramic packings at any rate).

Paper B7 shows that in the case of $1\frac{1}{2}$ in. Raschig rings, for instance, rates of absorption into inert and reacting solutions can indeed be calculated from the surface-renewal models in various ways using a single value of a for a single superficial liquid velocity. Paper B8 goes on to present effective interfacial areas for other $1 - 1\frac{1}{2}$ in. packings. The absorption of carbon dioxide into buffer solutions, catalysed by hypochlorite ions (Paper B7) and the absorption of oxygen into sulphite solutions, catalysed by cobalt sulphate (B8) give compatible results and I consider that the principles laid down for design on this scale are sound; but of course it would be desirable to extend the experiments to higher temperatures and pressures and to reversible reactions.

Paper B3 was a painstaking attempt to reconcile the observed rates of transient absorption of carbon dioxide into alkaline solutions (using an improved rotating drum) with the available physicochemical data. The difficulty of doing so, in spite of the fact that more published data are probably available for this set of reactions than for any other, strongly suggests that the empirical use of laboratory absorbers is a better basis for design: e.g. stirred vessel - Papers 7,8,9,15,16,17; laminar jet - Papers B13, 14; wetted wall column - Paper B4; string of spheres - Paper B10.

The use of catalysts, which can increase the rate of reaction without changing the physical properties of the system, has proved a powerful tool: see Papers B4,5,7,8,10,15,16,17.

Papers B15,16 deal with the absorption of carbon dioxide into ammonia and ethanolamine solutions, with and without alkali, and the effect of adding arsenite which catalyses the interconversion of carbamate and bicarbonate in the bulk of the solution (in a stirred cell or bubble plate). It is interesting that when catalyst is added to an aqueous solution of amine alone the hydrolysis of the carbamate formed by the absorption of carbon dioxide is accelerated by the catalyst and thus the latter results in a higher concentration of free amine and a higher rate of absorption; whereas when alkali carbonate is also present the interconversion process may reduce the concentration of free amine so that the catalyst actually decreases the rate of absorption except at high carbonation ratios. These experiments were not done under industrial conditions with regard to either temperature or composition but provoke a useful train of thought which I am currently exploring.

Paper B11 describes a visual observation of the Marangoni effect which is well known to occur when carbon dioxide is absorbed into an ethanolamine solution. We also observed, although it is not recorded here, that the effect does not occur when hydrogen sulphide is absorbed, the reason being that the reaction in that case is instantaneous. Paper B13 confirms this and gives strong support to the applicability of the surface-renewal model to a small packed column as well as providing reliable values for the diffusivities of various amines.

My book "Gas-Liquid Reactions" was an attempt to sum up the state of knowledge of the subject to about the end of 1968.

B1

Significance of Liquid-Film Coefficients in Gas Absorption

THE conventional picture of the process of absorption of a gas into an agitated liquid, under conditions such that "the liquid film is controlling," is that there exists a stagnant "film" of liquid at the interface, similar to the laminar sublayer formed when a fluid flows past a solid. The concentration of the bulk of the liquid beneath this film is kept uniform by turbulent mixing, while the surface concentration has at all times the saturated value c^*. The rate of absorption (per unit area) will then be given by the expression

$$R = \frac{D}{x_L}(c^* - c_o) \qquad (1)$$

where R is the rate of absorption per unit area, c_o is the bulk concentration of the solute, D is its diffusivity, and x_L is "the effective film thickness" (*9*).

It seems doubtful, however, whether the conventional picture bears a very close relationship to the actual mechanism of absorption. The conditions required to maintain a stagnant film at the free surface of an agitated liquid appear to be lacking, and it seems more probable that turbulence extends to the surface and that there is no laminar boundary layer. In particular, if we consider the liquid flowing over the packing in an absorption tower, it appears most unlikely that the surface layer of liquid maintains its identity throughout all the discontinuities of flow.

The fictitious nature of the "liquid film" is probably widely suspected; nevertheless, it is constantly referred to as though it actually existed. This may be regarded, for many purposes, as a harmless and convenient usage, as measured absorption rates appear to conform to the expression

$$R = k_L(c^* - c_o) \qquad (2)$$

where k_L, the liquid-film mass-transfer coefficient, is constant for a given liquid and gas under given conditions. However, if the film model is in fact an unrealistic one, it may lead to erroneous results if it is used as the basis for theories which seek to relate k_L to the conditions of operation.

It may be of value, therefore, to show that the proved usefulness of Equation 2 in no way provides support for the stagnant-film hypothesis, as this expression can be derived from what is considered to be a more realistic picture of the processes occurring during absorption into an agitated liquid. The treatment is here extended to cases in which gas-film resistance or surface resistance plays a part. Provided no chemical reactions occur, the expressions derived for the rate of absorption are similar in form to those based on the film theory, but a different significance is attached to the liquid-film coefficient. On the other hand, when the absorbed gas undergoes a reaction in solution, the two methods of treatment give different expressions for the relationship of the rate of absorption to such quantities as the liquid-film coefficient without reaction, the reaction-velocity constant, and the diffusivities of the components of the system.

The expressions derived in this paper for the rate of absorption, R, under various conditions contain a quantity, s, which relates to the rate of renewal of liquid surface. A complete development of the subject would require further discussion of the relation of s to the hydrodynamics and geometry of the system. For the present s is regarded as a quantity which must be determined experimentally for any given system. The various expressions for R in terms of s are thus likely to be of use mainly in comparing rates of absorption in systems of equal s. For instance, suppose the rate of absorption of carbon dioxide into water is measured in a packed tower. With the same packing and flow rate, and with a liquid of virtually the same density and viscosity, s will have the same value, and under these conditions it may prove possible to use the measurements on carbon dioxide and water to predict the rate of absorption of, say, sulfur dioxide into water or carbon dioxide into dilute alkali. (The possibility of doing so depends in part on factors discussed under "Nonhomogeneous Distribution of Surface Ages.")

No attempt has been made to compare the expressions derived here with published experimental measurements. In spite of the enormous number of these, very few are suitable for an analysis of this sort, and in any case a knowledge of some of the essential physical constants is at present lacking (particularly of k_S, the "surface resistance" coefficient).

The theory must be regarded as speculative until it has been tested by experiment. It may well be found that the processes occurring in gas-absorption equipment are too complicated to be treated successfully by such simple methods as those presented here, but the author hopes that these may constitute a step in the right direction.

ABSORPTION INTO A STAGNANT LIQUID

Consideration is limited in the first place to the case where the surface of the liquid is at all times saturated with the solute gas, and there is no chemical reaction between solute and solvent.

If the normal diffusion law is obeyed in the liquid, the rate of absorption (quantity per unit area per unit time) into a stagnant liquid of infinite depth is given by

$$\psi(\theta) = (c^* - c_o)\sqrt{\frac{D}{\pi\theta}} \tag{3}$$

where θ is the time of exposure to the gas and the initial concentration is uniformly equal to c_o (see discussion of derivations of R below).

It can be seen that unless the surface is renewed by stirring or by convection, the rate of absorption becomes very slow after a time. Stagnant water will normally absorb in 1 hour a quantity of gas equivalent to a saturated layer about 2 mm. thick (taking $D \sim 10^{-5}$ sq. cm. per second). If the "depth of penetration" is arbitrarily defined as the depth at which the rise in concentration is $^1/_{100}$ that at the surface, it will be equal to $3.6\sqrt{D\theta}$ (see Equation 27), with a value of about 6 mm. at 1 hour. (In the case, for instance, of carbon dioxide and water, ψ will conform to Equation 3 only for small values of θ; thereafter convection currents will arise because the density of a solution of carbon dioxide is greater than that of water, and the predicted values of ψ will be exceeded.)

Under certain conditions Equation 3 can be applied as a close approximation to: (*a*) liquid layers of restricted depth, and (*b*) liquid moving parallel to the surface with a velocity that varies with the depth. The necessary condition for (*a*) is that the time of exposure should be so short that the depth of penetration is less than the depth of the liquid; for (*b*) it must be so short that the depth of penetration is less than the depth at which the velocity is appreciably different from that at the surface.

ABSORPTION INTO SURFACE OF TURBULENT LIQUID

When a liquid is in turbulent motion it is a mass of eddies which incessantly change their conformation and position. These eddies are pictured as continually exposing fresh surfaces to the gas, while sweeping away and mixing into the bulk parts of the surface which have been in contact with the gas for varying lengths of time. The assumption will be made that during the time of exposure of any portion of the liquid it absorbs gas at a rate given by Equation 3, as in case (*b*) above. This is equivalent to assuming that the "scale of turbulence" is much greater than the "depth of penetration" of the solute diffusing from the surface, so that relative motion of the liquid at different levels close beneath the surface may be disregarded. Velocity gradients no doubt exist beneath the surface, but while an increase in intensity of turbulence will accentuate these, it will at the same time decrease the depth of penetration by shortening the period for which any part of the liquid is exposed to the gas before being submerged once more.

Consider a liquid which is maintained in turbulent motion by stirring at a steady rate. The total area of the surface which is exposed to the gas will be taken as equal to unity and the average rate of absorption is uniform over the area. The motion of the liquid will continually replace with fresh surface those parts which are older, in the sense that they have been exposed for a finite length of time. The mean rate of production of fresh surface will be constant and equal to s, and the chance of an element of surface being replaced within a given time is assumed to be independent of its age; hence the fractional rate of replacement of the elements belonging to any age group is equal to s.

Let the area of surface comprising those elements having ages between θ and $(\theta + d\theta)$ be $\phi(\theta)d\theta$. At steady state this does not vary with time. Hence in a short time interval equal to $d\theta$ the area entering the age group $\theta \ldots\ldots (\theta + d\theta)$ from the age group $(\theta - d\theta) \ldots \theta$ will be $\phi(\theta)d\theta$. This is also equal to the area in age group $(\theta - d\theta) \ldots\ldots \theta$ less the portion of this which is replaced by fresh surface in a time equal to $d\theta$—i.e.,

$$\phi(\theta)d\theta = \phi(\theta - d\theta)\, d\theta(1 - s d\theta)$$

Therefore $$\phi(\theta) = \phi(\theta) - \frac{d\phi}{d\theta}\, d\theta - s\phi(\theta)d\theta$$

and $$\frac{d\phi}{d\theta} = -s\phi$$

Since also

$$\int_0^\infty \phi d\theta = 1$$

we have

$$\phi = se^{-s\theta} \tag{4}$$

For areas other than unity, $\phi(\theta)d\theta$ is equal to the fraction of the area which at steady state is in the age group $\theta \ldots\ldots (\theta + d\theta)$. $\phi(\theta)$ is referred to here as the surface-age distribution function.

The rate of absorption into those elements of surface having age θ and combined area $se^{-s\theta}d\theta$ is found from Equation 3 to be

$(c^* - c_o)se^{-s\theta}\sqrt{\frac{D}{\pi\theta}}\,d\theta$. Hence the mean rate of absorption per unit area of turbulent surface is

$$R = (c^* - c_o)\sqrt{D}\int_0^\infty \frac{se^{-s\theta}}{\sqrt{\pi\theta}}\,d\theta$$

$$= (c^* - c_o)\sqrt{Ds} \qquad (5)$$

Comparing this with Equation 2 we see that $\sqrt{Ds}$ can be identified with k_L.

It has been tacitly assumed in the above calculation that immediately underlying all freshly formed surface is liquid with the same concentration, c_o, of dissolved gas. In identifying $\sqrt{Ds}$ with k_L, c_o is identified with the mean bulk concentration of the liquid; this implies that the mean concentration does not change with time, and that the bulk of the liquid is of uniform concentration. The first condition is automatically fulfilled in a continuous process operating at steady state. The extent to which the second is fulfilled will depend on the depth and flow characteristics of the liquid.

Consider a case in which at steady state the mean concentration at a level A, close to the surface, is c_A, while that at a greater depth, B, is maintained at the bulk or discharge concentration, c_o (the depth of A being comparable with the scale of turbulence). Solute is carried from A to B by eddy diffusion, and for steady state

$$(c^* - c_A)\sqrt{Ds} = (c_A - c_o)k_E = (c^* - c_o)k_L$$

or

$$\frac{1}{k_L} = \frac{1}{k_E} + \frac{1}{\sqrt{Ds}} \qquad (6)$$

where k_E is the coefficient of mass transfer from A to B by eddy diffusion. s and k_E will be functions only of the physical properties and flow conditions of the liquid and the geometry of the vessel. Hence, provided these remain constant, we see from

Equation 6 that for a series of solutes $1/k_L$ should be linear in $1/\sqrt{D}$.

Further complications will be introduced if the solute undergoes chemical reaction in the liquid, or if there is a gas-film resistance or surface resistance. These are considered in connection with absorption in packed columns, and the expressions obtained are in general applicable to stirred liquids also.

PACKED ABSORPTION TOWERS

Observation of the behavior of filaments of colored liquid suggests that the liquid running over the packing in a tower is in some places flowing in a thin laminar film, while at others it is in turbulent motion caused by discontinuities in the wetted surface. Most of the exposed surface appears to belong to liquid in laminar rather than turbulent flow—at least provided the rate of flow is not too great.

We consider first the case in which virtually all absorption is assumed to take place into the surface of those areas in which the liquid is in laminar flow. The system is assumed to be at steady state.

Consider a thin horizontal section of the packing. The section is supposed to have a volume great enough to offer a representative sample of flow conditions, but its height must be sufficiently small to allow the bulk concentration of the liquid to be regarded as uniform. Where a stream of liquid in laminar flow encounters a discontinuity in the packing, turbulent conditions prevail and the liquid is assumed to be thoroughly mixed before passing on to another laminar region. Fresh liquid surface, with a concentration equal (subject to local variations which will be ignored) to the mean concentration of the liquid stream, is thus being generated at the top of these laminar regions and destroyed at the same rate at the points of discontinuity. The time taken by the surface of the liquid to travel from the preceding point of discontinuity to some fixed point on the wetted surface of the packing is termed the "age" of the surface at this point.

It is assumed that the local rate of absorption of gas is the same as for a stagnant liquid of infinite depth which has been exposed to the gas for a time equal to the local age of the surface. A complete analysis of the assumption cannot be given here, but the following considerations show that it probably represents a reasonable approximation. A solution has been obtained (by graphical methods) for the rate of absorption of gas into a laminar layer of liquid flowing under the influence of gravity down a vertical wall. It is assumed that the parabolic velocity profile is fully developed, and that the surface is constantly saturated. The ratio of the amount of gas absorbed by an element of moving surface within a time θ of its formation to the amount absorbed in time θ by an equal area of surface of stagnant, infinitely deep liquid depends only on the quantity $L = \frac{lD\mu}{g\rho d^4}$ where d is the thickness of the liquid layer and l is the distance which the element of surface moves in time θ. The difference between the quantities absorbed in the two cases amounts to less than 5%, provided $L < 0.1$. To give an idea of the significance of this condition, d will be given the rather low value of 0.01 cm., with $l = 5$ cm., and $D = 10^{-5}$ sq. cm. per second. L is then about 0.05 (for water).

Although this analysis ignores the fact that the parabolic velocity distribution is established only after a considerable distance of travel, and that the surfaces of the packing will not generally be vertical, it does indicate that the assumption is likely to be justified under conditions of practical importance. While the discrepancy increases with the age of the surface, absorption into "older" surfaces is relatively unimportant in determining the mean rate of absorption.

The packing is supposed to be arranged in a completely random manner, so that there is no correlation between the age of any element of moving surface and the probability of its being destroyed within a given time (the validity of this assumption is discussed under "Nonhomogeneous Distribution of Surface Ages"). Under these circumstances the age-distribution function $\phi(\theta)$ is easily shown to be the same as that already found for the stirred liquid (Equation 4)—namely, $se^{-s\theta}$. The surface-renewal factor, s, will in this case depend on the physical properties and rate of flow of the liquid and the geometry and size of the packing. If the rate of absorption into a stagnant liquid, which has been exposed to the gas for a time θ, is denoted by $\psi(\theta)$, the mean rate of absorption per unit area of wetted surface in the section of packing will be

$$R = s \int_0^{\infty} e^{-s\theta}\psi(\theta)d\theta \tag{7}$$

The function ψ will take different forms according to the behavior of the liquid and gas which are under consideration. In principle it may be found by solving the appropriate diffusion equations with the required boundary conditions (see discussion of derivation of expressions for R). However, the necessity for this can usually be avoided, and the mathematical treatment simplified, by making use of the fact that the expression for R/s, as can be seen from Equation 7, is the Laplace transform of ψ. The transform is considerably easier to find than ψ itself.

Mean Absorption Rates for Various Types of System. The derivation of the expressions given here is discussed at the end of this paper.

SURFACE CONSTANTLY SATURATED, NO CHEMICAL REACTION IN SOLUTION.

$$R = (c^* - c_o)\sqrt{Ds} \tag{8}$$

As in the analogous case of absorption into a turbulent liquid (when k_E is large), $\sqrt{Ds}$ can be identified with k_L. If the solvent and the flow conditions remain unchanged, k_L should be proportional to $\sqrt{D}$ for a series of solutes, provided they behave toward the solvent in the manner postulated.

GAS-FILM RESISTANCE, NO CHEMICAL REACTION.

$$R = \frac{c^* - c_o}{\dfrac{1}{\sqrt{Ds}} + \dfrac{H}{k_G}} \tag{9}$$

In deriving this expression it has been assumed that the system obeys Henry's law and that the gas-film mass-transfer coefficient, k_G, has a constant value independent of the local rate of absorption into the liquid. The latter is almost certainly an oversimplification, but the approximation is probably no more drastic than that involved in the normal treatment of gas-film coefficients, and the result should give at least a general indication of the behavior to be expected from this type of system. It is interesting to find that Equation 9 is similar in form to that derived by a very different path from the simple two-film theory. It is shown in the discussion of derivation of expressions for R that the conventional rule for the addition of "resistances in series" can be applied to all the types of systems here considered.

Surface Resistance, No Chemical Reaction. Even when there is no effective gas-film resistance to absorption—for instance, when a pure gas is absorbed—there is some evidence that the gas and the liquid may not attain equilibrium at the interface immediately on being brought into contact (*4*, *7*). Such a phenomenon may be explained by assuming that only a very small fraction of the gas molecules striking the surface enters the liquid, while the remainder is reflected (this hypothesis has been discussed more fully elsewhere, *4*). An alternative explanation might be that there is a region in the immediate neighborhood of the liquid surface in which surface forces restrict the mobility of the solute molecules, so that their diffusivity is much lower than the bulk of the liquid. In either case, the rate of absorption at any time will be given by an expression of the form $(c^* - c_s)k_S$, where c_s (the concentration at the surface or immediately beneath it) is a function of the time of exposure. Only in the case of carbon dioxide and water is there any experimental evidence as to the value of k_S (*7*). The apparent value (ca. 0.05 cm. per second) is in this case of the same order of magnitude as normal values of k_L, but this requires confirmation. Further investigation may show the phenomenon to be one of considerable importance in determining rates of gas absorption.

Until more information is available, it is not possible to say whether k_S is a function of the age of the surface, although this seems unlikely. Assuming it to be constant, we have by analogy with Equation 9

$$R = \frac{c^* - c_o}{\dfrac{1}{\sqrt{Ds}} + \dfrac{1}{k_S}} \tag{10}$$

if there is no gas-film resistance, and

$$R = \frac{c^* - c_o}{\dfrac{1}{\sqrt{Ds}} + \dfrac{H}{k_G} + \dfrac{1}{k_S}} \tag{11}$$

when there is gas-film resistance of appreciable magnitude.

Solute Destroyed by First-Order Reaction with Solvent.

$$R = \left[c^* - c_o\left(\frac{s}{r + s}\right)\right]\sqrt{D(r + s)} \tag{12}$$

where r is the first-order reaction velocity constant. This may be put in the form

$$R = \left[c^* - c_o\left(\frac{s}{r + s}\right)\right]\sqrt{k_L^2 + Dr} \tag{13}$$

if $\sqrt{Ds}$ is identified with k_L, the hypothetical liquid-film coefficient for the absorption of the same gas in the same solvent without reaction. The value of s might be determined from measurements of k_L for a gas which did not react with the solvent.

If there is a gas-film resistance to absorption, Equation 12 becomes

$$R = \frac{c^* - c_o\left(\dfrac{s}{r + s}\right)}{\dfrac{H}{k_G} + \dfrac{1}{\sqrt{D(r + s)}}} \tag{14}$$

H/k_G should be replaced by $1/k_S$ if only surface resistance is important, and by $(1/k_S + H/k_G)$ if both are of appreciable magnitude.

The term involving c_o may be omitted when conditions are such that the bulk concentration of unreacted absorbed gas is negligibly small. This will be the case if at all points in the absorption tower $AR \ll Vrc^*$, where A is area of wetted surface and V is volume of liquid, per unit volume of packing.

Equation 13 may be compared with the expressions derived from the conventional film theory (*11*).

INSTANTANEOUS REACTION BETWEEN DISSOLVED GAS AND REAGENT IN SOLUTION. When there is no surface or gas-film resistance

$$R = \frac{c^* \sqrt{Ds}}{erf[\beta/\sqrt{D}]} \tag{15}$$

where β is defined by

$$\frac{c^*}{\sqrt{D'}} e^{\beta^2/D'} erfc[\beta/\sqrt{D'}] - \frac{c_o'}{\sqrt{D}} e^{\beta^2/D} erf[\beta/\sqrt{D}] = 0 \tag{16}$$

c_o' being the bulk concentration of the reagent in the solution and D' its diffusivity. c^* and c_o' are expressed in chemical equivalents per unit volume. Graphs which will help in the solution of Equation 16 have been published (*6*).

If the diffusivities of the dissolved gas and of the reagent are equal (which will often be approximately true) we find from Equations 15 and 16

$$R = (c^* + c_o') \sqrt{Ds} \tag{17}$$

which is similar to the expression derived (with $D = D'$) from the conventional hypothesis (*10*).

If a surface or gas-film resistance of appreciable magnitude, $1/k$, is present Equation 15 becomes:

$$R = \frac{c^*/erf[\beta/\sqrt{D}]}{\frac{1}{\sqrt{Ds}} + \frac{1}{k}} \tag{18}$$

SECOND-ORDER REACTION BETWEEN DISSOLVED GAS AND REAGENT IN SOLUTION. The exact solution of this problem offers considerable mathematical difficulties. Under certain conditions, however, Equation 12, 13, or 14 may be used as an approximation if r is replaced by $r'c_o'$, where c_o' is the bulk concentration of the reagent and r' is the second-order velocity constant—i.e., the rate of reaction at any point in the liquid is $r'cc'$, where c' is the local concentration of the reagent. The error involved depends on the ratios $s/r'c_o'$ and c_o'/c^*. We are interested mainly in cases where the former is of the order of magnitude of unity—i.e., diffusion and chemical reaction both play appreciable parts in controlling the rate of absorption. Putting $s/r'c_o' = 1$ for purpose of illustration, we find that the error will be less than 10% when c_o'/c^* is equal to 50. (In other words, the concentration of the reagent in the absorbent solution must be about 50 times as great as the concentration of a saturated solution of the gas in the pure solvent.) The accuracy of the approximation will be lessened if either of the above-mentioned ratios is diminished. (The bulk concentration of unreacted absorbed gas is assumed to be much less than c^*.)

NONHOMOGENEOUS DISTRIBUTION OF SURFACE AGES

The expression previously derived (Equation 7) in terms of ψ for the mean rate of absorption in a section of a packed column was based on the assumption that the probability of an element of moving surface being destroyed by remixing was independent of its age. However, there are two types of system in which this condition will not be fulfilled.

Nonrandom Packing. Any regularity in the arrangement of the packing is likely to affect the age-distribution function, $\phi(\theta)$. For instance, if the packing consists of vertically stacked rings, virtually none of the surface will reach an age greater than the time, θ_c, taken for the surface of the liquid to move from the top to the bottom of a ring, and there is no chance of destruction before this age has been reached. If we ignore the acceleration of the surface during the descent from one discontinuity to another, we have approximately

$$\phi(\theta) = 1/\theta_c, \quad 0 < \theta < \theta_c$$

$$\phi(\theta) = 0, \quad \theta > \theta_c$$

In a simple case such as this it is, of course, easy if ψ is known to determine R from the expression

$$R = \int_0^\infty \phi(\theta) \, . \, \psi(\theta) \, . \, d\theta \tag{19}$$

which is of general applicability.

In a dumped packing, even of elements of uniform size, ϕ may be expected to approximate much more closely to the form of Equation 4, since the linear rate of flow of the liquid will vary widely with the inclination of the surface, and there will be considerable variation in the distance of travel between discontinuities. It may be objected that there is bound to be a finite upper limit to the age of any element of surface. However, this is not likely to lead to serious error in itself, provided the distribution of the "younger" parts of the surface does not depart too widely from that given by Equation 4. This is because most of the absorption takes place into surface which has been exposed for only a short time, and the upper limit of the integral in Equation 7 can be given a finite value without greatly altering the value of the integral. For instance, in the simplest type of system, in which there is no surface or gas-film resistance or chemical reaction, the integral is (cf. Equation 5):

$$\int_0^\infty se^{-s\theta} \sqrt{\frac{D}{\pi\theta}} \, d\theta = \sqrt{Ds}$$

If the upper limit of this integral is changed to θ_c, it becomes on rearrangement $\sqrt{Ds}\ erf\ \sqrt{s\theta_c}$, which is within 10% of $\sqrt{Ds}$ if $s\theta_c > 1.5$. Tentative values of s, calculated from published values of k_L for carbon dioxide and water in packed towers (*8*) are of the order of magnitude 5 sec.$^{-1}$, indicating that less than one tenth of the observed absorption rate is due to surfaces older than 0.33 second.

Nonuniform Value of s. If an appreciable fraction of the absorption in a packed column takes place into the surface of turbulent liquid at the points of discontinuity, the form of ϕ will again be modified. We may picture the effective area of surface as being made up of a number of areas, a, each having a surface-age distribution similar to Equation 4 but with its own value, s_a, for s. This leads to the expression

$$R = \frac{1}{\Sigma a} \int_0^\infty \psi(\theta) \sum_a a s_a e^{-s_a \theta} d\theta \qquad (20)$$

If for either of the reasons discussed above, the function expressing the distribution of surface ages in an absorption system differs from $se^{-s\theta}$, and if its actual form is undetermined, systematic investigations or attempts to predict absorption rates in various types of system may be rendered difficult.

In the simplest case, in which there is no chemical reaction or gas-film or surface resistance, we find from Equations 8 and 19

$$R = (c^* - c_o) \sqrt{D} \int_0^\infty \frac{\phi(\theta)}{\sqrt{\pi\theta}} d\theta \qquad (21)$$

for nonrandom packing, and from Equations 8 and 20

$$R = (c^* - c_o) \sqrt{D} \frac{\Sigma a \sqrt{s_a}}{\Sigma a} \qquad (22)$$

for nonuniform values of s. Similarly, the analogs of Equation 15 are

$$R = \frac{c^*\sqrt{D}}{erf[\beta/\sqrt{D}]} \int_0^\infty \frac{\phi(\theta)}{\sqrt{\pi\theta}}\, d\theta \tag{23}$$

$$R = \frac{c^*\sqrt{D}}{erf[\beta/\sqrt{D}]} \frac{\Sigma a \sqrt{S_a}}{\Sigma a} \tag{24}$$

Equations 21 to 24 resemble Equations 8 and 15 in that they express R as the product of a term containing known (or independently measurable) physical constants, and another which depends only on flow conditions, etc., and which must be determined empirically. Thus, provided the latter term remains constant, there should be a simple and predictable relationship between the value of R and the values of the physical constants in the first term.

However, if there is any gas-film or surface resistance or if a first- or second-order reaction takes place, we find that the resulting expressions no longer have the relative simplicity which marks the equations referring to systems in which $\phi = se^{-s\theta}$. For instance, Equation 9 (simple gas-film resistance) becomes

$$R = (c^* - c_o) \frac{k_G}{H} \int_0^\infty e^{k_G^2\theta/HD} erfc\left[\frac{k_G}{H}\sqrt{\frac{\theta}{D}}\right] \phi(\theta)d\theta \tag{25}$$

when ϕ is undetermined (cf. Equation 29), and

$$R = \frac{c^* - c_o}{\Sigma a} \sum \frac{a}{\left(\frac{1}{\sqrt{Ds_a}} + \frac{H}{k_G}\right)} \tag{26}$$

when the value of s is not homogeneous. Thus even a system with constant flow conditions, and for which k_G is known, will show no simple relationship between R and D. All that can be said is that $R/(c^* - c_o)$ will be a unique function of D for a series of solutes if H/k_G is kept constant, and that it will be a unique function of H/k_G for a series of systems in which D is constant (liquid-flow conditions assumed constant throughout). Such considerations indicate a basis for the correlation of experimental results, but are not of great practical value. Similar conclusions may be reached regarding systems of the types referred to in Equations 10 to 14.

DERIVATION OF EXPRESSIONS FOR *R* IN VARIOUS TYPES OF SYSTEM

Two methods may be used. One is to solve the appropriate partial differential equation to find $\psi(\theta)$, the rate of absorption per unit area of surface into a stagnant liquid which has been exposed to the gas for time θ. $\psi(\theta)$ may then be multiplied by $se^{-s\theta}$ and integrated, as already explained, to give R. This method is used in cases 1, 2, and 4 below. Alternatively, the Laplace transform may be applied to the diffusion equation and the resulting "subsidiary equation" solved for $\bar{c}(x, s)$, the Laplace transform of $c(x, \theta)$ (*3*). R is equal to $-Ds(d\bar{c}/dx)_{x=0}$. The latter method, which is illustrated in cases 3 and 6 below, is generally considerably simpler than the former, but is applicable only when the diffusion equation is linear.

A general rule for the "addition of resistances" is derived below, so that it is not necessary to consider the effect of an added gas-film or surface resistance for each type of system separately.

1. No Reaction. Surface Constantly Saturated.

$$\frac{\partial c}{\partial \theta} = D\frac{\partial_2 c}{\partial x^2} \qquad \begin{array}{l} c = c_o,\ x > 0,\ \theta = 0 \\ c = c^*,\ x = 0,\ \theta > 0 \\ c = c_o,\ x = \infty,\ \theta > 0 \end{array}$$

The solution to this is (*1*)

$$c = c_o + (c^* - c_o)erfc\left[\frac{x}{2\sqrt{\theta D}}\right] \tag{27}$$

whence

$$\psi(\theta) = -D\left(\frac{\partial c}{\partial x}\right)_{x=0} = (c^* - c_o)\sqrt{\frac{D}{\pi\theta}}$$

$$R = (c^* - c_o)s \int_0^\infty e^{-s\theta} \sqrt{\frac{D}{\pi\theta}}\, d\theta$$

$$= (c^* - c_o)\sqrt{Ds} \tag{28}$$

2. Surface or Gas-Film Resistance. This system illustrates the rule for the addition of resistances given in Case 6.

The differential equation is the same as in case 1, but the boundary conditions become

$$c = c_o, \quad x > 0, \quad \theta = 0$$

$$-D\left(\frac{\partial c}{\partial x}\right) = k(c^* - c), \; x = 0, \; \theta > 0$$

$$c = c_o, \quad x = \infty, \quad \theta > 0$$

The second condition states that the rate of absorption is at all times proportional to the difference between the surface concentration and the saturated concentration. k stands for k_S or $k_{G/H}$. The solution is known (*2*) and may be put in the form

$$\psi(\theta) = k(c^* - c_s) = k(c^* - c_o)e^{k^2\theta/D} erfc\left[k\sqrt{\frac{\theta}{D}}\right] \tag{29}$$

whence

$$R = s\int_0^\infty e^{-s\theta}\psi(\theta)d\theta = \frac{c^* - c_o}{\frac{1}{k} + \frac{1}{\sqrt{Ds}}} \tag{30}$$

3. First-Order Reaction between Dissolved Gas and Solvent.

$$\frac{\partial c}{\partial \theta} = D\frac{\partial^2 c}{\partial x^2} - rc \qquad \begin{array}{l} c = c_o, x > 0, \theta = 0 \\ c = c^*, x = 0, \theta > 0 \\ c < c_0, x = \infty, \theta > 0 \end{array}$$

If conditions are such that the concentration of unreacted absorbed gas in the bulk of the liquid is negligibly small ($c_o = 0$), the solution of the above is, as has been shown elsewhere (*5*),

$$\psi(\theta) = c^* \sqrt{Dr}\left[erf\sqrt{r\theta} + \frac{e^{-r\theta}}{\sqrt{\pi r\theta}}\right] \tag{31}$$

whence

$$R = c^* \sqrt{D(r + s)} \tag{32}$$

This result could have been reached much more directly by the Laplace transform method, which will be illustrated by its application to the general case where $c_o \neq 0$.

Multiplying each term in the differential equation by $e^{-s\theta}$, integrating with respect to θ between 0 and ∞, and making use of the first boundary condition, we find (*3*)

$$s\bar{c} - c_o = D\frac{d_2\bar{c}}{dx^2} - r\bar{c}$$

while the second and third boundary conditions become

$$\bar{c} = \frac{c^*}{s},\ x = 0$$

$$\bar{c} < \frac{c_o}{s},\ x = \infty$$

where $\bar{c}$, the Laplace transform of c, is defined by

$$\bar{c}(x, s) = \int_0^\infty e^{-s\theta}c(x, \theta)d\theta$$

The solution of the above is

$$\bar{c} = \frac{c_o + \left[c^*\left(\frac{r+s}{s}\right) - c_o\right]e^{-x\sqrt{\frac{r+s}{D}}}}{(r + s)} \tag{33}$$

Now $\psi(\theta)$, the rate of absorption at time θ, is equal to $-D\left(\frac{\partial c}{\partial x}\right)_{x=0}$ whence it can be seen from the definition of $\bar{c}$ that

$$R = s\int_0^\infty e^{-s\theta}\psi(\theta)d\theta = -sD\int_0^\infty e^{-s\theta}\left(\frac{\partial c}{\partial x}\right)_{x=0} d\theta$$

$$= -sD\left[\frac{d}{dx}\int_0^\infty e^{-s\theta}.c.d\theta\right]_{x=0} = -sD\left(\frac{d\bar{c}}{dx}\right)_{x=0}$$

so that R is immediately found from Equation 33:

$$R = \left[c^* - c_o\left(\frac{s}{r+s}\right)\right]\sqrt{D(r+s)} \qquad (34)$$

4. Instantaneous Reaction between Absorbed Gas and Reagent in Solution. The method of solution for the stagnant liquid has been given (*6*). There is a steadily deepening zone beneath the surface which contains absorbed gas but no reagent. This is bounded by a plane at which the concentration of both is zero, and beneath this the liquid contains reagent alone. The absorbed gas and the reagent obey the normal diffusion equations, but the boundary conditions are somewhat complicated. The solution is

$$\psi(\theta) = \frac{c^*}{erf[\beta/\sqrt{D}]}\sqrt{\frac{D}{\pi\theta}} \qquad (35)$$

from which Equation 15 follows.

5. Second-Order Reaction between Absorbed Gas and Reagent in Solution. Attention will be confined to the case in which the concentration of unreacted absorbed gas in the bulk of the liquid is negligibly small compared to c^*.

Putting $c(x, \theta)$ = concentration of unreacted absorbed gas, $c'(x, \theta)$ = concentration of unreacted reagent (both in chemical equivalents per unit volume), and r' = the second-order reaction velocity constant, we have for a stagnant liquid

$$\frac{\partial c}{\partial \theta} = D\frac{\partial_2 c}{\partial x^2} - r'cc'$$

$$\frac{\partial c'}{\partial \theta} = D'\frac{\partial_2 c'}{\partial x^2} - r'cc'$$

$$c = c^*,\ x = 0,\ \theta > 0$$
$$c = 0,\ x > 0,\ \theta = 0$$
$$c = 0,\ x = \infty,\ \theta > 0$$
$$c' = c_o',\ x > 0,\ \theta = 0$$
$$c' = c_o',\ x = \infty,\ \theta > 0$$
$$\frac{\partial c'}{\partial x} = 0,\ x = 0,\ \theta > 0$$

These equations are nonlinear and no solution has so far been found. However, Equations 31 and 32 can be used as approximations, under certain conditions, if $r'c_o'$ is substituted for r. Let us call the expressions for R and $\psi(\theta)$ obtained in this way R_1 and $\psi_1(\theta)$, while the exact solutions of the above equations would lead to other expressions, R_2 and $\psi_2(\theta)$. Now it can be shown (*5*) that $\psi_2 \approx \psi_1$ if $r'c^*\theta \ll 1$. If $r'c^*\theta$ is taken to be 0.05—i.e., $\theta = 0.05/r'c^*$—the difference between ψ_2 and ψ_1 is less than 5%, hence the approximation

$$s\int_0^{0.05/r'c^*} e^{-s\theta}\psi_2(\theta)d\theta \approx s\int_0^{0.05/r'c^*} e^{-s\theta}\psi_1(\theta)d\theta$$

has an error less than 5%. If in addition it is stipulated that

$$s\int_0^{0.05/r'c^*} e^{-s\theta}\psi_1(\theta)d\theta > 0.95s\int_0^{\infty} e^{-s\theta}\psi_1(\theta)d\theta \qquad (36)$$

the following approximation can be used with an error of less than 10%:

$$R_1 = s\int_0^{\infty} e^{-s\theta}\psi_1(\theta)d\theta \approx s\int_0^{\infty} e^{-s\theta}\psi_2(\theta)d\theta = R_2$$

But from Equations 31 and 32

$$\frac{s\int_0^{0.05/r'c^*} e^{-s\theta}\psi_1(\theta)d\theta}{s\int_0^{\infty} e^{-s\theta}\psi_1(\theta)d\theta} = erf\sqrt{\frac{0.05(r'c_0' + s)}{r'c^*}} - \sqrt{\frac{r'c_0'}{r'c_0' + s}}\, e^{\frac{-0.05s}{r'c^*}}\, erf\sqrt{\frac{0.05c_0'}{c^*}}$$

The value of the ratio is seen to depend both on c_0'/c^* and on $r'c_0'/s$, but because interest centers on systems in which s and $r'c_0'$ are of comparable magnitude—i.e., diffusion and reaction play roughly equal parts in determining the absorption rate—we may put $r'c_0'/s = 1$ for purposes of illustration. It is then found that $c_0'/c^* > 50$ if Equation 36 is to be complied with.

6. Rule for Addition of Resistances. In all types of systems considered here the equation for $\bar{c}(x,s)$, the Laplace transform of the concentration at x, has the following form:

$$\frac{d_2\bar{c}}{dx^2} = \frac{a\bar{c}}{D} - \frac{b}{D} \tag{37}$$

where a and b are constants. In the case where there is no surface or gas-film resistance, so that the surface of the liquid is constantly saturated with the gas, the boundary condition at the surface becomes

$$\bar{c} = \frac{c^*}{s}, \; x = 0$$

whence

$$\bar{c} = \left(\frac{c^*}{s} - \frac{b}{a}\right) e^{-x\sqrt{\frac{a}{D}}} + \frac{b}{a}$$

and

$$R = -Ds\left(\frac{d\bar{c}}{dx}\right)_{x=0} = \left(c^* - \frac{bs}{a}\right)\sqrt{aD}$$

If, however, there is a constant resistance at the surface, the boundary condition is changed to

$$-D\left(\frac{\partial c}{\partial x}\right) = k(c^* - c), \; x = 0, \; \theta > 0$$

where k is the appropriate mass-transfer coefficient. This transforms to

$$-D\left(\frac{d\bar{c}}{dx}\right) = k\left(\frac{c^*}{s} - \bar{c}\right), \; x = 0$$

The new solution for $\bar{c}$ is

$$\bar{c} = \frac{\left(\frac{c^*}{s} - \frac{b}{a}\right) e^{-x\sqrt{\frac{a}{D}}}}{\frac{1}{k}\sqrt{Da} + 1} + \frac{b}{a} \tag{38}$$

whence

$$R = \frac{c^* - \dfrac{bs}{a}}{\dfrac{1}{k} + \dfrac{1}{\sqrt{Da}}} \tag{39}$$

In other words, if R has the form

$$R = (c^* - A)\, B\, \sqrt{D}$$

(A and B being constants) when the surface is constantly saturated, then for the case where there is a resistance $1/k$ at the surface

$$R = \frac{c^* - A}{\dfrac{1}{B\sqrt{D}} + \dfrac{1}{k}}$$

In cases such as Equation 15 this rule is ambiguous, and recourse must be had to the equation for $\bar{c}$ to determine the effect of an added resistance.

These arguments are easily extended to the case where there are two resistances at the surface.

CONCLUSION

The considerations advanced in this paper raise two distinct questions, which can be answered only by experiment.

1. Is the surface-renewal mechanism here postulated closer to the truth than the conventional picture of an undisturbed layer at the surface of the liquid?
2. If so, does the distribution of surface ages in absorption equipment of practical interest approximate to Equation 4?

It should be possible to shed some light on the first by qualitative observations on stirred and flowing liquids, without reference to absorption. If the answer should prove to be "yes," then the conventional picture is misleading, and should not be used as the basis for theories of absorption although its terminology would probably be retained as a matter of convenience. As regards the second, careful measurements of rates of absorption, and of the relevant physical quantities, such as k_S, would be required to determine the answer; it would be convenient if this should prove to be "yes," because a precise mathematical treatment could then be employed, as described above. Should the answer to the second question prove to be "no," the predictions of the theory lose a great deal of their precision and utility, but this in itself would not be a valid reason for preferring the "stagnant film" hypothesis as a basis for further theoretical developments.

NOMENCLATURE

A = area of wetted surface per unit volume of packing
a = area of those parts of absorbing surface having value s_a for s
c = concentration of absorbed gas in liquid at x, θ
c^* = saturated concentration of gas in liquid
c_o = initial or bulk concentration of gas in liquid
c' = concentration of reagent in solution
c'_o = initial or bulk concentration of reagent
$\bar{c}$ = Laplace transform of $c = \int_0^\infty ce^{-s\theta}d\theta$
D = diffusivity of absorbed gas
D' = diffusivity of reagent in solution
d = thickness of liquid layer on packing or other solid surface
H = Henry's law constant
k = generalized mass-transfer coefficient
k_L = liquid-film mass-transfer coefficient
k_G = gas-film mass-transfer coefficient (partial-pressure units)
k_S = mass-transfer coefficient for surface resistance (concentration units)
k_E = mass-transfer coefficient for eddy diffusion
l = distance traveled in time θ by element of surface of moving liquid
R = mean rate of absorption per unit area of nonstagnant liquid
r = velocity constant for first-order reaction between absorbed gas and liquid
r' = velocity constant for second-order reaction between ab-

sorbed gas and reagent in solution

s = fractional rate of renewal of surface of liquid

s_a = value of s for area a

V = volume of liquid per unit volume of packing

x_L = "effective thickness of liquid film"

x = distance beneath surface of liquid

β = quantity defined by Equation 16

θ = time for which a liquid surface has been exposed to gas, "age" of surface

ψ = rate of absorption into unit area of surface of stagnant liquid

ϕ = surface–age distribution function

μ = viscosity of liquid

ρ = density of liquid

$erf(z) = 1 - erfc(z) = \frac{2}{\sqrt{\pi}}\int_0^z e^{-y^2}dy$ (numerical values may be found in tables)

LITERATURE CITED

(1) Carslaw, H. S., and Jaeger, J. C., "Conduction of Heat in Solids," p. 43, Oxford University Press, 1947.

(2) *Ibid.*, p. 53.

(3) *Ibid.*, p. 240.

(4) Danckwerts, P. V., *Research*, **2**, 494 (1949).

(5) Danckwerts, P. V., *Trans. Faraday Soc.*, **46**, 300 (1950).

(6) *Ibid.*, p. 701.

(7) Higbie, R., *Trans. Am. Inst. Chem. Engrs.*, **31**, 65 (1935).

(8) Perry, J. H., "Chemical Engineer's Handbook," pp. 1179, 1184, New York, McGraw-Hill Book Co., 1941.

(9) Sherwood, T. K., "Absorption and Extraction," p. 61, New York, McGraw-Hill Book Co., 1937.

(10) *Ibid.*, p. 196.

(11) *Ibid.*, p. 202.

RECEIVED August 8, 1950.

B2

KINETICS OF LIQUID-FILM PROCESS IN GAS ABSORPTION. PART I: MODELS OF THE ABSORPTION PROCESS

By P. V. DANCKWERTS, G.C., M.B.E., M.A., S.M.,* and A. M. KENNEDY B.E., B.Sc.*

SUMMARY

Three models of the processes involved in the absorption of gas into agitated liquids are compared. These are: steady-state diffusion through a stagnant film; transient absorption into surfaces which are systematically replaced by fresh liquid; and transient absorption into surfaces which are randomly replaced. It is shown that the three models lead to closely similar predictions about the effect of physico-chemical variables (solubility, diffusivity, reaction rate, etc.) on the rate of absorption. Hence if one of the models is correct the effects of changes in these variables will not distinguish between them, but on the other hand the formulae derived for any of the models can be used for prediction.

The absorption of a gas by a liquid which is subject to agitation by stirring or by film-flow over solid surfaces is influenced, so far as the processes occurring on the liquid side of the interface are concerned, by two distinct sets of factors :

(*a*) Physico-chemical factors : solubility and diffusivity of gas in liquid ; concentration of reagent (if any) ; reaction-velocity constant, reaction-equilibrium constant, etc.

(*b*) Hydrodynamic factors : geometry and scale of equipment ; viscosity, density and flow-rate of liquid, etc.

The hydrodynamics of agitated liquids with free surfaces has received little attention. In particular it is not known whether turbulence, when maintained in the body of the liquid, extends to the surface or whether it is damped-out in the neighbourhood of a gas-liquid interface as it is near a solid-liquid interface. In the former case one would expect liquid which had been in contact with the gas to be mixed bodily into the interior, its place being taken by liquid from the interior. In the latter case, material transferred from one phase to another would have to cross a region in the neighbourhood of the surface by molecular diffusion alone, and there would be a fairly close analogy with mass-transfer between a solid and a fluid.

In the absence of definite information about the behaviour of the liquid near the surface, it is instructive to compare the consequences of making various assumptions. We shall consider a system (such as liquid running over packing in an absorption tower) in which the hydrodynamic factors are kept constant while the physico-chemical factors may be changed. There will be a relation between the rate of absorption and the magnitudes of the various physico-chemical quantities, and a valid theory of the absorption-process would enable us to predict this relation. In particular, if the rate of absorption of a particular gas into a liquid with which it does not react were known ("physical absorption") it would be possible to predict the rates of physical absorption of other gases and the rates of absorption of gases into reacting solutions, provided the required physico-chemical data were available.

Theories based on three distinct hypotheses about the behaviour of the liquid will be considered. It will be shown that the predictions of the three theories are quantitatively very close to one another under the conditions considered, in spite of their divergent physical and mathematical foundations. A number of the expressions given here are derived from *Absorption and Extraction*, by Sherwood and Pigford.[14]

The Three Models

The three physical models of the liquid behaviour will be referred to, for convenience, by the names of those who seem first to have suggested them :

(a) *Whitman : laminar film*[23]

The liquid at the surface is assumed to be in laminar flow parallel to the surface, while liquid below the surface is in turbulent motion. The rate of absorption is determined mainly by molecular diffusion in the surface layers. Although the relative importance of transport by diffusion and by turbulence will presumably vary continuously with depth below the surface, it is convenient to take as a model a completely stagnant layer of effective thickness x_L, overlying liquid of uniform composition. The thickness of the film is assumed to be small enough for the

* Department of Chemical Engineering, Tennis Court Road, Cambridge.

absorption-process to be treated as one of steady-state diffusion through the stagnant layer.

(b) *Higbie : systematic surface renewal*[6]

Liquid is assumed to run down a packing in a film that is in laminar flow throughout its thickness (not merely at the surface), until it is mixed up on meeting a discontinuity, or by some other cause. Each element of liquid surface is assumed to be exposed to the gas for the same length of time, and to absorb gas during this time at the same changing rate as though it were a stagnant layer of infinite depth. At any level in the column the mean absorption rate per unit wetted area will be the average of the rates of absorption into a large number of elements of liquid which have been exposed to the gas for various lengths of time. If therefore an element of surface absorbs an amount $Q\{\theta\}$ of gas per unit area during its time of exposure θ, the average absorption rate is $Q\{\theta\}/\theta$.

(c) *Danckwerts : random surface renewal*[1]

As in (*b*), the liquid is assumed to run down the packing in laminar flow until it is remixed by a discontinuity or other cause, and to absorb at the same rate as a stagnant surface during its exposure. It is assumed, however, that there is no correlation between the time for which an element of surface has been exposed and its chance of being remixed. The rate of surface renewal (fraction per unit time) is s. The average absorption rate per unit area is then :

$$s \int_{\theta=0}^{\infty} e^{-s\theta}\, dQ\{\theta\}$$

It can be shown[1, 11] that the Higbie and Danckwerts models, as well as the Whitman model, give rise to the conventional rule for " resistances in series " when a gas-film resistance occurs.

Prediction of Effect of Physico-Chemical Factors

The predictions which may be made concerning various absorption processes on the basis of these models will now be compared. The quantity R is the mean absorption rate per unit area of wetted surface in each case. (R will of course generally vary with height in a packed column.)

(i) *Physical absorption*

The liquid-film transfer coefficient for physical absorption, k_L, is defined by

$$R_1 = (C^* - C_0)\, k_L \qquad . \qquad . \qquad (1)$$

The physical significance given to k_L is different in each of the three models, as can be seen from the following expressions for the rate of absorption :

Whitman :[23] $$R_1 = (C^* - C_0)\,.\,\frac{D}{x_L} : \qquad . \qquad (2)$$

Higbie :[6] $$R_1 = (C^* - C_0)\,.\,2\sqrt{\frac{D}{\pi\theta}}\,; \;. \qquad (3)$$

Danckwerts :[1] $$R_1 = (C^* - C_0)\,\sqrt{Ds}\,; \qquad . \qquad (4)$$

where C^* = concentration of dissolved gas at surface (assumed constant), C_0 = concentration in bulk of liquid, D = diffusivity.

All three models lead to the same prediction concerning the effect of C^* and C_0 on the absorption rate. The Higbie and Danckwerts models predict that k_L is proportional to $\sqrt{D}$, which agrees reasonably well with observation.[14] In the Whitman model x_L must therefore be proportional to $\sqrt{D}$. This does not disprove the existence of a stagnant film, since the additional assumption of a completely stagnant layer of finite thickness would in any case be an over-simplification. (For the same reason, we find that the effective film-thickness for transfer from solid to fluid also depends on the diffusivity.[9])

(ii) *Very fast irreversible reaction with reagent in absorbing liquid*

There is a zone beneath the surface which contains dissolved gas and in which all the reagent has reacted ; this zone is bounded by a plane at which the concentrations of dissolved gas and reagent are both zero, and beyond this plane the concentration of reagent rises to its bulk value. Both reagent and dissolved gas diffuse towards the dividing plane.[15, 16]

Whitman :[15] $$R_2 = \left(C^* + \frac{D'}{D}\, C'_0\right) k_L \qquad . \qquad (5)$$

Higbie :[16] $$R_2 = C^* \frac{k_L}{\text{erf}\,[\beta/\sqrt{D}]} \qquad . \qquad (6)$$

Danckwerts :[1] $$R_2 = C^* \frac{k_L}{\text{erf}\,[\beta/\sqrt{D}]} \qquad . \qquad (7)$$

In each case k_L is the hypothetical liquid-film coefficient for absorption without reaction. D' is the diffusivity of the reagent, C'_0 its concentration in the bulk of the solution, and β is defined by

$$C^* \sqrt{D}\, e^{\beta^2/D'}\,.\,\text{erfc}\,[\beta/\sqrt{D'}] = C'_0 \sqrt{D'}\, e^{\beta^2/D}\,.\,\text{erf}\,[\beta/\sqrt{D}] \qquad . \qquad (8)$$

The condition which must be fulfilled for a second-order reaction to fall into this class is that $\sqrt{Dr' C_0'}$ should be much greater than $k_L \left(1 + \frac{C'_0 D}{C^* D'}\right)$, r' being the reaction-velocity constant.

The predicted rates of absorption with reaction may be compared with those for physical absorption (with $C_0 = 0$). If D' is taken as equal to D (this will often be approximately true) the predicted ratios R_2/R_1 (increase in absorption rate due to reaction) are found to be identical in the three cases, being equal to $[1 + (C_0'/C^*)]$. For other values of D/D', R_2/R_1 will of course have identical values for the Higbie and Danckwerts models; what is more surprising, these are also very close to the Whitman values, agreeing within a few per cent. over a large range of (C_0'/C^*) when, for example, $D/D' = 2$.[19]

(iii) *Reversible first-order reaction with the absorbent (equilibrium assumed in bulk of solution)*

The dissolved gas will be present partly as the unreacted species, A, partly as the product of the reaction, B. C^* is the concentration of A at the surface, C_0 its concentration in the bulk. Equilibrium is assumed to prevail in the bulk, so that

$$b_0 = K\,C_0 \quad . \quad . \quad . \quad (9)$$

where b_0 is the bulk concentration of B, and K the equilibrium constant. The total bulk concentration of A and B is therefore $C_0(1 + K)$. The velocity constant for the conversion of A to B (*i.e.*, for the forward reaction alone) is r.

Whitman:

$$R_3 = (C^* - C_0)\,k_L \cdot \frac{1 + K}{1 + \dfrac{K \tanh \beta}{\beta}}, \quad (10)$$

where

$$\beta = \frac{\sqrt{Dr(K+1)/K}}{k_L}.$$

Higbie:[17]

$$R_3 = (C^* - C_0)\,k_L \cdot f\left\{K, \frac{4Dr}{\pi k_L^2}\right\} \quad . \quad (11)$$

(The function $f\left\{K, \frac{4Dr}{\pi k_L^2}\right\}$ is given by equation 389, reference [17], where $k_c\theta$ is equal to $4Dr/\pi k_L^2$ in the present notation. On p. 332, Fig. 150 shows $f\{K, k_c\theta\}$ as a function of $k_c\theta$ for $K = 1$, 2 and ∞. It should be noted that these authors use C_0 to mean the total bulk concentration of A and B.)

Danckwerts:

$$R_3 = (C^* - C_0)\,k_L \cdot \frac{(K+1)\sqrt{1 + Dr(K+1)/Kk_L^2}}{K + \sqrt{1 + Dr(K+1)/Kk_L^2}} \quad . \quad (12)$$

(This expression was obtained by the Laplace transformation method described in reference [1].) The ratio R_3/R_1 for the three models has been calculated for $K = 1$ and $K = 2$, and for a range of values of Dr/k_L^2. It can be seen from Table I that here again the ratios predicted by each theory are within 5% of one another. Equations (10), (11) and (12), with $K = \infty$, are identical with equations (13), (14) and (15) with $C_0 = 0$.

(iv) *Irreversible first-order reaction*

The bulk concentration of unreacted solute is C_0, and the reaction velocity constant is r.

Whitman:[18]

$$R_4 = \left[C^* - \frac{C_0}{\cosh(\alpha\sqrt{\pi}/2)}\right] \cdot k_L \frac{\alpha\sqrt{\pi}/2}{\tanh(\alpha\sqrt{\pi}/2)}; \quad . \quad (13)$$

Higbie:[22]

$$R_4 = \left[C^* - \frac{C_0}{\alpha} \cdot \frac{\operatorname{erf}\alpha}{(\alpha + 1/2\alpha)\operatorname{erf}\alpha + e^{-\alpha^2}/\sqrt{\pi}}\right] k_L \cdot \frac{\sqrt{\pi}}{2}\left[(\alpha + 1/2\alpha)\operatorname{erf}\alpha + \frac{e^{-\alpha^2}}{\sqrt{\pi}}\right]; \quad . \quad (14)$$

Danckwerts:[1]

$$R_4 = \left[C^* - \frac{C_0}{1 + \pi\alpha^2/4}\right] k_L \cdot \frac{\alpha\sqrt{\pi}}{2}\sqrt{1 + 4/\pi\alpha^2}; \quad (15)$$

TABLE I.—*Values of* R_3/R_1

	$K = 1$					$K = 2$				
Dr/k_L^2	0	1	10	100	∞	0	1	10	100	∞
Whitman	1	1·23	1·63	1·87	2	1	1·26	1·98	2·58	3
Higbie	1	1·27	1·68	1·89	2	1	1·32	2·05	2·65	3
Danckwerts	1	1·27	1·64	1·87	2	1	1·33	2·00	2·58	3

where

$$\alpha = \frac{2}{k_L}\sqrt{\frac{Dr}{\pi}}.$$

The ratios, R_4/R_1, of the absorption-rates with and without chemical reaction are seen to depend in each case on the ratio C_0/C^*. The ratios R_4/R_1 have been evaluated for a range of values of α with $C_0 = 0$ and $C^* = 2C_0$ for each expression. As may be seen from Table II, there is agreement within 10% over the whole range of α when $C_0 = 0$, although the discrepancy between the Whitman expression and the others may approach 20% when $C^* = 2C_0$.

solubilities, diffusivities, reaction-velocity constants, etc., were known.

Of the three models, that of Whitman is the simplest to visualise, and its terminology has become a familiar part of the chemical engineer's vocabulary. However, the Higbie and Danckwerts models have one great advantage. If it is possible to measure the rate of absorption of gas into a stagnant liquid experimentally, it should be possible to use either model to predict absorption rates in a packed column without the necessity for knowing explicitly the values of any of the physico-chemical quantities involved, or even the orders of any chemical reactions which may take place. In view of the frequent difficulty of assigning values to C^*, D and reaction-velocity constants, the uncertainty of the mechanisms of many reactions, and the mathematical difficulties attending the analysis of the more complicated processes, this seems the most promising line to follow.

A method of measuring absorption-rates into stagnant liquids over short periods of time is discussed in Part II.

TABLE II.—*Values of* R_4/R_1

	$C_0 = 0$							$C_0 = C^*/2$						
α	0	0·5	1	2	3	5	$\gg 1$	0	0·5	1	2	3	5	$\gg 1$
Whitman	1	1·06	1·25	1·88	2·70	4·45	$\alpha\sqrt{\frac{\pi}{2}}$	1	1·15	1·63	3·14	5·03	8·80	$\alpha\sqrt{2\pi}$
Higbie	1	1·08	1·31	2·01	2·82	4·54	,,	1	1·24	1·87	3·58	5·35	8·90	,,
Danckwerts	1	1·09	1·34	2·03	2·84	4·55	,,	1	1·27	1·92	3·58	5·35	8·90	,,

Discussion

It has been shown that the three models lead to closely similar predictions about the effect of chemical reaction and other physico-chemical factors on absorption-rates, at any rate over considerable ranges of the variables concerned. It seems unlikely, therefore, that a study of the effect of these variables will indicate which of the three models is closest to the actual mechanism in any given case. Whatever the mechanism is, however, the formulae based on any one of the three models may be approximately applicable in practice. If this were so, a single determination of the transfer coefficient for the physical absorption of any gas would enable absorption-rates to be predicted under other physico-chemical conditions, provided that the hydrodynamic conditions were identical and that the values of the relevant

References

For references to the literature, see references to Part II, page S59.

The manuscript of this paper was received on 1 December, 1953.

KINETICS OF LIQUID-FILM PROCESSES IN GAS ABSORPTION. PART II: MEASUREMENTS OF TRANSIENT ABSORPTION RATES

By P. V. DANCKWERTS, G.C., M.B.E., M.A., S.M.,* and A. M. KENNEDY, B.E., B.Sc.*

SUMMARY

Values of the physico-chemical quantities discussed in Part I are often not available, and are difficult or impossible to measure. The second and third models considered do not require explicit knowledge of these quantities, but only measurements of transient rates of absorption for short times of contact between gas and liquid. Such transient rates have been measured in an apparatus incorporating a rotating drum. Carbon dioxide was absorbed into water, solutions of salts, buffer solutions and sodium hydroxide solution. Times of exposure were less than 0·25 sec. Comparison with rates of absorption in a packed column suggests that none of the three models gives an adequate description of the absorption process.

There is prima facie evidence of a substantial surface resistance to the absorption of carbon dioxide in water.

Introduction

It has been shown in Part I that the three models proposed for the liquid-film absorption process lead to closely similar predictions about the effect of changes in the physico-chemical variables under constant hydrodynamic conditions. Since the three models represent a wide range of postulated mechanisms, this leads one to hope that the predictions may be approximately applicable in practice. If this were so, a knowledge of the quantities k_L and a for the physical absorption of any gas under given hydrodynamic conditions would enable absorption-rates of various gases into solutions of various reagents to be predicted. In order to test this hypothesis, it is possible either to study absorption processes which can be subjected to mathematical analysis, and measure or estimate the appropriate values of C^*, D, r, etc.; or alternatively, to attempt to measure transient absorption-rates into stagnant liquid for very short times of contact, and base the predictions on the Higbie or Danckwerts models.

It is not possible to measure directly the solubility, C^*, or the diffusivity, D, of a dissolved gas in a solution with which it reacts. Solubilities may be estimated by empirical formulae, such as that of van Krevelen and Hoftijzer.[21] However, the universal validity of such approximations can never be certain; for instance, the method of van Krevelen and Hoftijzer predicts a value of 0·476 for the Bunsen absorption coefficient for carbon dioxide in 1 M. magnesium sulphate solution at 25° C., whereas the measured value is 0·414.[10] Data on diffusivities are scarce, even for non-reacting solutions. In the absence of measured values, the best estimate which can be made is that $D\mu/T$ is constant.[24] The rates of reactions involving ions (such as that between carbon dioxide and the hydroxyl ion) are in general strongly affected by the total ionic concentration and the species of ions present.

It is also possible that the "accommodation coefficient" at the surface of the liquid may be very low, giving rise to a "surface resistance," and that the absorption of the gas may cause the temperature at the surface to rise appreciably; these possibilities are mentioned later.

We shall give here a brief description of some work in progress designed to measure transient rates of absorption and thus avoid the uncertainties discussed above. The transient measurements have been compared with measurements in a packed column, to see whether the Higbie or Danckwerts models can be used as a basis for prediction.

The Rotating Drum Method

The apparatus used to measure transient absorption rates consists of a cylindrical drum, 3 in. × 3 in., which can be rotated at speeds up to 8·7 rev. per sec. (Fig. 1). The drum carries a film of liquid through the gas, and by varying

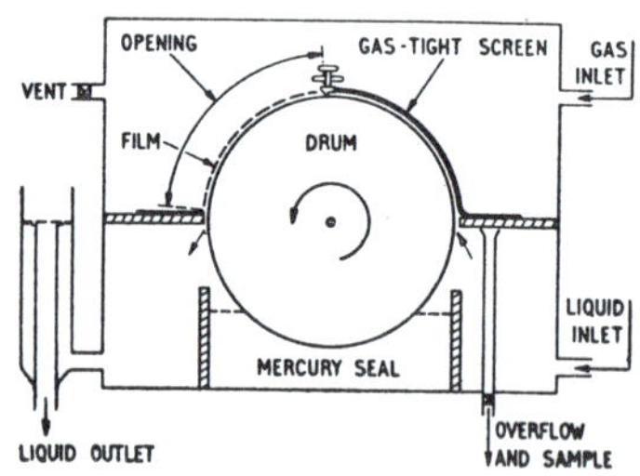

Fig. 1.—*Sketch of rotating drum*

* Department of Chemical Engineering, Tennis Court Road, Cambridge.

the speed of rotation and the fraction of the circumference of the drum which is exposed, the time of contact between liquid and gas can be varied from 0·0075 to 0·25 sec.

The chief problems in design were to dispose of the film without entraining gas, and to produce a film of which the surface accelerated to the speed of the drum very rapidly after formation. There was also the possibility that turbulence or other convective movements in the film might affect the results. However, the measurements are believed to be reasonably reliable ; the most convincing evidence is the compatibility of absorption rates obtained with equivalent exposure-times, but with different velocities and areas exposed. Even so it would be desirable to check at least one set of results by an independent method.

The gas (pure carbon dioxide in all experiments so far) is admitted to the space above the drum through a positive-displacement soap-bubble flowmeter, the whole gas system being at atmospheric pressure. The gas is kept saturated with water-vapour. The solutions are de-aerated, and the gas system thoroughly purged of air. All experiments are carried out in a constant-temperature room at 25° C.

The average rate of absorption per unit area of liquid surface exposed on the drum (for a given setting of speed and area exposed) is equal to $Q\{t\}/t$, where $Q\{t\}$ is the quantity of gas absorbed by unit area of stagnant liquid surface in the time t for which each element of surface is exposed. Some typical results are shown in Fig. 2.

Interpretation of Results

The absorbing liquids used were water, 1·05 M. sodium sulphate, 0·72 M. magnesium sulphate, 2 M. sodium hydroxide, and the carbonate-bicarbonate buffer solutions shown in Table I.

Water.—Since the solubility and diffusivity of carbon dioxide in water are known from independent experiments to be

$C^* = 0{\cdot}338 \times 10^{-4}$ gm. mol./cc. water (under 1 atm. p.p.) ;[7]

$D = 2{\cdot}02 \times 10^{-5}$ sq. cm./sec. (extrapolated from 20° C.) ;[8]

at 25° C., it is possible to calculate the absorption-rate if the surface is assumed to be always saturated with carbon dioxide. For a stagnant liquid of infinite depth

$$Q = 2C^* \sqrt{\frac{Dt}{\pi}}$$

$$= 1{\cdot}71 \times 10^{-7} \sqrt{t} \text{ gm. mole per sq. cm.} \quad (t \text{ in sec.}) \quad . \quad (1)$$

In Fig. 4 this expression is plotted against $\sqrt{t}$ and compared with the experimental results,

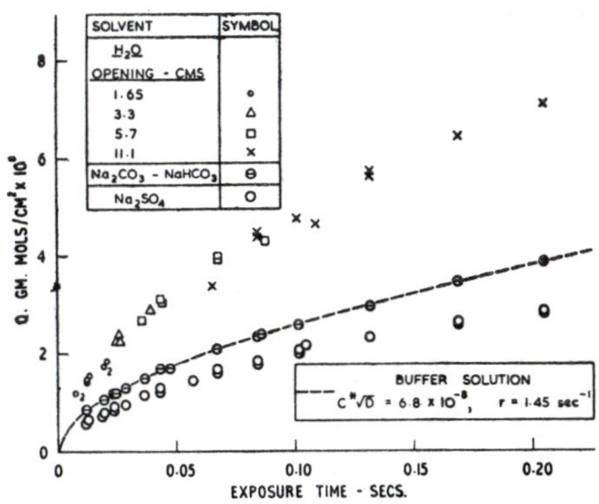

Fig. 2.—*Transient absorption of carbon dioxide at 25° C. into water, 1·05 molar sodium sulphate, and a buffer solution 0·68 molar in sodium carbonate and 0·40 molar in sodium bicarbonate*

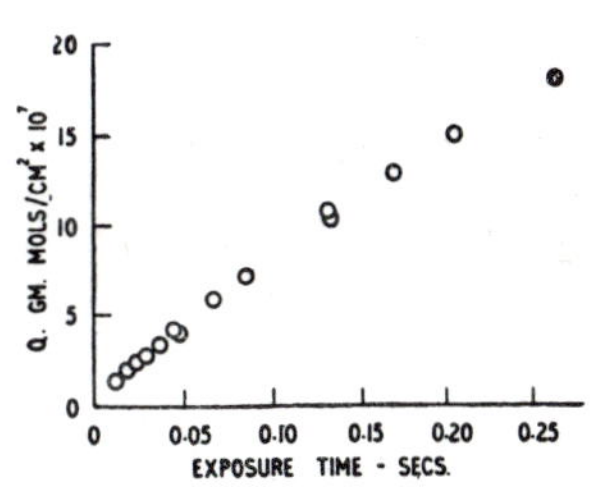

Fig. 3.—*Transient absorption of carbon dioxide into 2 molar caustic soda at 25° C.*

TABLE I.—*Buffer Solutions*

$[Na_2CO_3]$ (gm. mole/litre)	$[NaHCO_3]$ (gm. mole/litre)	$\frac{[CO_3'']}{[HCO_3']}$	(i) $[OH'] \times 10^4$ (gm. ion/litre)	$C^* \cdot \sqrt{D} \times 10^8$ gm. mole per sq. cm. sec.$^{\frac{1}{2}}$ exp.	est. (ii)	r sec.$^{-1}$ exp.	est. (iii)
0·763	0·228	3·35	7·15	6·9	6·6	2·55	4·3
0·735	0·297	2·47	5·26	6·7	6·6	2·25	3·2
0·679	0·400	1·70	3·63	6·8	6·6	1·45	2·2
0·489	0·818	0·60	1·28	6·7	6·6	0·70	0·8

Notes : (i) [OH′] estimated from equation (4).
(ii) C^* from formula of van Krevelen and Hoftijzer.[21] D from value in pure water by taking $D\mu/T$ = const.[24]
(iii) r taken as 6000 [OH′] — equation (3).

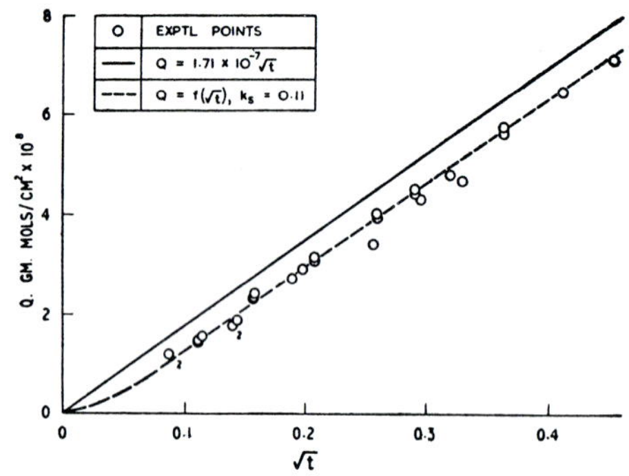

Fig. 4.—*Transient absorption of carbon dioxide into water at 25° C. plotted against square root of the exposure time; also theoretical curves for absorption into water*
(a) *without surface resistance*
(b) *assuming a surface transfer coefficient of 0·11 cm./sec.*

which are seen to be significantly lower. It is possible that subsequent work may show this to be due to a systematic error in the measurements. The effect is not due to the finite depth of the liquid. A calculation of the rise in temperature of the liquid surface due to the heat of solution of the carbon dioxide[2] shows this to be negligible, (as it is for all the other solutions studied except 2 M. sodium hydroxide, the surface temperature of which was calculated to rise 5·8° C. per sec.$^{\frac{1}{2}}$). Accepting the results at their face-value, we should be forced to conclude that there is a substantial " surface-resistance " to the absorption of carbon dioxide into water, the concentration of dissolved carbon dioxide at the surface being initially less than C^*. The dotted line in Fig. 4 shows the curve which should relate Q to $\sqrt{t}$ if there is a surface resistance $1/k_s$, where k_s (the " surface mass transfer coefficient ") has a value of 0·11 cm./sec. The equation of the curve (obtained by integrating an expression developed elsewhere[3]) is

$$Q = 2C^* \sqrt{\frac{Dt}{\pi}} + \frac{C^*D}{k_s}\left\{\text{erfc}\left[k_s\sqrt{\frac{t}{D}}\right]\exp[k_s^2\, t/D]-1\right\}. \quad (2)$$

Higbie's measurements[6] on carbon dioxide and water also indicated a surface resistance of this order of magnitude. In the present experiments the liquid-film coefficient for physical absorption of carbon dioxide in the packed tower was lower than 0·1 cm./sec. by a factor of 10 or more, so that the surface-resistance did not play a significant part in determining the rate of absorption. If the surface resistance really exists, however, it sets an upper limit, k_sC^*, to the rate of absorption, no matter how effective the agitation of the liquid or how fast the absorbed gas may be destroyed by reaction. However, the value of k_s will not necessarily be the same for concentrated solutions as for water ; the high rate of absorption of carbon dioxide into 2 M. sodium hydroxide solution (Fig. 3) indicates that if there is a surface resistance in this case it is much lower than that for water.

A possible cause of surface resistance would be a very low accommodation coefficient. Thus if only one gaseous carbon dioxide molecule in 10^5 striking the surface were able to penetrate it, the others being reflected, the apparent surface resistance would be explained.

Magnesium sulphate and sodium sulphate solutions

These were used as non-reacting liquids having the same viscosity as the alkaline solutions. It is not possible to confirm the presence of the apparent " surface resistance " effect in these cases, as the values of D had to be estimated from the viscosity and were therefore not sufficiently reliable. If the surface resistance had the same value as the apparent value for water, its effect would in any case be more difficult to detect owing to the lower rates of absorption into these solutions.

Buffer solutions

The predominating reaction in solutions of the pH used in these experiments is a virtually irreversible second-order reaction of dissolved carbon dioxide with hydroxyl ions to form bicarbonate ions.[13] Provided the concentration of the latter formed anywhere in the liquid is not too large in proportion to their initial concentration, the reaction is pseudo-first order, since the concentration of hydroxyl ions remains nearly constant.

$$CO_2 + OH' \rightarrow HCO_3'$$

$$\text{Reaction rate} = r'\,[OH'][CO_2] = r\,[CO_2]\ . \quad (3)$$

$$[OH'] \approx 2\cdot 13 \times 10^{-4}\ .\ \frac{[CO_3'']}{[HCO_3']}$$

$$\text{gm. ion per litre (25° C.)}\ . \quad (4)$$

The quantities in equation (4) should be activities, but it is very roughly true for concentrations. The value of r' may be expected to change with changes in the total ionic strength, and with marked changes in the ratio

$$[CO_3'']/[HCO_3'],$$

but its order of magnitude at 25° C. would be expected to be about 6000 litres per gm. mole-sec.[22] The value of r will probably be almost constant for any one buffer solution, under the conditions of the experiments carried out with the rotating drum and with the packed tower.

The equations for absorption with irreversible first-order reaction, with[5] and without[4] surface-resistance, have been given elsewhere. Since we do not know the magnitude of the surface-resistances, if any, of these solutions, we have for the moment analysed the results as though the resistance were zero. Values of $C^*\sqrt{D}$ and r deduced by fitting calculated curves to the experimental points are included in Table I. Values of $C^*\sqrt{D}$ and r estimated by other methods are included for comparison.

Caustic soda

Here again, the rate-determining reaction in the solution is between carbon dioxide and hydroxyl ions, the product in this case being initially carbonate ions. The reaction will only exhibit pseudo-first order kinetics so long as the concentration of hydroxyl ions near the surface is not substantially depleted by the reaction. Preliminary calculations, based on the experimental measurements (Fig. 3), suggest that when carbon dioxide is absorbed into 2M. sodium hydroxide this depletion may become appreciable after a few hundredths of a second at most. On the other hand the course of the absorption over a period of 0·25 sec. is not compatible with the equations for "instantaneous" reaction between carbon dioxide and hydroxyl ion with a deepening zone free from hydroxyl ion at the surface.[16] If neither of these approximations is applicable, the results can only be analysed on the basis of a second-order reaction between reagents which are diffusing into one another. This is a notoriously difficult mathematical problem, but some progress is being made by computational methods.[12] In the present case the situation is complicated by the shifting equilibrium between bicarbonate and carbonate ions in the regions of depleted hydroxyl ion concentration, and by the rise in temperature near the surface, referred to above.

It has not yet been possible to deduce any of the physico-chemical constants from the experimental observations on sodium hydroxide owing to the complexity of the kinetics of the absorption process.

Comparison with Absorption in Packed Column

Rates of absorption in a packed column, using similar solutions, have been measured for comparison. The column was a 4 in. i.d. Pyrex tube randomly packed with ½ in. porcelain Raschig rings to a depth of 10 in., at a density of about 72 rings per inch of depth. The liquid was distributed uniformly over the top of the packing through 40 inlets. The liquid flow-rate in all experiments was 11·8 cc./sec. (1100–1200 lb. per hr./sq. ft.). The viscosity of all solutions used was 1·57 times that of water at the same temperature, and the densities varied from 1·08 to 1·12 gm. per cc. The liquid hold-up was 4·8 cc. per cm. depth, or 6% of the total volume. The lower part of the column was filled with liquid, and the effective height of the packing could be varied by altering the liquid level. It was established that the rate of absorption into the horizontal liquid surface was sufficiently small to be ignored. The gas used was pure carbon dioxide at atmospheric pressure ; a few runs were made with nitrous oxide. The solutions were de-aerated, and the gas-system carefully and frequently purged with carbon dioxide. The absorption-rates of carbon dioxide were measured by means of a soap-bubble flow-meter, and all experiments took place in a constant-temperature room at 25° C.

If the average absorption rate per unit wetted area at a given height in the column is R, and the total absorption-rate is M when the effective height of the packing is h, then

$$\frac{dM}{dh} = AaR \quad . \quad . \quad . \quad (5)$$

where A is the cross-sectional area of the column (81·1 sq. cm.) and a the wetted surface-area per unit volume. Thus (aR) can be determined from the slope of a graph of M against h. In general, of course, (aR) varies with height in the column, because the bulk concentration of the liquid changes as absorption proceeds. The differential method of finding the value of (aR) for a specific level in the column, and hence for a specific liquid composition, has obvious advantages over the integral method normally used, in which the average absorption rate over a considerable range of compositions is measured.

In the case of absorption without reaction, the bulk concentration of absorbed gas in the liquid at any level is simply equal to the total rate of absorption into the column above this level, divided by the liquid flow-rate. In the present experiments, therefore,

$$\frac{C_0}{C^*} = \frac{M}{11{\cdot}8\,C^*} \quad . \quad . \quad (6)$$

and

$$-\frac{d}{dh}\left[\log_{10}\left(1 - \frac{M}{11{\cdot}8\,C^*}\right)\right] = \frac{81{\cdot}1}{11{\cdot}8 \times 2{\cdot}303} \,.\, (k_L a) \quad . \quad (7)$$

Typical results for absorption into a non-reacting solution are shown in Fig. 5. It may be noticed that the transfer coefficient is apparently abnormally high at the extreme top of the packing, although sensibly constant elsewhere. The value of $(k_L a)$ may be deduced from the slope of the

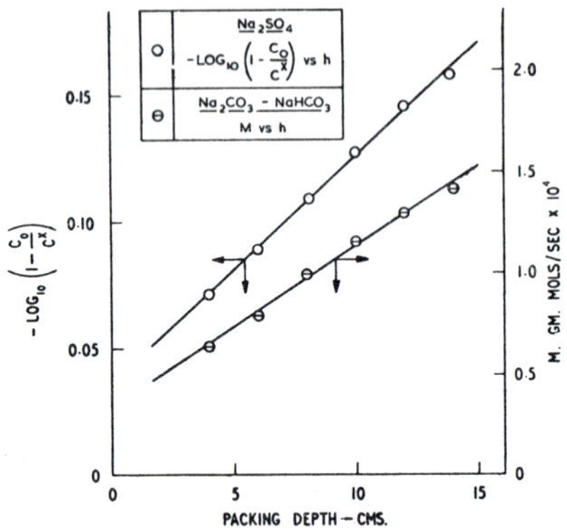

Fig. 5.—*Packed column results for the absorption of carbon dioxide at 25° C. into 1·06 molar sodium sulphate and into a buffer solution 0·64 molar in sodium carbonate and 0·50 molar in sodium bicarbonate*

logarithmic plot according to equation (7). The value of (aR) for absorption into pure water is $(k_L a C^*)$.

The experimental values of $(k_L a)$ for physical absorption are shown in Table II ; they are only about 70% as great as those predicted from the formula of Sherwood and Holloway.[14] This is probably partly due to the lower packing density in the present experiments (10,000 as against 12,000 rings per cu. ft.).

For reacting solutions, the calculation of the bulk composition at various levels is less simple ; if the reaction of dissolved carbon dioxide is sufficiently slow, there may be an appreciable concentration of unreacted carbon dioxide in the bulk of the liquid. However, calculation shows that in the present experiments the concentration of unreacted absorbed carbon dioxide never reached 5% of C^* even in the slowest-reacting

TABLE II.—*Liquid-film Transfer Coefficients for Physical Absorption*

System	$k_L a \times 10^3$ (sec.$^{-1}$)
$CO_2 - H_2O$	4·9
$CO_2 - 1{\cdot}05$ M. Na_2SO_4	3·1
$CO_2 - 0{\cdot}72$ M. $MgSO_4$	3·2
$N_2O - 1{\cdot}05$ M. Na_2SO_4	3·2

buffer solution. As expected, the absorption into the buffer solutions did not appear to alter (aR) appreciably (Fig. 5). In the case of 2 M. sodium hydroxide solution the curve of (dM/dh) vs. h was extrapolated to $h = 0$ to find (aR) for pure caustic soda solution ; the correction involved, however, was not large.

If the Danckwerts model (see Part I) of the absorption process gives valid results we have :

$$aR = as \int_{t=0}^{\infty} e^{-st}\, dQ\{t\} \quad . \qquad (6)$$

The quantities (aR) and $Q\{t\}$ are determined from measurements in the column and on the drum respectively, the bulk composition of the liquid in the column being the same as that of the liquid fed to the drum. The quantities a and s should (since the hydrodynamic conditions in the column do not vary appreciably) have the same values for all the column experiments.

To determine the values of a and s, a trial value of s is taken, the integral

$$s \int_0^{\infty} e^{-st}\, d\, Q\{t\}$$

evaluated for the transient absorption measurements with a given solution, and hence the value of a determined from equation (6). A set of trial values of s can be plotted against the corresponding values of a (it is more convenient in practice to plot against $1/a^2$). The same is done for each solution investigated and the various curves should converge at a point, giving the true values of a and s in the column. The calculation does not require any of the physico-chemical constants for the solutions, such as C^*, D, and r, to be found explicitly from the observations on the rotating drum.

When this procedure was applied to the experimental observations so far accumulated it was found that the curves for the non-reacting solutions and buffer solutions converged in the neighbourhood of $s = 0{\cdot}8$ sec.$^{-1}$, $a = 1{\cdot}1$ cm.$^{-1}$, while those for the non-reacting solutions and for 2M. sodium hydroxide converged in the neighbourhood of $s = 8$ sec.$^{-1}$, $a = 0{\cdot}3$ cm.$^{-1}$ Analysis on the basis of Higbie's model displays a similar discrepancy. In effect, 2M. sodium solution absorbs too slowly in the column in view of its performance on the drum, or else the buffer solutions absorb too fast in the column.

It would be premature to conclude from these relatively few results that the Higbie and Danckwerts models are unreliable guides. For instance, the high rates of absorption into sodium hydroxide solution may introduce some source of error into the experiments which has not yet been detected.

However, a possible explanation of the results is that the distribution of " ages " of parts of the surface (that is, the times for which they have been exposed to the gas) postulated in either the Higbie or Danckwerts models gives too little weight to the older parts of the surface to describe what occurs under the conditions reigning in the packed column in these experiments. If a considerable fraction of the wetted surface were in fact covered by relatively sluggishly-moving liquid, having a rate of surface-renewal less than the average, the effects described above would be expected, since the rate of absorption into sodium hydroxide solution may be expected to fall off relatively more rapidly than the rate of absorption into a buffer solution. If this explanation is right, it is to be expected that the actual distribution of surface ages will vary with the hydrodynamic conditions and that results obtained, for instance, with wetted-wall columns or "totem-pole" columns[20, 22] will be of little value for making predictions about industrial absorption columns except in special circumstances.

Conclusions

The rotating-drum method of measuring transient rates of gas-absorption promises to be a useful tool for investigating mechanisms of absorption, and for supplying the necessary physico-chemical data for industrial design purposes. It may also provide information of purely scientific interest about reaction kinetics and surface effects.

Comparison of transient absorption rates on the drum with steady-state absorption rates in a packed column leads to the tentative conclusion that neither the Higbie nor the Danckwerts models of the gas-absorption process can safely be used for design purposes when gases are absorbed by reacting liquids.

Acknowledgment

One of the authors (A. M. K.) wishes to thank the New Zealand Defence Science Authorities for an award which has enabled this work to be carried out.

Nomenclature

Symbols

a = area of wetted surface per unit packed volume.
A = cross-sectional area of column.
b_0 = bulk concentration of reacted dissolved gas (in equilibrium with C_0).
C^* = saturated concentration of (unreacted) dissolved gas in liquid.
C_0 = concentration of (unreacted) dissolved gas in bulk of liquid.
C_0' = bulk concentration of reagent in absorbing liquid.
D = diffusivity of dissolved gas.
D' = diffusivity of reagent in absorbing liquid.
h = height of packing exposed to gas.
k_L = liquid-film transfer coefficient for physical absorption.
k_s = surface mass-transfer coefficient.
K = equilibrium constant between reacted and unreacted dissolved gas.
M = total rate of absorption in height h of packed column.
$Q\{t\}$ = quantity of gas absorbed into unit area of surface of stagnant liquid in time t.
r = first-order or pseudo-first-order reaction velocity constant.
r' = second-order reaction velocity constant between dissolved gas and reagent.
R = mean rate of absorption per unit area of wetted surface (*e.g.*, in packed column).
s = fractional rate of renewal of surface (Danckwerts' model).
t = time.
T = absolute temperature.
x_L = effective thickness of liquid film (Whitman theory).
α $= \frac{2}{k_L}\sqrt{\frac{Dr}{\pi}}$.
β = constant defined by equation (8), Part I; or $\frac{1}{k_L}\sqrt{Dr(1+K)/K}$.
μ = viscosity.
θ = time of exposure of liquid surface (Higbie model).

$$\text{erf}\,(a) = 1 - \text{erfc}\,(a) = \frac{2}{\sqrt{\pi}}\int_0^a e^{-y^2}\,dy.$$

Subscripts 1, 2, 3, 4 refer to absorption into solutions which interact with the dissolved gas in different ways.

References

1 Danckwerts, P. V. *Industr. Engng. Chem.*, 1951, **43**, 1460.
2 Danckwerts, P. V. *Appl. Sci. Research* (A), 1952, **3**, 385.
3 Danckwerts, P. V. *Research*, 1949, **2**, 494.
4 Danckwerts, P. V. *Trans. Faraday Soc.*, 1950, **46**, 300.
5 Danckwerts, P. V. *Trans. Faraday Soc.*, 1951, **47**, 1014.
6 Higbie, R. *Trans. Amer. Inst. Chem. Engrs.*, 1935, **31**, 365.
7 "Int. Crit. Tables," Vol. III, p. 260.
8 "Int. Crit. Tables," Vol. V, p. 65.
9 Linton, W. H. and Sherwood, T. K. *Chem. Engng. Progr.*, 1950, **46**, 258.
10 Markham, A. E., and Kobe, K. A. *J.A.C.S.*, 1941, **63**, 449.
11 Marshall, W. R., and Pigford, R. L. *Application of Differential Equations to Chemical Engineering Problems*, 1947. (University of Delaware.)
12 Perry, R. H., and Pigford, R. L. *Industr. Engng. Chem.*, 1953, **45**, 1247.
13 Pinsent, B. R. W., and Roughton, F. J. W. *Trans. Faraday Soc.*, 1951, **47**, 263.
14 Sherwood, T. K., and Pigford, R. L. "Absorption and Extraction," p. 288. New York, McGraw-Hill, 1952.
15 Sherwood, T. K., and Pigford, R. L. "Absorption and Extraction," p. 321. New York, McGraw-Hill, 1952.
16 Sherwood, T. K., and Pigford, R. L. "Absorption and Extraction," p. 332. New York, McGraw-Hill, 1952.
17 Sherwood, T. K., and Pigford, R. L. "Absorption and Extraction," p. 328. New York, McGraw-Hill, 1952.
18 Sherwood, T. K., and Pigford, R. L. "Absorption and Extraction," p. 324. New York, McGraw-Hill, 1952.

[19] Sherwood, T. K., and Pigford, R. L. "Absorption and Extraction," p. 338, Fig. 153. New York, McGraw-Hill, 1952.
[20] Stephens, E. J., and Morris, G. A. *Chem. Engng. Progr.*, 1950, **46**, 258.
[21] van Krevelen, D. W., and Hoftijzer, P. J. *Chim. et Industr.*, XXIe *Congres Intern. Chimie Ind.*, 1948, p. 166.
[22] Wagstaffe, F. J. Unpublished work quoted by permission of I.C.I.
[23] Whitman, W. G. *Chem. metall. Engng.*, 1923, **24**, 146.
[24] Wilke, C. R. *Chem. Engng. Progr.*, 1949, **45**, 218.

The manuscript of this paper was received on 1 *December*, 1953.

B3

The kinetics of absorption of carbon dioxide into neutral and alkaline solutions

P. V. DANCKWERTS* and A. M. KENNEDY†

(*Received* 7 *October* 1957)

Abstract—The transient rates of absorption of carbon dioxide into water, non–reactive solutions, sodium hydroxide and alkaline buffer solutions have been measured at 25°C. The liquid was carried through the gas in the form of a thin film on a rotating drum, the time of exposure varying from 0·01 to 0·25 sec. The liquid was free from convection currents, so that the rates of absorption were controlled by diffusion and chemical reaction in the liquid. The experimental results are analysed in this light and values obtained for solubilities, diffusivities and reaction–velocity constants in the various solutions.

The relationship between the rate of absorption into reacting solutions and the homogeneous reaction kinetics is in general not a simple one, because of the depletion of reactant near the surface and other factors. Caution should be exercised against oversimplification when predicting or explaining the results of gas absorption under industrial conditions.

Résumé—Les vitesses d'absorption, en régime transitoire, du bioxyde de carbone dans l'eau, des solutions non réactives, des solutions caustiques et des solutions tampon alcalines ont été mesurées à 25°C. Le liquide était transporté à travers le gaz sous forme d'un film mince sur un tambour tournant. La durée du contact variait de 0,01 à 0,25 sec. Les courants de convection étaient évités dans la phase liquide, de telle manière que les vitesses d'absorption étaient controlées par diffusion et réaction chimique dans la phase liquide. Les résultats expérimentaux sont analysés à partir de ces considérations et les constantes de solubilité, de diffusion et de vitesse de réaction sont obtenues pour différentes solutions. La relation entre la vitesse d'absorption dans les solutions réactionnelles et la cinétique des réactions homogènes n'est pas simple en général, à cause des diminutions de concentration en surface, ainsi que de l'influence d'autres facteurs. Il faudra prendre des précautions quant aux simplifications utilisées quand on prédit ou explique les résultats de l'absorption des gaz dans des conditions industrielles.

Zusammenfassung—Die nichtstationären Geschwindigkeiten der Absorption von Kohlendioxyd in Wasser, nichtreaktiven Lösungen, Natriumhydroxyd und alkalischen Pufferlösungen wurden bei 25°C gemessen. Die Flüssigkeit wurde durch das Gas in Form eines dünnen Films auf einer rotierenden Trommel geleitet, die Kontaktzeit lag zwischen 0,01 bis 0,25 sec. Die Flüssigkeit war frei von Konvektionsbewegungen, so dass die Geschwindigkeiten der Absorption durch die Diffusion und die chemische Reaktion in der Flüssigkeit bestimmt wurden. Die Versuchsergebnisse wurden in dieser Hinsicht analysiert und es wurden Werte für die Löslichkeiten, die Diffusionskoeffizienten und die Geschwindigkeitskonstanten der Reaktion in den verschiedenen Lösungen erhalten.

Die Beziehung zwischen der Geschwindigkeit der Absorption in reagierende Lösungen und der homogenen Reaktionskinetik ist im allgemeinen nicht einfach wegen der Erschöpfung des Reaktanten in der Nähe der Oberfläche und wegen anderer Faktoren. Die Vorgänge dürfen daher nicht allzu stark vereinfacht werden bei der Vorausberechnung oder der Deutung von Ergebnissen der Casabsorption unter industriellen Bedingungen.

* Imperial College of Science and Technology, London.

† Dominion Laboratories, Chemical Engng. Section, Private Bag, Lower Hutt, Wellington, N.Z.

THE experiments described here were carried out in connection with an investigation of the mechanism of gas-absorption in packed columns, which has already been the subject of a perliminary report [1], and will be further reported elsewhere. An interpretation of the results in physico-chemical terms, although not necessary for the purpose of the original investigation, may be of some interest and the technique used may find a wider application in the study of fast liquid–phase reactions.

It was necessary to measure the rate of absorption of carbon dioxide into various aqueous solutions in the first fraction of a second of contact. A number of previous workers have had the same objective. Their methods, which were recently reviewed by CULLEN and DAVIDSON [2], have probably all been subject to error. Since the work described in this paper was done, the latter authors [2, 3] have carried out meticulous measurements of the absorption of gases into jets and into water running over the surface of a sphere. The shortest exposure-time attained was about 0·01 sec, as with the rotating drum described here. The internal consistency of their results suggests that the methods used were highly reliable, and the value for the diffusivity of carbon dioxide in water at 25°C deduced from their work (namely $1{\cdot}92 \times 10^{-5}$ cm^2/sec) has been adopted here as being probably the most trustworthy in the literature. The rotating drum is mechanically more complicated than the falling film or jet used by CULLEN and DAVIDSON, but it yields results which are more easily interpreted when the absorbed gas reacts with the liquid, as the complication introduced by the " stretching " of liquid surface is avoided.

The compositions of the aqueous solutions used were dictated by the requirements of the parallel series of experiments on absorption in packed columns. It was necessary for the viscosities and, as far as possible, the densities of all solutions to be the same. In the case of sodium hydroxide this was achieved by adding varying concentrations of sodium sulphate to the solutions. The sodium hydroxide solutions also contained some carbonate, for reasons connected with the packed column experiments.

EXPERIMENTAL

A diagrammatic view of the apparatus used for the principal series of experiments is shown in Fig. 1. The drum, of stainless steel, was 11·4 cm dia. and 12·4 cm long. The drum surface was ground smooth to give a cylinder truly concentric with the spindle. The drum itself was inside a Perspex box, through which the spindle passed via glands stuffed with graphitised asbestos. The spindle was supported externally on ball bearings and driven by an electric motor through a system of gears.

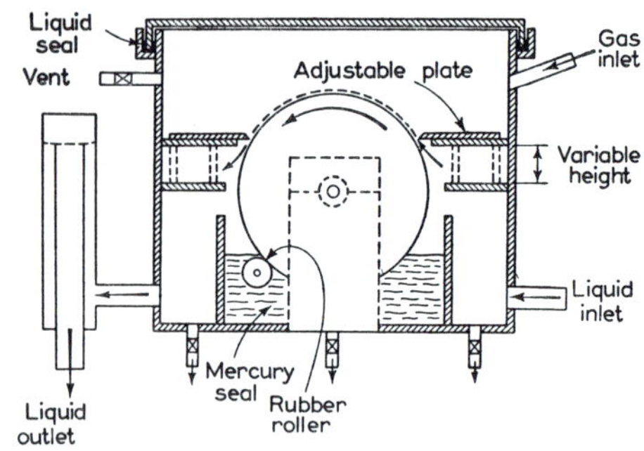

FIG. 1. Rotating drum apparatus.

In order to separate the inflowing and outflowing liquid, the lower part of the drum dipped into a trough of mercury, and a rubber roller bore against the drum so as to squeeze off any film of liquid which might otherwise have been carried through the mercury. The gap between the end faces of the drum and the sides of the Perspex box was so narrow that the rate of liquid leakage was negligible compared to the rate of transfer of liquid by the intended route. Two horizontal plates separated the gas from the horizontal liquid surface. These plates could be set at different heights, and each carried an adjustable plate which could be slid forward to cover the gap between the edge of the fixed plate and the drum.

The liquid level on either side of the drum was adjusted to conform with the level at which the horizontal plates were set. It was thus possible to expose varying proportions of the drum's surface to the gas. The adjustable horizontal plates were knife-edged, and by means of feeler gauges they could be aligned close to and parallel to the drum surface.

The drum speed could be varied by eight steps from 1·9 to 5·1 rev/sec. By varying also the fraction of the surface exposed, the same exposure time could be obtained with different drum speeds.

The thickness of the liquid film on the drum must be sufficient for it to be in effect "infinitely deep". It is easily shown from the diffusion equations [4] that the finite depth, l, of a liquid layer will not affect the amount absorbed in time t by more than $\frac{1}{2}$ per cent providing $Dt/l^2 < 0{\cdot}25$, where D is the diffusivity of the dissolved gas. For carbon dioxide in water, this corresponds to $l > 4 \times 10^{-3}$ cm for $t = 0{\cdot}25$ sec. The film thickness was kept well above this value; it could be controlled by varying the clearance of the adjustable knife edge on the inlet side, and determined by measuring the rate of carryover of liquid by the drum. The gap between the knife edge and the drum was in practice set at $5\text{–}7{\cdot}5 \times 10^{-3}$ cm, and variation within these limits did not affect the absorption rate. (The thickness of the film was greater than the width of the gap.)

With the drum running, the knife edge on the outlet side was advanced until it touched the liquid surface, a fine ripple being formed parallel to the knife edge; entrainment of gas on the outlet side then stopped. The actual width of the gap was not critical.

The box was closed by a lid which rested in a mercury seal. The box was provided with a gas inlet and a vent for purging the gas space, and with pipes for the entry and exit of the liquid. Although the water used for the solutions was de-aerated by boiling under vacuum, it was found that the rate of absorption gradually fell off after purging, due to the accumulation of insoluble gases in the box.

The carbon dioxide was held at atmospheric pressure in a large balloon, and allowed to flow into the box through a soap-bubble flow-meter. The carbon dioxide was taken from a cylinder, and contained 0·3 per cent of relatively insoluble gas. However, in some comparative runs using carbon dioxide from dry ice, and containing less than half as much impurity, the rates of absorption were not significantly different. The carbon dioxide was humidified by passing it through water at 25°C before storage in the balloon.

If the partial pressure of water at 25°C is p and the carbon dioxide is saturated with water-vapour, the measured rate of passage of carbon dioxide and water-vapour through the flow-meter will be $[P/(P-p)]\,q$ moles/sec, where q is the rate of absorption of carbon dioxide and P the total pressure. Now in general, subject to a qualification mentioned below, theoretical considerations indicate that the rate of absorption into a particular solution is proportional to its physical solubility. Since the solubility obeys Henry's law under the conditions of these experiments, it will be equal to $[(P-p)/P]\,C^*$, where C^* is the solubility at a partial pressure P.

Thus the effect of the dilution of carbon dioxide with water-vapour was to increase its apparent bulk and decrease its rate of solution in the same ratio, so that the observed volumetric rate of passage of moist carbon dioxide was equal to the rate at which dry carbon dioxide would have been absorbed at the same total pressure P. All experimental results were corrected to a standard total pressure of 760 mm, and the whole apparatus was housed in a constant temperature room at 25°C.

In interpreting the experimental results it is assumed that the liquid surface moves at the speed of the drum, and that end effects can be ignored. This assumption was justified by the fact that for a given exposure time there was generally no systematic difference between results obtained with different drum speeds, exposed areas and film thicknesses. However, at the lowest drum speeds it was noticed that the capillary ripples formed by the knife edge at the outlet side sometimes had their origin several millimetres upstream of the knife edge. Particles of talc scattered on the surface collected in the zone between the knife edge and the ripples, where the the surface seemed to be stagnant. This effect is undoubtedly due to the accumulation of surface active contaminants, and has been noticed elsewhere [5]. This is a potential source of error, but precautions were taken to keep the stagnant zone, in the worst cases, down to about 2 per cent of the exposed surface.

Technical grade sodium hydroxide, sodium carbonate and sodium bicarbonate were used because of the large quantities required for the packed column experiments. The composition of the sodium carbonate - sodium bicarbonate buffer solutions was determined by titrating a sample with N/10 hydrochloric acid to an end-point with brom–cresol green, thus determining $([NaHCO_3] + 2\,[Na_2\,CO_3])$. To another sample a known excess of sodium hydroxide was added, the carbonate in the solution precipitated with barium chloride, and the remaining excess sodium hydroxide titrated with N/10 hydrochloric acid, using equal amounts of phenolphthalein and thymol blue as indicator; this gave

$[NaHCO_3]$. In practice it was found best to neutralize almost all the remaining excess of sodium hydroxide with acid before adding barium chloride.

Results and Discussion

The results given here supersede some reported earlier [1], which were obtained with a more primitive apparatus, and before the possibility of error from the "stagnant region" had been appreciated.

Suppose the drum has a width w and its surface a linear velocity v, and the length of the periphery of the drum exposed to the gas is l. Then the strip of liquid surface formed in time δt has an area $wv \,.\, \delta t$, and it is exposed to the gas for a length of time l/v. If the amount of gas absorbed by unit area of surface in time t is $Q\{t\}$, the amount absorbed by the strip is $wv \,.\, \delta t \,.\, Q\{l/v\}$. Since $1/\delta t$ such strips are formed in unit time, the amount of gas absorbed in unit time by the total exposed area of the drum is $wv \,.\, Q\{l/v\}$.

Absorption into water. If the surface is saturated for $t > 0$ with carbon dioxide at a concentration C^*, the amount Q absorbed in time t by unit area is $2C^*\sqrt{Dt/\pi}$, where D is the diffusivity of the dissolved gas. A plot of Q against $\sqrt{t}$ should then be a straight line of slope $2C^*\sqrt{D/\pi}$. The experimental results are plotted in Fig. 2. The full line is the calculated relation, taking $C^* = 3{\cdot}37 \times 10^{-5}$ g mole/ml [6] and $D = 1{\cdot}92 \times 10^{-5}$ cm^2/sec. The latter is the value found by Davidson and Cullen [2, 3], which is considered more reliable than other values to be found in the literature. It will be seen that the observed results are systematically lower than the calculated line. It has previously been suggested that this might be caused by a very low "invasion coefficient" of carbon dioxide into water. However, Cullen and Davidson [2] absorbed carbon dioxide into jets of water, using exposure times down to about 0·01 sec, and found no trace of such an effect. The rise in surface temperature caused by the condensation of carbon dioxide and its associated water-vapour can be calculated [7] and is much too small to account for the discrepancy; in any case, this effect would also have been operative in Cullen and Davidson's experiments. The effect is no doubt a systematic experimental error; this is probably caused by end effects, such as the initial adjustment of the liquid velocity to that of the drum or "stagnant" surface near the exit. The magnitude of the error gives a measure of the reliability of the other results, which cannot be so accurately checked.

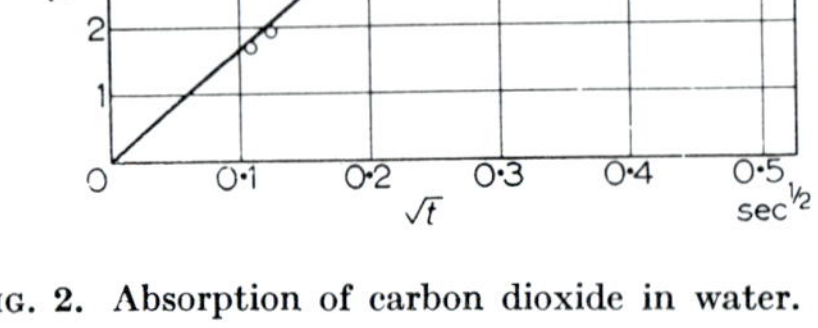

Fig. 2. Absorption of carbon dioxide in water.

Absorption into inert salt solutions. The results for absorption into 1·05 M sodium sulphate and 0·72 M magnesium sulphate solutions are shown in Fig. 3. The solubilities of carbon dioxide in these solutions are known [8] so the diffusivities can be calculated. For sodium sulphate, with $C^* = 1{\cdot}7 \times 10^{-5}$ g mole/ml, D is found to be $1{\cdot}1 \times 10^{-5}$ cm^2/sec. For magnesium sulphate, with $C^* = 2{\cdot}2 \times 10^{-5}$ g mole/ml, D is $1{\cdot}2 \times 10^{-5}$ cm^2/sec. When estimating diffusivity it is often assumed [9] that $D\mu$ is constant for a series of aqueous solutions at a particular temperature. Since the viscosities of both solutions were 1·57 times that of water, this rule would predict the diffusivity in both solutions to be $(1{\cdot}92 \times 10^{-5}/1{\cdot}57)$, or $1{\cdot}22 \times 10^{-5}$ cm^2/sec. It is assumed in subsequent calculations that D had the value $1{\cdot}2 \times 10^{-5}$ cm^2/sec in all other solutions used, since these all had the same viscosity.

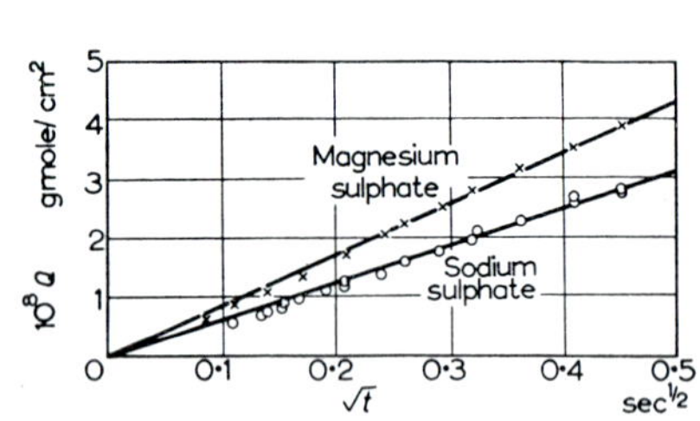

Fig. 3. Absorption of carbon dioxide in 1·05 M sodium sulphate and 0·72 M magnesium sulphate.

Absorption into sodium hydroxide solutions. Typical plots of Q against t and $\sqrt{t}$, for 0·085 M and 1·5 M sodium hydroxide solutions, are shown in Figs. 4 and 5. In general Q is initially proportional to t and later becomes linear in $\sqrt{t}$. The intercept on the $\sqrt{t}$ axis increases with increasing concentration.

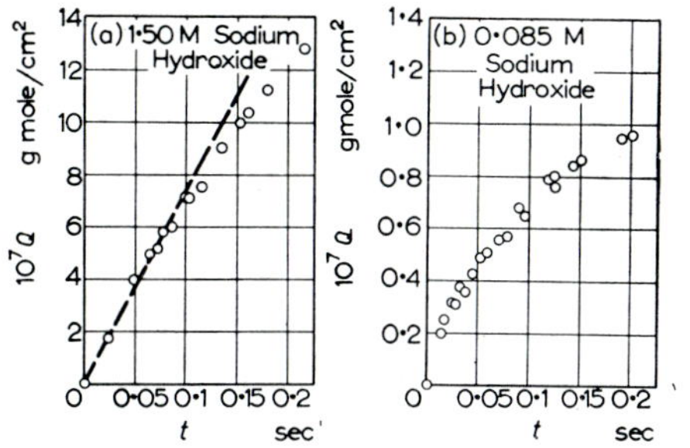

FIG. 4. Absorption of carbon dioxide in 1·50 M and 0·085 M sodium hydroxide (Q v. t).

In the earlier stages of absorption carbon dioxide is diffusing into a solution containing an almost uniform concentration of OH′ ions, with which it undergoes the reaction :

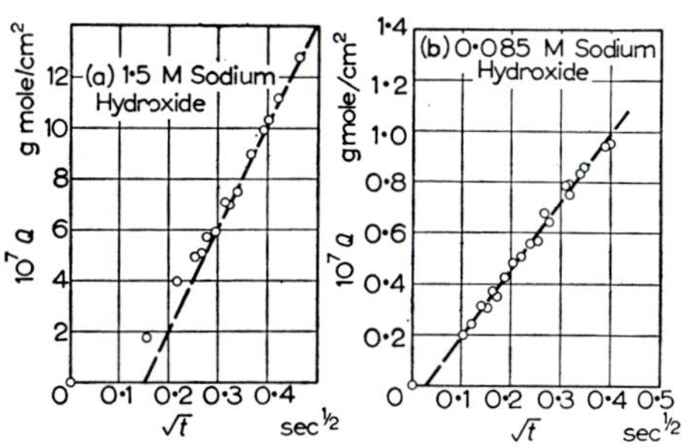

FIG. 5. Absorption of carbon dioxide in 1·50 M and 0·085 M sodium hydroxide (Q v. $\sqrt{t}$).

$$CO_2 + OH' \rightarrow HCO_3', \tag{1}$$

which is first-order with respect to CO_2 and to OH′. This reaction is followed by

$$HCO_3' + OH' \rightarrow CO_3'' + H_2O, \tag{2}$$

which has a much higher velocity constant than reaction (1). Thus, provided the concentration of OH′ is sufficiently high ($> 0{\cdot}1$ M) for the equilibrium concentration of HCO_3' to be negligible, the overall reaction is

$$CO_2 + 2OH' \rightarrow CO_3'' + H_2O, \tag{3}$$

being first-order with respect to CO_2 and to OH′, and having the velocity constant of reaction (1). So long as the OH′ ions are not substantially depleted in the neighbourhood of the surface, the reaction will be pseudo first-order with a velocity constant $k = k'\,[OH']$. The quantity absorbed in time t by diffusion accompanied by chemical reaction is given by [10]

$$Q = C^* \sqrt{\frac{D}{k}}\left[(kt + \tfrac{1}{2})\,\mathrm{erf}\,\sqrt{kt} + \sqrt{\frac{kt}{\pi}}\,e^{-kt}\right]. \tag{4}$$

When $kt > 2{\cdot}5$, Q can be closely approximated by :

$$Q = C^* \sqrt{\frac{D}{k}}\,(kt + \tfrac{1}{2}). \tag{5}$$

As the order of magnitude of k' is 10^7 ml/g mole sec [11,12], the value of kt is greater than 2·5 for the most dilute solution used, 0·045 M, and for exposure times greater than 0·01 sec. One would therefore expect the Q v. t curves to be linear with a slope of $C^* \sqrt{kD}$, until depletion of OH′ ions near the surface disturbs the first-order character of the reaction.

In the case of 1·5 M sodium hydroxide the expected order of magnitude of k is $1{\cdot}5 \times 10^{-3} \times 10^7$, or $1{\cdot}5 \times 10^4$ sec^{-1} ; the intercept of equation (3) on the Q axis would therefore be very small, and in effect the equation would represent a straight line through the origin. As may be seen from Fig. 4 it is possible to draw a straight line through the origin and the experimental points for times less than 0·1 sec. The slope of the line is $7{\cdot}5 \times 10^{-5}$ g mole/cm² sec. If the solubility of carbon dioxide in this solution is assumed to be $1{\cdot}7 \times 10^{-5}$ g mole/ml (estimated by the method of VAN KREVELEN and HOFTIJZER [13] for sodium hydroxide solution with an ionic strength of 2·25) and its diffusivity $1{\cdot}2 \times 10^{-5}$ cm²/sec, k is found to be about $1{\cdot}6 \times 10^4$ sec, whence k' is about $1{\cdot}1 \times 10^7$ ml/g mole sec, which is the expected order of magnitude. The results with other solutions, down to 0·15 M sodium hydroxide yield apparent values of k' within 20 per cent of this.

Although at first sight this analysis seems plausible enough, it is shown in Appendix I that if the value of k were in fact $1{\cdot}6 \times 10^4$ sec^{-1} in the 1·5 M sodium hydroxide solution, the concentration of OH′ at the surface would be depleted to the extent of 10 per cent within about $1{\cdot}5 \times 10^{-3}$ sec ; in the more dilute solutions the fractional rate of depletion is somewhat more rapid. Since the experimental observations refer to exposure times of 0·01 sec and more, it must be supposed that the depletion of OH′ ions renders the above analysis invalid, and that the true value of k' is substantially larger than 10^7 ml/g mole sec. Only if the effective diffusivity of OH′ ions was very much greater than that of dissolved carbon dioxide would the depletion be sufficiently small to be ignored. This is not the case, as is shown below.

If the reaction between the dissolved gas and the reactant in the solution were instantaneous, the rate of solution would be controlled by the diffusion of CO_2 and OH′ to a reaction site situated at a depth below the surface which increases with time. The concentration distribution in this case is shown in Fig. 6. The quantity

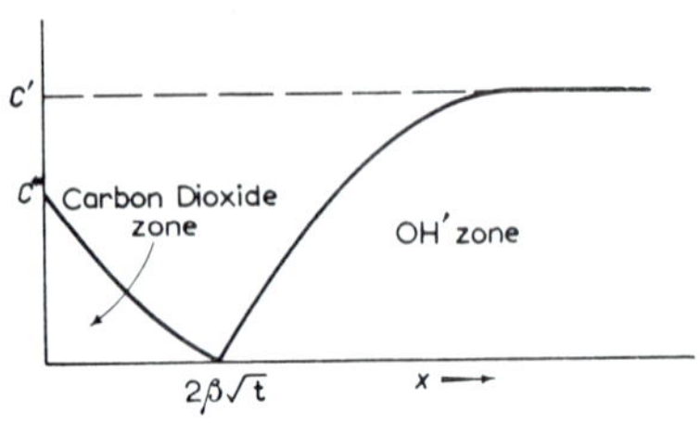

FIG. 6. Instantaneous reaction. Concentration distribution at time t.

of carbon dioxide absorbed in time t would then be given by [14]

$$Q = 2\,\alpha\, C^* \sqrt{\frac{Dt}{\pi}}, \tag{6}$$

where

$$\alpha = 1/\mathrm{erf}\,(\beta/\sqrt{D}) \tag{7}$$

and β is defined by

$$\frac{C'/n}{\sqrt{D}} \exp\,[\beta^2/D]\,\mathrm{erf}\,[\beta/\sqrt{D}] = \frac{C^*}{\sqrt{D}} \exp\,[\beta^2/D']\,\mathrm{erfc}\,[\beta/\sqrt{D'}], \tag{8}$$

where C' is the concentration of the reactant, and n the number of moles which react with one mole of dissolved gas. When $D = D'$,

$$\alpha = (1 + C'/nC^*). \tag{9}$$

A rigorous treatment of the kinetics of absorption of a gas which undergoes a second order reaction of *finite* speed with a reactant solute involves the solution of non-linear differential equations. PERRY and PIGFORD [15] have used a computer to obtain solutions covering a certain range of parameters in the special case where $D = D'$. Their results are not expressed in the form used here, but show implicitly that after a sufficient time of exposure Q approaches the form :

$$Q = 2C^* \left(1 + \frac{C'}{nC^*}\right) \sqrt{\frac{Dt}{\pi}} - A, \tag{10}$$

Here A is independent of t. It may be expected that, by analogy, when $D \neq D'$, Q will approach the form :

$$Q = 2\,\alpha\, C^* \sqrt{\frac{Dt}{\pi}} - B = S\sqrt{t} - B, \tag{11}$$

with B independent of t, α having the value given by equations (7) and (8). This view is supported by dimensional analysis, which shows that the relation between Q and t may be put in the form :

$$\frac{1}{2C^*}\sqrt{\frac{\pi}{D}}\,\frac{dQ}{d\sqrt{t}} = f\left\{\frac{D'}{D}, \frac{C'}{C^*}, k'\,C^*\,t\right\}. \tag{12}$$

As k' is increased indefinitely the behaviour must become that of the "instantaneous" reaction, and the left hand side becomes equal to α for any value of t, however small. It follows that for any value of k' the left-hand side becomes equal to α for sufficiently large values of t, so that a plot of Q v. $\sqrt{t}$ will approach asymptotically a straight line of slope S.

It can be shown that when $\alpha \gg 1$ equations (7) and (8) become :

$$\alpha \approx \frac{C'}{nC^*}\cdot\sqrt{\frac{D'}{D}} + \sqrt{\frac{D}{D'}}, \tag{13}$$

whence from equation (11) :

$$S \approx \frac{2C'}{n}\sqrt{\frac{D'}{\pi}} + 2C^*\sqrt{\frac{D}{\pi}}\sqrt{\frac{D}{D'}}, \tag{14}$$

where S is the limiting slope of the Q v. $\sqrt{t}$ curve. The error in α is about 1 per cent when $\alpha = 10$.

Experimental values of S are plotted against the sodium hydroxide concentration C' in Fig. 7. Assuming $C^*\sqrt{D}$ and D' to have the same value in all the solutions, the curve should tend to a straight line of slope $\sqrt{D'/\pi}$ (since $n = 2$) and intercept $2C^*\sqrt{\frac{D}{\pi}}\sqrt{\frac{D}{D'}}\cdot$ The limiting slope of Fig. 7 is $2{\cdot}8 \times 10^{-3}$ cm/sec$^{\frac{1}{2}}$, whence $D' = 2{\cdot}5 \times 10^{-5}$ cm^2/sec.

Using this value of D', S can be calculated from equations (7), (8) and (11) for the whole range of C', and compared with the experimental results. Since the intercept of equation (14) is so small,

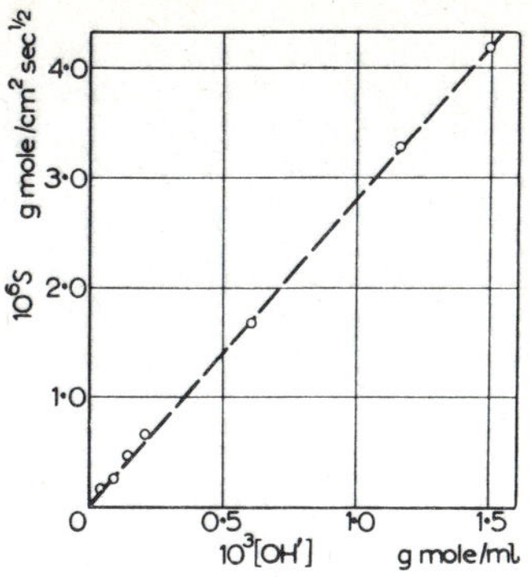

FIG. 7. Absorption of carbon dioxide in sodium hydroxide (S v. [OH′]).

the calculated values for the more alkaline solutions are virtually independent of C^*, and it is therefore appropriate to use a value of C^* estimated for the three or four least alkaline solutions. The compositions and ionic strengths of the solutions are shown in Table 1. The major

Table 1. *Sodium hydroxide*

Run	*Initial concn.* (g mole/l.)			$Q \rightarrow S\sqrt{t} - B$ (g mole/cm²)	
	OH′	CO_3''	SO_4''	10^6S	10^6B
115	0·045	0·03	0·96	0·17	0·004
116	0·085	0·07	0·91	0·26	0·006
117	0·15	0·13	0·78	0·48	0·02
118	0·21	0·19	0·70	0·68	0·04
119	0·62	0·30	0·42	1·7	0·16
120	1·16	0·38	0	3·3	0·40
122	1·50	0·25	0	4·2	0·61

contribution to the ionic strength of the less alkaline solutions is made by sodium sulphate, added to increase the viscosity; treating them as 1M solutions of sodium sulphate, of ionic strength 3M, one finds $C^* = 1{\cdot}7 \times 10^{-5}$ g mole/ml. Using the method of VAN KREVELEN and HOFTIJZER [13] to predict the solubility in 1M sodium carbonate one finds the same value, whereas if one uses the figures of NIJSING and KRAMERS [12] one finds $C^* = 1{\cdot}8 \times 10^{-5}$.

These figures suggest that the substitution of CO_3'' for SO_4'' will have little effect on the solubility. We shall adopt here the value $1{\cdot}7 \times 10^{-5}$ g mole/ml. (It may be noted that during absorption of carbon dioxide the OH′ at the surface is rapidly converted to CO_3''; the resultant increase in the ionic strength is, however, small in the less alkaline solutions).

Taking $D = 1{\cdot}2 \times 10^{-5}$ cm²/sec equations (7), (8) and (11) are used to calculate S as a function of C'. In Fig. 8 the calculated curve showing $10^6\left(S - 2C^*\sqrt{\frac{D}{\pi}}\sqrt{\frac{D}{D'}}\right)$ as a function of $10^3C'$ is compared with the experimental results. The agreement is good, but it happens that over the range of experimental conditions the closeness of the agreement is not very sensitive to the value assigned to C^* even in the less alkaline solutions.

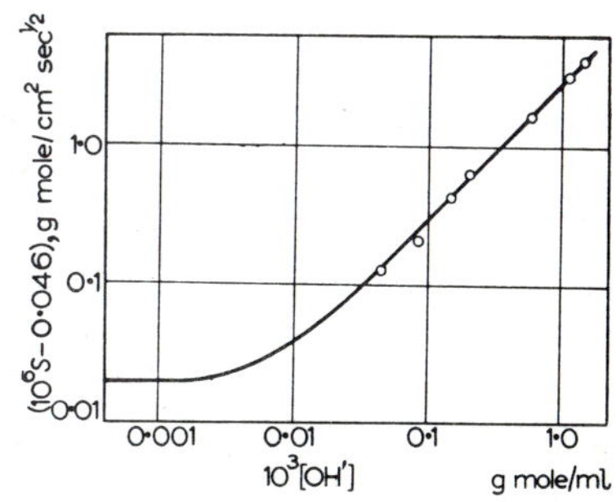

FIG. 8. Comparison of experimental and calculated values of S.

The agreement therefore gives little information about the accuracy of the estimate of C^*. However, it may be noted that the experimental value of S for 1·05 M sodium sulphate is $0{\cdot}063 \times 10^{-6}$ compared to the calculated value of $0{\cdot}066 \times 10^{-6}$ g mole/cm² sec½ for infinitely dilute sodium hydroxide. Since C^* in the former solution is very nearly equal to that estimated for the weaker sodium hydroxide solutions, this indicates that the latter estimate was correct.

The rate of absorption of carbon dioxide into sodium hydroxide solutions is not proportional to C^* as it is in the case of buffer solutions and non-reacting liquids. Strictly speaking, therefore, both the solubility of carbon dioxide and the observed rates of gas absorption should be corrected for the diluting effect of the water-vapour present in the carbon dioxide. Since the vapour-pressure of water at 25 C is 23·8 mm, the true absorption rate of carbon dioxide is 3 per cent less than the observed rate of passage of moist carbon dioxide, and the solubility is 3 per cent less than that of dry carbon dioxide at atmospheric pressure. In view of the experimental errors and other uncertainties, these corrections are small enough to be ignored, particularly since they tend to counteract one another.

The rate of absorption of carbon dioxide into the stronger sodium hydroxide solutions is so fast that the temperature at the surface may be suspected of rising appreciably and invalidating arguments based on an isothermal system. This temperature rise may be calculated approximately by assuming that $Q = S\sqrt{t}$ and that the heat of reaction, as well as the heat of solution, is liberated at the geometrical surface of the liquid. Both of these assumptions exaggerate the rise. If H is the sum of the heats, then the rise in temperature in time t will be [7].:

$$\Delta T = \frac{SH}{2\rho\sigma\sqrt{\pi\alpha}} \int_0^t \frac{d\theta}{\sqrt{\theta(t-\theta)}} = \frac{SH}{2\rho\sigma}\sqrt{\frac{\pi}{\alpha}}, \quad (15)$$

where ρ = density, σ = specific heat, and α = thermal diffusivity of the liquid. Putting $\rho\,\sigma = 1$ cal/°C cm^3, $\alpha = 1{\cdot}46 \times 10^{-3}$ cm^2/sec, H = 6,260 cal/g mole [7], we find for 1·5 M sodium hydroxide where $S = 4{\cdot}2 \times 10^{-6}$ g mole/cm^2 sec$^{\frac{1}{2}}$, $\Delta T = 0{\cdot}61$°C. The rise in temperature is therefore not large enough to exert any substantial effects.

From the foregoing it appears that after a certain exposure time the limiting rate of absorption into sodium hydroxide solutions is controlled by the diffusion of CO_2 and OH′ to the reaction site, and that the effective diffusivity of the latter is about twice that of CO_2. However, the actual situation is more complicated than the model on which the "instantaneous reaction" equations are based. In the first place, OH′ ions cannot diffuse independently of the other ions present, as the requirements of electrical neutrality must be satisfied.

Fig. 9 shows the way in which the various solute species might be momentarily distributed in the neighbourhood of the surface. In every element of solution the concentration of Na$^{\bullet}$ must

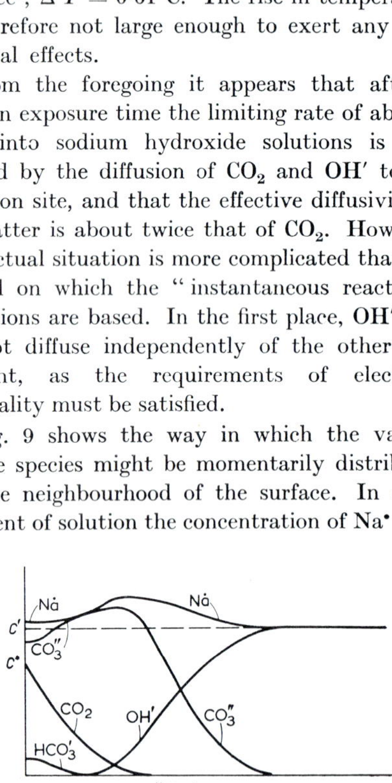

Fig. 9. Concentration distribution during absorption of carbon dioxide in sodium hydroxide solution.

be equal to the combined concentration (in equivalents) of CO_3'', OH′ and HCO_3'. The fact that the mobility of OH′ is greater than that of CO_3'' means that OH′ can diffuse to the reaction zone more rapidly (equivalent for equivalent) than CO_3'' can diffuse away from it.

As a result, the concentration of negative and hence of positive ions is greater in the reaction zone than in the bulk of the solution. The resulting concentration gradient of Na$^{\bullet}$ ions is opposed by an electrical potential gradient, the diffusion of Na$^{\bullet}$ ions into the bulk of the liquid being opposed by their bodily transport towards the surface. This potential gradient also opposes the transport of OH′ to the reaction zone (and expedites the transport of CO_3'' away from it).

The net effect will be to reduce the effective diffusivity of OH′ to less than its "self-diffusivity" in the sodium hydroxide solution. The value of the latter quantity may be estimated roughly as follows. The experimental evidence available concerning the self-diffusion of individual ions [16] shows that the value, D_0^*, of the self-diffusivity at infinite dilution is given by :

$$D_0^* = \frac{RT}{F}u_0, \quad (16)$$

where u_0 is the ionic mobility at infinite dilution. The value of u_0 for OH′ ions at 25°C is $2{\cdot}06 \times 10^{-3}$ cm^2/V sec [17], whence $D_0^* = 5{\cdot}3 \times 10^{-5}$ cm^2/sec. The effect of finite concentration can be partly accounted for by changes in viscosity ; thus the quantity $(D^*\,\mu/\mu_0)$ varies less than D^* itself. Other effects, due to interionic attraction, cannot be accurately allowed for in concentrated solutions. The measurements of the self-diffusivities of halide ions in alkali halide solutions made by Mills and Kennedy [16] show that values of $(D^*\,\mu/\mu_0)$ are some 10–22 per cent less in 2·3 M solutions than at infinite dilution. For the purposes of rough estimation it is reasonable to assume that it is the ionic strength of the solution which is of primary importance in determining the extent of this reduction.

The ionic strength in the 1·5 M sodium hydroxide solution was about 2·3 M (owing to the presence of sodium carbonate), so the estimated value of $(D^*\,\mu/\mu_0)$ is $4{\cdot}8 - 4{\cdot}1 \times 10^{-5}$ cm^2/sec. Since $\mu/\mu_0 = 1{\cdot}57$, the estimated value of D^* is $3{\cdot}1$–$2{\cdot}6 \times 10^{-5}$ cm^2/sec, compared with the value $2{\cdot}5 \times 10^{-5}$ cm^2/sec found for the "effective" diffusivity D' in gas absorption.

The method of estimating D^* is so rough that this results lends only slender support to the statement that $D' < D^*$. However it seems likely that D^* can be predicted in this way (when the requisite data are available) and used in place of D' with a precision sufficient for many purposes, particularly as the square root of D' is the factor determining the rate of absorption.

A further complication is the reaction of dissolved CO_2 with CO_3'' to form HCO_3' in the region between the reaction zone and the surface. The reaction which actually occurs is reaction (1) between CO_2 and the OH' ions which are present in a solution of CO_3'' by virtue of the equilibrium :

$$CO_3'' + H_2O \rightleftharpoons HCO_3' + OH'. \quad (17)$$

If the non-ideal behaviour of the solutes is ignored it is easily shown that :

$$[OH'] = \frac{K_w}{K_2} \cdot \frac{[CO_3'']}{[HCO_3']}, \quad (18)$$

where K_2 is the second dissociation constant of H_2CO_3. Thus the local rate of reaction is given by :

$$\frac{d}{dt}[HCO_3'] = 2k' [OH'] [CO_2]$$

$$= 2\frac{k' K_w}{K_2} \frac{[CO_2][CO_3'']}{[HCO_3']}. \quad (19)$$

The method of integrating this equation to give the total amount of HCO_3' formed in a given time is detailed in Appendix 2. The fraction of the CO_2 dissolving in 0·20 sec which reacts to form HCO_3' is greatest in the most dilute solution, namely 0·045 M, in which it is less than 7·5 per cent.

To sum up the discussion of the experiments with sodium hydroxide, although the results are readily explicable, quantitative prediction would be difficult in many circumstances because of the complexity of the processes involved. The initial homogeneous reaction to produce CO_3'' is followed by depletion of OH' at the surface, leading in time to a diffusion-controlled reaction ; superimposed on this is the secondary reaction leading to the formation of HCO_3'. These observations should be taken as a caution against over-simplification when predicting or explaining the results of gas absorption under industrial conditions.

Absorption into buffer solutions. Sodium carbonate–bicarbonate buffer solutions were used; their compositions are given in Table 2. The pH was in all cases greater than 10, and in these circumstances [18] dissolved carbon dioxide is removed by reaction (1). Provided that the reaction does not deplete the buffer sufficiently to alter the concentration of OH' appreciably, the rate of reaction, R, of carbon dioxide per unit volume of solution will be given by :

$$R = k' [CO_2] [OH'] = k [CO_2], \quad (20)$$

with k constant for a given buffer solution. The value of k' is likely to vary somewhat from one solution to another, even though the ionic strengths of the solutions were all about the same (2·3–2·5 M), since specific salt effects can be expected to occur at these concentrations. In addition the concentration of OH' is not likely to be strictly proportional to the buffer ratio $[CO_3'']/[HCO_3']$ [as implied by equation (18)]. We should not therefore expect the experimentally-observed value of k to be necessarily proportional to the buffer ratio ; however, the recently published work of NIJSING and KRAMERS [12] suggests that departure from proportionality is in fact small, perhaps owing to a cancellation of effects.

Neglecting for the moment the depletion of the buffer by the reaction, the relation between Q and t is given by equation (4). If erf $\sqrt{kt}$ and

Table 2. Buffer solutions

Run	*Concn.* (g mole/l.) CO_3''	*Concn.* (g mole/l.) HCO_3'	*Ionic strength* I (g ion/l.)	*Buffer ratio* $\frac{[CO_3'']}{[HCO_3']}$	$10^8 C^* \sqrt{D}$ (g mole/cm^2 sec$^{\frac{1}{2}}$)	k (sec^{-1})	$k\,[HCO_3']/[CO_3'']$ (sec^{-1})
107	0·49	0·815	2·29	0·60	6·7	0·70	1·16
108	0·62	0·535	2·40	1·16	6·8	1·51	1·30
109	0·68	0·40	2·44	1·70	6·8	1·45	0·85
110	0·735	0·295	2·50	2·47	6·7	2·25	0·91
111	0·765	0·225	2·52	3·40	6·8	2·55	0·75
112	0·78	0·175	2·52	4·45	7·6	3·74	0·84

e^{-kt} are expanded in powers of kt this becomes :

$$Q = 2C^* \sqrt{\frac{Dt}{\pi}} \times$$

$$\times \left[1 + \frac{1}{3} kt - \frac{1}{30} (kt)^2 + \frac{1}{210} (kt)^3 - - -\right]. \quad (21)$$

It transpires that the maximum value of kt encountered in these experiments was less than 1·5 so that the fourth term contributes less than 2 per cent of the total and may be ignored in view of the experimental errors. The values of $C^* \sqrt{D}$ and of k may be found in principle by fitting the first three terms of equation (21) to the experimental results by the method of least squares. The first term corresponds to absorption without reaction, and the greater the value of kt – that is, the more alkaline the solution and the longer the time of exposure – the greater is the influence of the reaction and the accuracy with which k can be determined.

However, the range of kt within which equation (21) can be expected to apply is limited by depletion of the buffer solution. It is shown in Appendix 1 that for the most and least alkaline solutions used (Runs 112 and 107) the values of kt leading to a change of 10 per cent in the buffer ratio at the surface are about 0·85 and 3·2 respectively. If a preliminary estimate can be made of the value of k it is possible to estimate the time of exposure above which depletion exceeds, say, 10 per cent and which may be arbitrarily taken as the limit above which equation (21) ceases to be applicable.

There are various ways in which k can be estimated, starting from the observation of Nijsing and Kramers [12] that its value is closely proportional to the buffer ratio and independent of the ionic strength, over a range of conditions not very different from those of the present work. Thus they report one value of k at 25°C, from which its values for the present Runs 112 and 107 may be estimated as 3·5 and 0·5 sec^{-1} respectively. Taking their value for $k \cdot [HCO_3']/[CO_3'']$ at 20°C, namely 0·55 sec^{-1}, and employing the temperature coefficient for the reaction determined by Pinsent, Pearson and Roughton [11], we find the values for Runs 112 and 107 at 25°C again to be 3·5 and 0·5.

On the other hand, if we assume that in the present work the degree of depletion in the runs with the lowest alkalinity (107 and 108) was small enough to be ignored (as will subsequently be shown) we can use the value of k in these solutions to predict the value of k in the more alkaline solution. The results for the least alkaline solutions cannot be expected to give the value of k with great accuracy, because of the small contribution made to Q by the reaction.

It will be seen that the value of $k\ [HCO_3']/[CO_3']$ for the less alkaline solutions is considerably greater than the value derived from the work of Nijsing and Kramers, which was about 0·8 ; however, in case there was a systematic discrepancy between their results and ours it was considered appropriate here to extrapolate from Run 108, which gave the largest value of $k\ [HCO_3']/[CO''_3]$, namely 1·30 sec^{-1}, giving the value of 5·7 sec^{-1} for k in the most alkaline solution (Run 112), and 0·8 sec^{-1} for the least alkaline (Run 107). For Run 107 the maximum time of exposure was 0·263 sec, giving a maximum value of 0·21 for kt ; this is so much less than the value, 3·2, at which the surface depletion reaches 10 per cent that we can take it that equation (21) applies to the entire course of the run.

For Run 112, with a maximum exposure time of 0·215 sec, on the other hand, the estimated maximum value of kt is 1·2, which is considerably greater than the value of 0·85 for 10 per cent depletion ; only the observations below 0·15 sec would be expected to conform to equation (21) with reasonable accuracy if the estimate of the value of k were correct. Similar calculations show that the change in the buffer ratio was less than 10 per cent in all runs except 112.

When the first three terms of equation (21) were fitted to the experimental values of Run 112 for times less than 0·15 sec, k was found in fact to be about 3·7 sec^{-1}. It was concluded that, since the value of kt for the maximum exposure time of 0·215 sec was therefore in fact below 0·85, the whole range of experimental values should conform to equation (21). The values of $C^* \sqrt{D}$ and k shown in Table 2 were obtained from the whole range of Run 112, and of the other runs.

In Fig. 10 the values of k are plotted against the buffer ratio. The results are compatible with the relationship :

$$k = 0{\cdot}8 \frac{[CO_3'']}{[HCO_3']} \text{ sec}^{-1}. \quad (22)$$

This is in agreement with the single result of Nijsing and Kramers [12] at 25°C, and also with their results at 20°C if corrected by the

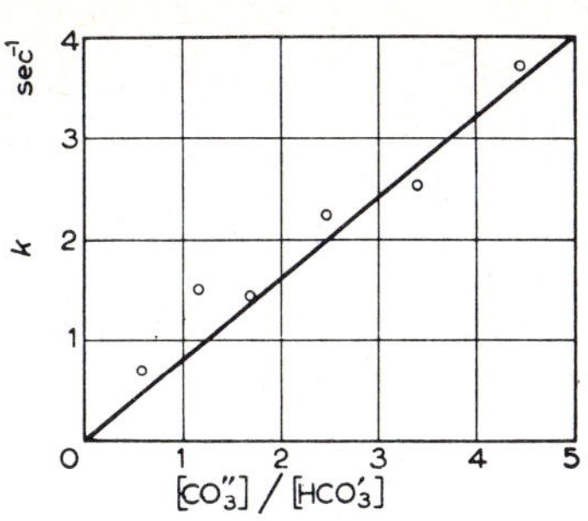

FIG. 10. Absorption into buffer solutions (k $v.$ buffer ratio).

temperature coefficient of PINSENT, PEARSON and ROUGHTON [11].

No direct comparison is possible between the values of $C^* \sqrt{D}$ obtained by NIJSING and KRAMERS and by ourselves, since the ionic strengths and viscosities of the solutions were different in the two cases. Taking D in our case to have been $1{\cdot}2 \times 10^{-5}$ cm^2/sec, we find $C^* = 2{\cdot}0 \times 10^{-5}$ g mole/ml as the average of runs 107–111, in which the ionic strengths varied from 2·29–2·52 M.

Using the method of VAN KREVELEN and HOFTIJZER [13], in conjunction with the coefficients found by NIJSING and KRAMERS [12], the predicted solubility in these solutions varies from $2{\cdot}13$–$2{\cdot}02 \times 10^{-5}$ g mole/ml. This agreement is good, and provides some support for the validity of the method of estimating solubilities and for the conclusion of NIJSING and KRAMERS that the solubility in carbonate–bicarbonate buffers depends on the ionic strength and nature of the cation, but not on the buffer ratio. However the present findings represent a relatively small extrapolation from their experimental results. The much higher apparent value of C^* in Run 112 suggests that the relationship may not hold at high buffer ratios.

CONCLUSIONS

The results obtained for the absorption of carbon dioxide in water are consistently somewhat lower than the values calculated from the most reliable values of C^* and D. In view of the work of CULLEN and DAVIDSON [2, 3], which showed no such discrepancy, this is thought to be due to systematic experimental error rather than to a low accommodation coefficient, as previously suggested.

The results for absorption into inert salt solutions yield values of D consistent with the assumption that D varies inversely with the viscosity of the solution.

The early stages of absorption into sodium hydroxide solution are presumably accompanied by a pseudo first-order reaction between CO_2 and a virtually uniform concentration of OH'. However, calculation shows that depletion of OH' near the surface will become appreciable in a time much less than the shortest exposure time observed in these experiments; this destroys the first order character of the reaction and makes it impossible to deduce the reaction velocity constant from the observations.

At higher exposure times the rate of absorption is governed by the diffusion of OH' and CO_2 to the reaction site; the results agree quantitatively with predictions, and the effective diffusivity of OH' is found to be about twice that of CO_2. The effective diffusivity is expected to be less than the self-diffusivity of OH' in the same solution; a rough estimate of the latter quantity, based on the ionic mobility, supports this conclusion.

At still higher exposure times, reaction of the CO_2 with the CO_3'' formed near the surface to give HCO_3' will become important and will complicate the calculation of absorption rates. In these experiments it is calculated that less than $7\frac{1}{2}$ per cent of the CO_2 absorbed was converted to HCO'_3 in the worst case.

The absorption of CO_2 into CO_3''/HCO_3' buffer solutions is accompanied by a first order reaction so long as the buffer is not appreciably depleted near the surface. Calculation shows that the depletion did not become substantial during these experiments. The calculated values of the velocity constant were compatible with those found by other workers.

Acknowledgement—One of the authors (A.M.K.) wishes to thank the New Zealand Defence Science Authorities for an award which enabled this work to be carried out.

APPENDIX I

THE DEPLETION OF SODIUM HYDROXIDE AND BUFFER SOLUTIONS BY REACTION

Sodium hydroxide solutions. Let the initial uniform concentration of OH' be $C(O)$, and its concentration at a distance x from the surface at time t be $C(x, t)$. The concentration of CO_2 at the surface is C^* at all times; it diffuses into the solution, in which it undergoes a second order reaction with OH'. During the initial stages of the process, when the depletion of OH' is small, the concentration of OH' can be regarded as uniform and the reaction is first order in the local concentration of CO_2, the velocity constant $k = k' C(O)$ having the same value everywhere.

The method to be described makes use of this approximation, and thus only allows small degrees of depletion to be calculated.

Putting

$$V = 1 - \frac{C(x, t)}{C(O)} \quad (23)$$

the simultaneous diffusion and reaction of OH′ are expressed by the equation :

$$\frac{\partial V}{\partial t} = D' \frac{\partial^2 V}{\partial x^2} + 2k \frac{\chi}{C(O)}, \quad \left\{ \begin{array}{l} V = 0,\ t = 0,\ x > 0 \\ V = 0,\ t > 0,\ x = \infty \\ \dfrac{\partial V}{\partial x} = 0,\ t > 0,\ x = 0 \end{array} \right\}, \quad (24)$$

where D' is the diffusivity of OH′ and χ is the concentration of unreacted CO_2 at (x, t) and is determined by the equation for diffusion accompanied by a first order reaction, the solution of which has been given elsewhere [10]. If the expression for χ is inserted into equation (24) it becomes mathematically analogous to that for the changes of temperature produced by a first order reaction accompanying gas absorption ; this problem has been considered elsewhere [7] and a comparison with the present problem shows that if $V(O, t)$ is the value of V at the surface at time t, and $L\{V(O, t)\}$ its Laplace transform, then :

$$\frac{C(O)}{2kc^* D} L\{V(O, t)\} = \frac{\sqrt{\dfrac{D'}{D}\left(1 + \dfrac{k}{p}\right)} - 1}{p\{p(D' - D) + kD'\}} \cdot \quad (25)$$

Here p is the parameter of the Laplace transformation, and D the diffusivity of dissolved CO_2. In order to simplify the expression, it will be assumed that the diffusivities of CO_2 and OH′ are equal, so that $D = D'$. (This will overestimate the extent of depletion, since in fact $D < D'$ as shown in the body of the paper ; the magnitude of the error so introduced is discussed later). Equation (25) then becomes :

$$\frac{C(O)}{2C^*} L\{V(O, t)\} = \frac{1}{p}\sqrt{1 + k/p} - \frac{1}{p} \cdot \quad (26)$$

The inverse transform of the first term on the right has been found elsewhere [7] ; using this result :

$$\frac{C(O) - C(O, t)}{2C^*} = \sum_{n=1}^{\infty} \frac{\left(\frac{1}{2}\right) \cdots \left(\frac{3}{2} - n\right)}{n!} \cdot \frac{(kt)^n}{n!}, \cdot \quad (27)$$

$$= f_1(kt).$$

The function $f_1(kt)$ can be read from Fig. 1 of [7] by subtracting 1 from the ordinate ; however, it can be approximated with sufficient accuracy for the present purpose by :

$$f_1(kt) \approx kt/2,\ kt < 1 ; \quad (28)$$

$$f_1(kt) \approx 2\sqrt{kt/\pi} - 1,\ kt > 5. \quad (29)$$

In order to calculate the time required for 10 per cent depletion at the surface, we put :

$$\frac{C(O) - C(O, t)}{C(O)} = 0{\cdot}1 = \frac{2C^*}{C(O)} f_1(kt). \quad (30)$$

Taking $C^* = 1{\cdot}7 \times 10^{-5}$ g mole/ml, we find for 1·5 sodium hydroxide, $f_1(kt) = 4{\cdot}4$. Using the approximation given in (29) we find $kt = 21$. If we take the value of k to be about $10^7 C(O)\ \text{sec}^{-1}$, t is found to be about $1{\cdot}5 \times 10^{-3}$ sec. A similar calculation shows that for 10 per cent depletion of 0·045 M sodium hydroxide, $f_1(kt) = 0{\cdot}22$, and using approximation (28) t is found to be about $6{\cdot}7 \times 10^{-4}$ sec.

If $D' \gg D$ the inverse transform of equation (25) takes the form :

$$\frac{C(o)}{2C^*}\sqrt{\frac{D'}{D}}\, V(O, t) = 1 - e^{-kt} + e^{-kt} \sum_{n=1}^{\infty} \mu_n (kt)^{n+1}, \quad (31)$$

$$= f_2(kt).$$

The coefficients μ_n and a graph of $f_2(kt)$ are given in [7]; $f_2(kt)$ tends to $2\sqrt{kt/\pi}$ for large values of kt, and to kt for small values. If the ratio D'/D were equal to 10, calculation shows that the time required for 10 per cent depletion at the surface would be about 9×10^{-3} sec. for 1·5 M sodium hydroxide and $1{\cdot}7 \times 10^{-3}$ sec for 0·045 M sodium hydroxide. Actually since the effective value of D'/D is about 2, as shown in the body of the paper, the times required will be considerably less than this.

Buffer solutions. As before, it is assumed that the change in composition of the solution is small enough for the OH′ concentration to be treated as uniform, so that the CO_2 undergoes a pseudo first-order reaction with a velocity constant $k = k'[\text{OH}']$. If the initial concentration of CO_3'' is $C(O)$ and its concentration at x, t is $C_1(x, t)$, and :

$$V_1 = 1 - \frac{C_1(x_1 t)}{C_1(O)} \quad (32)$$

then

$$\frac{\partial V_1}{\partial t} = D_1 \frac{\partial^2 V_1}{\partial x^2} + \frac{k\chi}{C_1(O)}, \quad \left\{ \begin{array}{l} V_1 = 0,\ x > 0,\ t = 0 \\ V_1 = 0,\ x = \infty,\ t > 0 \\ \dfrac{\partial V_1}{\partial x} = 0,\ x = 0,\ t > 0 \end{array} \right\}, \quad (33)$$

where D_1 is the diffusivity of CO_3''. Assuming the diffusivities of CO_2 and of CO_3'' to be equal, comparison with [7] shows the solution to be :

$$1 - \frac{C_1(O, t)}{C_1(O)} = \frac{C^*}{C_1(O)} \cdot f_1(kt), \quad (34)$$

where $f_1(kt)$ is given by equation (25). If the initial concentration of HCO_3' is $C_2(O)$, and its value at x, t is $C_2(x, t)$, the equation expressing the change in concentration of HCO_3' is :

$$\frac{\partial V_2}{\partial t} = D_2 \frac{\partial^2 V_2}{\partial x^2} + \frac{2k\chi}{C_2(O)}, \quad \left\{ \begin{array}{l} V_2 = 0,\ x > 0,\ t = 0 \\ V_2 = 0,\ x = \infty,\ t > 0 \\ \dfrac{\partial V_2}{\partial x} = 0,\ x = 0,\ t > 0 \end{array} \right\} \quad (35)$$

where

$$V_2 = \frac{C_2(x, t)}{C_2(O)} - 1, \quad (36)$$

and D_2 is the diffusivity of HCO_3'. If this is equal to the diffusivity of CO_2 :

$$\frac{C_2(O, t)}{C_2(O)} - 1 = \frac{2C^*}{C_2(O)} \cdot f_1(kt). \quad (37)$$

The fractional surface depletion, Δ, of the initial buffer ratio $C_1(O)/C_2(O)$ is thus :

$$\Delta = 1 - \frac{C_2(O)}{C_1(O)}\left[\frac{C_1(O) - C^* f_1(kt)}{C_2(O) + 2C^* f_1(kt)}\right]. \qquad (38)$$

Considering the most and least alkaline buffer solutions used, and putting $C^* = 2{\cdot}0 \times 10^{-5}$ g mole/ml, it is found that the values of kt required for Δ to reach a value of 0·1 are 0·85 and 3·2 respectively.

Appendix 2

The Formation of Bicarbonate in Sodium Hydroxide Solutions

In order to calculate the amount of CO_2 which undergoes the secondary reaction with CO_3'' to form HCO_3', the actual situation is simplified by assuming " instantaneous " primary reaction between CO_2 and OH' to give CO_3''. If the secondary reaction is left out of account, there will be a layer of depth $2\beta\sqrt{t}$ [β being defined by equation (8)], in which all the OH' has been converted to CO_3'', and in which the concentration χ of unreacted CO_2 is given by :

$$\frac{\chi}{C^*} = \frac{\operatorname{erfc}(x/2\sqrt{Dt}) - \operatorname{erfc}(\beta/\sqrt{D})}{\operatorname{erf}(\beta\sqrt{D})}, \qquad (39)$$

x being the depth below the surface [14]. Thus at a depth x the concentration of CO_2 is zero at $t = (x/2\beta)^2$ and thereafter rises progressively to C^*. The concentration of CO_3'' rises from C'' for $t < (x/2\beta)^2$ to $(C'' + \frac{1}{2}C')$ for $t > (x/2\beta)^2$, C' being the initial concentration of OH' and C'' the initial concentration of CO_3'' present in the solution. In the present calculation the diffusion of both CO_3'' and HCO_3' ions is neglected, the changes in their concentrations at a given depth being regarded as being caused solely by the local secondary reaction.

The errors introduced by these two approximations are opposite in sign ; that due to neglecting the diffusion of CO_3'' is probably the greater, and will lead to an overestimate of the amount of HCO_3' formed. The depletion of the CO_2 concentration by the secondary reaction is ignored, and its concentration is assumed to be given by equation (39) ; this also gives an overestimate of the HCO_3' formed, as does the assumption of an " instantaneous " primary reaction. The calculated extent of reaction will, however, be of the right order of magnitude.

If in equation (19) we put $[HCO_3'] = a$, $[CO_2] = \chi$, $[CO_3''] = \frac{1}{2}(2C'' + C' - a)$, the equation can be integrated with respect to t for constant x :

$$\int_{a_0}^{a} \frac{a\,da}{(2C'' + C' - a)} = C^* \frac{k' K_w}{K_2} \int_{(x/2\beta)^2}^{t} \frac{\operatorname{erfc}(x/2\sqrt{Dt}) - \operatorname{erfc}(\beta/\sqrt{D})}{\operatorname{erf}(\beta/\sqrt{D})}\,dt. \qquad (40)$$

The concentration a_0 is the equilibrium concentration of $[HCO_3']$ in a solution of sodium carbonate alone in concentration $(C'' + \frac{1}{2}C')$; a_0 was negligibly small in all cases, and may be ignored. Equation (40) then becomes on integration :

$$(2C'' + C')\ln\left(\frac{2C'' + C'}{2C'' + C' - a}\right) = \frac{k' K_w C^*}{K_2 \operatorname{erf}(\beta/\sqrt{D})} \times$$
$$\times \left[\left(t + \frac{x^2}{2D}\right)\left\{\operatorname{erf} c\left(\frac{x}{2\sqrt{Dt}}\right) - \operatorname{erfc}\left(\frac{\beta}{\sqrt{D}}\right)\right\}\right.$$
$$\left. - x\sqrt{\frac{t}{\pi D}}\cdot e^{-\frac{x^2}{4tD}} + \frac{x^2}{2\beta\sqrt{\pi D}}\, e^{-\beta^2/D}\right]. \qquad (41)$$

If we take t, the exposure time, to be 0·20 sec, a can be evaluated as a function of x, and a graphical integration from $x = 0$ to $x = 2\beta\sqrt{t}$ gives the total amount of HCO_3' formed.

For the purposes of the calculation we take $C^* = 1{\cdot}7 \times 10^{-5}$ g mole/ml, $D = 1{\cdot}2 \times 10^{-5}$ cm²/sec, $k' = 10^7$ ml/g mole sec. The total amount of HCO_3' formed is about 7·5 per cent of the total amount of CO_2 absorbed in the case of 0·045 M sodium hydroxide, and less for more concentrated solutions.

Notation

a = concentration of HCO_3'
A = defined by equation (10)
B = defined by equation (11)
C = (with various subscripts and superscripts) concentration of substances other than CO_2
C^* = saturated concentration of CO_2 in water or solution at 25°C, partial pressure 760 mm.
D = diffusivity of CO_2 in water
D', D_1, D_2 = diffusivity of solutes other than CO_2 in water
D^* = self-diffusivity of OH' ions in sodium hydroxide solutions
F = Faraday's number
H = sum of heat of solution and heat of reaction of CO_2 in sodium hydroxide solution
k = pseudo first-order reaction velocity constant
k' = second order reaction velocity constant
K_w = dissociation constant of water
K_2 = second dissociation constant of H_2CO_3
l = depth of liquid film ; length of exposed periphery of drum
n = no. of moles of solute reacting with one mole of CO_2
p = vapour-pressure of water ; parameter of Laplace transformation
P = total pressure
q = rate of absorption of CO_2
$Q\{t\}$ = amount of CO_2 absorbed per unit surface area in time t
R = gas constant ; rate of reaction per unit volume
S = defined by equation (11)
t = time of exposure
T = temperature
ΔT = temperature rise at surface
u_0 = ionic mobility of OH' ions at infinite dilution
V = velocity of drum surface.

v, v_1, v_2 = quantities defined by equations (23), (32), (36)
w = width of drum
x = depth below liquid surface
α = defined by equation (6); or thermal diffusivity of liquid
β = defined by equation (8)
Δ = fractional change in buffer ratio
μ = viscosity of liquid
μ_0 = viscosity of water
μ_n = coefficient in equation (31)
ρ = density of liquid
σ = specific heat of liquid
χ = concentration of CO_2 in liquid

References

[1] Danckwerts P. V. and Kennedy A. M. *Trans. Inst. Chem. Engrs.* 1954 **32** Supplement No. 1 p. S53.

[2] Cullen E. J. and Davidson J. F. *Trans. Faraday Soc.* 1957 **53** 113.

[3] Cullen E. J. and Davidson J. F. *Trans. Inst. Chem. Engrs.* 1957 **35** 51.

[4] Carslaw H. S. and Jaeger C. J. *Conduction of Heat in Solids* p.84. The Clarendon Press, Oxford 1948.

[5] Lynn S., Straatemeier J. R. and Kramers H. *Chem. Engng. Sci.* 1955 **4** 58.

[6] Seidell A. and Linke W. F. *Solubilities of Inorganic and Organic Compounds. Supplement* p.89. D. van Nostrand and Co., New York 1952.

[7] Danckwerts P. V. *Appl. Sci. Res.* 1952 **A3** 389.

[8] Markham A. E. and Kobe K. A. *J. Amer. Chem. Soc.* 1941 **63** 449.

[9] Wilke C. R. *Chem Engng. Progr.* 1950 **46** 258.

[10] Danckwerts P. V. *Trans. Faraday Soc.* 1950 **46** 300.

[11] Pinsent B. R. W., Pearson L. and Roughton F. J. W. *Trans. Faraday Soc.* 1956 **52** 1512.

[12] Nijsing R. A. T. O. and Kramers H. Absorption of CO_2 in carbonate bicarbonate buffer solutions in a wetted-wall column. *Chemical Reaction Engineering* p. 81. Permagon Press, London 1958.

[13] Van Krevelen D. W. and Hoftijzer P. J. *Chim. Ind.* XXIme *Congres Int. Chim. Ind.* 1948 p. 168.

[14] Danckwerts P. V. *Trans. Faraday Soc.* 1950 **46** 701.

[15] Perry H. and Pigford R. L. *Industr. Eng. Chem.* (*Industr.*) 1953 **45** 1247.

[16] Mills R. and Kennedy J. W. *J. Amer. Chem. Soc.* 1953 **75** 5696.

[17] Conway B. E. *Electrochemical Data* p. 145. Elsevier Pub. Co., Amsterdam 1952.

[18] Pinsent B. R. W. and Roughton F. J. W. *Trans. Faraday Soc.* 1951 **47** 263.

B4

Kinetics of CO_2 absorption in alkaline solutions—I

Transient absorption rates and catalysis by arsenite

D. ROBERTS† and P. V. DANCKWERTS‡

Imperial College of Science and Technology, London

(*Received* 20 *July* 1962)

Abstract—A method of eliminating the "stagnant film" end-effect on a wetted-wall column has been devised and appears to be satisfactory.

The transient rates of absorption of CO_2 into NaOH, $Na_2CO_3/NaHCO_3$ and Na_2SO_4 solutions have been measured for times up to one second. The results are to be used for comparison with absorption in packed columns, as will be reported in Part II. The kinetics of the various processes are discussed, particularly the marked influence of ionic strength on reaction rates.

The catalytic effect of arsenite ions on the reaction between CO_2 and water has been measured.

IN AN attempt to test the "surface-renewal" theory of gas-absorption, DANCKWERTS and KENNEDY [1] measured the transient rate of absorption of CO_2 into various solutions by means of a rotating drum which carried a film of liquid through the gas. These results were compared with measurements of the rate of absorption of CO_2 into the same solutions in a packed column. The comparison suggested that a decisive test of the theory required transient measurements over times extending up to 1 sec or more, rather than about 0·25 sec, which was the upper practical limit with the rotating drum. A wetted-wall column has accordingly been used to extend the measurements. This is a convenient apparatus for making transient measurements for times of exposure extending up to 1 sec or more, but precautions are necessary regarding entrance effects, ripples and the stagnant region which forms at the bottom of the column.

The primary purpose of the experiments reported in this paper was to obtain data for comparison with absorption rates in the packed column. However, points of interest relating to the kinetics are discussed where they arise, and the catalytic effect of arsenite ions has been investigated.

EXPERIMENTAL METHOD

Apparatus

The apparatus used is shown in Fig. 1. The falling film was formed on the outside of a 19 mm. O.D. stainless steel tube. The liquid passed up the inside of the tube and was distributed around its circumference by a cap fitted over the top of it. The liquid flowed from the inside of the tube, through twelve holes distributed around its circumference, into a small annular chamber, and from there it flowed through the annulus (0·04 cm wide) formed between the cap and the tube, finally falling freely down the outside of the tube. An air vent was provided in the top of the cap.

The falling film was removed from the column by means of a channelled polyethylene collar for the reasons given below. This collar is shown in Fig. 2. The stainless steel tube passed through the centre of dual steel base-plates and was gripped by a rubber ring held firmly between them. This arrangement allowed the tube to be moved through an angle of about 5° in any direction, and it was adjusted to the vertical position by a set of three centering screws acting at the lower end of the tube. The gas was stored in a balloon and the gas-absorption rate measured by a soap-film flow-meter. The solutions used were carefully de-aerated and the absorption apparatus was purged at frequent intervals. The liquid flow-rate was measured by a rotameter. The whole apparatus was kept in a room where the temperature was maintained at 25°C. The gas was saturated with water vapour at this temperature.

Entrance effect

The velocity of the newly-formed surface at the upper end of the column is zero, and the terminal velocity is approached asymptotically further down the column. An error is introduced by assuming that the surface has the terminal velocity at all points. This effect is well understood and can be rendered unimportant by suitable design. Using the equation given by WILKES and NEDDERMAN [2] we have calculated that in our case the error in the effective height of the column was less than 1 per cent for all heights and flow-rates.

Ripples

Ripples, which enhance the absorption rate, can be eliminated by adding surface-active agents. In this work we used Lubrol W (I.C.I.) which has been shown by NYSING and KRAMERS [3] not to give rise to a surface resistance to gas absorption.

Stagnant film

The most serious remaining cause of error is the rigid film of surface-active material which builds up on the surface at the bottom of the column. This tends to immobilize the surface here and thus reduces the local rate of absorption. NYSING and KRAMERS [3] made a correction for this effect,

† Present address: E. I. Du Pont de Nemours and Co. Inc., Eastern Laboratory, Gibbstown, New Jersey, U.S.A.
‡ Present address: Department of Chemical Engineering, Pembroke Street, Cambridge.

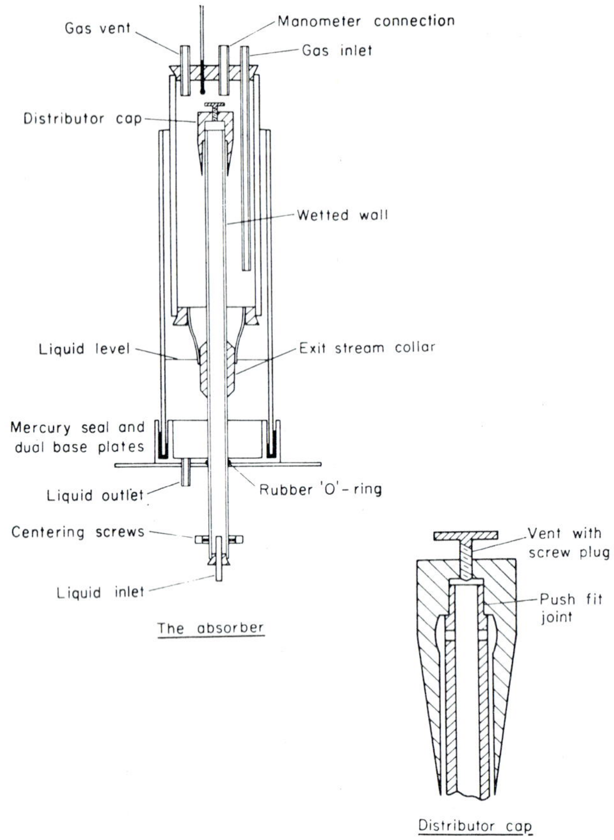

FIG. 1. Apparatus.

which amounted to assuming that the bottom 2 cm or so of the film absorbed no gas at all. However, we made some measurements of absorption rates with very short columns which showed that the rate of absorption into the region covered by the rigid film was not negligible, and that the height of the film was apt to vary considerably. Thus in our experiments the contribution of this region to the absorption rate varied in an unpredictable manner (particularly since a variety of types of reactant solution were used), and NYSING and KRAMER's method of correction could not be regarded as entirely satisfactory. We therefore tried to eliminate the effect as far as possible.

A collar made of polyethylene (Fig. 2) was fitted over the lower end of the tube, on which it was a sliding fit. Three vertical grooves, each about 2·5 mm wide, were cut in the outer surface of the collar. On reaching the top of the collar the liquid flowed away down the grooves, the rest of the surface remaining unwetted. The rigid film was confined to these grooves, which extended 1–2 cm above the standing liquid level. The area of the liquid surface exposed in the grooves was never more than about 2 per cent of the area of the falling film proper, and the fraction of the observed absorption rate which could be ascribed to this region was proportionately less because the specific absorption rate is always least at the bottom end of the column, and also because of the effect of the rigid film. Thus absorption into the liquid in the grooves could be ignored in interpreting the results.

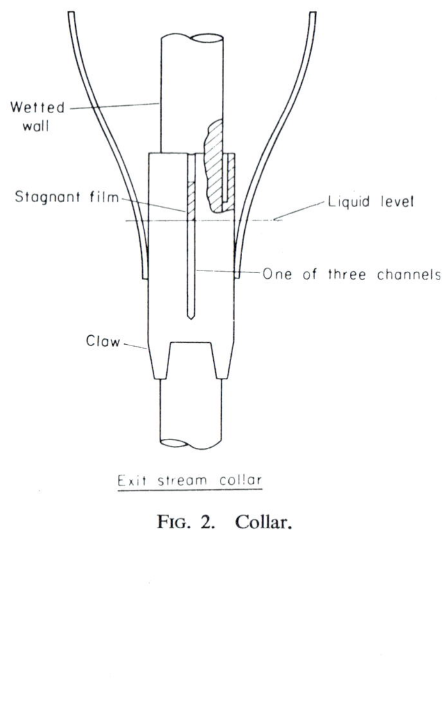

FIG. 2. Collar.

The effect of the grooves in withdrawing the rigid film from the bottom of the column proper is shown in Fig. 3. As the standing liquid level is reduced and a greater length of grooves is exposed, the rate of absorption increases owing to the progressive withdrawal of the rigid film into the grooves, freeing the surface of the falling film proper. When the liquid level is about 1 cm below the top of the grooves the whole film has been withdrawn into the grooves. Further increase in the height merely introduces a small additional area of "clean" surface in the grooves, the resulting increase in absorption rate being very small. The area of the film is much reduced when it is withdrawn into the grooves, both because its width is reduced and because the increase in liquid velocity reduces its height.

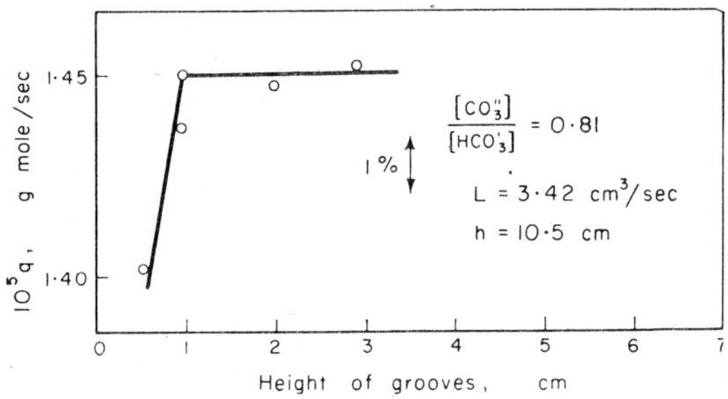

FIG. 3. Effect of collar in removing stagnant zone.

Fig. 4 shows as circles the results obtained with CO_2 and Lubrol solution when the exit collar was not in use. The effect of the rigid film clearly varies from one run to another. The crosses denote the results obtained with the exit collar in use and may be compared with the "theoretical" curve calculated using the value for the diffusivity found by DAVIDSON and CULLEN [4]. It appears that the measures taken to reduce errors from various sources have been satisfactory.

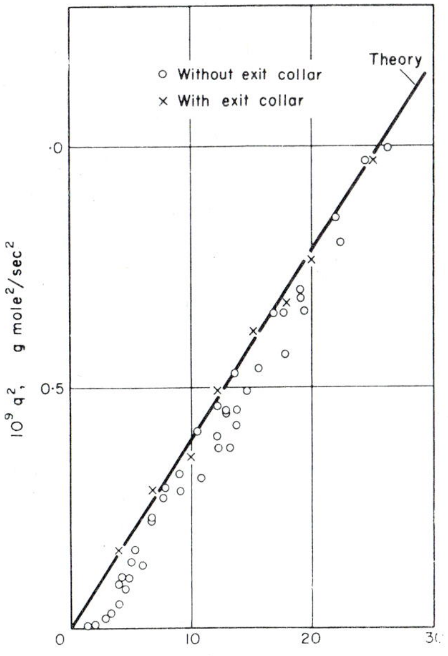

FIG. 4. Absorption of CO_2—effect of collar.

Solutions used

Solutions of sodium carbonate-bicarbonate buffers (some of which contained sodium arsenite), sodium hydroxide and sodium sulphate were used. In order that the compositions of the solutions should correspond as closely as possible to those used in the parallel series of experiments in a packed column (which will be discussed in Part II of this series), sodium sulphate and carbonate were added to the sodium hydroxide solutions. The solutions were analysed by the methods used by DANCKWERTS and KENNEDY [1].

RESULTS

The time of exposure of the liquid surface on the wetted-wall column is

$$t = \frac{2h}{3}\left(\frac{3\nu}{g}\right)^{1/3}\left(\frac{\pi d}{L}\right)^{2/3} \tag{1}$$

where h is the height of the column, ν the kinematic viscosity of the liquid, d the column diameter, L the volumetric liquid flow rate (the expression is accurate when the film thickness is much less than the column diameter, as in our work). If the total rate of absorption into the surface of the liquid on the column is q, the amount $Q(t)$ of gas absorbed by unit area of surface during an exposure time t can be calculated from

$$Q(t) = \frac{qt}{\pi hd} \tag{2}$$

The amount absorbed into unit area of surface of the falling film in a given time will be the same as the amount absorbed into a stationary liquid exposed for the same length of time provided the depth to which the absorbed gas penetrates by diffusion is small compared to the thickness of the film. Using the equation developed by JOHNSTONE and PIGFORD [5] it is calculated that the error, which is due to the finite thickness of the film and the velocity gradients in the liquid, was never more than 0·5 per cent in our work.

Na_2SO_4 *solutions*

The solubility C^* of CO_2 in these solutions was taken from SEIDELL and LINKE [6], and the diffusivity, D, of dissolved CO_2 calculated from

$$Q = 2C^*\sqrt{\left(\frac{Dt}{\pi}\right)}$$

where t, the time of exposure, is given by equation (1). The resulting values of D are plotted against the concentration of Na_2SO_4 in Fig. 5, and com-

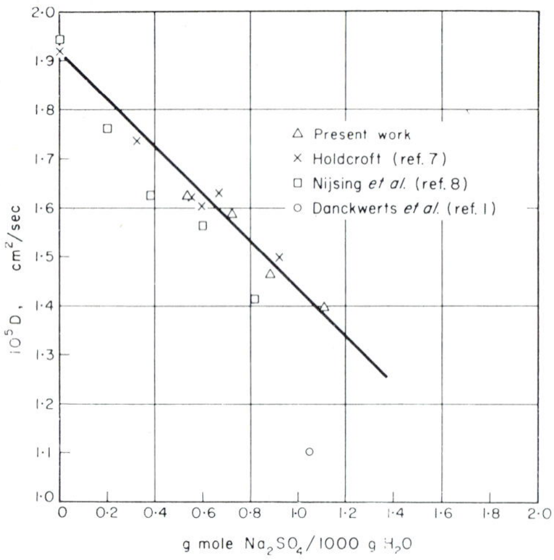

FIG. 5. Diffusivity of CO_2 in Na_2SO_4 solutions at 25°C.

pared with those of other workers. There is good agreement with the values found by HOLDCROFT [7], who used a method in which the liquid ran over a sphere and the stagnant film was eliminated (this method has been described by DAVIDSON and CULLEN [4]). The values reported by NYSING *et al.* [8] are by comparison low, and that of DANCKWERTS and KENNEDY very low.

Buffer solutions

The reactions which dissolved CO_2 may undergo in the buffer solutions are

$$CO_2 + OH' \rightarrow HCO_3' \quad (3)$$

$$CO_2 + H_2O \rightarrow HCO_3' + H^{\cdot} \quad (4)$$

(the latter being followed by very rapid reaction between $H^{\cdot}$ and OH'). Both reactions are first order in the concentration of CO_2 and will thus display pseudo first-order over-all kinetics in the buffer solutions, in which the concentration of OH' remains constant (except in the circumstances mentioned below). The second-order rate constant, $r_{OH'}$, of reaction (3) at 25°C is about 8500 l/g mole sec [9], while the pseudo first-order constant, r_{H_2O}, of (4) is 0·0257 sec^{-1} [9, 10] (both values refer to vanishingly small concentrations of electrolyte). Thus in the buffer solutions used, in whch the concentration of OH' was not less than about 10^{-4} g ion/l., reaction (4) did not make a substantial contribution to the rate of reaction of CO_2. (The catalysis of reaction (4) is discussed later.)

The equation for absorption accompanied by first-order reaction has been given by DANCKWERTS [11]:

$$Q = C^*\left(\frac{D}{r}\right)^{1/2}\left\{(rt + \tfrac{1}{2})\operatorname{erf}\sqrt{rt} + \exp(-rt)\sqrt{\frac{rt}{\pi}}\right\} \quad (5)$$

At values of rt greater than about 3, this becomes

$$Q \approx C^*\left(\frac{D}{r}\right)^{1/2}(rt + \tfrac{1}{2}) \quad (6)$$

The experimentally determined values of Q were fitted to equation (5) by the method of NYSING and KRAMERS [3], yielding the values of $C^*\sqrt{D}$ and r which are shown in Fig. 6 and Table 1.

The concentration of OH' (g ion/l.) in the buffer solutions is given by the expression

$$[OH'] = \frac{K_W'}{K_2'}\frac{[CO_3'']}{[HCO_3']} \quad (7)$$

where

$$K_2' = \frac{[H^{\cdot}][CO_3'']}{[HCO_3']} \quad (8)$$

and

$$K_W' = [H^{\cdot}][OH']$$

the quantities in square brackets being concentrations. The values of the "stoichiometric" dissociation constants K_2' and K_W' at any temperature

Table 1.
25°C. (Concentrations in g ion/l.)

$[CO_3'']$	$[HCO_3']$	I	$\frac{[CO_3'']}{[HCO_3']}$	$10^8 C^*\sqrt{D}$ $\left(\frac{\text{g mole}}{\text{cm}^2\,\text{sec}^{1/2}}\right)$	r (sec^{-1})	$r\frac{[HCO_3']}{[CO_3'']}$ (sec^{-1})
0·531	0·765	2·36	0·694	7·3	0·70	1·01
0·625	0·530	2·40	1·18	7·3	1·01	0·86
0·716	0·412	2·56	1·74	7·1	1·60	0·92
0·740	0·295	2·50	2·51	7·4	2·02	0·80
0·760	0·214	2·49	3·55	7·6	2·50	0·70
0·193	0·482	1·16	0·40	10·05	0·35	0·88
0·207	0·215	0·94	0·96	10·75	0·75	0·78

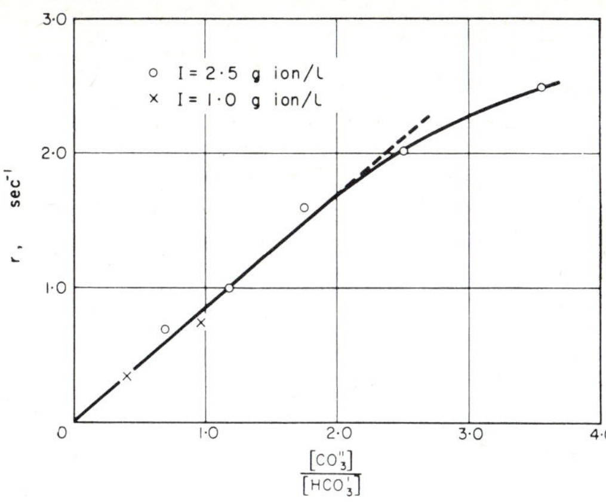

FIG. 6. Reaction rate constant in carbonate–bicarbonate buffer solutions at 25°C.

depend on the ionic strength of the solution and the nature of the ions present. These quantities have not, so far as we know, been measured in carbonate-bicarbonate buffers. We shall assume for the moment that they depend only on the ionic strength. The total pseudo first-order velocity constant may then be written

$$r \approx r_{H_2O} + r_{OH'} \frac{K'_W}{K'_2} \frac{[CO_3'']}{[HCO_3']} \tag{9}$$

and thus a plot of r against $[CO_3'']/[HCO_3']$ in solutions of constant ionic strength might be expected to yield a straight line of slope $r_{OH'}K'_W/K'_2$. It can be seen from Fig. 6 that such a relationship is obeyed within the limits of error by the less alkalin. solutions, but that there is a falling-off in the reported values of r in the most alkaline solutione Using the method of DANCKWERTS and KENNEDY [1], it can be calculated that significant depletion of the buffer took place near the surface of the absorbing liquid at higher exposure times in this solution, and may have reached 17 per cent. It is not possible to calculate the effect of this on the estimated value of r, but it seems likely that the falling-off in the more alkaline solutions is due to this depletion. For the primary purpose of the measurements, which was the comparison with the packed-tower results, the effect was of no importance.

For the less alkaline solutions, the value of $r_{OH'}K'_W/K'_2$ is found from the slope of the curve in Fig. 6 to be 0·86 sec^{-1}. At infinite dilution (where $K'_W = 1{\cdot}01 \times 10^{-14}$ (g ion/l.)^2, $K'_2 = 4{\cdot}69 \times 10^{-11}$ g ion/l. [12], $r_{OH'} = 8500$ l/g mole sec [9] $r_{OH'}K'_W/K'_2$ is 1·83 sec^{-1}. Thus, for a given buffer ratio, the reaction of CO_2 was less than half as fast in our solutions as at infinite dilution. The results of NYSING and KRAMERS [3] yield the following values of $r_{OH'}K'_W/K'_2$ in CO_3''/HCO_3' buffers of ionic strengths 1·2–2·1 g ion/l.; sodium buffer at 25°C, 0·80 (compared to 0·86 in our experiments); sodium buffer at 20°C, 0·55; potassium buffer at 20°C, 0·73 sec^{-1}, whereas the value at infinite dilution at 20°C is 0·95. It will be seen that quite large errors arise if the values of $r_{OH'}$, K'_W and K'_2 appropriate to infinite dilution are used to calculate the rates of reaction in solutions in the range of concentration encountered by industry; also that the rate depends markedly on the species of ion present as well as on the ionic strength. PINSENT *et al.* [9] report the variation of $r_{OH'}$ with ionic strength at 20°C in various salt solutions. The results of NYSING *et al.* [8] in NaOH and KOH, which are less accurate, show the same general behaviour (Fig. 7). NÄSÄNEN [13] gives the following formula for the effect of ionic strength on K'_W/K'_2 at 25°C in solutions of KCl and NaCl with ionic strengths up to 2 g ion/l:

$$\log_{10} \frac{K'_W}{K'_2} = -3{\cdot}677 - \frac{1{\cdot}01\sqrt{I}}{1 + \alpha\sqrt{I}} - \beta I \tag{10}$$

where $\alpha = 1{\cdot}49$, $\beta = 0{\cdot}061$ for KCl, and $\alpha = 1{\cdot}27$, $\beta = 0{\cdot}125$ for NaCl. We have assumed the *relative* variation to be the same at 20°C, and have extrapolated to 3 g ion/l. We have combined the resulting values of K'_W/K'_2 with the values of $r_{OH'}$ at 20°C given by PINSENT *et al.* [9] to estimate the value in KCl and NaCl solutions at 20°C of the kinetic factor $r_{OH'}K'_W/K'_2$. The results are shown in Fig. 8, and compared with those of NYSING and KRAMERS.

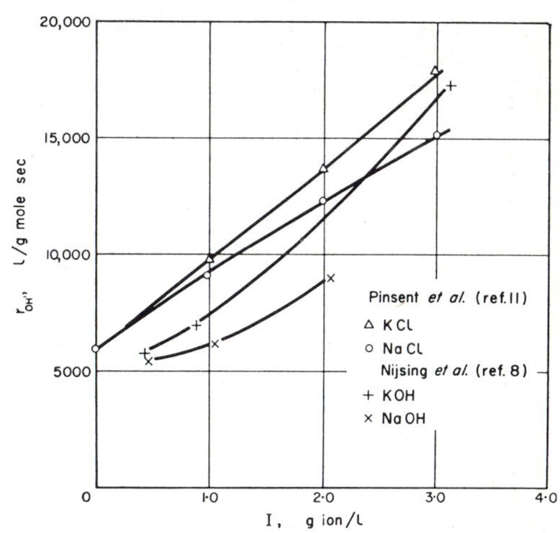

FIG. 7. Effect of ionic strength on rate of reaction of CO_2 with OH′ at 20°C.

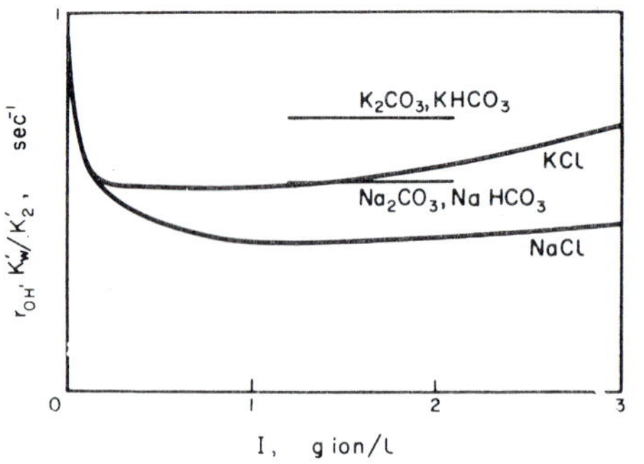

FIG. 8. Effect of ionic strength on rate of reaction in carbonate–bicarbonate buffer solutions at 20°C.

It is clear that the nature of the anions as well as of the cation can make a considerable difference to the rate of reaction. In general, the rate is greater in potassium solutions because both K_W'/K_2' and $r_{OH'}$ are greater, and there is little variation in the range 1–2 g ion/l.

Reaction (4) is probably catalysed by carbonate ions, although the catalytic power is known not to be large [14]. The effect might be appreciable in the more alkaline buffer solutions used, but is likely to have been masked by the depletion which occurred in these solutions.

Sodium hydroxide solutions

In this case the dissolved CO_2 undergoes what is effectively an irreversible second-order reaction with hydroxyl ions. The absorption rate arising from a process of this kind has been the subject of discussion elsewhere [1, 8] and recently BRIAN *et al.* [15] have presented numerical solutions for certain ranges of the parameters. As t increases, $Q(t)$ becomes linear in $\sqrt{t}$:

$$Q(t) \rightarrow S\sqrt{t} - B \qquad (11)$$

The experimental values of S and B are plotted in Fig. 9. The results obtained by DANCKWERTS and KENNEDY [1] are included for comparison. The subsequent discussion relates only to the present results. The viscosities of all the solutions were about the same, and the ionic strengths lay between 2·3 and 3·1 g ion/l. The theoretical expression (1) for S shows that if D' were constant, S should be linear in C_0, the concentration of NaOH, for values of the latter greater than about 10^{-5} g mole/cm^3, becoming equal to $C_0\sqrt{(D'/\pi)}$, where D' is the effective diffusivity of OH' in the solution. It can be seen from Fig. 9 that the curve appears to be linear with a slope of about $2{\cdot}7 \times 10^{-3}$ cm/sec$^{1/2}$, which

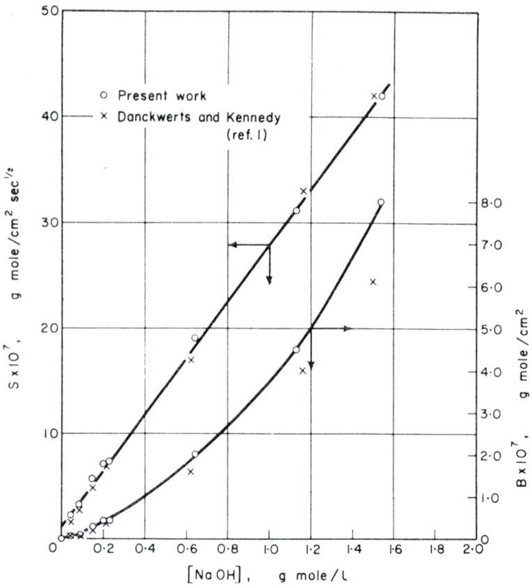

FIG. 9. Quantities S and B for rate of absorption of CO_2 into NaOH solutions.

indicates a value of $2{\cdot}3 \times 10^{-5}$ cm^2/sec for D'. The value calculated from the results of DANCKWERTS and KENNEDY [1] was $2{\cdot}5 \times 10^{-5}$ cm^2/sec.

For the calculations in Part II it is necessary to know the initial rate of absorption, K, into the hydroxide solutions, although not with great accuracy. The rate can be estimated more reliably from the results of KENNEDY [16] with the rotating drum that from those obtained with the falling film; the discrepancy between the two sets of results is not large enough to be important in this context. The values of K obtained from KENNEDY'S results are shown in Fig. 10. It can be seen from equation (6) that the initial rate of absorption should be $C^*\sqrt{(Dr_{OH'}[OH'])}$; Fig. 10 shows that a relationship of this kind is not very closely obeyed by the experimental results, perhaps owing to systematic error.

A rough calculation indicated that the temperature rise associated with the absorption of CO_2 would not have been great enough to affect the absorption rate. (For the purpose of comparison with absorption in packed towers, the temperature effect is immaterial, as it will occur to about the same extent in the packed tower also.)

Catalysis by arsenite ions

ROUGHTON [17] pointed out in 1943 that the reaction between CO_2 and water is catalysed by arsenite ions, among other bases. The fact is the

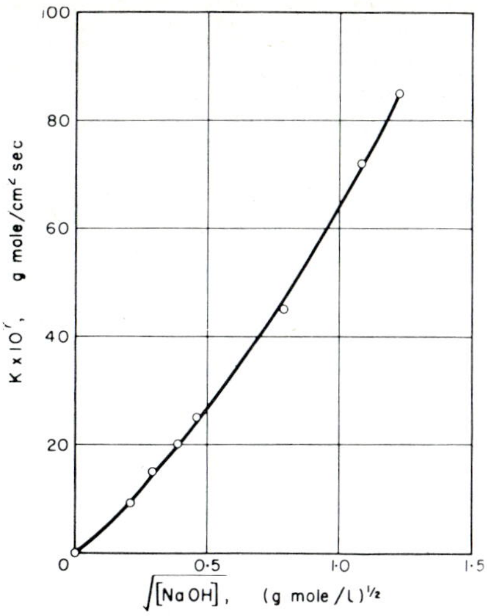

FIG. 10. Initial rates of absorption into NaOH solutions.

basis of one form of a process patented by S. P. A. VETROCOKE [18], in which CO_2 is absorbed in a solution containing potassium carbonate and potassium arsenite. In the absence of arsenite the CO_2 reacts predominantly with the hydroxyl ions present, which are in low concentration (e.g. 10^{-4} g ion/l.) in carbonate–bicarbonate mixtures so that the reaction is relatively slow, although faster than

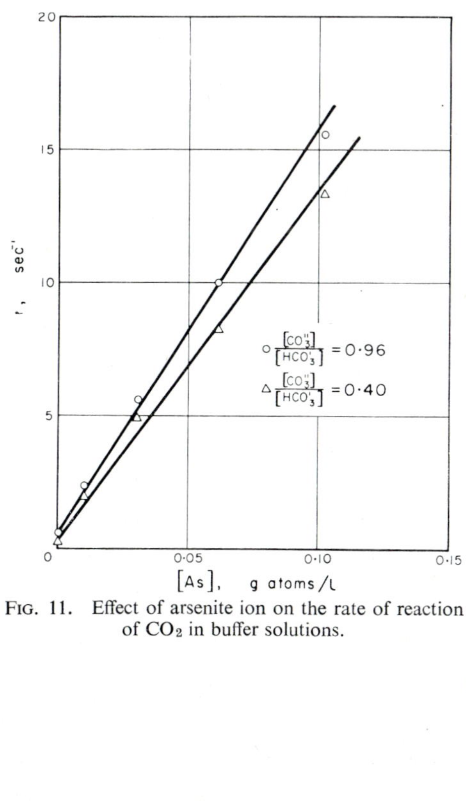

FIG. 11. Effect of arsenite ion on the rate of reaction of CO_2 in buffer solutions.

the reaction with water. The arsenite ions catalyse the latter, and thus accelerate the reaction of CO_2 and hence the processes of absorption and desorption.

To obtain a quantitative measure of the catalytic effect, we absorbed CO_2 into carbonate–bicarbonate buffer solutions with sodium arsenite added, and determined the apparent rate constants for the pseudo first-order reaction of CO_2. The ionic strength in these solutions was about 1 g ion/l. The results are shown in Fig. 11, which shows the increase in rate to be proportional to the concentration of arsenic added.

Assuming the increase in rate to be proportional to the concentration, [A′], of arsenite ions,

$$r = r_{OH'}[OH'] + r_{H_2O}\{1 + l_A[A']\} \tag{12}$$

$$= r_{OH'}[OH'] + r_{H_2O} + r_A[A'] \tag{13}$$

$$= r_{OH'}[OH'] + r_{H_2O} + \frac{r\,C}{1 + (K_1'[HCO_3']/K_2'[CO_3''])} \tag{14}$$

where C is the total concentration of arsenic and K_1' the first (stoichiometric) dissociation constant of arsenious acid. There is the usual difficulty about evaluating K_1'/K_2' in the solutions used. NÄSÄNEN [13] gives K_2', and ANTIKAINEN and ROSSI [19] K_1', as functions of ionic strength in KCl solutions at 25°C, but it is not certain that the value of K_1'/K_2' in 1M KCl is any closer to the value in the buffer solutions used than is the value at infinite dilution, which is 10·6. To make equation (14) fit the experimental results exactly the value would have to be 11. For present purposes we shall use the value 10·6; then the arsenite would be 81 per cent ionized when the buffer ratio $[CO_3''^-]/[HCO_3']$ is 0·40 and 92 per cent ionized when it is 0·96. It can be calculated from Fig. 10 that r_A is about 164 l/g mole sec, and since r_{H_2O} is 0·0257 sec^{-1} at 25°C [9, 10] the "multiplying factor" l_A is 6400 l/g mole.

Thus the addition of 1M arsenic (about 10 per cent by weight As_2O_3) to an equimolal CO_3''/HCO_3' mixture would increase the rate of reaction of CO_2 from about 1·8 to about 140 sec^{-1} at 25°C. (The temperatures used in the Vetrocoke process are much higher.)

Considerable depletion of the buffer solutions must have occurred near the surface of the absorbing liquid, but this will introduce much less error than in the case of the uncatalysed reaction. The resultant fall in the concentration of hydroxyl ions makes little direct difference to the total rate. The effect on the degree of ionization of arsenious acid to arsenite is more serious, but a fall in the buffer ratio from 0·96 to 0·40 (for example) would cause a reduction in the concentration of arsenite ions of only 11 per cent. Experimental plots of Q vs. t remained linear at the higher exposure times, as required by equation (6), and showed no signs of falling-off; this also suggests that depletion was unimportant. The reported values of the total rate constant are therefore probably not far wrong.

Notation

Symbol	Meaning	Units
B	Constant in equation (9)	
C_0	Concentration of NaOH	g mole/cm^3
C^*	Saturated concentration of CO_2 (Chemical symbols in square brackets denote concentrations in g mole/l.)	g mole/cm^3
d	Diameter of wetted column	cm
D	Diffusivity of dissolved CO_2	cm^2/sec.
D'	Effective diffusivity of hydroxyl ions	cm^2/sec
g	Gravitational acceleration	cm/sec^2
h	Height of wetted column	cm
I	Ionic strength	g ion/l.
K	Initial rate of absorption of CO_2 into NaOH	g mole/cm^2 sec.
K'_w	Stoichiometric dissociation "constant" of water	(g ion/l.)^2
K_1'	Stoichiometric first dissociation "constant" of arsenious acid	g ion/l.
K_2'	Stoichiometric second dissociation "constant" of carbonic acid	g ion/l
L	Liquid flow rate	cm^3/sec.
l_A	Catalytic multiplying factor of arsenite ions	l/g mole
q	Rate of absorption on column	g mole/sec
Q, $Q(t)$	Amount of gas absorbed per unit area in time t,	g mole/cm^2
r	Total first-order reaction-rate constant	sec^{-1}
r_A	Catalytic constant for arsenite ions	l/g mole sec
r_{OH}'	Rate constant for reaction of CO_2 with OH'	l/g mole sec
r_{H_2O}	Rate constant for reaction of CO_2 with H_2O	sec^{-1}
S	Constant in equation (9)	g mole/cm^{2}sec$^{\frac{1}{2}}$
t	Time of exposure	sec
ν	Kinematic viscosity	cm^2/sec

$$\text{erf}\, x = \frac{2}{\sqrt{\pi}} \int_0^x \exp(-y^2)\, dy$$

References

[1] Danckwerts P. V. and Kennedy A. M., *Chem. Engng. Sci.* 1958 **8** 201.
[2] Wilkes J. O. and Nedderman R. M., *Chem. Engng. Sci.* 1962 **17** 177.
[3] Nysing R. A. T. O. and Kramers H., *Chem. Engng. Sci.* 1958 **8** 81.
[4] Davidson J. F. and Cullen E. J., *Trans. Inst. Chem. Engs.* 1957 **35** 51.
[5] Johnstone H. F. and Pigford R. L., *Trans. Amer. Inst. Chem. Engrs.* 1941 **37** 5.
[6] Seidell A. and Linke W. F., *Solubilities of Inorganic and Organic Compounds* (3rd Ed. and Supplement). Van Nostrand, New York 1952.
[7] Holdcroft G. Ph.D. Thesis, Cambridge University 1959.
[8] Nysing R. A. T. O., Hensriksz R. H. and Kramers H., *Chem. Engng. Sci.* 1959 **10** 88.
[9] Pinsent B. R. W., Pearson L. and Roughton F. J., *Trans. Faraday Soc.* 1956 **52** 1512.
[10] Pinsent B. R. W. and Roughton F. J. W., *Trans. Faraday Soc.* 1951 **47** 263.
[11] Danckwerts P. V., *Trans. Faraday Soc.* 1950 **46** 300.
[12] Harned H. S. and Owen B. B., *The Physical Chemistry of Electrolytic Solutions* (3rd Ed.). Reinhold, New York 1958.
[13] Näsänen R., *Suom. Kem.* 1946 B **19** 90.
[14] Roughton F. J. W. and Booth V. H., *Biochem. J.* 1938 **42** 2049.
[15] Brian P. L. T., Hurley J. F. and Hasseltine E. H., *Amer. Inst. Chem. Engrs. J.* 1961 **7** 22.
[16] Kennedy A. M. Ph.D. Thesis, Cambridge University 1954.
[17] Roughton F. J. W., *Harvey Lect.* 1943 **39** 96.
[18] *Brit. Pat.* 786669; 791150; 798856; Mullowney J. F., *Petrol. Refin.* 1957 **36** 149; Riesenfeld F. C. and Mullowney J. F., *Petrol. Refin.* 1959 **38** 161; Guntermann V. W., *Gas-u. Wasserfach (Gas)* 1959 **100** 1225; Kohl A. L. and Riesenfeld F. C., *Chem. Engng.* 1959 **66** 127.
[19] Antikainen P. J. and Rossi V. M. K., *Suom. Kem.* 1959 B **32** 185.

Résumé—On a conçu une méthode qui s'est avérée satisfaisante pour éliminer l'effet terminal du film stagnant dans une colonne à paroi mouillée.

Les vitesses d'absorption de CO_2 en des solutions de NaOH, $Na_2CO_3/NaHCO_3$ et de Na_2SO_4 ont été mesurées par des expériences dans lesquelles les temps de contact étaient inférieurs à une seconde.

Dans une suite à cet article ces résultats seront comparés à ceux concernants l'absorption dans une colonne à garnissage. On discute la cinétique des différents processus et plus particulièrement l'influence, très nette, de la force ionique. On a mesuré également l'effet catalytique des ions d'arsénite sur la réaction entre le CO_2 et l'eau.

Zusammenfassung—Es wurde eine Methode zur Verhinderung des "stagnierenden Film"-Endeffektes in einer Dünnschicht-Absorptions-kolonne entwickelt, die befriedigend zu sein scheint.

Die Anfangs-Absorptionsgeschwindigkeiten (oder kurzzeitigen Absorptionsgeschwindigkeiten) von CO_2 in NaOH-, $Na_2CO_3/NaHCO_3$- und Na_2SO_4-Lösungen wurden bis zu Zeiten von einer Sekunde gemessen. Die Resultate können zum Vergleich mit Absorptionen in gepackten Kolonnen herangezogen werden, worüber in einem II. Teil noch berichtet werden soll. Die Kinetik verschiedener Prozesse wurde diskutiert, wobei vor allem dem markanten Einfluss der Ionenstärke auf die Absorptionsgeschwindigkeit Beachtung geschenkt wurde.

Der katalytische Einfluss von Arsenit-Ionen auf die Reaktion zwischen CO_2 und Wasser wurde gemessen.

B5

Kinetics of CO_2 absorption in alkaline solutions—II

Absorption in a packed column and tests of surface-renewal models

P. V. DANCKWERTS, A. M. KENNEDY† and D. ROBERTS‡

Department of Chemical Engineering, Pembroke Street, Cambridge

(*Received* 15 *October* 1962)

Abstract—In Part I transient rates of absorption of CO_2 into NaOH, $Na_2CO_3/NaHCO_3$ and Na_2SO_4 solutions were measured. These results are here compared with the rates of absorption of CO_2 into similar solutions in a column packed with ½ in. Raschig rings. It is found that the Higbie and Danckwerts surface-renewal models enable these rates to be predicted to within about 10 per cent. The Delta model (all surface has same age) is less accurate.

INTRODUCTION

Models of the absorption mechanism

IN ITS general form, as applied to a packed absorption column, the surface-renewal theory can be stated thus. The surface of the liquid in a small but representative section of the packing is comprised of a mosaic of elements which have been exposed to the gas for different lengths of time, and which are themselves replaced in due course by fresh liquid having the local average bulk composition. The rate of absorption of gas per unit area into an element of surface which has been exposed for a length of time t is $\phi(t)$ (which depends on the nature of the gas and the local composition of the liquid, and in general decreases as t increases). The fraction of the surface which has been exposed for (has "ages" equal to) times between t, $t + dt$ is $\psi(t)dt$. The average rate of absorption per unit area of wetted surface in the section of packing is then

$$R = \int_0^\infty \phi(t)\psi(t)\, dt \tag{1}$$

or, if $Q(t)$ is the amount absorbed per unit area in time t,

$$R = \int_0^\infty \psi(t)\, dQ(t) \tag{2}$$

The "HIGBIE" model [1] assumes that all elements of surface are replaced after the same time of exposure, θ, giving

$$\begin{aligned} \psi(t) &= \frac{1}{\theta}, \qquad 0 < t < \theta \\ \psi(t) &= 0, \qquad t > 0 \end{aligned} \tag{3}$$

R is then simply the average rate of absorption over time θ, namely $Q(\theta)/\theta$. The "DANCKWERTS" model [2] assumes that the chance of an element of surface being replaced is independent of its age, giving

$$\psi(t) = s\exp(-st) \tag{4}$$

where s is the fractional rate of surface-renewal.

DAVIDSON [3] has suggested models leading to two other possible forms for $\psi(t)$. He considers liquid running in a film over surfaces inclined at different angles to the horizontal, each angle having an equal probability of occurrence. The liquid is in laminar flow until it meets a discontinuity, where it is completely mixed. In the first model every surface has the same length, d, in the direction of flow. In the second there is an equal probability of all lengths up to some maximum value, d. The length d is identified with the principle dimension of the packing particle, without regard to its geometry. Experimental results with Raschig rings and Berl saddles, over quite a wide range of sizes and flow-rates, agree fairly well with DAVIDSON'S second model, which seems the more plausible of the two. The observed values of the absorption rate tend to be lower than those predicted at low flow-rates, perhaps owing to incomplete surface-renewal at discontinuities, and higher at high flow-rates, perhaps owing to ripples or turbulence in the film.

The surface-age distributions for the HIGBIE, DANCKWERTS and DAVIDSON (random lengths and angles) models are compared in Fig. 1. (The ordinates have been put into dimensionless form, so that the comparison can be made independent of the absolute rates of surface renewal; k_L is the coefficient for physical absorption of a gas which has a diffu-

† Present address: Department of Chemistry, University of Otago, Dunedin, New Zealand.

‡ Present address: E. I. Du Pont de Nemours and Co. Inc., Eastern Laboratory, Gibbstown, New Jersey, U.S.A.

sivity D in solution.) The Fig. also shows the distribution for the Delta model, discussed later.

In all forms of the surface-renewal model the quantity k_L^2/D, which has the dimensions of reciprocal time, is a measure of the fractional rate of renewal of the surface or the reciprocal of its average age. It is a quantity which is determined by the hydrodynamics of the absorption system and is not dependent on its physico-chemical properties.

DANCKWERTS [4] and ANDREW [5] have discussed the possibility of "surface rejuvenation", whereby liquid underlying the surface below a certain depth may be thoroughly mixed when the liquid flows over a discontinuity, leaving a layer on the surface which flows on undisturbed. A model of a similar type has been proposed by TOOR and MARCHELLO [6].

Finally, there is the familiar Whitman film model, which may be regarded as a limiting case of surface rejuvenation.

It is clear that the actual behaviour of the liquid running over a packing is likely to be very complicated. Experimental evidence can be adduced [7, 8] both for mixing and for rejuvenation at discontinuities, as well as for enhancement of the absorption rate on smooth surfaces by ripples. Surface-active agents may have a considerable effect [8]. The behaviour depends on the circumstances. We regard it as unprofitable to construct over-complicated models with many parameters, and prefer to see whether a surface-renewal model characterized by two parameters—the rate of renewal (or k_L^2/D) and the effective surface area—can be used to make predictions of sufficient accuracy for design purposes. The practical advantages of using a surface-renewal rather than the film model are set out in the Discussion.

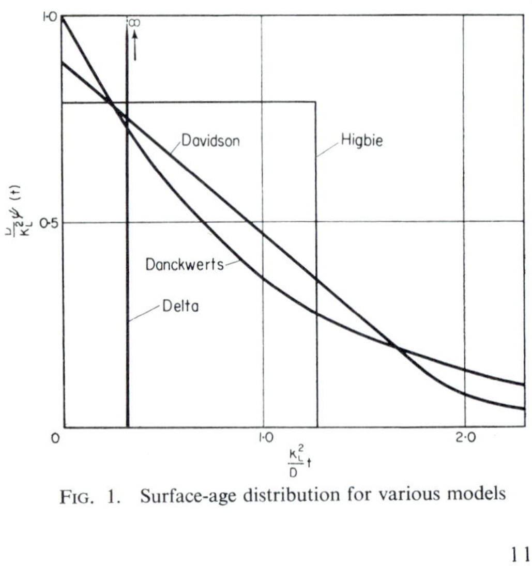

FIG. 1. Surface-age distribution for various models

Experimental tests

DANCKWERTS and KENNEDY [9] measured transient absorption rates of CO_2 with a rotating drum, and compared them with absorption rates in a packed column to see whether the surface-renewal models could be used for purposes of prediction. The tentative conclusion was that the surface-renewal models could not account for the effects of chemical reaction on rates of absorption in a packed column. However, further work by KENNEDY [10] suggested that a conclusive test would require measurement of the transient rates up to exposure-times of about 1 sec, as against about 0·25 sec which was the maximum which could be attained with the drum. The work reported by ROBERTS and DANCKWERTS in Part I of this series [11] extended the transient measurements with CO_2 up to 1·3 sec, using a wetted-wall column and solutions covering the same range of compositions as those used by KENNEDY in his packed column. In this paper we compare these measurements with the rates of absorption of CO_2 in a packed column measured by KENNEDY [10].

EXPERIMENTAL

All experiments were carried out at 25°C.

The packed column

The technique used for measuring absorption rates in the packed column and eliminating end-effects has been described by DANCKWERTS and KENNEDY [9]. The column was packed with $\frac{1}{2}$-in. (12·7 mm) porcelain Raschig rings. The internal diameter of the column was 4 in. (10·2 cm) and the liquid flow-rate 11·8 cm^3/sec (1100–1200 lb/hr ft^2). The volume of liquid held up was 7 per cent of the packed volume of the column.

The solutions

The compositions were similar to those of the solutions dealt with in Part I. Carbonate–bicarbonate buffers, sodium hydroxide and inert solutions of sodium sulphate were used. To ensure that the hydrodynamic conditions in the packed column were the same with each solution, sodium sulphate was added to the NaOH solutions to control density and viscosity. The viscosities of all solutions used were 1·57 times that of water, while their densities ranged from 1·08 to 1·12 g/cm^3. To economize in the use of reagents, solutions were sometimes re-used after they had absorbed some CO_2. In such cases they were re-analysed before re-use. It was considered that the presence of

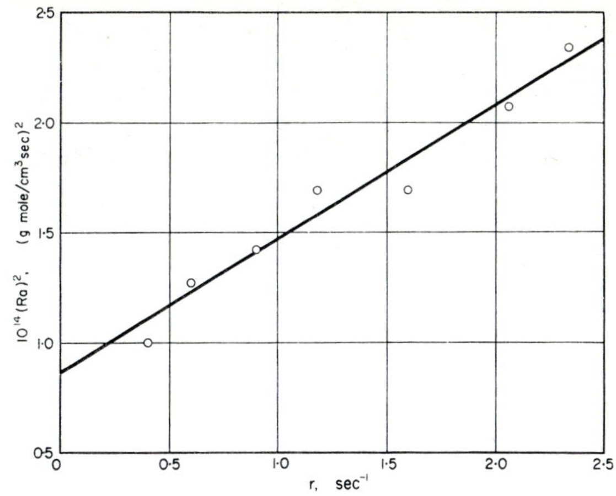

FIG. 2. Absorption into buffer-solutions in packed column with first-order reaction.

sodium carbonate in the sodium hydroxide solutions would not affect the reaction rate, which would be determined only by the residual hydroxide concentration.

RESULTS

Water

Trial runs with water, at a flow-rate of 11·8 cm^3/sec, gave a value for $k_L a$ of 6·4 × 10^{-3} sec^{-1}. The value predicted from the correlation given by SHERWOOD and HOLLOWAY [12] is 7·6 × 10^{-3} sec^{-1}, but in their experiments the packing-density of ½-in. Raschig rings was 420,000 pieces per m^3, whereas in our experiments it was 370,000. If $k_L a$ were proportional, other things being equal, to the packing density, a value of 6·65 × 10^{-3} would be predicted for our experiments by appropriate correction of the correlation.

1·05 M Na_2SO_4 *solution*

Five separate determinations gave values of the volumetric absorption rate, Ra, ranging from 7·50 × 10^{-8} to 7·95 × 10^{-8} g mole/cm^3 sec, the mean value being 7·60 × 10^{-8}; the corresponding value of $k_L a$ was 4·55 × 10^{-3} sec^{-1}. The difference between this and the value of $k_L a$ for water is due to the higher viscosity of the Na_2SO_4 solution and the lower diffusivity of dissolved CO_2.

Buffer solutions

The transient absorption into these solutions followed the course to be expected for first-order kinetics, the apparent rate-constant, r, for the reaction of CO_2 being related to the composition of the solutions as shown in Fig. 6, Part I. According to the DANCKWERTS model (2),

$$\mathrm{Ra} = ac^*\sqrt{[D(r + s)]} \qquad (5)$$

Thus a plot of $(\mathrm{Ra})^2$ against r should give a straight line of slope $(ac^*)^2 D$ and intercept $(ac^*)^2 Ds$. The ratio of the intercept to the slope yields s, and a can be calculated from the value 0·73 × 10^{-7} g mole/cm^2 sec$^{1/2}$ reported for $C^*\sqrt{D}$ for these solutions in Part I. The plot is shown in Fig. 2, together with a straight line which gives s = 1·45 sec^{-1}, a = 1·06 cm^{-1}. These values may be used to predict the value of $k_L a$ in 1·05 M Na_2SO_4 solution. According to the DANCKWERTS model, $k_L = \sqrt{(Ds)}$; the value of D in this solution is estimated from Fig. 5, Part I, to be 1·42 × 10^{-5} cm^2/sec. The predicted value of $k_L a$ or $a\sqrt{(Ds)}$ is then 4·8 × 10^{-3} sec^{-1}, as against the mean experimental value of 4·55 × 10^{-3}.

The results of the experiments on buffer solutions, taken together with those on an inert solution, therefore appear to be compatible with the DANCKWERTS surface-renewal model and hence, by implication, with the DAVIDSON and HIGBIE models.

NaOH *solutions*

The results are shown in Fig. 3 and the relationship between Ra and the concentration of NaOH can be approximated by a straight line of slope about 2·5 × 10^{-3} sec^{-1}. Both the DANCKWERTS and HIGBIE surface-renewal models predict that when [OH′] is substantially greater than C^* (i.e. about 0·01 M),

$$\mathrm{Ra} = k_L a\left[C^*\sqrt{\left(\frac{D}{D'}\right)} + \tfrac{1}{2}[\mathrm{OH}']\sqrt{\left(\frac{D'}{D}\right)}\right] \qquad (6)$$

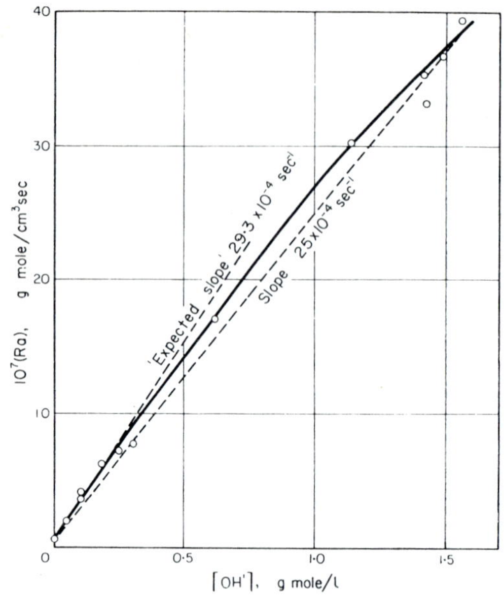

FIG. 3. Absorption into NaOH solutions in packed column.

provided that the reaction between CO_2 and OH′ is effectively instantaneous (here D' is the effective diffusivity of OH′). Using the value $4{\cdot}55 \times 10^{-3}$ sec^{-1} for $k_L a$, the apparent value of the ratio D'/D corresponding to a slope of $2{\cdot}5 \times 10^{-3}$ sec^{-1} would be only 1·1. This is an unlikely finding and contradicts previous findings [13] which indicate that the ratio is about 1·7. The measurements of transient absorption in Part I show that in fact the reaction in these solutions departs significantly from the instantaneous, in the sense required by equation (6). The transient rate of absorption at the higher values of t tends to

$$Q = S\sqrt{t} - B \tag{7}$$

both B and S increasing with NaOH concentration. If we adopt the nomenclature of the HIGBIE model, we can say that the reaction will only be effectively "instantaneous" in the absorption column experiment if $Q \simeq S\sqrt{t}$ when $t = \theta$, the exposure time in the HIGBIE model. This condition may be written $S\sqrt{\theta} \gg B$. As will be shown later, θ is apparently about 1·6 sec. An inspection of Fig. 9, Part I, shows that the above condition is not fulfilled in the case of the more concentrated NaOH solutions. It may be concluded that under our experimental conditions the kinetics of the absorption into NaOH solutions corresponded not to "infinitely fast reaction", but to a second-order reaction of finite speed. No simple analytical expression for the dependence of Ra on [OH′] is available in these circumstances.

The upper dotted line in Fig. 3 has the expected slope for an instantaneous reaction with $D'/D = 1{\cdot}7$; the results with the more dilute solutions seem to conform to this, as would be expected. The results with the more concentrated solutions are best fitted by the curve (full line).

COMPARISON OF TRANSIENT AND PACKED-COLUMN MEASUREMENTS

For the DANCKWERTS model

$$\boldsymbol{R} = s\int_0^\infty \exp(-st)\, dQ(t) \tag{8}$$

and for the HIGBIE model

$$R = Q(\theta)/\theta \tag{9}$$

In addition to the HIGBIE and DANCKWERTS models, we have tested a third, in which all the surface has the same age, τ. Since the surface-age distribution in this model is an impulse or delta function, we have called this the "Delta" model. It leads to

$$R = \left(\frac{dQ}{dt}\right)_{t=\tau} \tag{10}$$

Each model was tested by seeing whether it was possible to choose values for the parameter a and for s, θ or τ which were compatible with the transient and packed-column results for all the solutions. A guide to the best values for these parameters can be found by the following method, illustrated by reference to the DANCKWERTS model. For a solution of a particular composition, R was calculated for a number of values of s by the methods described in the next section, using smoothed figures for the transient data taken from the full curves in Figs. 6, 9 and 10, Part I. The values of Ra in the packed column for solutions of the same compositions were read from the smoothed (full) curves in Figs. 2 and 3 of this paper. For each chosen value, s', of s the corresponding value, a', of a was calculated from

$$a' = (\text{Ra})/R' \tag{11}$$

R' being the value of R calculated for s'. The resulting values of a' were plotted against s' for each absorbent. If the DANCKWERTS model represented the facts exactly and if there were no experimental errors the curves for all the absorbents would converge at a single point, and the corresponding values of a' and s' could be identified with the values of the parameters a and s in the packed column. The HIGBIE and Delta models can be tested similarly.

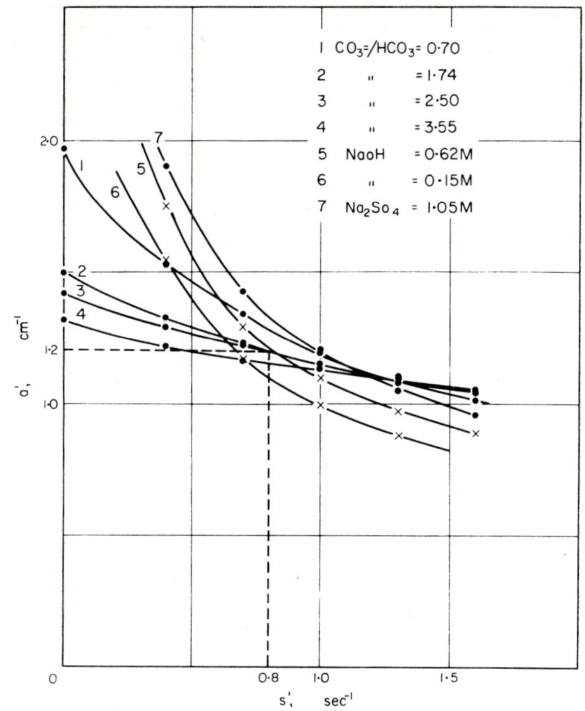

FIG. 4. Test of Danckwerts model.

The plot for the DANCKWERTS model is shown in Fig. 4. (The compositions for which the calculations have been done do not necessarily represent actual experimental solutions, the absorption-rate data having been taken from smoothed curves as described above.) The curves shown represent the limits of the region of convergence; curves for other solutions having compositions in the range investigated would not extend this region.

The DANCKWERTS model is clearly not exactly compatible with the smoothed experimental results, but there is convergence in an area centred roughly about $a = 1{\cdot}2$ cm^{-1}, $s = 0{\cdot}80$ sec^{-1}. Since in the DANCKWERTS model $k_L = \sqrt{(Ds)}$, the value of k_L^2/D corresponding to $s = 0{\cdot}80$ is $0{\cdot}80$ sec^{-1}. The plots for the HIGBIE and Delta models are similar, although the convergence is somewhat less well-defined in the former and considerably less in the latter. The expressions for k_L^2/D corresponding to these two models are $4/\pi\theta$ and $1/\pi\tau$ respectively. If we assign the value 0·80 to k_L^2/D in both we get $\theta = 1{\cdot}6$ and $\tau = 0{\cdot}40$ sec respectively. The point corresponding to $a = 1{\cdot}2$, $\theta = 1{\cdot}6$ is in fact fairly centrally located in the region of convergence in the plot for the HIGBIE model, and $a = 1{\cdot}2$, $\tau = 0{\cdot}40$ similarly agrees fairly well with the Delta model.

PREDICTION OF ABSORPTION RATES

The most direct test of the usefulness of the surface-renewal concept to the process designer is to use the smoothed data on transient absorption to predict the rates of absorption in the packed column, and to compare these predictions with the actual measured rates. We have therefore calculated Ra as a function of composition for the inert solution and the two sorts of reacting solution, using the methods of calculation given below, and the values $a = 1{\cdot}2$ cm^{-1}, $s = 0{\cdot}80$ sec^{-1}, $\theta = 1{\cdot}6$ sec, $\tau = 0{\cdot}40$ sec.

In the case of the buffer solutions we have taken r from the full curve in Fig. 6, Part I, and have given $C^*\sqrt{D}$ the value $7{\cdot}3 \times 10^{-8}$ g mole/cm^{3}sec$^{1/2}$ for all solutions; for the HIGBIE model Q has been calculated from equation (5), Part I, while for the Delta model this was differentiated to give

$$\frac{dQ}{dt} = C^* \sqrt{\left(\frac{D}{r}\right)} \left\{ r \operatorname{erf}\sqrt{rt} + \sqrt{\left(\frac{r}{\pi t}\right)} \exp(-rt) \right\} \quad (12)$$

For the Danckwerts model we have the relation $R = C^*\sqrt{[D(r+s)]}$ [2].

In the case of the NaOH solutions, the rates of solution at sufficiently high values of t are given by equation (7). The quantities B and S have been taken from the curves in Fig. 9, Part I. Since this relation holds for the values of t equal to the assumed values of τ and θ, and also over the range of values of t required for the calculations described in the previous section, equation (7) could be used to evaluate

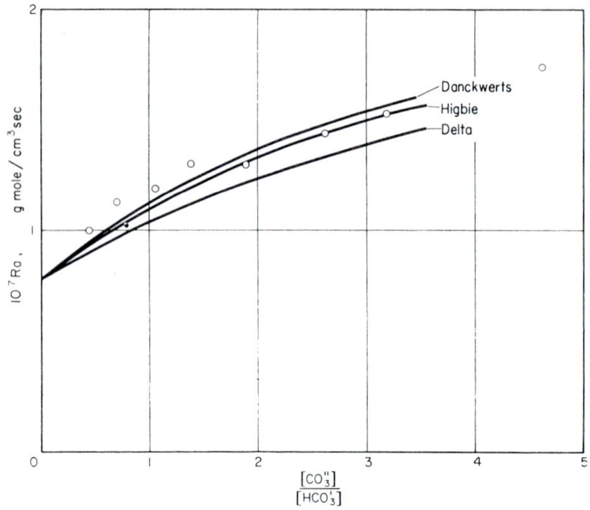

FIG. 5. Comparison of measured and predicted rates of absorption in packed column:
(a) carbonate-bicarbonate buffers

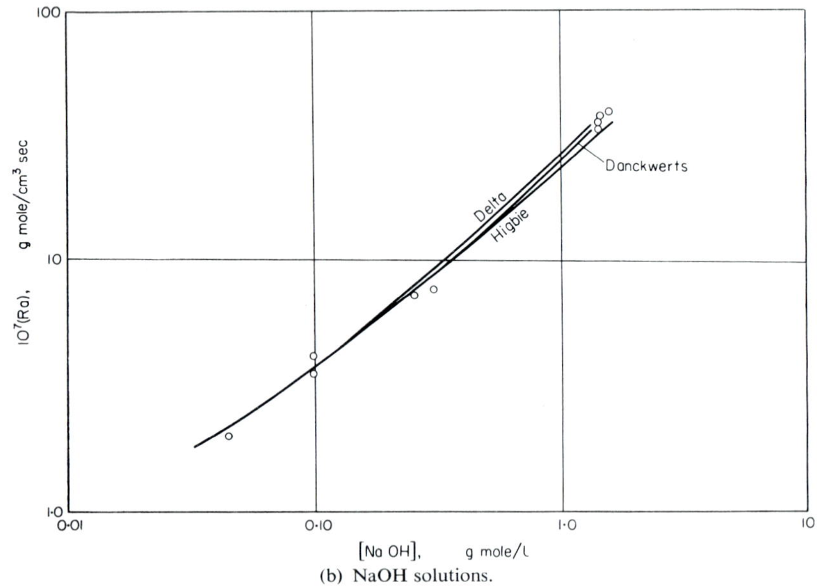

(b) NaOH solutions.

Q and dQ/dt for the HIGBIE and Delta models. In order to evaluate the integral in equation (8) account must be taken of the absorption rate at times smaller than those to which equation (7) applies. This need not be done accurately, as the contribution to the integral from this region is not large. Initially Q is proportional to t:

$$Q = Kt \tag{13}$$

Approximate values of K can be taken from the curve in Fig. 10, Part I. The actual relationship between Q and t can be approximated by combining equations (7) and (13), the transition occurring at the value of t which satisfies both. This enables the integral in equation (8) to be evaluated algebraically (comparison with numerical integration of some of the experimental curves showed the error involved to be small). The expression used is

$$R = s\int_0^\infty \exp(-st)\, dQ(t) = K[1 - \exp(-st')] + \\ + \frac{S}{2}\sqrt{(\pi s)}\operatorname{erfc}\sqrt{(st')} \tag{14}$$

where

$$t' = \frac{S^2}{2K^2} - \frac{B}{K} + \frac{S}{2K^2}\sqrt{(S^2 - 4BK)} \tag{15}$$

In the case of physical absorption into 1·05 M Na_2SO_4 we have taken $C^* = 1{\cdot}67 \times 10^{-5}$ g mole/cm³ [14] and $D = 1{\cdot}42 \times 10^{-5}$ cm³/sec (from Fig. 5, Part I). Thus

$$Q = 2C^*\sqrt{\left(\frac{Dt}{\pi}\right)} = 7{\cdot}1 \times 10^{-3}\sqrt{t} \text{ g mole/cm}^2 \quad (16)$$

The predictions are compared with the experimental values of Ra for the NaOH and buffer solutions in Figs. 5(a) and (b). The agreement with the predictions of the DANCKWERTS and HIGBIE models is quite good, although there is a systematic discrepancy in the case of the buffer solutions. The Delta model cannot be made to fit both sets of results simultaneously so well as the other two models. For absorption into 1·05 M Na_2SO_4 solution all three models lead to a predicted value of $6{\cdot}8 \times 10^{-8}$ g mole/cm³ sec for (Ra), whereas the observed values lay in the range $7{\cdot}5 - 8{\cdot}0 \times 10^{-8}$, with a mean of $7{\cdot}6 \times 10^{-8}$. This discrepancy and that observed with the buffer solution correspond to the fact, established above, that the results for these solutions are best fitted by $a = 1{\cdot}06$, $s = 1{\cdot}45$. The maximum discrepancy could probably be reduced by using a value of a rather less than 1·2 and of s rather larger than 0·80; a trial has shown, however, that the adoption of $a = 1{\cdot}15$, $s = 1{\cdot}0$ gives unacceptably large discrepancies in the case of the NaOH solutions, so the values $a = 1{\cdot}2$, $s = 0{\cdot}80$ can probably not be greatly improved on.

It appears, therefore, that the experimental results are roughly compatible with all three models. The HIGBIE and DANCKWERTS models lead to agreement with the experimental results to within about 10 per cent, the Delta model being less accurate. (The chosen values of s, θ and τ are so related that all three models give rise to the same predictions concerning rates of physical absorption.)

DISCUSSION

The average vertical velocity of the liquid (calculated from the flow rate and liquid hold-up) was 2·1 cm/sec, so that the average time taken for the liquid to travel vertically a distance equal to the diameter of one packing-particle was 0·6 sec. In fully-developed laminar flow the surface velocity of the liquid would be 2·1 × 3/2, and the time of passage of the surface over the height of one particle 0·4 sec. This cannot be compared rigorously with the "average exposure time", θ, of the HIGBIE model which was found to be 1·6 sec. However, the rate of surface-renewal seems to be less than might be expected from the flow rate; this is probably due to incomplete mixing at the discontinuities between packing particles. DAVIDSON [3] also found that at low Reynolds numbers, comparable to those used in our experiments, the liquid–film coefficients predicted by his flow model were greater than those actually observed.

It is interesting that the value of the effective interfacial area, a, is much larger than that reported by previous investigators, amounting as it does to about 30 per cent of the surface area of the dry packing. The conclusions of previous workers, summarized by YOSHIDA and KOYANAGI [15] would lead one to expect a figure of about 10 per cent or less for gas absorption on ½-in. Raschig rings with an irrigation rate of 1100 lb/hr ft². However, the methods used to determine the "effective" interfacial area by these workers probably measure something quite different from the quantity a which enters into our calculations. This question will be the subject of a future communication.

The results bear out the suggestion [16], based on cases amenable to mathematical analysis, that there would be little to choose between the DANCKWERTS and HIGBIE models as regards accuracy in predicting the effects of chemical reactions on gas-absorption. The distribution function of the DAVIDSON model is much closer to the DANCKWERTS model than that of the HIGBIE model, and would be expected to give rise to predictions almost indistinguishable from those based on the DANCKWERTS model. The fair success of the Delta model shows how remarkably insensitive the predictions are to the age distribution assumed. It has been shown [16] that predictions based on the film model are quantitatively similar to those based on the HIGBIE and DANCKWERTS models, in cases where the comparison can be made.

It must be emphasized that neither calculation from first principles, nor the use of the film model, could have taken the place of the combination used here of the surface-renewal models and measured transient absorption rates.

(a) *Calculation from first principles*

Supposing that k_L^2/D and a are known, it is possible in principle to calculate rates of absorption accompanied by chemical reactions of certain types, provided certain physico-chemical data relating to the system are available. Formulae for the HIGBIE and DANCKWERTS models and the WHITMAN film model, with reversible or irreversible first-order reaction, or instantaneous second-order reaction, are given by DANCKWERTS and KENNEDY [16]. The effects of a second-order reaction of finite speed have been computed for the film model by VAN KREVELEN and HOFTIJZER [17] and for the HIGBIE model by BRIAN *et al.* [18]. There are, however, two difficulties about this procedure. The first is to assign values to such quantities as $C^*\sqrt{D}$ and the diffusivity of the reactant. It is difficult to make a reliable estimate of the solubility and diffusivity of a solute in a solution with which it reacts, or of the effective diffusivity of a single ion in an electrolyte solution (e.g. OH′ in the CO_2–NaOH system). Moreover, as pointed out in Part I, ionic equilibria (such as determine, for instance, the pH of the solution) and reaction-rate constants are much affected by the ionic strength and composition of the solution; the same is true of solubilities and diffusivities. Substantial inaccuracies in prediction are likely to arise from these sources. A second type of difficulty arises when the kinetics of the reaction are unknown or are more complicated than those for which formulae or computed results are available. This frequency occurs in industrial practice (e.g. the absorption of CO_2 in K_2CO_3; or the depletion of the carbonate-bicarbonate buffers near the surface, referred to in Part I, which led to a falling-off in the apparent value of r which was not quantitatively predictable).

(b) *The film model*

This model suffers from a disadvantage relative to the surface-renewal models, because it is not possible to combine empirical transient absorption measurements with the parameters of the model to make predictions. If one makes use of one of the surface-renewal models it is only necessary to measure Q as a function of t for the gas and liquid in question, and to know k_L^2/D and a for the absorption equipment in question. The mechanism of the reaction, the possibility of temperature rise at the surface of the liquid and so on need not be taken into account. If the film model is to be used, the reaction must be kinetically simple, and a number of physico-chemical quantities must be determined explicitly.

To sum up, therefore, if predictions must be made by calculation from first principles the choice of model is merely a matter of convenience. It has been shown [16] that all give rise to quantitatively similar predictions. The computations of BRIAN *et al.* [17] enable the HIGBIE model to deal with somewhat more complicated systems than the film model (diffusivities of solute and reactant unequal). However, the uncertainties of prediction can be much reduced by laboratory measurements of transient absorption rates, coupled with some form of the surface-renewal model. It matters little which is used, but the HIGBIE model is likely to prove the most convenient. The tests reported in this paper suggest that predictions made in this way may be accurate to about 10 per cent, but these tests should be extended to cover a wider range of flow rates, packing sizes and reaction types before the method can be regarded as trustworthy.

CONCLUSIONS

In the particular hydrodynamic conditions chosen for the experiment either the HIGBIE or DANCKWERTS models can be used to predict within about 10 per cent the rates of absorption in a packed column under the influence of chemical reactions of widely-different natures and speeds. The Delta model gives somewhat less accurate results.

These findings contradict the earlier tentative conclusions of DANCKWERTS and KENNEDY [9], based on extrapolation of transient results obtained with shorter exposure times, that the surface-renewal models could not be used for predicting the effects of chemical reactions.§

NOTATION

a	Effective interfacial area per unit packed volume	cm^{-1}
B	Constant in equation (7)	g mole/cm^2
C^*	Saturated concentration of CO_2 in solution	g mole/cm^3
D	Diffusivity of dissolved CO_2	cm^2/sec
D'	Effective diffusivity of OH′ ions	cm^2/sec
K	Initial rate of absorption of CO_2 into NaOH solution	g mole/cm^2sec
k_L	Liquid–film mass-transfer coefficient	cm/sec
Q, $Q(t)$	Quantity of gas absorbed per unit area in time t	g mole/cm^2
R	Rate of absorption per unit area of interface	g mole/cm^2sec
r	First-order reaction-rate constant	sec^{-1}
S	Constant in equation (7)	g mole/$cm^2sec^{1/2}$
s	Fractional rate of surface renewal	sec^{-1}
t	Time of exposure	sec
θ	Time of exposure in HIGBIE model	sec
τ	Surface age in Delta model	sec
$\phi(t)$	Rate of absorption per unit area after exposure for time t	g mole/cm^3sec
$\psi(t)$	Surface-age distribution function	sec^{-1}

§ A preliminary report of these results was included in a communication to the 25th Anniversary Congress of the Society of Chemical Engineers, Japan (November 1961). In the present paper the original experimental figures have been treated somewhat differently, leading to slightly different results.

References

[1] Higbie R., *Trans. Amer. Inst. Chem. Engrs.* 1935 **31** 365.
[2] Danckwerts P. V., *Industr. Engng. Chem.* 1951 **43** 1460.
[3] Davidson J. F., *Trans. Inst. Chem. Engrs.* 1959 **37** 131.
[4] Danckwerts P. V., *Amer. Inst. Chem. Engrs. J.* 1955 **1** 456.
[5] Andrew S. P. S., *Alta Technologia Chimica; Processi di Scambia.* Academia Nazionale dei Lincei, Rome, 1961 p. 153.
[6] Toor H. L. and Marchello J. M., *Amer. Inst. Chem. Engrs. J.* 1958 **4** 97.
[7] Wint A., Thesis, Cambridge University 1960.
[8] Davidson J. F., Cullen E. J., Hanson D. and Roberts D., *Trans. Inst. Chem. Engrs.* 1959 **37** 122.
[9] Danckwerts P. V. and Kennedy A. M., *Trans. Inst. Chem. Engrs.* 1954 **32** S53.
[10] Kennedy A. M., Thesis, Cambridge University 1954.
[11] Roberts D. and Danckwerts P. V., *Chem. Engng. Sci.* 1962 **17** 961.
[12] Sherwood T. K. and Holloway F. A. L., *Trans. Amer. Inst. Chem. Engrs.* 1940 **36** 39.
[13] Nijsing R. A. T. O.. Hendriksz R. H. and Kramers H., *Chem. Engng. Sci.* 1959 **10** 88.
[14] Seidell A. and Linke W. F., *Solubility of Inorganic and Organic Compounds* (3rd Ed. and Suppl.), Van Nostrand, New York 1952.
[15] Yoshida F. and Koyanagi T., *Amer. Inst. Chem. Engrs. J.* 1962 **8** 309.
[16] Danckwerts P. V. and Kennedy A. M., *Trans. Inst. Chem. Engrs.* 1954 **32** S49.
[17] Hoftijzer P. J. and Van Krevelen D. W., *Trans. Inst. Chem. Engrs.* 1954 **32** S60.
[18] Brian P. L. T., Hurley J. F. and Hasseltine E. H., *Amer. Inst. Chem. Engrs. J.* 1961 **7** 226.

Résumé—Dans la première partie de cette publication, les vitesses d'absorption de CO_2 par des solutions de NaOH de $Na_2CO_3/NaHCO_3$ et de Na_2SO_4 ont été mesurées. Les résultats de ces mesures sont comparés avec les vitesses d'absorption de CO_2 par les mêmes solutions dans des colonnes garnies avec des anneaux Rashig de ½ pouce.

On constate que la théorie de Higbie et Danckwerts permet de prévoir ces vitesses avec une précision d'environ 10%. La théorie de Delta est moins précise.

Zusammenfassung—Die Resultate der Absorption von CO_2 in NaOH, $Na_2CO_3/NaHCO_3$ und Na_2SO_4 in einer Kolonne mit 12,5 mm Raschig-Ringen werden mit den durch die von Higbie und Danckwerts aufgestellten Oberflächen-Erneuerungs-Theorie vorausgesagten Werten verglichen; es wird eine gute Uebereinstimmung gefunden.

B6

Kinetics of CO_2 absorption—III

First-order reaction in a packed column

G. M. RICHARDS,† G. A. RATCLIFF and P. V. DANCKWERTS

Department of Chemical Engineering, Pembroke Street, Cambridge

(*Received* 1 *August* 1963; *in revised form* 12 *November* 1963)

Abstract—CO_2 has been absorbed in a packed column into carbonate-bicarbonate buffer solutions containing arsenite as a catalyst. In these solutions CO_2 undergoes a pseudo-first-order reaction. The results confirm that rates of absorption can be predicted on the basis of the Danckwerts surface-renewal model. Values of the effective interfacial area, *a*, and the surface-renewal rate, *s*, are reported for $\frac{1}{2}$ in. Raschig rings and various liquid flow-rates.

A TEST of surface-renewal models for gas-absorption in a packed column has been made by DANCKWERTS, KENNEDY and ROBERTS [1]. They found that both the Higbie and Dankwerts models enabled predictions of the effects of chemical reactions on rates of absorption to be made with an accuracy of about 10 per cent. The work reported here extends the test to much faster first-order reactions and to various liquid flow-rates.

EXPERIMENTAL

The absorbent was a buffer solution containing 0·207 M Na_2CO_3, 0·215 M $NaHCO_3$ (buffer-ratio 0·96) and varying amounts of arsenious oxide, which forms arsenite ions ($H_2AsO_3^-$) which catalyse the reaction of dissolved CO_2 to bicarbonate [2,3]. When the arsenic was added to the solution, sufficient NaOH was also added to restore the pH to that of the original buffer. Varying amounts of sodium chloride were also added to the solutions to maintain the ionic strength approximately constant at 0·94 g ions per 1000 g water. This also helped to minimise changes in density and viscosity and hence to minimise changes in the hydrodynamic conditions in the column. Densities ranged from 1·035 to 1·045 g/cm^3, and viscosities were between 1·16 and 1·18 times that of water.

Pure CO_2 saturated with water vapour and at a total pressure of 1 atm was absorbed at 25°C in a packed column of 4 in. (10·2 cm) internal diameter containing $\frac{1}{2}$ in. (1·3 cm) Raschig rings at a packing density of 28,200 rings/M^3. The apparatus and technique were almost identical to those described by DANCKWERTS and KENNEDY [4].

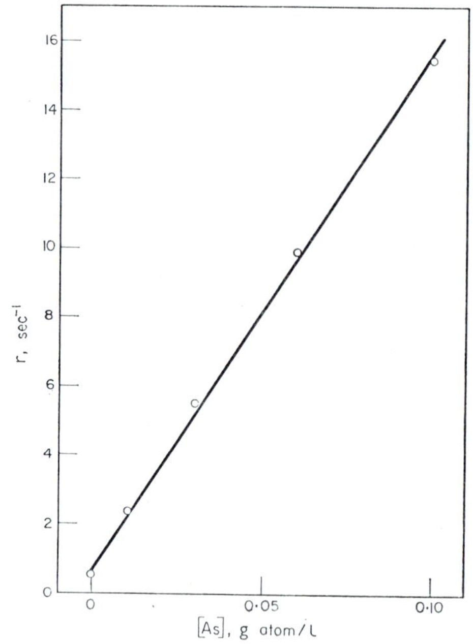

FIG. 1. Rate constants for reactions of CO_2 (25°C). Data of ROBERTS [2].

RESULTS AND DISCUSSION

ROBERTS and DANCKWERTS [2] have shown that the transient rates of absorption into arsenic-containing buffers of the type used here indicate that the dissolved CO_2 undergoes a pseudo-first-order reaction with a rate-constant which is a linear function of the amount of arsenite added. Figure 1 shows the values of the rate-constant *r* obtained when varying amounts of arsenic were added to a

† Present address: Shell Trinidad Limited, Point Fortin, Trinidad.

buffer having the same composition as that used by us. Values of r in the presence of concentrations of arsenite greater than 0·1 M were estimated by extrapolation of Fig. 1.

If the time of contact between liquid and gas is too long, or the reaction-rate too fast, the pH in the neighbourhood of the surface will fall and reduce the degree of ionisation of arsenious acid; since the arsenite ion is the catalytic species, this will reduce the reaction-rate constant near the surface after a sufficient time of contact. In the work reported by ROBERTS and DANCKWERTS (exposure up to 1·2 sec, r up to 15·5 sec^{-1}) there was no sign of this occurring. In our work values of r up to 26 sec^{-1} were attained, and the rate may have fallen off in the older parts of the surface in the packed column. This possibility is discussed further below.

For the purpose of analysing the results, it is convenient to use the Danckwerts surface-renewal model, which leads to

$$Ra = ac^*\sqrt{[D(s + r)]} \qquad (1)$$

where Ra and a are the absorption rate and the effective interfacial area per unit packed volume respectively, s the surface-renewal rate, and D and c^* the diffusivity and solubility respectively of CO_2. ROBERTS and DANCKWERTS [2] determined the value of $c^*\sqrt{D}$ in buffers of this composition to be $1{\cdot}08 \times 10^{-7}$ g mole/cm^2 $sec^{1/2}$. Estimates of c^* using the method of VAN KREVELEN and HOFTYZER [5] and of D using the method of RATCLIFF and HOLDCROFT [6] give results which agree with ROBERTS and DANCKWERTS to within the accuracy of the predictions. We shall use the value $1{\cdot}08 \times 10^{-7}$ here.

Equation (1) indicates that a plot of $(Ra)^2$ against r should give a straight line of slope $a^2(c^*\sqrt{D})^2$ and with an intercept $a^2(c^*\sqrt{D})^2 s$. Thus a can be found from the slope and s is the ratio of intercept to slope. The experimental results for several liquid flow-rates are shown in Fig. 2. It will be seen that a linear relationship holds for the lower values of r, but that there is a slight falling-off in the values of $(Ra)^2$ at higher values of r. This may well be due to depletion of the buffer in the older parts of the surface, as discussed above.

The values of a and s deduced from these plots are shown in Figs 3 and 4. The values of a and s found for a flow-rate of 0·145 cm^3/cm^2 sec are compared in Table 1 with those found by DANCKWERTS *et al.* [1] for carbonate-bicarbonate buffer solutions at the same flow rate (but without catalyst). The densities of these solutions ranged from 1·08 to 1·12 g/cm^3. The viscosities were 1·57 times as great as that of water at the same temperature, as compared to 1·16–1·18 in the present work.

Table 1.

	a (cm^{-1})	s (sec^{-1})
Present work	0·97	1·63
DANCKWERTS *et al.* [1]	1·06	1·45

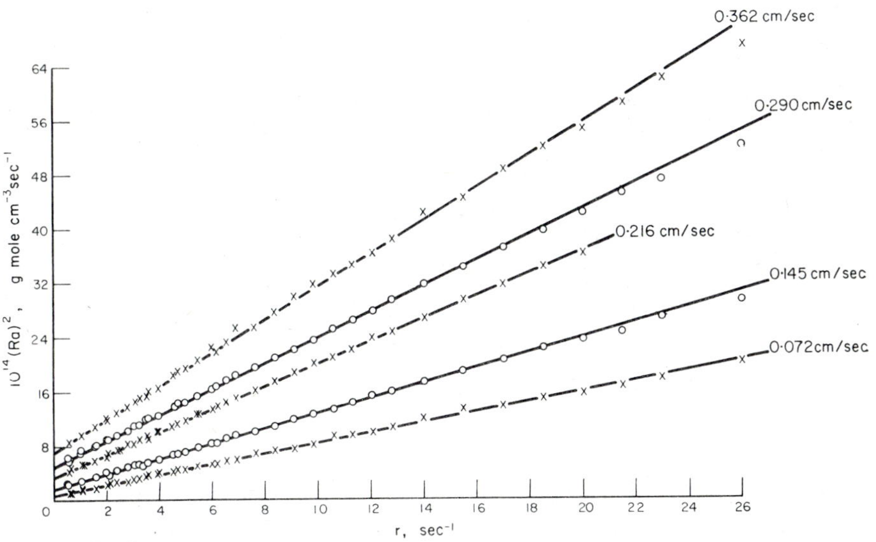

FIG. 2. Absorption into buffer solutions containing arsenite in a parched column. Superficial liquid flow-rates indicated.

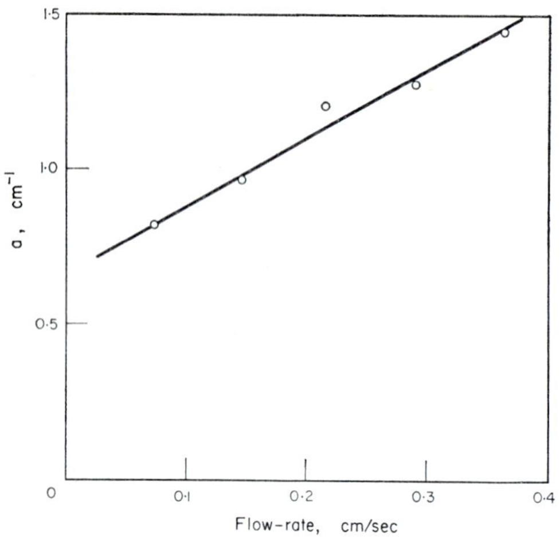

FIG. 3. Variation of effective interfacial area with superficial flow-rate.

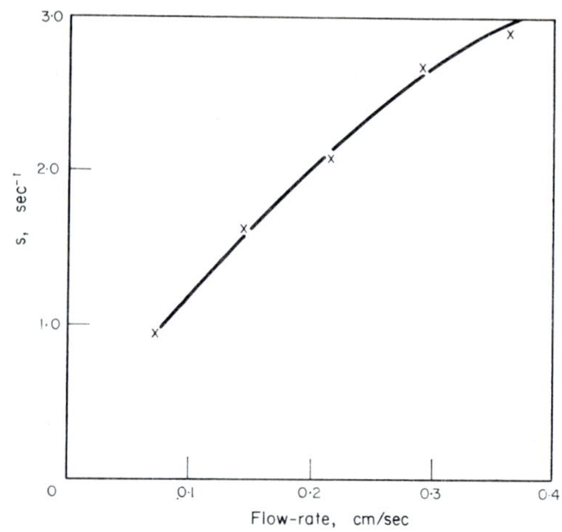

FIG. 4. Variation of surface-renewal rate with superficial flow-rate.

In view of the relatively small range of values of r in the work of DANCKWERTS *et al.* and the considerable scatter of their results, it would probably not be wise to draw conclusions from these figures about the effect of viscosity on a and s.

We have also measured rates of physical absorption into 1·78 molal NaCl solution, which has the same viscosity as and a density close to that of the reacting solutions. The measured values of $k_L a$ agree closely with those predicted by the Danckwerts model, which leads to $k_L a = a\sqrt{(Ds)}$. We have used the values of a and s found for reacting solutions at the same flow-rates, and a value of D taken from the data of RATCLIFF and HOLDCROFT [6] (see Table 2).

CONCLUSIONS

The Danckwerts surface-renewal model accounts well for the rates of absorption of CO_2 into non-reacting solutions and into solutions with which it undergoes a first-order reaction, over a considerable range of flow-rates and reaction-rate constants. However, since the predictions based on the Whitman film model and the Higbie surface-renewal model would be quantitatively very similar to those based on the Danckwerts model [7], the experiments do not provide strong support for one model rather than another but show rather that any one of the models can be used as the basis of design calculations.

It is possible in principle to deduce from the measurements reported here the actual surface-age distribution in the packed column. For an irreversible first-order reaction such as was used this involves the numerical inversion of a La ace transform, and will be the subject of a later communication.

NOTATION

a Effective interfacial area per unit packed volume cm^{-1}
c^* Saturated concentration of CO_2 g mole/cm^3

Table 2. Rates of physical absorption at 25°C compared with rates predicted from absorption into reacting solutions

Absorption into 1·78 molal NaCl solution; $D = 1{\cdot}63 \times 10^{-5}$ cm^2/sec [6]

Flow-rate (cm^3/cm^2 sec)	a (cm^{-1})	s (sec^{-1})	$10^3 a\sqrt{(Ds)}$ (cm/sec)	$10^3 k_L a$ (obs.) (cm/sec)
0.072	0·82	0·96	3·2	3·25
0·145	0·97	1·63	5·1	4·86
0·216	1·20	2·09	7·0	6·83
0·290	1·275	2·69	8·5	8·23
0·362	1·445	2·91	10·0	9·81

D	Diffusivity of dissolved CO_2	cm^2/sec	r	First-order reaction-rate constant	sec^{-1}
k_L	Liquid-film mass-transfer coefficient for physical absorption	cm/sec	s	Surface-renewal rate	sec^{-1}
R	Rate of absorption into unit interfacial area	g mole/cm^2 sec			

References

[1] Danckwerts P. V., Kennedy A. M. and Roberts D., *Chem. Engng. Sci.* 1963 **18** 63.
[2] Roberts D. and Danckwerts P. V., *Chem. Engng. Sci.* 1962 **17** 961.
[3] Sharma M. M. and Danckwerts P. V., *Trans. Faraday Soc.* 1963 **59** 386.
[4] Danckwerts P. V. and Kennedy A. M., *Trans. Inst. Chem. Engrs* 1954 **32** S53.
[5] van Krevelen D. W. and Hoftijzer P. J., *Chim. Industr.*; XXI Congr. Int. Chim. Industr. 1948 p. 168.
[6] Ratcliff G. A. and Holdcroft J. G., *Trans. Inst. Chem. Engrs* 1963 **41** 315.
[7] Danckwerts P. V. and Kennedy A. M. *Trans. Inst. Chem. Engrs* 1954 **32** S49.

Résumé—On a absorbé CO_2 dans une colonne à garnissage par des solutions tampon de carbonate et bicarbonate contenant de l'arsénie comme catalyseur. Dans ces solutions CO_2 subit une réaction du pseudo premier ordre. Les résultats confirment que la vitesse d'absorption peut être prévue grâce au schéma de renouvellement de surface de Danckwerts. L'auteur rapporte des valeurs de la surface de contact effective *a* et de la vitesse de renouvellement de surface *s*, pour des anneaux Raschig de 1/2 pouce et pour diverses vitesses d'écoulement des liquides.

B7

THE DESIGN OF GAS ABSORBERS

I—METHODS FOR PREDICTING RATES OF ABSORPTION WITH CHEMICAL REACTION IN PACKED COLUMNS, AND TESTS WITH $1\frac{1}{2}$ IN. RASCHIG RINGS

By Professor P. V. DANCKWERTS, G.C., M.B.E., M.A., S.M., C.Eng. (MEMBER),*
and A. J. GILLHAM, B.Sc., A.C.G.I., Ph.D.†

SYNOPSIS

The rate of absorption of carbon dioxide has been measured in a small stirred vessel and in an 18 in. dia. column packed with $1\frac{1}{2}$ in. Raschig rings. A variety of reacting solutions and a range of stirring speeds and flow-rates were employed.

The results appear to show that if the stirring-rate in the vessel is adjusted to give the same value of k_L for physical absorption as that in the column, the vessel can be used as a model of the column for the purpose of predicting the effects of chemical reactions in the latter.

The values of k_L and a for the column can also be determined with the aid of the stirred vessel, or by an independent method involving first-order reactions. These quantities have been determined by both methods for a range of flow-rates. They also can be used for predicting the effects of chemical reactions, either by the use of a laboratory model or by calculation from first principles.

The various methods of prediction have been tested by comparison with measured rates of absorption in solutions of sodium hydroxide in the packed column. The comparison suggests that the methods are sufficiently accurate for design purposes.

Introduction

The problem considered here is the prediction of the rate of absorption of a gas in a packed column, the absorbent being a liquid with which the gas reacts. The rate of absorption will be determined partly by hydrodynamic factors (density, viscosity, and flow-rate of liquid; shape and size of packing) and partly by physico-chemical factors (solubility of gas, kinetics of reaction, diffusivities of substances in solution). Various simple theoretical models of the absorption process can be used to calculate, from first principles, the effects of chemical reactions with simple kinetics. It has been shown[1] that the film model and various versions of the surface-renewal model lead to conclusions about the effects of chemical reactions which are numerically in very close agreement. It has also been shown experimentally[2, 3] that the surface-renewal models give reasonably accurate predictions of rates of absorption into reacting liquids in packed columns, although the range of conditions which has been studied is limited.

A tentative conclusion from the earlier work is that the absorption column resembles the theoretical models in so far as the effect of the hydrodynamic factors can be represented by two parameters, k_L (the mass-transfer coefficient in the absence of reaction) and a (the effective interfacial area per unit volume of packing); the latter is supposed to be unaffected by the type of reaction considered. (If a number of gases with different diffusivities, D, in solution are considered it seems probable, as far as the experimental evidence goes, that $k_L/\sqrt{D}$ is actually the relevant hydrodynamic parameter. In the work described here, however, the value of D was virtually constant.) It would follow that if a laboratory model of the packed column is to be constructed, the essential feature to be reproduced is the value of k_L. The rate of absorption in the laboratory model would then respond to chemical reactions in the same way as that in the column, even though the means of agitating the liquid in the two systems might be quite different.

If these conclusions are correct, it should be possible to set up small and convenient laboratory apparatus to simulate packed columns and to investigate the effects of chemical reactions. (In general, of course, the local absorption rate varies with height in the packed column, owing to changes in the compositions of the liquid and the gas. A series of measurements would then have to be made on the model, using gas and liquid compositions corresponding to those at different points in the column. The total rate of absorption in the column would then be predicted by integration of the model results, as described below.) Such direct model experiments would make it unnecessary to investigate the kinetics of the reaction explicitly, or to determine or estimate solubilities and diffusivities; in this way much laborious experimental work and a number of sources of error could be eliminated.

It is possible that similar techniques could be used to simulate other industrial absorbers, such as bubble plates.

Various types of laboratory models might be useful for simulating columns under different circumstances (*e.g.* jets, falling films, rotating drums, and strings of discs or spheres). The authors decided to use a stirred vessel, the gas being absorbed into the horizontal surface of the liquid (no bubbles being entrained). The object was to see whether this vessel, at a given rate of stirring, acted as a satisfactory model for a given packed column under given hydrodynamic conditions; and if so, to demonstrate its use for predicting absorption rates in the packed column, and also for determining values of k_L and a in the column.

* Department of Chemical Engineering, Pembroke Street, Cambridge.

† Imperial Chemical Industries Ltd., Mond Division, Runcorn, Cheshire.

Principles of Methods Used

The experiments consisted of absorbing pure carbon dioxide into a number of different solutions, both in a column about 18 in. (46 cm) dia. packed with $1\frac{1}{2}$ in. (3·8 cm) Raschig rings and in the stirred vessel, which was about 4 in. (10 cm) dia. The solutions all had about the same viscosity and density and thus the hydrodynamic behaviour of all of them was almost identical in the packed column at a given flow-rate, or in the stirred vessel at a given stirring speed. The results can be used as a basis for three methods of predicting absorption rates in the packed column. The object was to test the validity of these methods and to establish values of k_L (for physical absorption) and a for the packing. The discussion does not take into account the matter of gas-side resistance to transfer. It is hoped to discuss this in a later communication.

Method I: ratio matching

Consider absorption into a non-reacting (inert) solution and into a reacting solution, and let the stirring-rate in the vessel be specified. There will then be a particular flow-rate in the packed column (as yet undetermined) at which the value of k_L (the liquid-film coefficient in the absence of reaction) will be the same as that in the stirred vessel. Now, in general the ratio:

$$\gamma = \frac{\text{Rate of absorption into reacting solution}}{\text{Rate of absorption into } CO_2\text{-free inert solution}}$$

will depend on the value of k_L, as well as on the nature of the reacting solution. Thus, the flow-rate at which γ is observed to have the same value in the column as in the vessel (the composition of the reacting solution being the same in both cases) can be taken to be that which gives rise to the same value of k_L in the column as in the vessel. If the superficial flow-rate, L, in the column is systematically varied, the ratio $\gamma_{vessel}/\gamma_{column}$ (for a fixed stirring speed in the vessel) becomes equal to unity when L has such a value as to make the value of k_L in the column the same as that in the vessel. This procedure may be repeated with a number of different reacting solutions: if in each case the quantity $\gamma_{vessel}/\gamma_{column}$ is plotted against L, the result should ideally resemble Fig. 1,

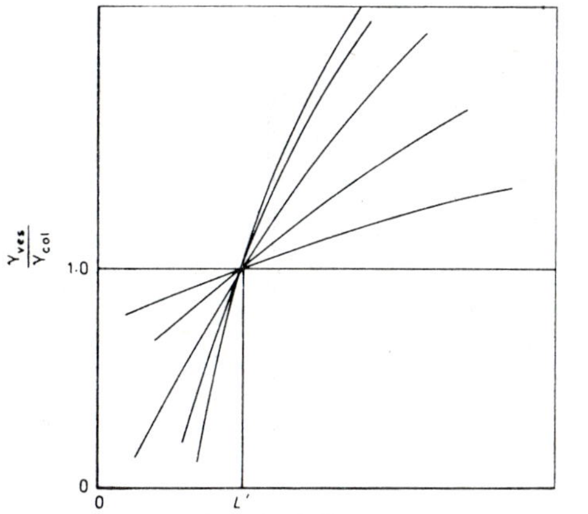

Fig. 1.—*Idealised plot*

if the basic assumptions of the method are justified. The various curves should intersect the line:

$$\frac{\gamma_{vessel}}{\gamma_{column}} = 1$$

at a single point, corresponding to the flow-rate, L', at which the value of k_L in the column is the same as that in the vessel at the specified stirring speed.

The values of γ_{vessel} for the various solutions can now be measured at a different stirring speed, and the procedure repeated. In this way one could determine and tabulate the stirring speed corresponding to any flow-rate in the column. The rates of absorption ($k_L C_0^*$) into the carbon-dioxide-free inert solution, at various stirring speeds in the vessel and flow-rates in the column, would also be tabulated.

The comparison of values of γ in the vessel and in the column is most conveniently done by using reacting solutions, such as buffer solutions, for which the rate of absorption does not change as carbon dioxide is absorbed. The rate of absorption per unit volume of packing is then the same at all points in the column. The procedure would be more complicated and less accurate if, for instance, sodium hydroxide solutions were used for the purpose.

If one wishes now to predict the rate of absorption into some novel absorbent in the column at a given flow-rate, one simply puts the absorbent into the vessel, measures the absorption rate at the appropriate stirring speed, and thus (from the previously obtained results with the inert solution) determines γ. The previously determined rate of absorption into the carbon-dioxide-free inert solution in the column is multiplied by γ to give the rate of absorption into the novel solution in the column.

The method can similarly be used with novel gases, as well as with novel liquids. The use of the method for predicting total absorption rates in columns where the local absorption rate varies with height is discussed below.

Once determined, the data relating the stirring speed in the vessel to the flow-rate in the column can be permanently recorded for use by designers. A possible disadvantage of the method is that experiments on new liquids or gases must be done with a vessel identical in design to that used to obtain the original data. This drawback is avoided by Method II.

Method II: matching k_L

The values of k_L and a for the column, at various flow-rates, are determined as described below and permanently recorded. A laboratory model (which might be the stirred vessel, or some other type of apparatus) is designed and operated in such a way as to have the appropriate value of k_L and a known value of the total interfacial area for absorption. The rate of absorption per unit interfacial area, using the appropriate gas and absorbent, is measured in the model; multiplied by a, it gives the rate of absorption per unit volume of the column.

This method has the advantage that all the data to be permanently recorded (k_L and a as functions of the flow-rate and of the physical properties of liquid) refer to the column, and not to an arbitrarily chosen laboratory model.

There are two methods whereby k_L and a for the column can be determined. The rate of absorption into an inert solution in the stirred vessel is $k_L A(C_0^* - C_0)$, where C_0^* is the saturated solubility of the gas and C_0 its bulk concentration. Provided C_0^* is known, $k_L A$ is easily determined by means of batchwise or steady-flow measurements. The value of A for the stirred vessel can be taken, with little error, as equal to the horizontal cross-section of the vessel (this point is further discussed below). Values of k_L corresponding to various stirring speeds are thus established for the particular vessel used.

The ratio-matching technique described under Method I, above, can now be used to determine the flow-rates in the column which give the same values of k_L as various stirring speeds in the vessel. The values of k_L in the column for various flow-rates are thus established once and for all. Values of a

are obtained from values of $k_L a$, determined by absorption into an inert solution in the column or from published information such as the correlation of Sherwood and Holloway.[4]

An alternative method of determining the values of k_L and a in the column is to absorb the gas into a series of solutions in which it undergoes a first-order reaction. If the absorption rate is determined as a function of the reaction-velocity constant, the values of k_L and a can be deduced. This method has been described by Richards, Ratcliff, and Danckwerts,[3] and is illustrated below.

Method III: calculation from kinetics

It is supposed that k_L and a for the column have been determined and recorded, by one of the methods described above. It is also supposed that the kinetics of the reaction are understood and that the relevant physico-chemical parameters (solubility, diffusivities, reaction-velocity constants, *etc.*) are known or can be estimated. The rate of absorption can then be calculated, on the basis of one of the models of gas absorption. The surface-renewal models (Higbie or Danckwerts) probably correspond more closely to the facts than the film model. Solutions of the relevant equations are available for first-order reactions,[1] for irreversible second-order reactions,[5] and for reversible first-/third-order reactions.[6] However, for many types of reaction of practical interest an elaborate computer programme is likely to be needed.

The Stirred Vessel

The vessel, shown in Fig. 2, was constructed from a Pyrex glass-column section of 10·5 cm i.d. Each end was closed with a Perspex plate sealed in position with rubber "O" rings. The stirrer shaft passed through a mercury seal on the top cover, and the base of the shaft was supported in a seating in the bottom of the vessel. The stirrer consisted of four Perspex strips, of rectangular cross-section, which were supported in the form of a cross at right angles to the stirrer rod. Static baffles were attached to the base of the vessel to promote bulk mixing. The level of the liquid flowing through the vessel was adjusted so that the bottom edge of the stirrer blades just cut the plane of the liquid surface. If the blades were too deeply immersed, splashing occurred at the higher stirring speeds. If insufficiently immersed, the value of k_L was reduced. For these reasons it was necessary to have the same liquid level, to within 1 mm, in different runs. The sensitivity of the performance to the liquid level is a disadvantage in this apparatus, although with practice reproducible results were easily obtained.

When the cell was filled with liquid up to the working level, the area, A, of liquid surface exposed was 80 cm^2 (cross-sectional area of vessel less that of stirrer).

The solution entered and drained from the vessel through pipes sealed into the bottom plate. The feed solution was stored in a 20 l. aspirator and delivered by gravity through a flow-measuring device, passing through a stainless-steel coil in a constant-temperature bath before entering the cell. The solution flowed out of the cell through a tube of 1 cm i.d., which led to a constant-level device.

The drive for the stirrer was transmitted from a synchronous motor through a variable-speed gear-box and a right-angled drive to the stirrer shaft.

The carbon dioxide, which was saturated with water vapour, was stored in a rubber balloon at a pressure close to atmospheric. When the vessel was in operation the gas passed from the balloon through a variable-area flow-measuring device before entering the stirred vessel through the top cover.

After the solution had started flowing through the vessel the gas space above the solution was purged with clean carbon dioxide by opening the gas vent, which was connected to a filter pump. The constant-level device was adjusted until the liquid surface was in the plane of the bottom edge of the cruciform stirrer while it was rotating. The total volume of solution in the vessel was then about 150 cm^3. Under this condition of operation the liquid surface remained reasonably flat, with no vortex formation, and with no entrainment of gas bubbles in the bulk of the solution. The highest value of k_L compatible with these conditions was 0·022 cm/s at a stirrer speed of 160 rev/min, using a solution of 0·53 M sodium sulphate. This value of k_L corresponds to the value obtained with the same solution in a column of 18 in. i.d. packed with 1½ in. ceramic Raschig rings at a superficial solution flow-rate of 1·7 cm/s (13 000 lb/h ft^2).

The stirred vessel could also be operated batchwise, with no flow of liquid. When an inert absorbent was used, the value of k_L could be determined by this method, making allowance for the changing concentration of absorbed gas. When a buffer solution was used as an absorbent, the rate of absorption into a batch of absorbent remained constant for long enough to allow it to be determined without difficulty. However, when solutions of sodium hydroxide were used as absorbents, the depletion of reagent caused the rate of absorption to change too rapidly to allow accurate measurements to be made batchwise, and the steady-flow method had to be used for these solutions and was in fact used for the other solutions too. The weakness of this method was that even at the fastest practicable flow-rates the solution of sodium hydroxide in the vessel was substantially depleted, and it was thus not possible to measure the rate of absorption into undepleted absorbent.

The Packed Column

This is shown diagrammatically in Fig. 3. The column had an i.d. of about 18 in. (its cross-sectional area, A', was 1655 cm^2) and was packed with 1½ in. (3·8 cm) Raschig rings to a depth of 12 ft (365 cm). The solutions entered through the top cover of the tower and flowed onto the liquid distributor supported immediately above the packing. This distributor was of the notched-chimney type, with 40 distribution chimneys, and gave excellent distribution over a wide range of liquid flows.

In order to alter conveniently the height of irrigated packing exposed to the gas, the lower part of the column was allowed to fill with liquid; the liquid level could be altered at will and was held steady by a constant-level device. This procedure introduced a small error, because some absorption took place into the horizontal liquid surface. The area of the horizontal surface corresponded to the effective surface area of about 1·3 cm depth of 1½ in. Raschig rings. Some absorption also took place into the surface of the liquid on the top distributor plate, the area involved being roughly the same, and it is possible that the flow over the packing immediately beneath the distributor was of a different character from that in the bulk of the packing. The errors introduced by these end-effects were of negligible magnitude in the experiments involving absorption into the inert solution and buffer solutions; in these cases the analysis of the results involved the slope of a line relating absorption rate to height of packing, rather than the total rate of absorption with any one height of packing exposed. In the case of absorption into solutions of sodium hydroxide, where the total rate of absorption was of interest, the area of the horizontal liquid surfaces on the distributor plate and at the bottom of the exposed packing constituted an appreciable fraction (say 25–30%) of the effective wetted surface of the smallest height of packing exposed (10 cm). The error, of course, was proportionally less when greater heights of packing were exposed.

In order to avoid, at this stage, the complications associated with gas-side resistance, pure carbon dioxide saturated with water vapour was used. The gas flowed into the top of the column and flowed out through a pair of outlet pipes immediately above the horizontal liquid surface. The absorption rate was the difference between the measured rates of inflow and outflow of gas, and was found to be independent of the rate of inflow over a wide range, thus demonstrating that any inert gases released from the liquid were being effectively purged from the column. For instance, when carbon dioxide was being absorbed at a rate of 500 cm^3/s, the total gas input to the tower was varied from 850 to 2300 cm^3/s. The rate of absorption into the solution was not influenced by this change in the total gas flow through the tower. This indicated that inert gases were being efficiently purged, and that the rate of gas flow did not significantly affect the behaviour of the liquid flowing over the packing. In view of this, it is not likely that the results would have been any different if the gas and liquid had flowed countercurrently, as in normal practice, rather than cocurrently, as in the experiments.

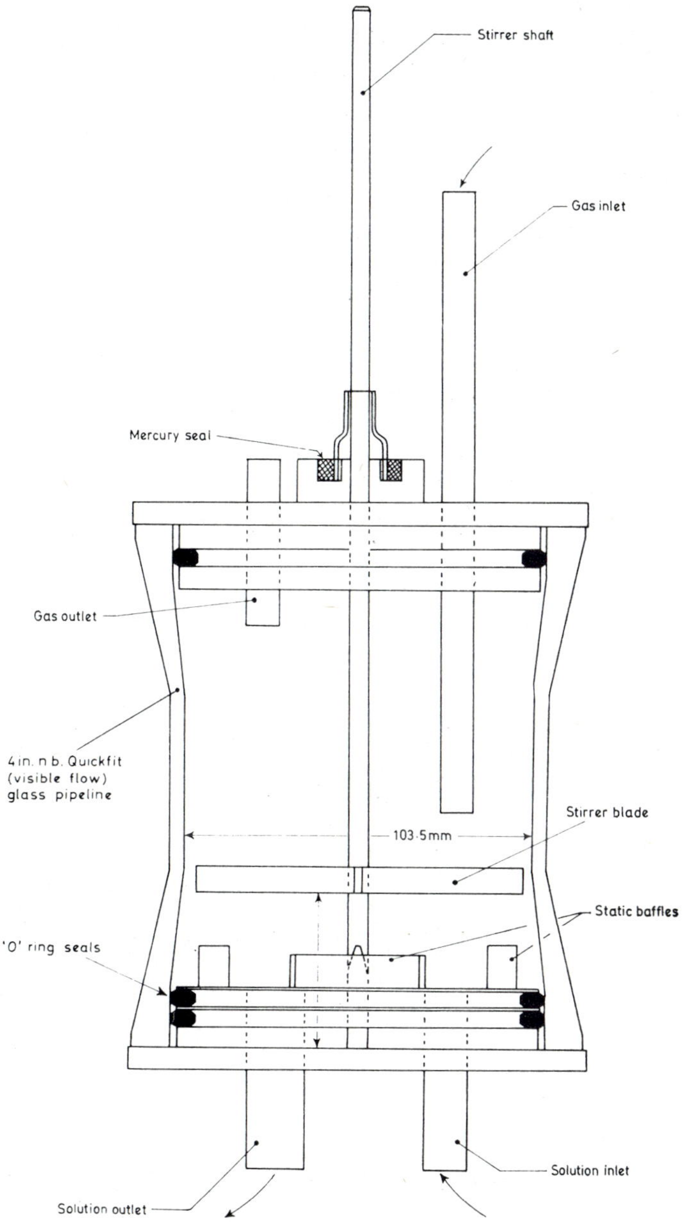

Fig. 2.—*The stirred vessel*

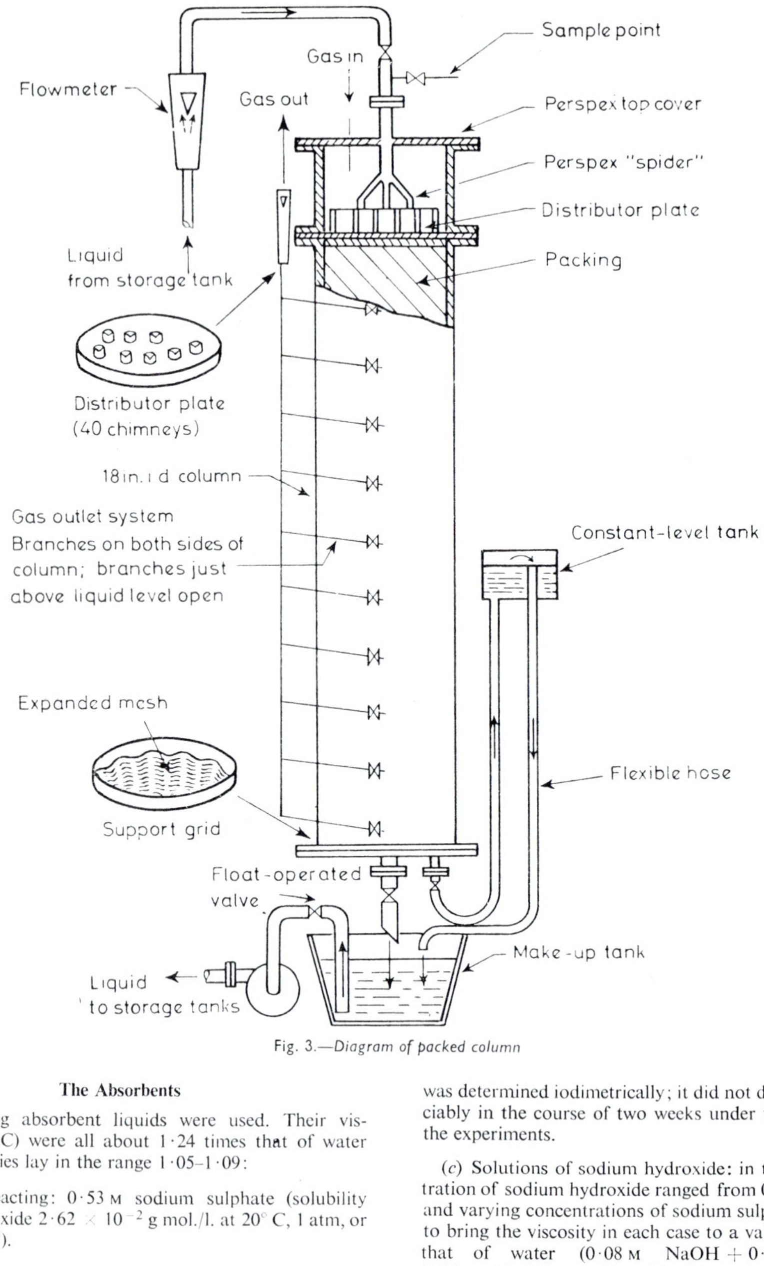

Fig. 3.—*Diagram of packed column*

The Absorbents

The following absorbent liquids were used. Their viscosities (at 20° C) were all about 1·24 times that of water and their densities lay in the range 1·05–1·09:

(*a*) Non-reacting: 0·53 M sodium sulphate (solubility of carbon dioxide $2 \cdot 62 \times 10^{-2}$ g mol./l. at 20° C, 1 atm, or 0·63 cm^3/cm^3).

(*b*) Buffer solutions: all contained 0·6 M potassium carbonate, 0·2 M potassium bicarbonate, together with concentrations of sodium hypochlorite ranging up to 0·05 M. In these solutions the dissolved carbon dioxide undergoes a first-order reaction with velocity constant:

$$k_1 = 1 \cdot 5 + 2700\,[\mathrm{ClO^-}]\ \mathrm{s^{-1}} \quad . \quad (1)$$

at 20° C, where $[ClO^-]$ is the concentration of hypochlorite in g ion/l. (the kinetics of the reaction are discussed further in Appendix I). The concentration of hypochlorite was determined iodimetrically; it did not decompose appreciably in the course of two weeks under the conditions of the experiments.

(*c*) Solutions of sodium hydroxide: in these the concentration of sodium hydroxide ranged from 0·08 M to 0·80 M, and varying concentrations of sodium sulphate were added to bring the viscosity in each case to a value of 1·24 times that of water (0·08 M NaOH + 0·52 M Na_2SO_4; 0·15 M NaOH + 0·51 M Na_2SO_4; 0·37 M NaOH + 0·2 M Na_2SO_4; 0·80 M NaOH + 0·1 M Na_2SO_4). In these solutions the dissolved carbon dioxide undergoes a second-order reaction with OH^-. The rate constant depends on the composition of the solution. The kinetics are discussed in Appendix II.

Measurement of Absorption Rates

The experiments were conducted at temperatures ranging from 20·5 to 23·0° C. With a given absorbent, measurements in the packed column and the stirred cell were conducted at the same temperature.

Stirred vessel

To make a measurement of absorption rate the stirred vessel was operated at a constant solution flow-rate, v, and stirrer speed. The gas space above the solution was purged at frequent intervals. The rate of absorption was measured when the rate had remained steady for about five minutes. In the case of non-reacting solutions the value of the mass-transfer coefficient, k_L, was determined from the relationships:

$$RA = k_L A(C_0^* - C_0) = vC_0 \quad . \quad . \quad (2)$$

where RA is the total rate of absorption (cm^3/s), A the area of the surface (cm^2), C_0^* the solubility of carbon dioxide at the operating temperature (cm^3 of gas per cm^3 of solution), and C_0 the concentration of carbon dioxide in the bulk of the liquid, which is assumed to be uniform. Thus:

$$k_L = \frac{RA}{A(C_0^* - RA/v)} \text{ cm/s} \quad . \quad . \quad (3)$$

The operation was repeated using the same stirrer speed but with different solution flow-rates, and thus different values of C_0. It was found that at a given rate of stirring the value of k_L did not change significantly with changes in solution flow-rate between 0 and 5 cm^3/s. Values of k_L and $k_L C_0^*$ calculated on the assumption that A was equal to the cross-sectional area of the vessel less that of the stirrer (80 cm^2) are shown in Table I.

Rates of absorption into a series of buffer solutions are given in Table II, expressed as values of γ, as defined under "*Method I: ratio matching*" above (*i.e.* ratio of absorption rate in buffer to value of $k_L A C_0^*$ with inert solution at same stirring speed). Values of the first-order reaction-velocity constant, k_1, calculated from equation (1), are also shown. (The value of k_1 in solution 1 was not accurately known.)

In the case of the sodium hydroxide solutions, the absorption rate not only depended on the concentration of sodium hydroxide and sodium sulphate in the inflowing liquid, but also had an unknown dependence on the concentration, C_0, of absorbed carbon dioxide in the liquid occupying the vessel. The latter could be calculated from the absorption rate and

TABLE I.—*Absorption into* 0·53 M *Sodium Sulphate in Stirred Vessel at* 20° *C*

($C_0^* = 0{\cdot}63$ cm^3/cm^3)

Stirrer speed (rev/min)	$10^3 k_L$ (cm/s)	$10^3 k_L C_0^*$ (cm/s)
160	21·6	13·6
141	18·9	11·9
125	16·3	10·3
110	13·1	8·3
95	11·1	7·0

the liquid flow-rate. Some of the experimental results are shown in Fig. 4; the inflowing solution contained 0·8 M sodium hydroxide and 0·1 M sodium sulphate. The concentration, C_0, of absorbed carbon dioxide is equal to the absorption rate divided by the flow-rate of the solution: $C_0 = RA/v$. Lines of Constant C_0 are shown in the figure. It is a simple matter to re-plot the data as R against C_0 for various stirring speeds ($A = 80$ cm^2); or, by dividing the data for each stirring speed by the relevant value of $k_L C_0^* A$ (see Table I) to re-plot as γ against C_0 for various stirring speeds. Values of R or γ for speeds other than those used in the experiments must be obtained by interpolation or cross-plotting.

Packed column

The observed absorption rate when a height, h, of packing is exposed to the gas is $\bar{R}ahA'$ where A' is the cross-sectional area of the column and $\bar{R}a$ the average absorption rate per unit volume of packed space. Dividing the absorption rate by A', one obtains $\bar{R}ah$. In the case of absorption into an

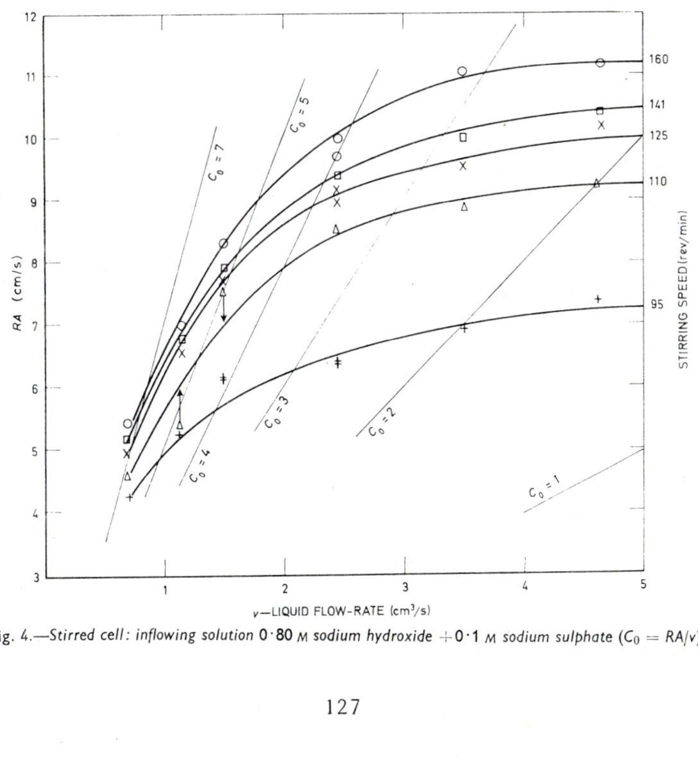

Fig. 4.—*Stirred cell: inflowing solution* 0·80 M *sodium hydroxide* +0·1 M *sodium sulphate* ($C_0 = RA/v$)

Table II.—*Absorption into Buffer Solutions in the Stirred Vessel—Values of* γ

($RA = \gamma k_L C_0^* A$ cm³/s; A = 80 cm²; k_1 in s^{-1})

Stirrer speed (rev/min)	$10^3 k_L C_0^*$ (cm/s)	Solution k_1 =	1	3 85	4 52·5	5 23	6 130	7 37	8 55	9 7	10 1·5
160	13·6		1·93	2·14	1·01	1·52	2·36	1·60	1·75	1·25	0·99
141	11·9		2·12	2·28	1·45	1·57	2·52	1·75	1·90	1·21	1·03
125	10·3		2·28	2·49	2·13	1·67	2·75	1·95	2·11	1·28	1·09
110	8·3		2·66	2·94	2·56	1·78	3·16	2·28	2·43	1·40	1·16
95	7·0		3·03	3·34	2·85	2·14	3·56	2·40	2·59	1·45	1·29

inert solution the value of $k_L a$ can be obtained by exposing different heights of packing, keeping the superficial volumetric flow-rate, L, constant.

Since: $$k_L a(C_0^* - C_0)\mathrm{d}h = L\mathrm{d}C_0$$

and: $$C_0 = \bar{R}ah/L$$

where C_0 is the concentration of dissolved carbon dioxide at the bottom of a height, h, of packing, therefore:

$$\frac{\mathrm{d}}{\mathrm{d}h}\ln\left(1 - \frac{\bar{R}ah}{LC_0^*}\right) = -\frac{k_L a}{L} \quad . \quad . \quad . \quad (4)$$

and $k_L a$ can be determined by plotting $\ln\left(1 - \frac{\bar{R}ah}{LC_0^*}\right)$ against h.

Values of $k_L a$ determined in this way with various flow-rates are shown in Fig. 5; heights of packing up to 100 cm were used in these measurements.

In the case of all the buffer solutions, the total absorption rate was linearly related to the height of packing (up to 80 cm, which was the maximum height used in this set of experiments). Thus, the local rate of absorption, Ra, was independent of height; that is, it was not affected by the variation in composition of the solution due to the carbon dioxide absorbed higher up the column. Values of Ra and of

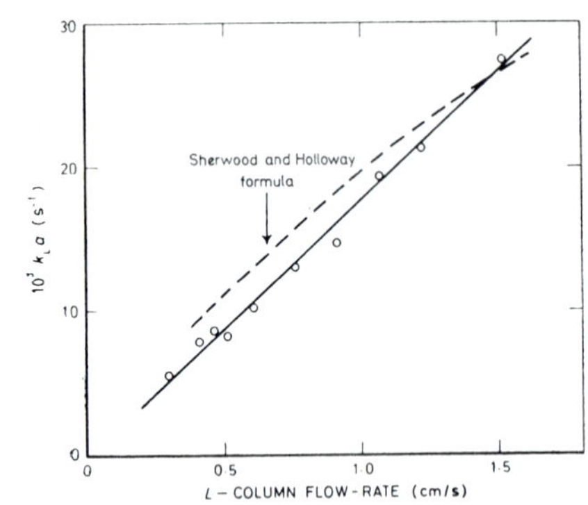

Fig. 5.—*Absorption into solution of 0·53 M sodium sulphate in packed column*

$\gamma = Ra/k_L aC_0^*$ for these solutions (the same as those referred to in Table II) are shown in Table III.

In the case of the solutions of sodium hydroxide, the local rate of absorption was less in the lower than in the higher parts of the packing, because of depletion of the OH^- by reaction with absorbed carbon dioxide. In Figs 6–9 the total absorption rate, $\bar{R}aA'h$, is plotted against the height, h, of exposed packing for several liquid inlet concentrations and flow-rates.

The experimental observed rates are compared in Figs 6–9 with rates predicted as described below.

Matching of Stirring Speeds with Flow-rates

For a given stirring speed, and for each buffer solution in turn, the ratio:

$$\frac{\gamma \text{ in stirred vessel}}{\gamma \text{ in packed column}}, \text{ or } \frac{\gamma_{ves}}{\gamma_{col}}$$

was calculated for the various liquid flow-rates employed in the column. This ratio was plotted against the flow-rate, L. The plot for a stirring speed of 95 rev/min is shown in Fig. 10. The lines do not all intersect at $\gamma_{ves}/\gamma_{col} = 1$, as in the ideal case (see Fig. 1), but there is a region of convergence. In order to determine the value of L at which k_L is the same as that in the stirred vessel at 95 rev/min, we have arbitrarily identified it with the value of L at which the average deviation from unity of the ratio $\gamma_{ves}/\gamma_{col}$ is zero; this occurs at L = 0·40 cm/s. A similar procedure was used for other stirring speeds. The results of the matching procedure are shown in Fig. 11.

Determination of the Value of k_L and a in Column

Since the value of $k_L a$ had already been determined for various flow-rates in the column, as described above, a separate determination of either k_L or a sufficed to give the values of both quantities. Two independent methods were used.

The first followed from the matching of stirring speeds with flow-rates, described above. It had been established that the value of k_L in the stirred vessel at 95 rev/min (for instance) was the same as that in the column at 0·40 cm/s. The value of k_L in the stirred vessel at 95 rev/min had been shown, from physical absorption measurements, to be 0·0111 cm/s (Table I); this is therefore the value of k_L in the column at a

Table III.—*Absorption Into Buffer Solutions in Packed Column*

Solution	1		3		4		5		6		7		8		9		10	
k_1 (s^{-1})=			85		52·5		23		130		37		55		7		1·5	
L (cm/s)	10^2Ra	γ	10^2Ra	γ	10^2Ra	γ	10^2Ra	γ	10^2Ra	γ	10^2Ra	γ	10^2Ra	γ	10^2Ra	γ	10^2Ra	γ
0·3025	1·16	3·41	1·18	3·47	1·07	3·15	0·79	2·32	1·39	4·09	0·87	2·57	0·91	2·65	0·58	1·70	0·42	1·24
0·605	1·69	2·60	1·82	2·80	1·57	2·42	1·26	1·96	1·96	3·02	1·29	1·98	1·41	2·17	0·98	1·52	0·74	1·14
0·9075	2·16	2·16	2·44	2·44	2·15	2·15	1·74	1·74	2·58	2·58	1·80	1·80	1·96	1·96	1·44	1·44	1·12	1·12
1·21	2·60	1·90	3·08	2·25	2·77	2·02	2·21	1·61	3·22	2·35	2·36	1·72	2·57	1·88	1·90	1·39	1·53	1·115
1·51	3·09	1·77	3·76	2·15	3·39	1·94	2·68	1·53	3·95	2·26	2·96	1·69	3·21	1·84	2·38	1·36	1·94	1·10

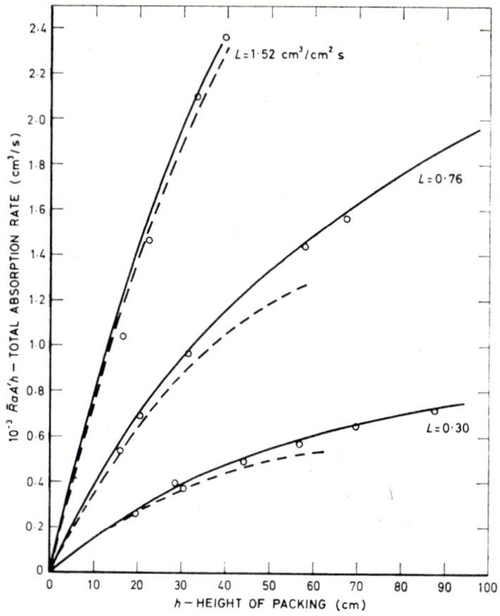

——— : predictions by Methods I and II
– – – : predictions by Method III
○ : experimental

Fig. 6.—*Feed:* 0·08 M *sodium hydroxide* +0.52 M *sodium sulphate*

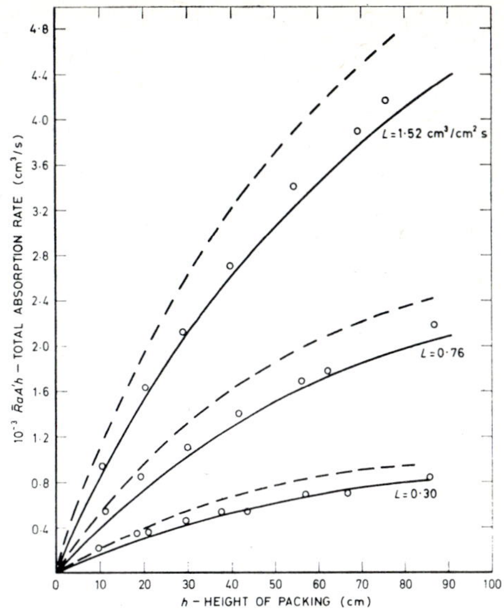

——— : predictions by Methods I and II
– – – : predictions by Method III
○ : experimental

Fig. 7.—*Feed:* 0·15 M *sodium hydroxide* + 0·51 M *sodium sulphate*

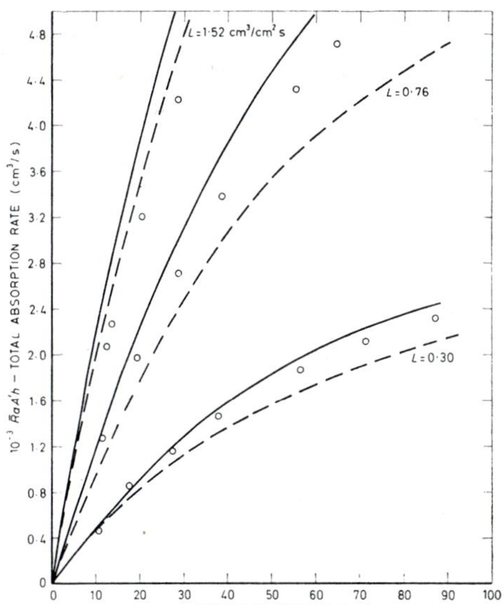

——— : predictions by Methods I and II
– – – : predictions by Method III
○ : experimental

Fig. 8.—*Feed* 0.37 M *sodium hydroxide* + 0.2 M *sodium sulphate*

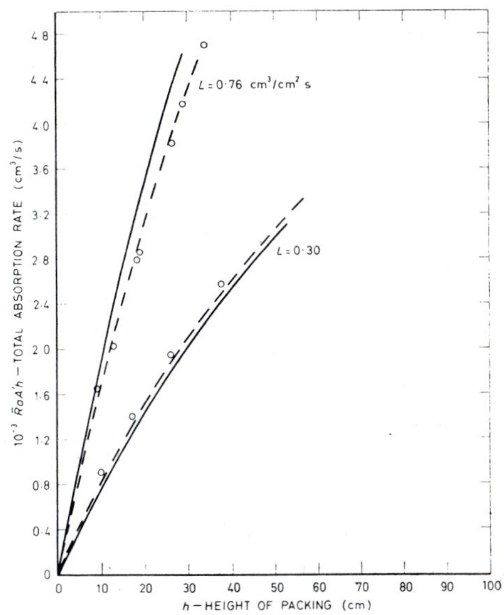

——— : predictions by Methods I and II
– – – : predictions by Method III
○ : experimental

Fig. 9.—*Feed:* 0.80 M *sodium hydroxide* + 0.1 M *sodium sulphate*

flow-rate of 0·40 cm/s. The value of $k_L a$ at this flow-rate was about $7{\cdot}0 \times 10^{-3}\ s^{-1}$ (Fig. 5), so the value of a is about $0{\cdot}63\ cm^{-1}$. Values of k_L and a determined by this method are shown in Fig. 12 (circles).

The second method, which did not involve the stirred vessel, was based on the measured rates of absorption into the buffer solutions in the column. In these solutions the dissolved carbon dioxide undergoes a pseudo-first-order reaction with known velocity constant, k_1. In these circumstances (according to the Danckwerts surface-renewal model):

$$Ra = aC^* \sqrt{(Dk_1 + k_L^2)} \quad . \quad . \quad (5)$$

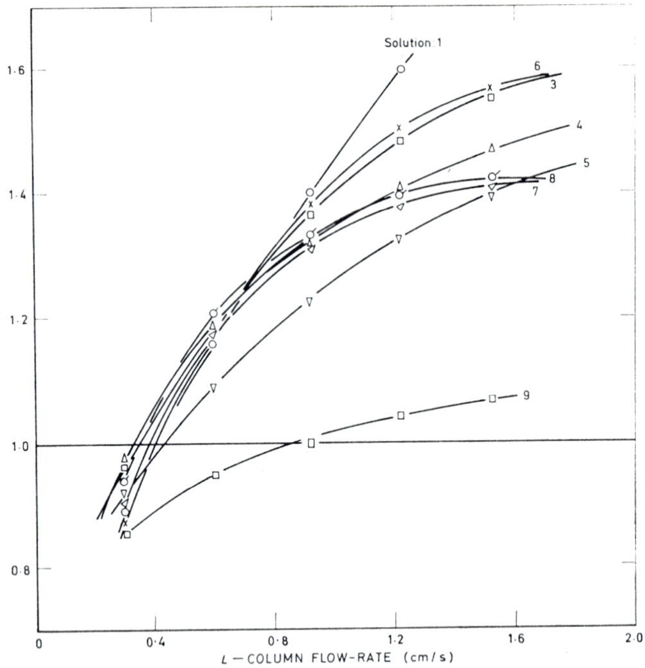

Fig. 10.—*Matching of column flow-rate to stirring speed of 95 rev/min*

Thus, a plot of $(Ra)^2$ against k_1 should give a straight line of slope $(aC^*)^2D$, from which a can be determined if $C^*\sqrt{D}$ is known; k_L is then calculated from the values of k_La already determined by physical absorption. (Values of k_L can also be calculated from the intercepts of the lines just mentioned, as can be seen from equation (5), but this is a less accurate method.)

The value of $C^*\sqrt{D}$ was determined in a laminar-jet absorption apparatus, as described by Sharma and Danckwerts.[7] Its value in the buffer solutions used by us was $10{\cdot}8 \times 10^{-8}$ g mol./cm^2 s$^{1/2}$ at 20° C, 1 atm, or $2{\cdot}60 \times 10^{-3}$ cm/s$^{1/2}$. Measurements of absorption rates in the column, using various buffer solutions (Table III) are plotted in the appropriate way in Fig. 14 (the results are shown for three flow-rates only). The slopes of the curves decrease at the higher values of k_1. This is no doubt due to the accumulation of bicarbonate near the surface of the absorbent, in sufficient

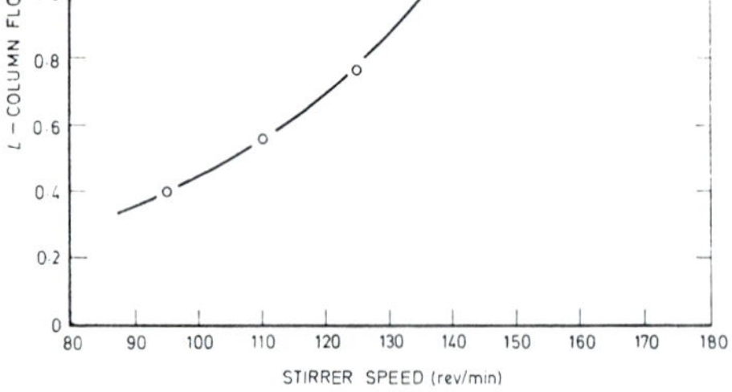

Fig. 11.—*Relation between column flow, L, and stirrer speed*

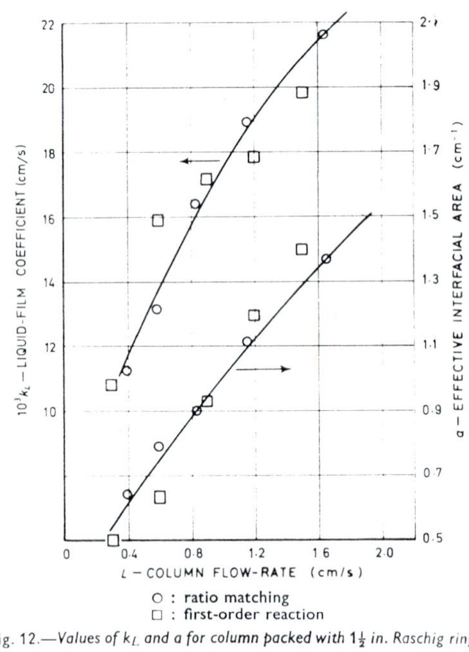

○ : ratio matching
□ : first-order reaction

Fig. 12.—*Values of k_L and a for column packed with $1\frac{1}{2}$ in. Raschig rings*

concentration to depress the real value of k_1 below that calculated from equation (1). The effect is discussed in Appendix I. The intial slopes of the curves were used to determine the value of a.

The values of a and k_L at different flow-rates found by this method are shown in Fig. 12 (squares), together with those found by the former method. The agreement between the two methods is reasonably good, and gives some confidence in the validity of the concepts employed, particularly as the ratio-matching method primarily determines k_L in the column, while the first-order method determines a. The first-order method appears to have the lower precision.

However, it will be noted that the value of a at the highest flow-rate (1·6 cm/s) was about 1·4 cm^{-1}, and the trend of the results suggests that greater flow-rates would lead to higher values of a. The geometrical surface area of $1\frac{1}{2}$ in. Raschig rings is about 1·3 cm^{-1}; the wall of the column contributed a further 0·085 cm^{-1} to the area. It seems, therefore, that the quantity, a, determined by these methods can at least equal, and possibly exceed, the geometrical surface area of the packing. (The work of Shulman *et al.* suggests that there is no drastic change in the trend of liquid hold-up with flow-rate when $L = 1·6$ cm/s. At this flow-rate the hold-up is only about 8% of the volume of the column.) It is desirable to extend the measurement of a to higher flow-rates (some modification of our apparatus would have been necessary). If values substantially greater than that of the geometrical surface arise it must be concluded that a is a quantity which can be used in empirically successful design calculations, but which cannot be identified with the area of an actual surface.

As a further check on the "first-order" method of determining the interfacial area, it was used to establish the value of this quantity in the stirred vessel. The results are shown in Fig. 13. The apparent value of the interfacial area

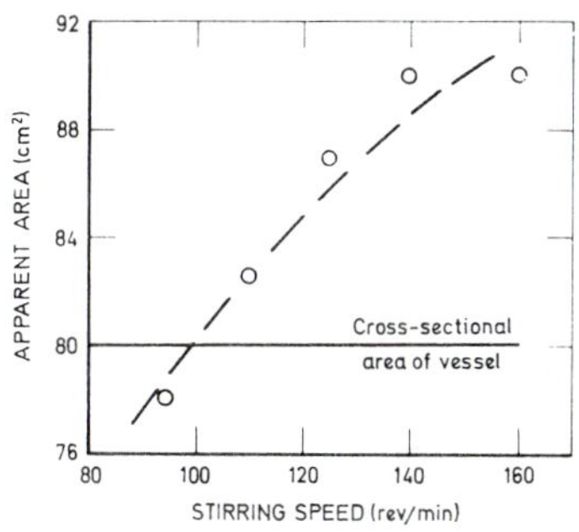

Fig. 13.—*Apparent interfacial area in stirred vessel, determined by "first-order" method*

varied by 14% over the range of stirrer speeds employed, possibly reflecting a true increase caused by ripples on the surface. (No central depression was observed to form even at the highest stirring speed.) The apparent area at the lowest stirring speed was some 2·5% less than the cross-sectional area of the vessel (less that of the stirrer). These results give an indication of the validity and accuracy of the method. (The cross-sectional area of the vessel less that of the stirrer was used as the value of A in all other calculations, *i.e.* 80 cm^2.)

Prediction of Rates of Absorption into Solutions of Sodium Hydroxide in Column

The three methods described earlier will now be used to predict the rates of absorption into solutions of sodium hydroxide in the packed column, and the predictions will be compared with the observations. The regime of absorption kinetics ranged, according to the operating conditions, from diffusion-controlled "instantaneous" reaction to a situation closer to pseudo-first-order reaction.

Method I

Consider a situation when a solution of 0·8 M sodium hydroxide is fed to the column at a rate of 0·76 cm/s. From Fig. 11 we see that this corresponds to a stirring speed of 124 rev/min. From Fig. 4, by interpolation or cross-plotting, we find the value of RA and hence the value of γ in the stirred vessel at this stirring speed with various concentrations, C_0, of absorbed carbon dioxide in the solution. To find the local value of (Ra) in the column at a level where C_0 has a specific value, we multiply the value of γ appropriate to the local value of C_0 by C_0^* and by the value of k_La found from Fig. 5 (the value of $k_LaC_0^*$ at a flow-rate of 1·52 cm/s is 0·0175 s^{-1}).

The differential equation relating height, composition and absorption-rate is:

$$LA'dC_0 = RaA'dh, \qquad (6)$$

h being the height of packing measured downwards from the top. If $C_0 = 0$ at the top (no carbon dioxide dissolved in feed solution):

$$\frac{h}{L} = \int_{C_0=0}^{C_B} \frac{dC_0}{Ra} = \frac{1}{k_LaC_0^*}\int_{C_0=0}^{C_B} \frac{dC_0}{\gamma}. \qquad (7)$$

The concentration, C_B, of absorbed carbon dioxide at the bottom of the column is given by:

$$C_B = \frac{\text{Total rate of absorption in column}}{LA'} \qquad (8)$$

Since γ has been determined as a function of C_0, the integral in equation (7) can be evaluated numerically for various values of the total absorption rate, and the corresponding values of h, the height of packing exposed, are thus determined. A curve representing the results of this series of calculations is compared with the experimental results for a flow-rate of 0·76 cm/s of a solution of 0·8 M sodium hydroxide in Fig. 9. Similar curves are presented for other sets of experimental conditions in Figs 6–9. It will be seen that the observed and calculated absorption rates are in reasonable agreement, the maximum discrepancy being less than 20%.

As explained above, it was not possible to determine γ in the stirred vessel at low values of C_0; this shortcoming is

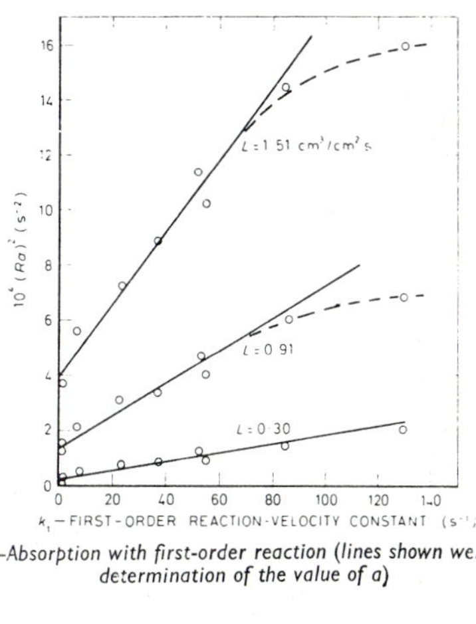

Fig. 14.—*Absorption with first-order reaction (lines shown were used for determination of the value of a)*

apparent in Fig. 4. There was consequently some uncertainty in carrying out the integrations between $C_0 = 0$ and the lowest values of C_0 for which γ had been determined. However, this was a minor source of error.

Method II

The values of k_L and a established experimentally are shown in Fig. 12. The smoothed curves shown there can be used for the purposes of Method II. Considering, once more, the solution of 0·8 M sodium hydroxide at a flow-rate of 0·76 cm/s, we see from Fig. 12 that the corresponding value of k_L is 0·0155 cm/s. Table I shows the relationship found to exist between stirring speeds and values of k_L in the stirred vessel; by interpolation, it is found that a stirring speed of 122 rev/min would give the required value of k_L. By interpolation in Fig. 4 we can find the value of RA, and hence of R, for a stirring speed of 122 rev/min, for a solution containing initially 0·8 M sodium hydroxide and for various concentrations, C_0, of absorbed carbon dioxide. We can then calculate the heights of column, h, corresponding to various absorption rates (*cf.* equation (7)):

$$h = \frac{L}{a}\int_{C_0=0}^{C_B} \frac{dC_0}{R} \quad . \quad . \quad . \quad (9)$$

(the value of a is found from Fig. 12 to be 0·86 cm^{-1}).

In the present case virtually the same data have been used for Methods I and II; the curves in Fig. 12 lie close to the experimental points determined by "ratio matching". Thus, the predictions made by the two methods will be essentially the same, the only difference being that introduced by smoothing the experimental points.

Method III: calculation from kinetics

The rate-determining reaction is that between OH^- and CO_2:

$$CO_2 + OH^- \rightarrow HCO_3^- \quad . \quad . \quad (10)$$

It is second order, the rate-constant k_2 being about $5{\cdot}9 \times 10^3$ l./g mol. s at 20° C and at infinite dilution; its value in electrolyte solutions varies markedly with the ionic strength of the solution. The reaction can be regarded as irreversible under the conditions of our experiments.

Brian, Hurley, and Haseltine[5] have carried out computations, based on the Higbie surface-renewal model, of the effect of an irreversible second-order reaction on the rate of absorption. In terms of the present system, they give $R/k_L C^*$ as a function of $\sqrt{\{Dk_2[OH^-]\}}$, with the quantity $\{[OH^-]/2[CO_2]^*\}\sqrt{(D'/D)}$ as a parameter. Here R is the rate of absorption per unit interfacial area (cm^3/cm^2 s), $[CO_2]^*$ the saturated solubility of carbon dioxide in the solution in question (g mol./l.), C^* its value in cm^3/cm^3, k_2 the reaction-rate constant (l./g mol. s), $[OH^-]$ the concentration of hydroxyl ion (g mol./l.), D the diffusivity of carbon dioxide (cm^2/s), and D' the effective diffusivity of OH^- in the solution. The value of $[OH^-]$ is known (original concentration of sodium hydroxide less equivalent concentration of absorbed carbon dioxide), and the value of k_L is taken to have been established by the methods described above. If values can be assigned to $[CO_2]^*$, D', D, and k_2 in the solution in question, the value of R can be determined from the relationship given by Brian *et al.* There is considerable uncertainty about the values of these quantities. Methods of estimating them are suggested in Appendix II. Using these methods, R has been calculated for the various solutions of sodium hydroxide, at different flow-rates and different degrees of depletion (*i.e.* values of C_0). Equation (9) has then been integrated to predict the height of packing required for various rates of absorption. The results are shown in Figs 6–9. The predictions are not much less accurate than those of Methods I and II; but it must be remembered that the extent of knowledge about the kinetics of the reaction and about the various physico-chemical parameters in the system carbon dioxide–sodium hydroxide is quite exceptional. In most cases of gas absorption accompanied by chemical reaction the use of Method III is likely to be beset with uncertainties.

Effect of Packing Height on Absorption Efficiency

It has often been suggested that the efficiency of a packing decreases as the height of the packed column is increased, the reason being that the liquid finds its way to the wall and runs down it; the flow-rate over the packing proper would then decrease, with an adverse effect on both k_L and a. Thus, if the liquid were properly distributed at the top of the packing, the packing would operate more efficiently near the top than lower down. Such behaviour might make the design methods suggested in this paper highly inaccurate or useless for tall columns, as no unique values of k_L and a could be used.

To test this we absorbed carbon dioxide into a buffer solution (0·6 M Na_2CO_3, 0·2 M $NaHCO_3$) with various heights of packing. The rate of chemical reaction in this

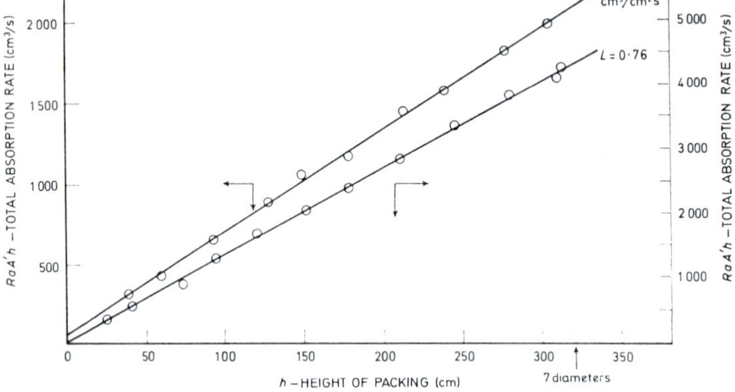

Fig. 15.—*Absorption into buffer solution without catalyst*

solution is not sufficient to have much effect on the absorption rate directly, but the carbon dioxide absorbed reacts in the bulk of the liquid to form bicarbonate, so that the back pressure of carbon dioxide in equilibrium with the solution remains negligible. Thus, the local rate of absorption would be the same at all levels of the column if the flow characteristics remained the same; on the other hand, a progressive decrease in k_L and a would be reflected in a progressive decrease in the local absorption rate.

As shown in Fig. 15, the local absorption rate (at two different flow-rates) was independent of the height of packing, at any rate between heights equal to 0·5 and seven column diameters. We conclude that if substantial redistribution of flow occurs it must be either complete within a distance equal to less than one column radius from the distributor; or it must take place so slowly that it has no appreciable effect on the irrigation of the packing within a distance of seven column diameters; or the increasing rate of absorption into liquid running down the wall must just make up for the decreasing rate of absorption in the bulk of the packing. A more likely interpretation of these results is that in the circumstances in which our measurements were made there is no substantial redistribution of flow, and that the liquid remains well distributed.

Conclusions

It appears that a stirred vessel (and hence probably other types of laboratory model) can be used to predict the effects of chemical reactions on rates of absorption in packed columns, with a degree of accuracy which would at any rate represent an improvement on present design procedures.

Ratio matching or the first-order method can be used to determine once and for all values of k_L and a for packed columns under various hydrodynamic conditions. These quantities can be used for prediction, either with the aid of measurements in models or by calculation from first principles. In general the data required for the last method are unlikely to be available.

APPENDIX I

Kinetics of the Catalysed Reaction with Water

In a carbonate–bicarbonate buffer solution, dissolved carbon dioxide reacts with OH^- ions, at a local rate $k_2[CO_2][OH^-]$; the value of $k_2[OH^-]$ at 20° C is about $1{\cdot}5\ s^{-1}$ in the buffer solution which we used (0·6 M potassium carbonate, 0·2 M potassium bicarbonate). The carbon dioxide reacts also with water at a rate $k_{H_2O}[CO_2]$, the value of k_{H_2O} being about $0{\cdot}02\ s^{-1}$ in these circumstances. In the presence of hypochlorite ions, there is also a catalysed reaction with water; the rate is $k_{OCl^-}[CO_2][OCl^-]$. The value of k_{OCl^-} varies greatly with the concentration and nature of the various ions in the solution. Its value in the buffer solution used by the authors was determined by means of a laminar-jet absorption apparatus,[7] and found to be 2700 l./g mol. s at 20° C. The rates of the three reactions are added together to give the total pseudo-first-order rate-constant for the reaction of carbon dioxide:

$$k_1 = k_2[OH^-] + k_{H_2O} + k_{OCl^-}[OCl^-] \quad . \quad (11)$$

(*cf.* equation (1)).

If the reaction is to be pseudo-first order, the value of k_1 must be the same at all times and in all regions of the liquid. During the time of exposure of an element of surface (to use the nomenclature of the surface-renewal model) the reaction of carbon dioxide near the surface converts some $CO_3^=$ into HCO_3^-. This change in local composition will bring about a change in k_{OCl^-}; Sharma[8] has shown that when, for instance, a solution containing 1·5 M K_2CO_3 and 0·5 M $KHCO_3$ is carbonated to 0·5 M K_2CO_3 and 2·5 M $KHCO_3$, the value of k_{OCl^-} decreases by a factor of 3·7. We might expect, therefore, that at higher values of k_1t (where t is the exposure time of the liquid surface) the value of k_1 in the neighbourhood of the surface would decrease. The result would be that rates of absorption would be lower than predicted by equations (1) and (5) at high values of k_1; a plot of $(Ra)^2$ against k_1, as calculated from equation (1), would show a diminishing slope at high values of k_1. This behaviour is shown by Fig. 14, and was shown by the corresponding plots for the stirred vessel.

APPENDIX II

Absorption Rates in Solutions of Sodium Hydroxide

The solutions contain, in general, sodium bicarbonate, sodium hydroxide, and sodium sulphate. We wish to estimate the solubility, $[CO_2]^*$, of carbon dioxide, the velocity constant, k_2, for the reaction of carbon dioxide with OH^-, the diffusivity, D, of carbon dioxide, and the effective diffusivity, D', of OH^-.

For the solubilities we use an extension of the method of van Krevelen and Hoftyzer.[9] It is assumed that the solubility is given by an expression of the form:

$$-\log_{10}\frac{[CO_2]^*}{[CO_2]^*_W} = m_1I_1 + m_2I_2 + m_3I_3 \quad . \quad (12)$$

where I_1, I_2, I_3 are the contributions to the ionic strength of sodium hydroxide, sodium sulphate, and sodium bicarbonate, respectively, and m_1, m_2, m_3 are constants which are not sensitive to temperature and which do not interact; $[CO_2]^*_W$ is the solubility of carbon dioxide in water at the temperature in question. Thus:

$$-\log_{10}\frac{[CO_2]^*}{[CO_2]^*_W} = m_1[NaOH] + 3\,m_2[Na_2SO_4] + 3\,m_3[Na_2CO_3] \quad . \quad . \quad . \quad (13)$$

the terms in brackets being concentrations (g mol./l.). The value of m_1 is given by Nijsing, Hendriksz, and Kramers,[10] and is 0·149 l./g mol. Nijsing and Kramers[11] have shown that the value of m_3 is 0·088 l./g mol. The value of m_2 can be found by comparing the solubilities of carbon dioxide in (for instance) water and a solution of 0·593 M sodium sulphate at 25° C;[10] it is found to be 0·101 l./g mol. Thus:

$$-\log_{10}\frac{[CO_2]^*}{[CO_2]^*_W} = 0{\cdot}149\,[NaOH] + 0{\cdot}303\,[Na_2SO_4] + 0{\cdot}264\,[Na_2CO_3] \quad . \quad . \quad (14)$$

The value of D in 0·52 M sodium sulphate at 25° C is $1{\cdot}58 \times 10^{-5}\ cm^2/s$ (Ref. 10); assuming the temperature coefficient to be the same as that in water, the value at 20° C will be $1{\cdot}375 \times 10^5\ cm^2/s$. From the same source we find the value of D in 0·80 M sodium hydroxide at 20° C to be $1{\cdot}47 \times 10^{-5}\ cm^2/s$. Thus, little error will probably be introduced by giving D the value $1{\cdot}4 \times 10^{-5}\ cm^2/s$, in all the solutions dealt with.

The value of k_2 can be estimated from the plot given by Pinsent, Pearson, and Roughton;[12] this method assumes that the value of k_2 depends only on the ionic strength and not otherwise on the composition of the solution. This is unlikely to be exactly true, but is as good an estimate as can be made with the data available.

The ratio D/D' has been determined in various solutions.[10, 13] It will be assumed here that it has the value 0·59 for all the solutions.

Suppose now that we wish to calculate the specific rates of absorption, R, into solutions which originally contained 0·8 M sodium hydroxide with 0·1 M sodium sulphate, the superficial flow-rate, L, over the packing being 0·76 cm/s. The value of k_L is then (Fig. 12) 0·0155 cm/s and of a 0·86 cm^{-1}. From equation (14) we calculate that the value of $[CO_2]^*$ is 0·029 g mol./l., with a variation of only about 1% over the range of carbonation covered in the absorption experiments (C_0 ranging up to 6 cm^3/cm^3). The ionic strengths of the solutions range only from 1·1 to 1·25 g ion/l., so that it is sufficiently accurate to take an average value of k_2, namely 10^4 l./g mol. s (ref. 12). The dimensionless variable $\sqrt{\{Dk_2[NaOH]\}}/k_L$ thus becomes 23·4 $\sqrt{[NaOH]}$. (In these calculations [NaOH] is the initial concentration of sodium hydroxide less the equivalent concentration of absorbed carbon dioxide.) The variable $\frac{[NaOH]}{[CO_2]^*}\sqrt{\left(\frac{D}{D'}\right)}$ becomes 23·3 [NaOH]. Thus, when the residual concentration, [NaOH], is 0·4 (corresponding to $C_0 = 4·8$ cm^3/cm^3) we find that $\sqrt{\{Dk_2[NaOH]\}}/k_L = 14·8$, $\frac{[NaOH]}{[CO_2]^*}\sqrt{\left(\frac{D}{D'}\right)} = 9·3$. From the numerical results of Brian *et al.*[5] we find that the value of R/k_LC^* is then about 7·8. Thus:

$$Ra = 7·8 \times k_L \times a \times [CO_2]^* \times 22·4 \times 293/273 = 0·73\ s^{-1}.$$

The calculation is repeated for other values of C_0, and the integral in equation (9) computed.

The calculations for other flow-rates and feed solutions are similar.

Acknowledgment

The authors are grateful to the Gas Council for financial support and technical assistance in the course of this work.

Symbols Used

A = cross-sectional area of stirred vessel less that of stirrer (80 cm^2).
A' = cross-sectional area of packed column (1655 cm^2).
a = effective interfacial area per unit volume of packing (cm^{-1}).
C_B = value of C_0 at bottom of column.
C^* = solubility of carbon dioxide in solution at 20°C, 1 atm (cm^3/cm^3).
C_0^* = solubility of carbon dioxide in inert (0·53 M sodium sulphate) solution (cm^3/cm^3).
C_0 = concentration of absorbed carbon dioxide in bulk of solution (cm^3/cm^3).
$[CO_2]^*$ = solubility of carbon dioxide in solution at 20° C, 1 atm (g mol./l.).
$[CO_2]_W^*$ = solubility of carbon dioxide in water at 20° C, 1 atm (g mol./l.).
D = diffusivity of carbon dioxide in solution (cm^2/s).
D' = effective diffusivity of OH^- in solution (cm^2/s).
h = height of packing exposed to gas (cm)
I_1, I_2, I_3 = ionic strengths (eqn. (12)) (g ion/l.).
k_1 = first-order reaction-velocity constant (s^{-1}).
k_2 = reaction-velocity constant for $CO_2 + OH^-$ (l./g mol. s).
k_{H_2O} = reaction-velocity constant for $CO_2 + H_2O$ (s^{-1}).
k_L = liquid film coefficient in absence of reaction (cm/s).
k_{OCl^-} = catalytic constant for OCl^- (eqn. (11). (l./g mol. s).
L = superficial liquid flow-rate in column (*i.e.* volumetric flow-rate divided by cross-sectional area) (cm/s).
m_1, m_2, m_3 = constants in equations (12) and (13) (l./g mol.).
R = rate of absorption per unit interfacial area (cm^3/cm^2 s).
$\bar{R}$ = average value of R in packed column.
v = liquid flow-rate through stirred vessel (cm^3/s).
γ = ratio of rate of absorption in reacting solution to rate of absorption in carbon-dioxide-free 0·53 M sodium sulphate.

Note:—In general, symbols in square brackets denote concentrations in g mol./l.

References

1 Danckwerts, P. V., and Kennedy, A. M. *Trans. Instn chem. Engrs*, 1954, **32**, 349.
2 Danckwerts, P. V., Kennedy, A. M., and Roberts, D. *Chem. Engng Sci.*, 1963, **18**, 63.
3 Richards, G. M., Ratcliff, G. A., and Danckwerts, P. V. *Chem. Engng Sci.*, 1964, **19**, 325.
4 Sherwood, T. K., and Holloway, F. A. *Trans. Am. Inst. chem. Engrs*, 1940, **36**, 21 and 39.
5 Brian, P. L. T., Hurley, J. F., and Haseltine, E. H. *A.I.Ch.E. Jl*, 1961, **7**, 226.
6 Brian, P. L. T., Vivian, J. E., and Habib, A. G. *A.I.Ch.E. Jl*, 1962, **8**, 205.
7 Sharma, M. M., and Danckwerts, P. V. *Chem. Engng Sci.*, 1963, **18**, 729.
8 Sharma, M. M. *Ph.D. Thesis*, 1964, p. 32. Cambridge University.
9 van Krevelen, D. W., and Hoftyzer, P. J. "*Chim. Ind. XXIme Congres Internationale de Chimie Industrielle*", 1948, p. 168 (Paris: Société de Productions Documentaires).
10 Nijsing, R. A. T. O., Hendriksz, R. H., and Kramers, H. *Chem. Engng Sci.*, 1959, **10**, 88.
11 Nijsing, R. A. T. O., and Kramers, H. *Chem. Engng Sci.*, 1958, **8**, 81.
12 Pinsent, B. R. W., Pearson, L., and Roughton, F. J. W. *Trans. Faraday Soc.*, 1956, **52**, 1512.
13 Danckwerts, P. V., and Kennedy, A. M. *Chem. Engng Sci.*, 1958, **8**, 201.
14 Shulman, H. L., Ullrich, C. F., and Wells, N. *A.I.Ch.E. Jl*, 1955, **1**, 247.

The manuscript of this paper was received on 12 November, 1965

THE DESIGN OF GAS ABSORBERS PART II: EFFECTIVE INTERFACIAL AREAS FOR SEVERAL TYPES OF PACKING

By Professor P. V. DANCKWERTS, G.C., M.B.E., M.A., C.Eng., F.R.S. (MEMBER)† and S. F. RIZVI, M.Sc.‡

SYNOPSIS

Effective interfacial areas for gas absorption were measured for a number of different types of packing and rates of irrigation. The method used was a chemical one, involving the absorption of oxygen from air by a solution of sodium sulphite containing a little cobalt sulphate; the latter catalyses the oxidation of sulphite by oxygen. The results confirm those of Danckwerts and Gillham[1] (Part I of this paper) obtained by different methods, using 1½ in Raschig rings.

Introduction

The object of the work described in this paper was to measure the effective interfacial area for gas absorption accompanied by chemical reaction when packings of various kinds were irrigated with liquid at various flow-rates. The results obtained duplicated and confirmed those obtained in Part I of this paper by Danckwerts and Gillham[1] for 1½ in Raschig rings, and extended them to other packings, but the method used was more direct and convenient.

Principle of Method

The method used was the measurement of the rate of absorption of oxygen from air into a solution of sodium sulphite containing a small amount of cobalt sulphate. The reaction of dissolved oxygen with sulphite to form sulphate is catalysed by the cobalt to the point at which it is so rapid that it all takes place within the diffusion film, the rate of absorption being independent of k_L. The kinetics of the reaction, in the range of conditions relevant to the present discussion, have been investigated by de Waal and Okeson;[2] however, these authors wrongly assumed that the reaction is first-order with respect to oxygen, whereas in fact it is second-order (Reith,[3] and Wesselingh and van't Hoog[4]), so that the rate of absorption, R, is proportional to the 3/2 power of the partial pressure of oxygen at the surface:

$$R = kp^{3/2} \quad \text{g mol/cm}^2\,\text{s} \qquad (1)$$

where p is the partial pressure of oxygen (atm). The value of k depends on the temperature, the pH, and the ionic strength but is independent of the extent of oxidation of sulphite provided that the pH is kept constant during the transformation of sulphite to sulphate by the addition of alkali. In our experiments the change in pH during one pass of the liquid through the packed column or stirred cell was negligible so that k had the same value at all points in the apparatus. The order of magnitude of k in our experiments (pH = 8·2–8·5) was $1{\cdot}5\times10^{-7}$ g mol/cm² s (atm)$^{1{\cdot}5}$; thus the specific rate of absorption of oxygen from air was about $1{\cdot}5\times10^{-8}$ g mol/cm² s.

The principle of the method was to measure the rate of absorption of oxygen from air in a packed column, and then, using the same liquid at the same temperature, to measure the rate of absorption of oxygen from air in a stirred vessel. The latter measurement yielded k from which it was possible to calculate the effective interfacial area required to give the observed rate of absorption in the column. The partial pressures of oxygen in all streams were measured by an oxygen meter which gave the oxygen partial pressure directly as a percentage of a standard atmosphere.

A steady stream of air was passed through the stirred cell, and the rate of absorption was calculated from the flow-rate and the partial pressures of oxygen in the in-flowing and out-flowing streams. The gas in the cell was well mixed and thus had the same composition as the out-flowing stream. If the meter-reading for the out-flowing stream was z the rate of absorption in the cell was:

$$RA = Ak\left(\frac{z}{100}\right)^{1.5} \quad \text{g mol/s} \qquad (2)$$

RA being the measured rate of absorption and A the known area of liquid surface in the stirred cell. Thus k could be calculated from equation (2).

For the packed column we can write the following differential equation:

$$\bar{G}\frac{dp}{P} = -Ra\,dh \qquad (3)$$

Here $\bar{G}$ is the superficial mass velocity of gas (g mol/cm² s); the maximum variation in the value of this quantity was about 12%, owing to the absorption of oxygen from the air and $\bar{G}$ was therefore treated as a constant in equation (3), being the mean between the mass velocity, G_i, at the inlet and that at the outlet, G_0. The two are related by:

$$G_0/G_i = \frac{0{\cdot}79}{1-0{\cdot}21y/x} \qquad (4)$$

† Department of Chemical Engineering, Pembroke Street, Cambridge.

‡ Warren Spring Laboratory, Stevenage, Herts.

where x is the meter reading for the air at point of entry and y the value for the outgoing depleted air. (x of course varies with the atmospheric pressure.)

In equation (3) p is the partial pressure of oxygen and P the total or atmospheric pressure (the pressure-drop in the column was negligible, the superficial velocity of gas being about one centimetre per second); h is the height measured downwards from the top of the column (the gas flowed downwards), and a the effective interfacial area per unit of packed volume. Since a was expected to vary with height (see below), we put:

$$dB = a\,dh \quad . \quad . \quad . \quad (5)$$

and substituting from equations (1) and (5), equation (3) becomes:

$$\bar{G}\frac{dp}{P} = -kp^{1 \cdot 5}\,dB \quad . \quad . \quad . \quad (6)$$

$$B = -\frac{\bar{G}}{kP}\int_{x/100}^{y/100}\frac{dp}{p^{1 \cdot 5}} \quad . \quad . \quad (7)$$

and since $P = x/21$,

$$B = \frac{420\bar{G}}{kx}\left(\frac{1}{\sqrt{y}} - \frac{1}{\sqrt{x}}\right) \quad . \quad . \quad (8)$$

The procedure was to carry out absorption experiments with various heights, h, of packing exposed and to determine B for each height. As may be seen from equation (5), the slope of a plot of B against h at a given h is the value of a, the specific effective interfacial area, at a distance h below the top of the packing.

The effect of gas side resistance was quite negligible in all experiments. At the rates of absorption observed, the gas side resistance would have had to be equivalent to a film of stagnant gas 10 cm thick to make the partial pressure of oxygen at the interface one per cent less than in the bulk of the gas.

Calculations showed that the rise in temperature at the surface of the liquid, above the bulk temperature, was less than 0·02 degC, while the bulk temperature in the most extreme case rose by about 0·5 degC. According to de Waal and Okeson[2] the reaction velocity constant rises by about 5·5% per degC, but the absorption rate only varies as the square-root of the reaction velocity constant, and the effect on the absorption rate is offset by the decrease with temperature of the solubility of oxygen in the liquid. It may be concluded that the value of k (equation (1) did not vary by more than two per cent between the top and bottom of the column.

Experimental

The stirred vessel was similar to that used by Danckwerts and Gillham[1] and described in Part I, having a liquid surface-area of about 76 cm^2. The liquid stirrer consisted of four flat blades which intersected the plane of the liquid surface. The stirrer shaft also carried a gas stirrer to ensure that the composition of the gas was uniform and thus equal to the out-flowing composition. The rate of air-flow was measured by a soap-bubble flow-meter and the partial pressure of oxygen by a Servomex Type O.A.137 magnetic oxygen meter with a digital voltmeter attached for greater accuracy. The gas was in all cases dried before the oxygen pressure was measured. It was confirmed that the rate of absorption was independent of the rate of stirring, from 90–130 rev/min, thus covering the range of k_L likely to be encountered in the packed column;[1] this showed that absorption was in the "fast" regime and independent of k_L, and that equation (1) applied. The stirred vessel was immersed in a tank of water and thus could be maintained at the same temperature as the liquid entering the top of the column, which ranged from 34°C to 11°C.

The column was essentially identical to that used by Danckwerts and Gillham and described in Part I, being about 4 m high and 0·5 m internal diameter. The air was blown into the top and flowed downwards, leaving through a pair of openings in the wall of the column. The liquid level was adjusted by an overflow device so as to be immediately below the openings through which the gas left. There were a number of such pairs of openings, and the effective height, h, of the column was varied by changing the exit from one pair to another and adjusting the liquid level accordingly. (See Part I of this paper.[1])

The air and liquid flow-rates were measured by rotameter. The partial pressure of oxygen was measured as with the stirred cell.

The absorbent solution was made up in batches of 10 m^3, using tap water. The initial composition was 0·8 M Na_2SO_3, 4×10^{-4} M $CoSO_4$. All free liquid surfaces except those in the column itself were blanketed with nitrogen. Under the most extreme conditions the sulphite in the solution was depleted by 0·005 M in one pass through the column. Since the solution was originally 0·8 M, and the absorption rate is undiminished by depletion down to 0·4 M,[2] the solution could be reused many times providing it was protected from air, and the pH was adjusted occasionally by the addition of sodium hydroxide. The interior metal parts of the plant were protected by epikote resin or rubber. (The solution must not be allowed to come into contact with copper or brass.[2])

A sample of the liquid was taken from the feed tank every time a run was made in the column and the absorption-rate measured in the stirred vessel, as described above, in order to determine the value of k for the run.

Some characteristics of the packings used are given in Table I.

TABLE I.—*Characteristics of Packings*

	Number per m^3 (a)	Number per m^3 (b)	Geometrical Surface Area (cm^{-1}) a
1½ in Raschig rings (ceramic)	14000	13000	1·3
1½ in Pall rings (polypropylene)	14000	15500	1·3
1½ in Intalox saddles (ceramic)	23000	21000	1·6
2 in Intalox saddles (polypropylene)	6500	6200	1·1

(a) Stated by manufacturer (Hydronyl Limited)

(b) Observed

Results and Discussion

Figure 1 shows a typical plot of B against h, for two-inch (50 mm) polypropylene Intalox saddles. In all cases such plots gave good straight lines with a positive intercept on the B-axis, indicating that there is an end-effect at the top of the column where the liquid is introduced. Otherwise there is no indication of a change of slope (*i.e.*, of specific interfacial area) with height. The implications of this are discussed later.

The values calculated for the specific interfacial area at different flow-rates in the case of 1½ in (38 mm) ceramic

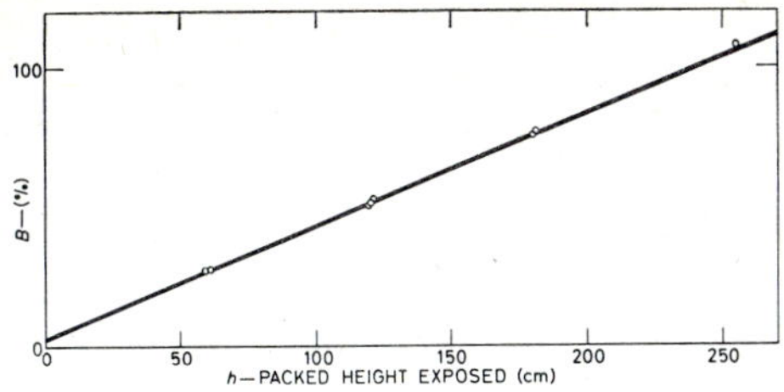

Superficial liquid velocity = 0·86 cm/s

Fig. 1.—*Plot of B against h for two-inch polypropylene Intalox saddles*

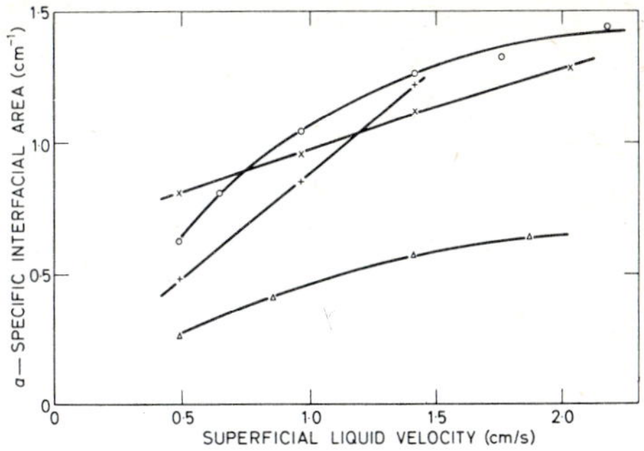

○: 1½ in ceramic Raschig rings
+: 1½ in polypropylene Pall rings
×: 1½ in ceramic Intalox saddles
△: 2 in polypropylene Intalox saddles
(1 cm/s is equivalent to about 7400 lb/ft² h)

Fig. 3.—*Plot of a against superficial liquid velocity for various packings*

Raschig rings are shown in Fig. 2, together with the best line through the results of Danckwerts and Gillham.[1] The agreement between them is reassuring, in view of the difference in the techniques used; however, the new results suggest that the value of the specific interfacial area *a*, levels-off at something over a value of $1{\cdot}4\ \text{cm}^{-1}$ at high flow-rates. This is somewhat higher than the geometrical surface-area of the packing (see Table I), suggesting that at high flow-rates not only is most of the surface of the packing effectively irrigated, but that drops may be falling through the gas as well. (The column wall makes only a very small contribution to the total area.)

Figure 3 summarises the results of the present work. There is not a great deal to choose, as regards effective interfacial area, between the various types of 1½ in packing. Unfortunately it was not possible to make a comparison between packings of the same shape and size made of ceramic and of polypropylene respectively. However, the ceramic Raschig rings and polypropylene Pall rings have the same geometrical surface area, and it is interesting to note that the two have the same effective surface area at a superficial liquid velocity of about 1·5 cm/s, although the effective surface area of the polypropylene packing is substantially less at lower velocities. In view of the large contact angle of water on polypropylene one might have expected the polypropylene packing to be decisively the worst at all flow-rates from the point of view of effective interfacial area.

In the case of Intalox saddles, the ratio of the geometrical surface area of the two inch (50 mm) polypropylene saddles to that of the 1½ in (38 mm) ceramic saddles is 0·85, whereas the ratio of the effective interfacial areas is 0·5 or less. This suggests that the wetting of the polypropylene is less effective than that of the ceramic. Coughlin[5] carried out experiments somewhat similar to those described above, using one inch (25 mm) Raschig rings made of ceramic, polyethylene and Saran. It is not possible to deduce from the published results the actual values of the effective interfacial area, but the results might be interpreted as showing that the areas were about 25% larger for the ceramic than for the plastic packings. As against this, the operating hold-up of liquid was the same for all the materials. It thus does not seem possible at this stage to make any general statement about the relative values of the effective interfacial area for ceramic packings and geometrically similar plastic packings.

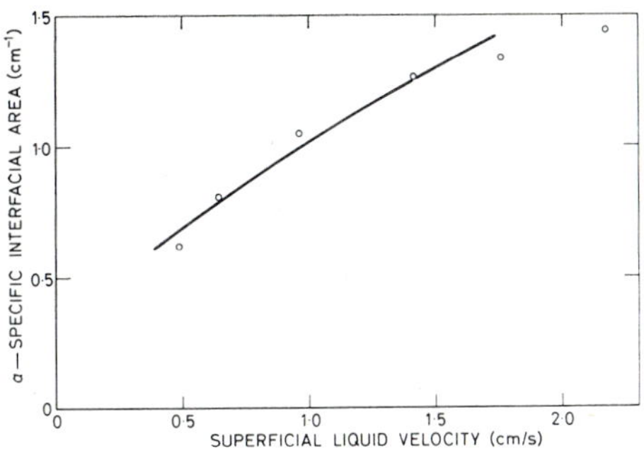

○: present results
——: best line through results of Danckwerts and Gillham[1]

Fig. 2.—*Plot of a against superficial liquid velocity for 1½ in Raschig rings*

The gas velocity in these experiments was too low to influence the interfacial area. It is known[6] that in counter-current flow the gas velocity does not influence the interfacial area until the loading-point is reached. There has been much discussion of the concept of the "effective" interfacial area (for a summary, see Danckwerts).[7] It is supposed that, in the case of absorption without reaction, thin and slow moving parts of the layer of liquid on the packing may become saturated with gas and cease to take part in the absorption process. The addition of a reactant which reacts at a finite rate generally increases the capacity of the liquid for gas proportionately more than the rate of absorption, so that it takes longer for a thin film to become saturated and more of the surface is effective. In the limit, with sufficiently concentrated solutions of reactant, we may expect virtually all the wetted surface to be effective, and the effective wetted area to be independent of the type of reaction. This expectation is borne out by the agreement of the present results with those of Danckwerts and Gillham,[1] using quite a different reaction, which suggests that the whole wetted surface was effective for gas absorption in both cases.

The liquid was distributed evenly across the top of the packing, but experiment showed that after traversing four metres of packing a substantial part of the liquid was flowing down the walls. Thus the actual rate of irrigation of the packing is highest at the top of the column and becomes less lower down. One might expect this to be reflected in a progressive diminution in the value of *a* down the column. No such trend was evident in the plots of B against h (*e.g.*, Fig. 1), suggesting that the wall-flow is established in a short distance at the top of the column; it may account for the end-effect referred to above. Danckwerts and Gillham[1] also found no variation of effective interfacial area, a, with height. These findings are in accord with the observation of Porter and Templeman,[8] that the equilibrium wall-flow is established close to the top of the column. It is concluded that the values of a reported here are less than those which would be obtained if the liquid were uniformly distributed at all levels. However, Porter and Templeman[8] concluded that the proportion of liquid flowing down the wall is insignificant in columns of large diameter. This implies that the values of a reported here, using a 50 cm column, are smaller than those to be expected in large industrial equipment, and thus the use of the values given here for design purposes is a conservative practice.

Acknowledgments

We are grateful to the Director of the Warren Spring Laboratory (Ministry of Technology), for accommodation and assistance and to Hydronyl Limited for gifts of packings.

Symbols Used

A = interfacial area in stirred cell (cm^2).
a = specific effective interfacial area (cm^{-1}).
B = quantity given by equation (8) (dimensionless).
$\bar{G}$ = average superficial mass-velocity of gas (g mol/cm^2 s).
G_i = superficial mass-velocity of gas at inlet (g mol/cm^2 s).
G_0 = superficial mass-velocity of gas at outlet (g mol/cm^2 s).
h = height of packing exposed to gas.
k = absorption-rate constant defined by equation (1) [g mol/cm^2 s $(atm)^{-\frac{3}{2}}$].
p = partial pressure of oxygen (atm).
x = partial pressure of oxygen in air entering column (% of 1 atm).
y = partial pressure of oxygen in air leaving column (% of 1 atm).
z = partial pressure of oxygen in air leaving stirred vessel (% of 1 atm).

References

1 Danckwerts, P. V., and Gillham, A. J. *Trans. Instn chem. Engrs*, 1966, **44**, T42.

2 de Waal, K. J. A.. and Okeson, J. C. *Chem. Engng Sci.*, 1966, **21**, 559.

3 Reith, T. *Thesis*, 1968. Delft Technical University.

4 Wesselingh, J. A., and van't Hoog, A. C. *Trans. Instn chem. Engrs*, 1970, **48**, T69.

5 Coughlin, R. W. C. *A.I.Ch.E. Jl*, 1969, **15**, 654.

6 Mannford Doble, M. G. *Ph.D. Thesis*, 1966, Cambridge University.

7 Danckwerts, P. V. "*Gas—Liquid Reactions*", 1970, p. 217. (New York: McGraw-Hill Book Co.).

8 Porter, K. E., and Templeman, J. J. *Trans. Instn chem. Engrs*, 1968, **46**, T86.

The manuscript of this paper was received on 5 *October*, 1970.

DESIGN OF GAS ABSORBERS: PART III—LABORATORY "POINT" MODEL OF A PACKED COLUMN ABSORBER

By P. V. DANCKWERTS (FELLOW)† and E. ALPER‡

SYNOPSIS

A stirred cell has been used as a model of a " point " in a packed column in which gas absorption is accompanied by reaction in the liquid. Values of k_L and k_G could be varied independently, and were in the range of industrial interest. Using data from the stirred cell, it was possible to predict the height of a packed column for a specified duty to within 9%. The system used was air–CO_2–NaOH–Na_2CO_3.

Using the physico-chemical data pertaining to the system, it was also possible to predict rates of absorption from first principles.

Introduction

The designing of a counter-current packed column absorber consists, firstly, of deciding on the size and material of packing, the values of L (the superficial liquid velocity) and $\bar{G}/L$ (the ratio of volumetric flow-rates of gas to liquid) and lastly, computing the required packing height. This paper is concerned with the last item for the case of gas absorption with chemical reaction.

If all the physico-chemical data relating to the system and the reaction kinetics are known it is possible to design a packed column from first principles. However, it is usually difficult to estimate the solubility and the diffusivity of a solute in a solution with which it reacts and the kinetics may be complicated. These difficulties justify the idea of making relatively simple laboratory models and carrying out experiments to obtain information in a form which could be applied directly to design.

Danckwerts and Gillham proposed a stirred cell as a model of a packed column and measured the rates of absorption of pure carbon dioxide into sodium hydroxide solutions in a stirred cell and a large packed column.[1] Jhaveri and Sharma absorbed oxygen into sodium hydrosulphite solution in a stirred cell and a small packed column.[2] Both groups of workers compared the measured absorption rates in their packed columns with the predictions obtained from their stirred cell experiments. Their comparisons indicate that a stirred cell can be used satisfactorily to design a packed column. However, in these experiments there was no gas side resistance. When there is resistance in the gas phase as well, both liquid side and gas side phenomena have to be considered.

Criteria for Simulating a " Point " in a Packed Column by Laboratory Absorber

A small but statistically representative volume element of packed space will be referred to as a " point " in the packed column. The values of k_L, the physical liquid side coefficient, k_G, the gas side coefficient, and a, the effective interfacial area per unit volume of packing, will be assumed to be the same at every " point " (i.e., in all parts of the column). If both fluids are in " plug flow ", the points at a particular height of packing behave identically and form a horizontal slice of packing of height dh.

In certain cases, for instance when the reaction between the dissolved gas and a reactant in solution is slow, there will be reactions proceeding in the bulk of the liquid and the rate of absorption in column or stirred cell will depend on the volume of bulk liquid available per unit area of interface. Since this quantity is quite different in the two pieces of equipment, the stirred cell cannot be used as a model of a " point " in a packed column in such cases. The technique discussed in this paper is therefore confined to cases in which all the reactions which determine the rate of absorption occur essentially in the diffusion film and the concentration of dissolved gas in the bulk of the liquid is zero. Under these conditions we can write a material balance equation for the case where the gas contains only one soluble component and the liquid only one reactant, thus relating the partial pressure of solute gas to the concentration of unreacted reactant at various points in the column:

$$\frac{(G_h\, p_h - G_o\, p_o)}{v_M\, P} = \frac{L}{z}(B_{in}{}^0 - B_h{}^0) \quad . \quad . \quad . \quad (1)$$

where P is the pressure in the column, G is the superficial velocity of the gas, p is the bulk partial pressure of solute gas, B^0 is the bulk concentration of reactant, L is the superficial velocity of the liquid, v_M is the molar volume of the gas, z is the number of moles of B reacting with each mole of gas, and subscripts $_o$, $_{in}$, and $_h$ denote the conditions at outlet, inlet, and level h, respectively.

For dilute gas systems, we can replace G_h and G_o by $\bar{G}$, the mean superficial velocity of the gas.

When a horizontal slice of packing of differential height, dh, is considered, we can write:

$$\bar{R}(C^*, B^0)a\,\mathrm{d}h = \frac{L}{z}\mathrm{d}B_h{}^0 = -\frac{\bar{G}}{Pv_M}\mathrm{d}p_h \quad . \quad . \quad (2)$$

where $\bar{R}$ is the rate of absorption per unit area.

Thus, integrating over the height, H, of the column:

$$H = \frac{L}{za}\int_{out}^{in}\frac{\mathrm{d}B_h{}^0}{\bar{R}} = \frac{\bar{R}}{aPv_M}\int_{out}^{in}\frac{\mathrm{d}p_h}{\bar{R}} \quad . \quad . \quad (3)$$

† Department of Chemical Engineering, Pembroke Street, Cambridge, CB2 3RA.

‡ Department of Chemical Engineering, Hacettepe University, Ankara, Turkey.

In order to simulate a " point ", firstly it is necessary to calculate the compositions of gas and liquid corresponding to " points " at various levels in the packed column from equation (1). The compositions at the top (or bottom) of a packed column are usually given.

Consider a laboratory stirred cell absorber in which the liquid and gas are agitated in such a way as to give rise to mass transfer coefficients, k_L and k_G, of the same magnitude as those occurring in the packing. The method of agitating gas and liquid in the laboratory absorber and the packed column are different. On the other hand, it is known that the enhancement factor due to the chemical reaction, E, differs little whether the film model, or the Higbie or Danckwerts surface renewal models are used as the basis of calculation.[3] Porter showed that, when various surface renewal models of extremely different contact time distributions but with the same mean contact time (i.e., the same k_L) were used to predict the absorption rate for a second order reaction, they all gave almost the same results.[4] If we also make the bulk compositions of gas and liquid in the laboratory absorber the same as those of a level in the packed absorber, we may expect that $\bar{R}$, the absorption rate per unit interfacial area at that level, will be the same as that in the laboratory model according to the above argument. Thus, $\bar{R}$ can be determined experimentally as a function of B^0 or p, and equation (3) integrated numerically to find the height of the column.

The laboratory " point " model can take various forms, the essential feature in each case being that the area of interface between gas and liquid is well defined and known. For instance, the string of discs or string of spheres absorbers might be used but their disadvantage is that the composition of the liquid and gas may vary substantially over the length of the string and it may not be possible to have a very short string to minimise these changes.[5]

The Stirred Cell as a " Point " Model

A stirred cell was first used by Danckwerts and Gillham but they only considered matching the values of k_L, not k_G.[1] Their work has here been extended so that the value of k_G in the cell can also be varied to match the value of k_G in a packed column; a somewhat similar device has been described by Levenspiel.[6]

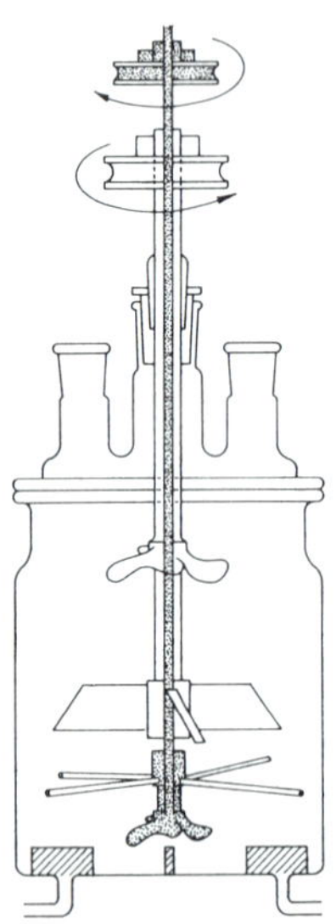

Fig. 1—*Stirred cell*

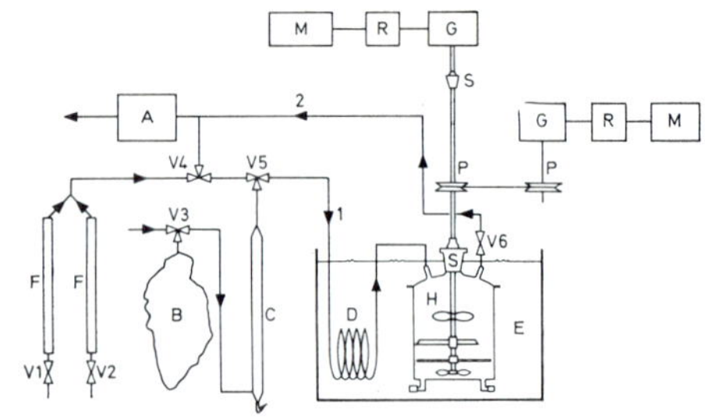

A: gas analyser
B: rubber balloon
C: soap film meter
D: heat exchanger
E: constant temperature bath
F: rotameter
G: reducing gear
H: stirred cell
M: motor
R: variator
P: pulley
S: mercury seal
V: valve (or stop-cock)
1: inlet gas stream
2: outlet gas stream

Fig. 2—*Schematic diagram of the stirred cell setup*

Figure 1 shows our cell which contains symmetrical baffles to prevent vortex formation in the liquid, a submerged turbine type stirrer to ensure bulk mixing of liquid, a stirrer with four blades skimming the liquid surface, a turbine type stirrer in the bulk of the gas phase, and a stirrer with three inclined blades (60° to the horizontal) just above the liquid surface. The clearance between the inclined blades and the liquid surface is about 0.75 cm. The shafts of the gas and liquid stirrers are concentric and each is connected to a reducing gear which is coupled to a variator and a synchronous motor (see Fig. 2). This arrangement makes it possible to vary the speed of the liquid side stirrers between 30 and 160 rev/min, and the speed of the gas side stirrers between 80 and 1000 rev/min, accurately and independently. Both shafts were sealed with mercury seals.

The volume of liquid in the cell was always exactly 250 cm^3, thus ensuring that the level of the liquid surface relative to the stirrers was always the same. During the operation of the cell there were no bubbles or splashes at the wall and the surface of the liquid remained about horizontal. Consequently, the gas–liquid interfacial area in the cell, A, was assumed to be the same as the cross-sectional area of the cell, less that of the blade type stirrer; the value of A was 89 cm^2. In all that follows it is assumed that the gas in the cell was perfectly mixed.

In order to determine k_L as a function of liquid side stirrer speed (η_L) pure CO_2 was absorbed into water and the following equation governed the process:

$$V \frac{d \ln}{dt} \left[\frac{C^*}{(C^* - q/V)} \right] = k_L A \quad . \quad . \quad . \quad (4)$$

where q is the total amount of gas absorbed and V is the volume of liquid in the cell (i.e., 250 cm^3). The total amount absorbed, q, was measured as a function of time, t, and $V \ln [C^*/(C^* - q/V)]$ was calculated and plotted against t:

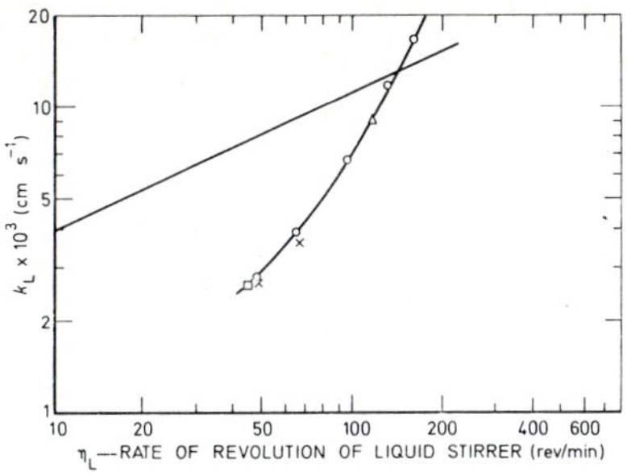

Gas stirrer revolutions

○ : —

× : 180

△ : 240

⊡ : 410

Fig. 3—*Liquid side physical mass transfer coefficients in the stirred cell (the straight line represents* $k_L = 2[D_A/\pi(60/4\eta_L)]^{\frac{1}{2}}$)

this plot is a straight line and the slope gives the value of $k_L A$. The same process was repeated at different liquid side stirrer speeds.

The values of k_L varied from about 2×10^{-3} to 20×10^{-3} cm s^{-1} as η_L varied from about 30 to 160 rev/min. (At higher speeds, the gas–liquid interfacial area became unknown due to the occurrence of bubbles and splashes at the wall of the cell.) Figure 3 shows the results of experiments in cells containing a stirrer, with four blades, which is just skimming the surface. The straight line in the same figure is the plot of $k_L = 2[D_A/\pi(60/4\eta_L)]^{\frac{1}{2}}$. This line, of course, represents the situation where elements of surface remain stationary until they have been removed from the surface by the action of the blades of the stirrer. Experimental values of k_L indicate that this is obviously not the physical situation. At low speeds the surface of the liquid moves in the same direction as the stirrer, thus resulting in contact times which are greater than $(60/4\eta_L)$. When the speed of the stirrer increases, eddying occurs resulting in high mass transfer coefficients. Both of these factors are affected considerably by the size and shape of the baffles and the submerged stirrer, and the depth to which the stirrers are immersed. It is clear why different cells of similar but not identical design may give different values of k_L.

In order to determine k_G as a function of gas side stirrer speed, η_G, sulphur dioxide (diluted with air) was absorbed into about 2M NaOH solution. The reaction between sulphur dioxide and OH^- ions is instantaneous and if:

$$k_G\, p < k_L \frac{B^0}{z}\left(\frac{D_B}{D_A}\right)^{\frac{1}{2}} \qquad (5)$$

(where D_B and B^0 are, respectively, the diffusivity and bulk concentration of OH^- ions) the absorption process is entirely controlled by the transport of SO_2 across the gas film and:

$$\bar{R} = k_G\, p \qquad (6)^3$$

(This condition was amply satisfied in our experiments.) Thus measurement of $\bar{R}$ and p gives the value of k_G for the gas absorbed (i.e., SO_2); the value of k_G for other gases under the same conditions can be estimated by correcting for the diffusivity of solute gas—and assuming that such quantities as the viscosity, the density, *etc.* of the gas stream (solute gas + inerts) are about the same. The dependence of k_G on the diffusivity of solute gas is not firmly established but the experimental evidence of Vidwans and Sharma[7] indicates that k_G varies as the square-root of the diffusivity, and this relationship was used; values of the diffusivities were taken from Reid and Sherwood.[8]

The following equation can be used to calculate k_g from the experimental observations:

$$k_G\, A p_o = g\left(\frac{p_{in}}{P - p_{in}} - \frac{p_o}{P - p_o}\right) \qquad (7)$$

where g is the molar flow-rate of air, p_{in} and p_o are the partial pressures of SO_2, and P is the pressure in the cell (atmospheric). The concentration of SO_2 was measured by means of a calibrated katharometer and the results are shown in Fig. 4 as a plot of k_G *versus* η_G. (The speed of the liquid side stirrer was kept at 129 rev/min.)

The values of k_G which were achieved in the cell ranged from 2×10^{-5} to 14×10^{-5} mol s^{-1} cm^{-2} atm. A typical value of k_G in a large packed column is around 10×10^{-5} mol s^{-1} cm^{-2} atm;[5] thus this value of k_G is in the range which could be achieved in the stirred cell. The gas side mass transfer coefficient for a given solute–inert gas system in the stirred cell relates to the rate of revolution in the following form:

$$k_G \propto \eta_G{}^n \qquad (8)$$

where η_G is the speed of the stirrer. The value of the exponent, n, was found to be 0.9 in this cell. The experimental results are shown in Fig. 4.

Absorption from Air into NaOH Solutions in the Stirred Cell and a Packed Column

Carbon dioxide from air was absorbed into sodium hydroxide solutions in the stirred cell and in a packed column. Conditions were such that there were resistances in both the gas and liquid phases. The solutions were made up from sodium hydroxide and sodium carbonate so that the concentration of Na^+ ion was always equal to 0.6 g ion l^{-1}. The object was to compare the predictions obtained from the stirred cell experiments with the results of packed column experiments.

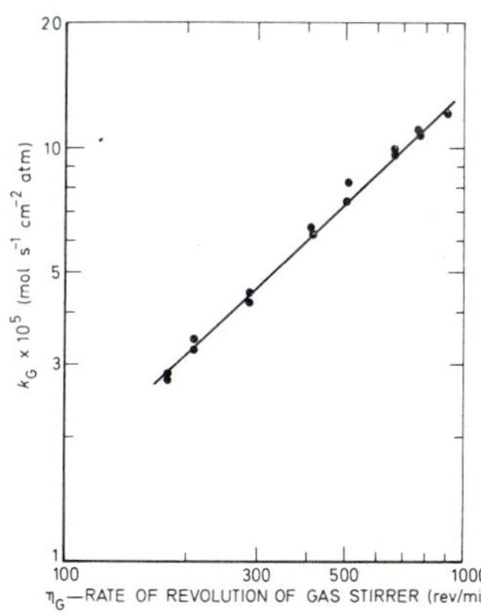

Fig. 4—*Gas side mass transfer coefficients in the stirred cell (SO_2—air system)*

TABLE I—*Results of CO_2–Air–Aqueous ($NaOH+Na_2CO_3$) Experiments (packed Column, 25°C and 1 atm)*

$L = 0.28$ cm s^{-1}
$G = 36.0$ cm s^{-1}
$a = 1.3$ cm^{-1} (Danckwerts[3])
$k_L = 5.7\times10^{-3}$ cm s^{-1} (Danckwerts[3])
$k_g = 4.2\times10^{-5}$ mol cm^{-2} s^{-1} atm (Mannford Doble[9])
$[Na^+] \approx 0.6$ mol l^{-1}
$[OH^-]+2[CO_3] \approx 0.6$ mol l^{-1}

$[OH^-]$ (mol l^{-1}) Inlet	Outlet	$p_{CO_2}\times10^2$ (atm) Inlet	Outlet	$\int_{in}^{out}\frac{dB}{\bar{R}} = \frac{aHz}{L}$ (s cm^{-1})	**Packing height** (cm)	**Predicted height** (cm)	**Difference** (%)	**Error in material balance** (%)
0.39	0.29	4.6	3.8	430	48.0	46.3	+3.5	8.4
0.52	0.42	4.1	3.3	410	48.0	44.1	+8.1	4.2
0.58	0.46	4.6	3.6	430	48.0	46.3	+3.5	3.6
0.28	0.10	5.2	3.6	970	108.0	104.0	+3.7	3.5
0.62	0.19	7.6	4.0	1430	163.0	154.0	+5.5	0.5
0.60	0.08	10.3	6.2	1430	163.0	154.0	+5.5	0.8
0.55	0.19	6.0	3.1	1420	163.0	153.0	+6.1	7.0
0.52	0.20	5.0	2.2	1430	163.0	154.0	+5.5	0.0
0.55	0.08	10.6	7.0	1320	143.0	142.0	+0.7	5.3
0.46	0.10	7.8	5.0	1310	143.0	141.0	+1.4	6.7
0.46	0.13	7.8	5.2	1120	123.0	121.0	+1.6	5.4
0.58	0.13	10.2	6.7	1130	123.0	122.0	+0.8	2.6
0.53	0.26	5.3	3.0	1120	123.0	121.0	+1.6	6.0

The packed column

The packed column was 183 cm high and 10.2 cm in internal diameter. An efficient liquid distributor was provided and there were two equally spaced redistributors (Fig. 6). In order to vary the height of packing exposed, the column was flooded up to various levels. The gas was introduced at three points around the circumference immediately above the reigning liquid level. The packing consisted of 1.27 cm ceramic Raschig rings, the packing density being 0.37 rings per cm^3, the same as that found by Mannford Doble[9] and given by Danckwerts.[3]

In all experiments the superficial velocity of the liquid was 0.2 cm s^{-1}, and that of the gas 36 cm s^{-1}; the temperature was kept at 25°C. Under these circumstances the values of k_L (for CO_2 in a liquid of similar density and viscosity) and a are 5.7×10^{-3} cm s^{-1} and 1.3 cm^{-1}, respectively;[3] the value of k_G is 4.2×10^{-5} mol s^{-1} cm^{-2} atm.[9] The gas side resistance contributed from 10% to 30% of the total "resistance" to absorption.

The gas and liquid feed and outlet compositions are shown in Table I. The feed consisted of mixed solutions of NaOH and Na_2CO_3, the concentration of Na^+ being always 0.6 g ion l^{-1}. The compositions of liquid streams were determined by titration[10] and the compositions of gas streams were determined by means of a katharometer.

The stirred cell

In the stirred cell, the values of k_L and k_G were adjusted to be 5.7×10^{-3} cm s^{-1} and 4.2 mol s^{-1} cm^{-2} atm, respectively. The absorption rate $\bar{R}A$ was measured for six different liquid compositions and various gas compositions (details can be found elsewhere[5]). The results of the experiments are given in Fig. 5 as plots of $\bar{R}$ against p, the bulk partial pressure of CO_2.

Comparison of actual and predicted packing heights

For each experiment with the packed column the relevant values of $\bar{R}$ from the stirred cell were used to integrate equation (3) between the inlet and outlet conditions of the packed column. The resulting values of H are shown as "Predicted height" in Table I. The predicted and actual heights differed by less than 9% in all cases, indicating that the method of using the stirred cell to design packed columns is sound. The range of error in the predicted heights is similar to that in the material balances.

Calculation of $\bar{R}$ from First Principles

As has been remarked before, it is possible in principle to calculate $\bar{R}$ if the values of k_L and k_G and adequate physico-chemical data are available. The Higbie surface renewal

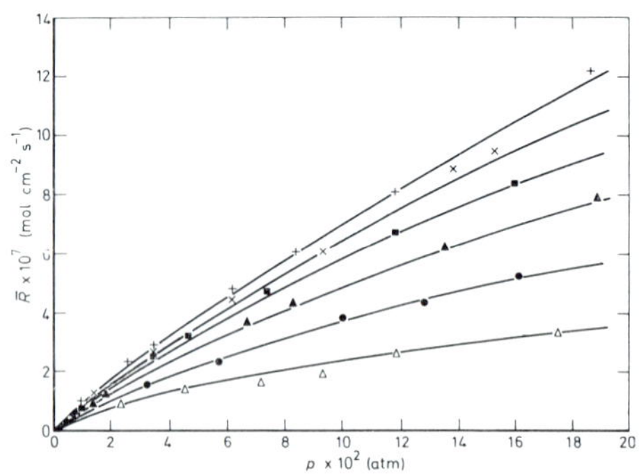

Liquid composition (g ion l^{-1})

	$[Na^+]$	$[OH^-]$	$[CO_3^{--}]$
+	0.6	0.6	0.00
×	0.6	0.5	0.05
■	0.6	0.4	0.10
▲	0.6	0.3	0.15
●	0.6	0.2	0.20
△	0.6	0.1	0.25

Fig. 5—*Experimental results in the stirred cell; the lines are calculated from first principles* ($k_L = 5.3\times10^{-3}$ cm s^{-1}; $k_G = 4.2\times10^{-5}$ mol s^{-1} cm^{-2} atm)

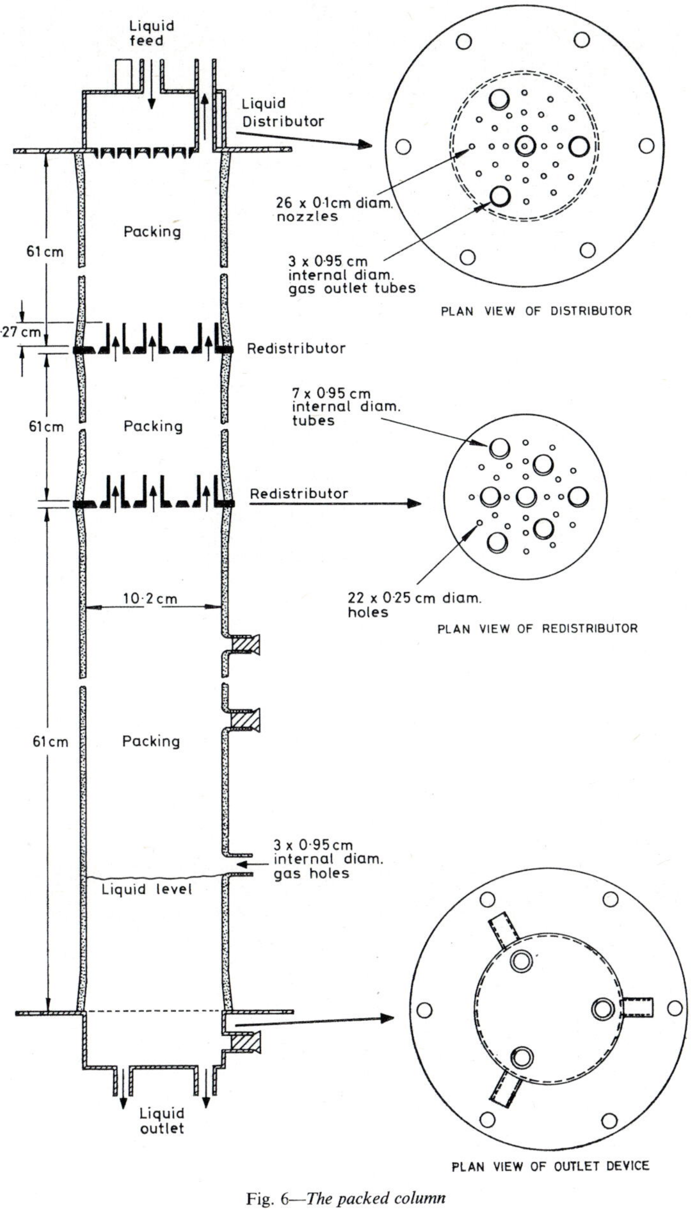

Fig. 6—*The packed column*

Table II—*Physico-chemical Data for CO_2–Air–Aqueous ($NaOH+Na_2CO_3$) System*

Characteristics	Method and the result	Remarks
D_A: diffusivity of CO_2 in the solution ($cm^2\,s^{-1}$)	$D_A\,\mu_L^{0.82}$ = constant (Joosten[12]) ($D_A = 1.8\times10^{-5}\,cm^2\,s^{-1}$)	The diffusivity of CO_2 in water at 25°C was taken as $1.92\times10^{-5}\,cm^2\,s^{-1}$ and the viscosity of liquid relative to water was measured by means of a Schurz–Immergut viscometer.
D_A/D_B: ratio of diffusivities of CO_2 to OH^-	$D_A/D_B = 1.7$ (Danckwerts *et al.*[13] and Nijsing *et al.*[14])	The dissolved reactant OH^- is an ion and some problems arise in assigning diffusivity to a single ionic species in such a situation.[3] The experimental data of various workers fit the theory best if D_A/D_B is taken as 1.7.
He: Henry's law constant (atm l mol^{-1})	$\log_{10}(He/He^0) = h_1 I_1 + h_2 I_2$ (Danckwerts[3])	In mixed electrolytes, this expression is probably the best estimate that can be made. The Henry constant for water, He^0, was obtained from Seidell and Linke.[15] (He^0 = 0.0296 atm l mol^{-1} at 25°C.)
k_2: the second order rate constant (l $mol^{-1}\,s^{-1}$)	k_2 / OH^- / CO_3^{--} 10200 0.6M + — 10300 0.5M + 0.05M 10400 0.4M + 0.10M 10500 0.3M + 0.15M 10600 0.2M + 0.20M 10700 0.1M + 0.25M	The value of k_2 at infinite dilution was calculated from $\log_{10} k_2 = 13.635 - 2895/T$.[16] Pinsent *et al.* state that the effect of Na_2CO_3 is about the same as NaCl at a given ionic strength.[17] This fact and the curve of Danckwerts and Sharma[16] which gives the effect of NaCl addition, were used to obtain the values of k_2.

model[3] will be used here for this purpose. The physico-chemical data for the experimental system used are unusually well known as the result of work by many experimenters.

The rate-determining reaction is that between CO_2 and hydroxyl ions:

$$CO_2 + OH^- \rightarrow HCO_3^- \quad (9)$$

This reaction is second order and under the conditions of our experiments can be regarded as irreversible and sufficiently fast to take place in the diffusion film, the bulk concentration of CO_2 being virtually zero. Reaction (9) is followed by:

$$HCO_3^- + OH^- \rightarrow CO_3^= + H_2O \quad (10)$$

which can be regarded as instantaneous. Thus the overall reaction is:

$$CO_2 + 2OH^- \rightarrow CO_3^= + H_2O \quad (11)$$

and the stoichiometric number z is 2 (moles of OH^- reacting with each mole of CO_2).

The method of calculating $\bar{R}$ for absorption with an irreversible reaction in the presence of gas side resistance has been described elsewhere.[3] The physico-chemical data required are the reaction-velocity constant, k_2, of reaction (9), the Henry's Law constant for CO_2 in the solution, the diffusivity, D_A, of dissolved CO_2 in the solution, and the effective diffusivity, D_B, of OH^-. All these quantities differ from the corresponding quantities at infinite dilution. Table II indicates the methods used to calculate these quantities and the figures adopted.

The continuous lines in Fig. 5 show the calculated values of $\bar{R}$ for different liquid and gas compositions. The calculated values agree very well with the experiments, showing that it is possible to design packed columns from first principles, provided that the physico-chemical data can be reliably estimated.

Merchuk, Nakanishi, and Yoshida have also compared calculated and observed rates of absorption in the same chemical system, and have obtained reasonable agreement.[11]

Discussion and Conclusions

A stirred cell such as that shown in Fig. 1 can be used satisfactorily to simulate the conditions existing at a "point" in an industrial packed column. The values of k_L and k_G are likely to depend critically on the shape, size, and position of corresponding stirrers.

The results of experiments in the packed column and the predictions obtained from the stirred cell indicate that a stirred cell, such as was used in this project, can be used to design a packed column satisfactorily for gas absorption with fast irreversible chemical reaction when there are both gas phase and liquid phase resistances. If the seals of the stirrers were suitably designed, a similar cell could be used to simulate a high pressure absorber.

In the system CO_2–NaOH–Na_2CO_3, the relevant physico-chemical data can be estimated with unusual certainty. In these circumstances it is also possible to calculate the height of a packed column from first principles with considerable precision.

Symbols Used

A = interfacial area in stirred cell (cm^2).
a = effective interfacial area per unit packed volume (cm^{-1}).
B^0 = bulk concentration of reactant, NaOH.
C^* = saturated concentration of CO_2 (mol l^{-1}).
D_A = diffusivity of dissolved CO_2 ($cm^2\,s^{-1}$).
D_B = effective diffusivity of OH^- ($cm^2\,s^{-1}$).
E = enhancement factor (ratio of chemical to physical absorption rates).
$\bar{G}$ = mean superficial velocity of gas in column (cm s^{-1}).
G = superficial velocity of gas at point in column (cm s^{-1}).
g = rate of flow of gas through stirred cell (mol s^{-1}).
H = height of packing (cm).
He = Henry's Law constant for CO_2 in solution (atm l mol^{-1}).
h = distance from bottom of packing (cm).
k_2 = second order rate constant (l $mol^{-1}\,s^{-1}$).
k_G = gas side mass transfer coefficient (mol $s^{-1}\,cm^{-2}$ atm).
k_L = liquid side mass transfer coefficient (cm s^{-1}).
L = superficial velocity of liquid.
n = exponent in equation (8).
P = atmospheric pressure (1 atm).
p = partial pressure of CO_2 (atm).
q = amount of gas absorbed in time t (mol).
$\bar{R}$ = rate of absorption per unit interfacial area at point in column (mol $cm^{-2}\,s^{-1}$).
t = time of absorption in stirred cell (s).

V = volume of liquid in stirred cell (l).
v_M = molar volume of gas (cm^3 mol^{-1}).
z = number of moles of NaOH reacting with one mole of CO_2.
η_G = rate of revolution of gas stirrer (rev/min).
η_L = rate of revolution of liquid stirrer (rev/min).

Subscripts

$_h$ = at level h in packing.
$_{in}$ = pertaining to inflowing stream.
$_o$ = pertaining to outflowing stream.

References

1 Danckwerts, P. V., and Gillham, A. J. *Trans. Instn chem. Engrs*, 1966, **44,** T42.
2 Jhaveri, A. S., and Sharma, M. M. *Chem. Engng Sci.*, 1968, **23,** 1.
3 Danckwerts, P. V. "*Gas-Liquid Reactions*", 1970. (London and New York: McGraw-Hill).
4 Porter, K. E. *Trans. Instn chem. Engrs*, 1966, **44,** T25.
5 Alper, E. *Ph.D. Thesis*, 1971. University of Cambridge.
6 Levenspiel, O., and Godfrey, J. H. *Chem. Engng Sci.*, 1974, **29,** 1723.
7 Vidwans, A. D., and Sharma, M. M. *Chem. Engng Sci.*, 1967, **22,** 673.
8 Reid, R. C., and Sherwood, T. K. "*The Properties of Gases and Liquids*", 1966 (2nd Edn). (New York: McGraw-Hill).
9 Mannford Doble, M. G. *Ph.D. Thesis*, 1966. University of Cambridge.
10 Vogel, A. I. "*A Textbook of Quantitative Analysis*", 1961 (3rd Edn). (London: Longmans).
11 Merchuk, J. C., Nakanishi, K., and Yoshida, F. J. *Journal of Chemical Engineering of Japan*, 1970, **3,** 62.
12 Joosten, G. E. H. *Ph.D. Thesis*, 1971. University of Cambridge.
13 Danckwerts, P. V., and Kennedy, A. M. *Chem. Engng Sci.*, 1958, **8,** 201.
14 Nijsing, R. A. T. O., Hendriksz, R. H., and Kramers, H. *Chem. Engng Sci.*, 1959, **10,** 88.
15 Seidell, A., and Linke, W. F. "*Solubilities of Inorganic and Organic Compounds*", 1952. (New York: Van Nostrand).
16 Danckwerts, P. V., and Sharma, M. M. *Chem. Engr, Lond.*, October 1966, p. CE 244.
17 Pinsent, B. R. W., Pearson, L., and Roughton, F. J. W. *Trans. Faraday Soc.*, 1956, **52,** 1512.

The manuscript of this paper was received 27 *March* 1974.

The revised manuscript of this paper was received 1 *July* 1974.

B10

LABORATORY SCALE-MODEL OF A COMPLETE PACKED COLUMN ABSORBER

E. ALPER† and P. V. DANCKWERTS

Department of Chemical Engineering, Pembroke Street, Cambridge CB2 3RA, England

(*Received* 15 *May* 1975; *accepted* 31 *July* 1975)

Abstract—A method has been established and tested for simulating in the laboratory the essential features of a complete packed column in which absorption with chemical reaction is occurring. The necessary rules for scaling-down have been established. These are: the same values of k_L, k_G, (aH/L), (vH/L) or (v/a) and $(L/\bar{G})$ in the model and the packed column which is being simulated.

It has been shown that a laboratory-scale string-of-spheres column can simulate a complete packed column as regards both gas-side and liquid-side phenomena and also reactions occurring in the bulk of the liquid.

INTRODUCTION

The design of a packed column for a given absorption duty consists firstly of deciding on the size and material of the packing and the values of L, the superficial liquid velocity and $(\bar{G}/L)$, the ratio of volumetric flowrates of gas to liquid and, secondly, computing the required packing height. If all the physico-chemical data and the reaction kinetics are known it is possible to design a packed column from first principles. Alternatively, a laboratory model which simulates a "point" in the packed column may be used for design purposes in certain cases[1]. The point model can only be used when it is

†Department of Chemical Engineering, Hacettepe University, Ankara, Turkey.

possible by means of a simple material balance to calculate the corresponding compositions of the gas and liquid streams at various points in the column. It is not applicable in cases where the rate of absorption is influenced by reactions occurring in the bulk of the liquid, when two gases are being absorbed simultaneously or when the liquid contains two or more reagents.

In such circumstances a different approach must be adopted. This paper describes a method of making a laboratory-sized scale model of a complete full-sized packed column. The height of the model is related to that of the full-sized column by a known scaling factor; the gas-liquid ratio and the inlet and outlet compositions are the same for the model as for the full-sized column of the height simulated. The model thus serves as a complete demonstration of the behaviour of the full-sized equipment. Gas-side and liquid-side phenomena and bulk liquid reactions are all taken into account.

CRITERIA FOR SIMULATION OF A FULL-SIZED PACKED COLUMN BY A LABORATORY ABSORBER

In the packed column, which is to be simulated, the values of k_L, k_G, a (the effective interfacial area per unit volume of packing) and v (the liquid hold-up) will be assumed to be the same in all parts. Under normal operating conditions, the deviations from the plug-flow of both fluids may be assumed to be negligible[2]. Consider the case where the liquid contains only one reactant B and the gas stream contains one solute gas C.

When a horizontal slice of packing of height dh is considered, one can write

$$\bar{R}(p^0, C^0, B^0)a\,dh = L\,\mathrm{d}C^0 + vr(C^0, B^0)\,dh \qquad (1)$$

$$-L\,\mathrm{d}B^0 = zvr(C^0, B^0)\,dh \qquad (2)$$

where $\bar{R}$ is the rate of absorption per unit interfacial area, p^0 is the partial pressure of soluble gas in the bulk gas stream, C^0 is the bulk concentration in the liquid of unreacted dissolved gas, B^0 is the bulk concentration of reactant, z is the number of mols of B reacting with each mol of C, v is the volume of liquid per unit volume of packed space and r is the rate of reaction per unit volume of liquid.

We can rearrange eqns (1) and (2) and integrate to give

$$\frac{aH}{L} = \int_{\text{in}}^{\text{out}} \frac{\mathrm{d}c}{\bar{R}(p^0, C^0, B^0)} - \int_{\text{in}}^{\text{out}} \frac{\mathrm{d}B^0}{z\bar{R}(p^0, C^0, B^0)} \qquad (3)$$

$$\frac{vH}{L} = \int_{\text{in}}^{\text{out}} \frac{\mathrm{d}B^0}{zr(C^0, B^0)}. \qquad (4)$$

Consider a hypothetical laboratory absorber (in which deviations from plug flow of gas and liquid are negligible) whose values of k_L, k_G, $(L/\bar{G})$, C^0_{in}, B^0_{in}, C^0_{out}, B^0_{out}, p^0_{in}, temperature and pressure are the same as in the packed column which is being simulated (the material balance will make p^0_{out} the same for both absorbers). We now make the assumption that, given the same gas and liquid compositions, the specific absorption rate $\bar{R}$ will be the same at corresponding points in the two absorbers provided that k_L and k_G are the same. We may now expect the right hand side of eqn (3) to be the same for the laboratory absorber and the packed column. (In general, ways and means of agitating liquid and gas in the laboratory absorber and the packed column will be different. However this is unlikely to be important in determining the relationship between k_G, k_L and the absorption rate, as long as in the laboratory absorber the liquid and gas are agitated in such a way as to give rise to mass transfer coefficients, k_L and k_G of the same magnitude as those occurring in the packing[1, 3, 4]). This means that (aH/L) must be made the same for both absorbers. This can be rewritten to give:

$$\frac{H}{H_1} = \frac{(L/a)}{(L_1/a_1)} \qquad (5)$$

where subscript 1 denotes the laboratory absorber. The ratio (H/H_1) may be defined as the "scaling ratio" and one desires this ratio to be larger than one (i.e. about 5–10). The right hand side of eqn (4) should also be the same in each absorber, thus necessitating the ratio (vH/L) to be the same. This ratio (vH/L) is, in fact, the mean residence time of liquid in the column and the requirement of the same residence time in both absorbers is obvious if there are bulk reactions.

$$\bar{t}_t = \frac{vH}{L} = \frac{v_1 H_1}{L_1}. \qquad (6)$$

Rearranging eqn (6) and using eqn (5) one obtains eqn (7) which requires the same interfacial area per unit volume of liquid and this is, of course, an obvious condition for bulk reactions.

$$\frac{a}{v} = \frac{a_1}{v_1}. \qquad (7)$$

Therefore the necessary conditions for simulation are the same k_L, k_G, (aH/L), (vH/L) or (v/a) and $(L/\bar{G})$ in the model and the packed column. The model now simulates all the essential features of the packed column and if the inlet compositions are the same so would be the outlet compositions. It must be noted that no assumption has been made about the reaction kinetics or transfer mechanism and the method should apply to any case (except possibly when reactions occur in the gas phase). In the Appendix it is shown that the above conditions are also sufficient when various gases are absorbed simultaneously or when a gas is absorbed into a solution which contains various reactants.

LABORATORY ABSORBERS FOR SIMULATION OF A LARGE PACKED COLUMN

In order to simulate the packed column, the values of k_G, k_L, (aH/L), $(L/\bar{G})$, (vH/L) or (v/a) in the laboratory absorber should be the same as those in the full-scale packed column. Furthermore it is desirable to make the "scaling ratio", (H/H_1) about 5–10 for convenience. This "scaling ratio" makes the value of the wetting rate (L_1/a_1) for the laboratory absorber small (see eqn 5). This, in turn, makes the flowrate of gas in the model small as the ratio $(L/\bar{G})$ is fixed. We shall consider here whether it is possible to build a model which will satisfy the necessary conditions.

THE WETTED-WALL COLUMN

Imagine a wetted wall column in which the liquid film is interrupted and the surface is renewed at a number of equally-spaced points. Then k_L would be inversely proportional to the square root of distance between two such "mixing devices" at a given liquid flowrate. Equations (3) and (4) can be rewritten for such a wetted wall column to give

$$\frac{a'H_1}{Q} = \int_{\text{in}}^{\text{out}} \frac{dC^0}{\bar{R}(p^0, C^0, B^0)} - \int_{\text{in}}^{\text{out}} \frac{dB^0}{z\bar{R}(p^0, C^0, B^0)} \quad (8)$$

$$\frac{v'H_1}{Q} = \int_{\text{in}}^{\text{out}} \frac{dB^0}{r(C^0, B^0)} \quad (9)$$

where a' is the interfacial area per unit height, H_1 is the height of wetted wall exposed to the gas and v' is the volume of liquid per unit height. As before one obtains

$$\left(\frac{H}{H_1}\right) = \frac{L/a}{Q/a'} \quad (10)$$

$$\frac{v}{a} = \frac{v'}{a'} \quad (11)$$

The value of k_L should, of course, be the same in the modified wetted wall column and the packed column and this might be achieved by adjusting the distance between successive mixers. (It might also be possible to incorporate hold-up in the mixing devices so that eqn (11) is satisfied). However, it was shown experimentally using a variety of mixing devices that it was not practicable to achieve mixing, and thus surface renewal, in such a wetted wall column[4]. The liquid seems to creep past such "flow-interruptors" as elastic bands.

THE STRING-OF-SPHERES COLUMN

On the other hand, when a liquid runs down a strong-of-spheres column[5] one might expect complete mixing at the junction between the spheres under certain conditions. Furthermore the spheres might have pools at the top of them (see Fig. 1) so that the mean residence time of liquid in this model could be matched to that of an industrial column. In general, one might expect that k_g could be varied by varying the diameter of the confining tube (see Fig. 1) and hence the velocity of the gas.

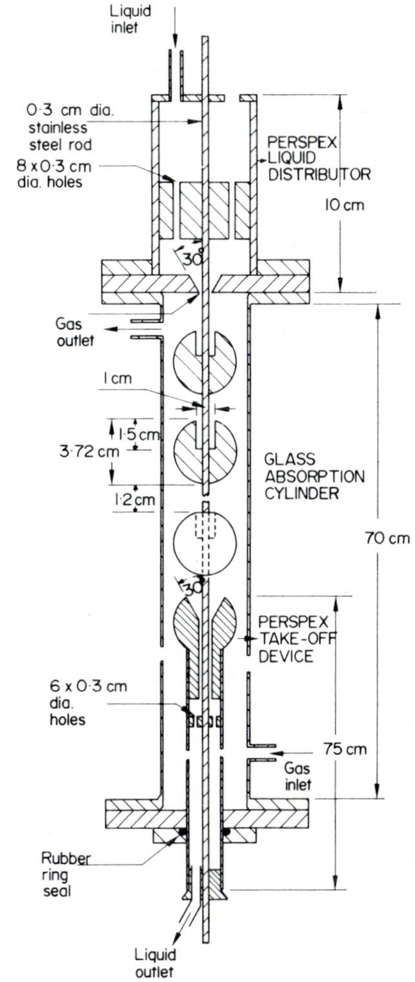

Fig. 1. String-of-spheres column.

The column used, which is shown in Fig. 1 schematically, consisted of spheres, each 3·72 cm in diameter with a cylindrical pool at the top of each sphere 1 cm diameter and 1·5 cm deep. The spheres were supported on a rod of 0·32 cm diameter and the distance between two consecutive spheres was 1·2 cm. The number of spheres exposed to the gas could be changed between 1 and 10 by removing (or putting back) some of the spheres and adjusting the position of the slidable take-off tube. A second column, which contained 1·89 cm dia. spheres with cylindrical pools of 0·5 cm dia. and 0·75 cm deep was also used. In these columns, the dimensions of the pools in the tops of the spheres were such that the value of (v'/a') was of the same order of magnitude as that of an industrial column.

In both columns ripples were observed even at the lowest liquid flowrates of all spheres.

Oxidation of aqueous sodium sulphite solution using cobalt as catalyst was used to determine the value of the effective interfacial area per sphere [2] and it was found that this area was only a few per cent more than the geometrical area (surface area of a dry sphere plus that of a 1·2 cm long rod) and remained constant throughout the column. The difference between dry and wet areas is no doubt due to the actual increase in surface area when the spheres and rods are covered by liquid.

(i) *The liquid side coefficient*

Pure carbon dioxide was absorbed into tap water to check whether k_L remained the same throughout the column and to determine its value. Considering the Nth sphere we can write (taking $(C^* - C^0_{N-1})$ as the driving force on the Nth sphere):

$$K_L{}^N A(C^* - C^0_{N-1}) = Q(C_N{}^0 - C^0_{N-1}) \qquad (12)$$

where Q is the liquid flowrate, A is the interfacial area per sphere, $C_N{}^0$ is the concentration of dissolved gas in the liquid which is leaving the Nth sphere. Assuming $k_L{}^N = k_L{}^{N-1} = \ldots k_L{}^1 = k_L$ (where $k_L{}^N$ is the liquid side coefficient for the Nth sphere) one can solve eqn (13) to give

$$\ln\left(\frac{C^* - C_0{}^0}{C^* - C_N{}^0}\right) = -N \ln\left[1 - \frac{k_L A}{Q}\right]. \qquad (13)$$

Total absorption rates, $(\bar{R}AN)$, were measured at various Q for different numbers of spheres which were exposed to the gas. As the inlet liquid contained no dissolved CO_2, $C_0{}^0$ was equal to zero and $C_N{}^0$ given by total absorption rate divided by the liquid flowrate. Taking C^* as 0·826 cm^3 of CO_2/cm^3 of water at 25°C and 1 atm [7] the values of $\ln(C^* - C_N)$ can be calculated. Figures 2 and 3 are the plots of $\ln(C^* - C_N)$ against N for 3·72 cm dia. spheres and 1·89 cm dia. spheres respectively. These plots are straight lines, indicating that k_L is independent of the length of the column. The slopes of the lines are $(1 - k_L A/Q)$ which, of course, provides the value of k_L. The intercepts should be $\ln C^*$ – measurement yields values of $C^* = 0{\cdot}826$, as given in [7]. Figure 4 shows the plots of k_L against Q for the two sphere-columns.

(ii) *The gas-side coefficient*

The NH_3-air-aq. H_2SO_4 system was used to determine the value of the gas-side coefficient in the sphere columns.

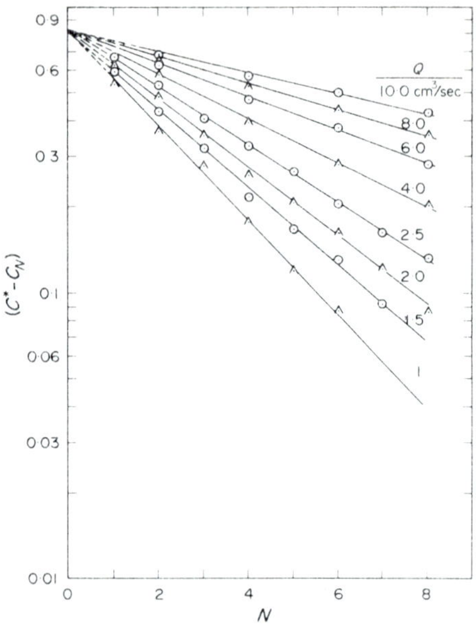

Fig. 2. Plots of $(C^* - C_N)$ vs N (3·72 cm dia sphere column).

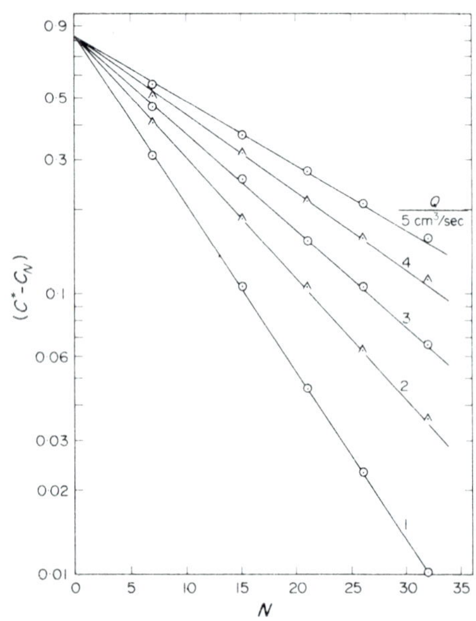

Fig. 3. Plots of $(C^* - C_N)$ vs N (1·89 cm dia sphere column).

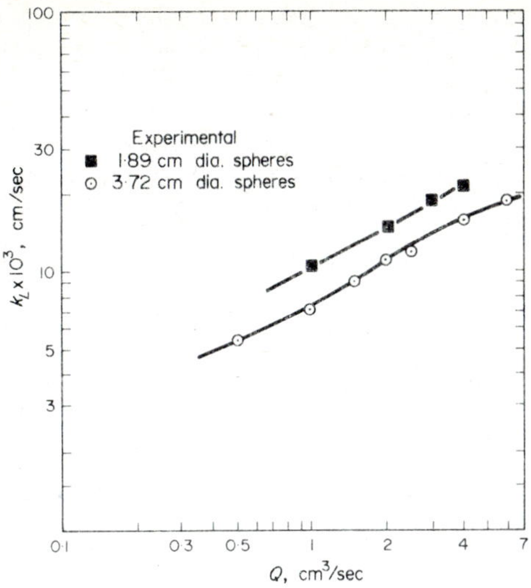

Fig. 4. Liquid side physical mass transfer coefficients in the 1·89 cm and 3·72 cm dia spheres columns.

The concentration of the acid solution was about 2 M and the concentration of ammonia in the gas stream was always less than 10%. In these experiments, $k_G p < k_L B^0/2\sqrt{(D_B/D_C)}$ so that the absorption rate per unit area was given by [8]

$$\bar{R} = k_G p. \tag{14}$$

Considering the nth sphere, one can write

$$k_G A p_{n+1} = \frac{g_{n+1}p_{n+1} - g_n p_n}{P} \tag{15}$$

where p_n is the partial pressure of solute gas, g_n is the molar flowrate of gas and P is the total pressure in the column. Making a mass balance on inerts, eqn (15) can be written in the following form:

$$p_n = P \frac{\dfrac{p_{n+1}}{P - p_{n+1}} - \left(\dfrac{k_G AP}{g_{\text{in}}}\right)\left(\dfrac{p_{n+1}}{P - p_{\text{in}}}\right)}{1 + \dfrac{p_{n+1}}{P - p_{n+1}} - \left(\dfrac{k_G AP}{g_{\text{in}}}\right)\left(\dfrac{p_{n+1}}{P - p_{\text{in}}}\right)}. \tag{16}$$

There is no analytical solution of this equation. However, if one can assume, as was actually the case, that the fraction of the total gas stream removed by absorption is small, so that $g_n = g_{n-1} = \ldots g_1 = g_{\text{in}} + g_{\text{out}}/2 = \bar{g}$, then eqn (16) becomes linear and an analytical solution results in

$$k_G = \bar{g}\frac{1 - (p_{\text{out}}/p_{\text{in}})^{1/N}}{AP} \tag{17}$$

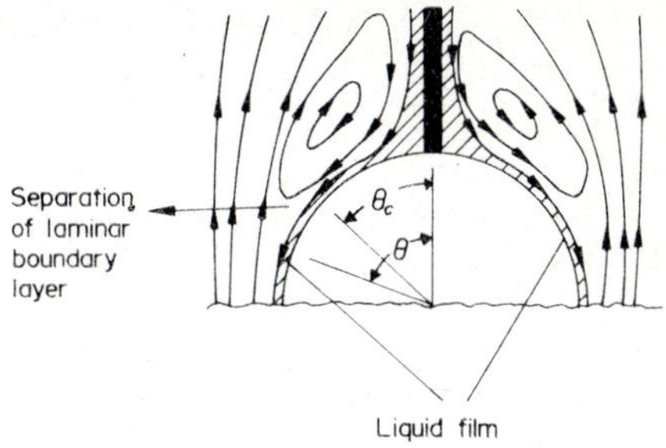

Fig. 5. Idealized flow patterns around a sphere.

where N is the total number of spheres. The following procedure can now be used to determine the value of k_G. From the measured values of p_{out} and p_{in} a first approximation for k_G is obtained using eqn (17). Insert this value of k_G into eqn (16) and obtain a value for p_{out} (where $p_{n+1} = p_{\text{in}}$ and $p_1 = p_{\text{out}}$) after N iterations. Compare the calculated value of p_{out} with the measured value. If they are not sufficiently nearly equal, change the value of k_G by a small amount. Insert this new value into eqn (16) and repeat until the calculated and measured values are almost equal.

Experiments were carried out with 10 × 3·72 cm dia. spheres using either 4·50 or 5·50 cm internal diameter glass tubes and (i) keeping the liquid flowrate constant and varying the gas flowrate or (ii) keeping the gas flowrate constant and varying the liquid flowrate. Similar experiments were carried out with 30 × 1·89 cm diameter spheres and using 4·50, 3·57, 3·00 and 2·58 cm internal diameter glass tubes. The air flowrate and the inlet and outlet concentration of ammonia were measured and the values of k_G were calculated using the above procedure. These results showed that k_G depended on the liquid flowrate as well as the gas flowrate. It is known that k_G can be correlated to $\bar{u}_r$, the mean relative velocity of the gas and the surface of the liquid, reasonably successfully in the case of disc columns [4, 9]. We may, therefore, expect k_G to be a function of a mean relative velocity.

The relative velocity u_r depends on the angle θ (see Fig. 5) for given gas and liquid flowrates and sphere and column diameters, and it is given approximately by

$$u_r = \frac{u_0}{1 - (d/D)^2 \sin^2\theta} + \frac{3Q \sin^{-1/3}\theta}{2\pi d \Delta_1} \tag{18}$$

where u_0 is the superficial gas velocity, d is the diameter of the sphere, D is the internal diameter of the column and Δ_1 is the liquid film thickness at $\theta = 90°$ and is equal to $(3Q\nu_l/dg)^{1/3}$. The first term of eqn (18) is the mean gas velocity (ignoring the liquid film thickness) and the second term is the velocity of liquid surface assuming the flow is laminar. The mean value of u_r was obtained by

$$\bar{u}_r = \frac{180/\pi}{(\theta_2 - \theta_1)} \int_{\theta_1}^{\theta_2} u_r \, d\theta \tag{19}$$

where θ_2 and θ_1 were taken as 40° and 100°. The choice of these limits is arbitrary but it appears from Fig. 6 that all the experimental data can be correlated by a single line within experimental errors.

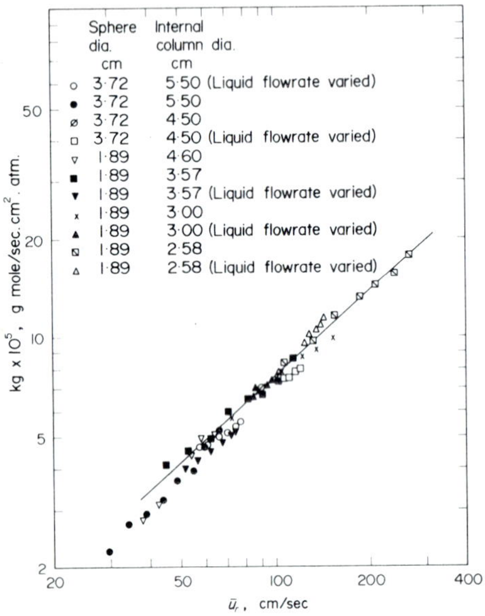

Fig. 6. Gas side mass transfer coefficients (3·72 cm and 1·89 cm dia spheres columns) (NH_3-air systems).

(iii) *The* (v/a) *ratio*

The volume of liquid per sphere, V, can be taken as the volume of the pool plus the volume of the film on the sphere. The latter can be calculated using eqn (20) (which assumes laminar flow of the liquid)

$$V_s = \frac{\pi d^2}{2} \left(\frac{3\nu_L Q}{\pi d g}\right)^{1/3} \int_0^{180°} \sin^{1/3}\theta \, d\theta \tag{20}$$

(iv) *Summary of the method*

The method for simulating a "complete" packed column of known L, $(L/\bar{G})$ and H (and also k_L, k_G, v and a), as regards both gas and liquid side phenomena and bulk reactions, using a string-of-spheres may be summarised as:

1. Determine the flowrate of liquid in the sphere column Q so that k_L will be the same as that of the packed column (e.g. use Fig. 4).

2. Calculate the flowrate of gas in the sphere column so that it is equal to $Q \times (\bar{G}/L)$.

3. Calculate the diameter of the tube confining the sphere column from the flowrate of liquid and gas and the diameter of spheres using eqn (19) and Fig. 6 to make the gas velocity such that k_G is the same in model and packed column.

4. Calculate the dimensions of the pool on the top of the spheres to make (v/a) the same by making use of eqn (20).

5. Determine the number of spheres, N, from $NA/Q = aH/L$ where A is the interfacial area per sphere.

ABSORPTION WITH CHEMICAL REACTION IN THE SPHERE AND THE PACKED COLUMNS—TESTS OF THE MODEL

The method of simulation which is given above has been tested by carrying out absorption experiments in the column consisting of 3·72 cm diameter spheres and a packed column. The packed column was made up by joining the 10·2 cm internal diameter Q.V.F. glass columns 61 cm long and packing them with 1·27 cm ceramic Raschig rings. Between the sections of the column redistributors were placed and the initial distribution of liquid was very good (see Ref.[1] for pertinent details). The following chemical systems were used (i) absorption of pure carbon dioxide into monoethanolamine solutions containing sodium arsenite, (ii) simultaneous absorption of CO_2 and NH_3 from air into water, (iii) simultaneous absorption of SO_2 and CO_2 from air into monoethanolamine solutions, (iv) absorption of CO_2 from air into 2,6 dimethyl morpholine solutions and (v) absorption of pure CO_2 into a solution which contains two amines.

(i) *Absorption of carbon dioxide into monoethanolamine solutions containing arsenite*

A fast reaction takes place near the interface according to:

$$CO_2 + 2RR'NH \rightleftharpoons RR'NCOO^- + RR'NH_2^+. \tag{21}$$

This reaction is followed by the following overall reaction in the bulk of the liquid[10]

$$RR'NCOO^- + H_2O \rightarrow RR'NH + HCO_3^-. \tag{22}$$

The hydrolysis of carbamate (22) is normally too slow to take place during the passage of liquid through the absorber but it can be catalysed by the addition of sodium arsenite.

A batch of amine solution (containing arsenite) was pumped into an overhead tank and this solution was fed to either the sphere column or the packed column. Pure CO_2 was absorbed in the absorbers, total absorption rates were measured and the samples were collected from the inlet and outlet liquid streams and the concentration of CO_2 in the liquid (i.e. $(RR'NCOO^-) + (HCO_3^-)$ was obtained by analysing them.

In these experiments, there was no gas-side resistance but the absorption rate was dependent on the bulk reaction (22). Therefore it was only necessary to make k_L, (aH/L) and (v/a) the same in both absorbers. Table 1 shows the results of the experiments. Table 1 also contains some runs (9 and 10) where the solutions did not have any catalyst. Runs 15–18 were only carried out in the sphere column (for economic reasons) and in these runs, only the concentration of arsenite was different and everything else was the same. According to runs 15–18, an increase in the arsenite concentration results in an increase in the total absorption rate. This suggests that the reaction (22) takes place appreciably so that the absorption rate increases (due to increased rate of supply of amine from the bulk of liquid to the zone near the gas-liquid interface). The conditions for other runs (e.g. concentration of arsenite, carbonation ratios etc.) were similar to those for the runs 15–18. Consequently they should involve the reaction (22) in the bulk. The predicted total absorption rate in the packed column, $(\bar{R}aH_xH)_p$ (from the results of sphere column experiments) is:

$$(\bar{R}aA_xH)_p = \frac{LA_x}{Q}\overline{(RAN)} \qquad (23)$$

where $\overline{(RAN)}$ is the total absorption rate in the sphere column and A_x is the cross sectional area of the packed column. The last column of Table 1 shows the percentage difference between the predicted and measured total absorption rates, $(\bar{R}aA_xH)$ and the maximum difference is less than ±7%.

(ii) *Simultaneous absorption of* NH_3 *and* CO_2 *into water*

When NH_3 and CO_2 are absorbed into water simultaneously, reactions take place in the liquid, leading to the formation of ammonium carbamate and bicarbonate. The rate of absorption of NH_3 will mostly depend on the value of the k_G but there will also be some liquid-side resistance. The rate of absorption of CO_2 will approximately depend on the concentration of NH_3 at the gas-liquid interface[11], thus on the value of k_G. In a tall column, most of the ammonia will be absorbed in the bottom part of the column and at the top of the column, CO_2 will be absorbed physically (as there is a negligible concentration of NH_3 here).

Table 2 shows the results of the experiments. In the packed column, the superficial gas (at inlet) and liquid velocities were 0·4 and 16 cm/s. The values of k_L (CO_2 into water) and k_G (NH_3 in air) were $6{\cdot}4\times10^{-3}$ cm/s[12] and $2{\cdot}1\times10^{-5}$ g mole/s . cm^2 atm[13]. In the sphere column, when the liquid and gas rates were 0·8 and 32 cm^3/s respectively, and the diameter of the column was 4·0 cm, the values of k_L, k_G and $(L/\bar{G})$ were the same in both absorbers. The height of packed column was 158 cm and when the height of the sphere column was 49 cm, the values of (aH/L) were the same for both absorbers. We may now expect the compositions of liquid and gas streams leaving the column to be the same if the inlet composition of gas is the same as that of the packed column. Comparisons of the sphere column runs 1, 2 . . . 15 with the packed column runs 16, 17 . . . 30 (the runs 1, 2, 7 etc. in the sphere column correspond to the runs 16, 17, 22 etc. in the packed column) indicate that if the inlet compositions of gas is the same, the outlet compositions of the gas and the liquid streams will also be the same.

(iii) *Simultaneous absorption of* CO_2 *and* SO_2 *into amine solutions*

CO_2 and SO_2 were absorbed simultaneously into aqueous monoethanolamine solutions in the packed and the sphere columns. The flowrates of gas and liquid were the same as those in (ii) and the values of k_L and k_G were the same in both absorbers.

Table 3 shows the results of the experiments in the sphere (run no. 1–5) and packed (run no. 6–8) columns. Runs in the sphere column (3–5) can be compared with the respective packed column runs (6–8). Comparisons of these runs indicates that the outlet compositions of the gas and the liquid will be closely the same if the inlet compositions are the same in the two absorbers.

(iv) *Absorption of* CO_2 *from air into 2,6 dimethylmorpholine solutions*

The reaction between CO_2 and 2,6 dimethylmorpholine is second order and the reaction rate constant is about 15,000 l/g mole s at 25°C[14]. When the flowrates of liquid and gas streams are the same as in (ii) and (iii), the values of k_L and k_G are $6{\cdot}4\times10^{-3}$ cm/s and $2{\cdot}1\times10^{-5}$ g mole/s cm^2 atm. At the top of the column, the reaction between the uncarbonated amine and CO_2 can be considered as effectively irreversible and when the concentration of amine and the partial pressure of CO_2 are 0·4 g mole/litre and 0·04 atm respectively the value of $(1 + Ek_L/Hek_g)$ is about 1·55. This means roughly that there is about 35% gas-side resistance[15]. Consequently, the rate of absorption of CO_2 depends not only on k_L but also on k_G.

CO_2 from air was absorbed into 0·4 M dimethylmorpholine solution in the sphere and packed columns. The inlet and outlet compositions of gas and the concentration of CO_2 in the leaving liquid (as carbamate etc.) were analysed.

Table 4 shows the results in the sphere column (runs 1, 2 and 3) and the packed column (run 4). The run 1 can be compared with the corresponding packed column run 4.

(v) *Absorption of* CO_2 *into a solution which contains two amines*

This is an example of absorption of a gas into a solution which contains two reactants. The data in Table 5 covered

Table 1. Experimental results of CO_2-aq. monoethanolamine (+ sodium arsenite) system. (Sphere column and packed column, 25°C and 1 atm). (Sphere column: liquid flowrate = 1·15 cm^3/s, height = 49 cm). (Packed column: superficial liquid velocity = 0·6 cm/s, height = 133 cm)

Run no.	Arsenite concn	$([RNH_2] + [RNHCOO^-] + [RNH_3^+])$	Carbonation ratio			Total absorption rate		$\frac{(RaA_xH)-(RaA_xH)_P}{(RaA_xH)} \times 100$
			Inlet	Outlet		$(\overline{RAN})$	$(\overline{RaA_xH})$	
				Sphere column	Packed column	Sphere column	Packed column	
	gmole/ℓ	gmole/ℓ				cc/s		%
1	0.110	1.09	0.00	0.52	0.52	15.10	640	-0.50
2	0.100	1.16	0.00	0.52	0.53	16.70	715	+0.40
3	0.100	1.16	0.40	0.59	0.59	5.85	261	+4.40
4	0.100	1.16	0.57	0.65	0.66	3.13	134	+0.50
5	0.150	1.16	0.69	0.76	0.77	2.39	99	+1.50
6	0.090	1.58	0.30	0.51	0.51	8.50	370	+0.25
7	0.090	1.58	0.48	0.55	0.56	3.27	145	+3.45
8	0.070	1.10	0.37	0.53	0.53	4.90	208	+1.0
9	-	1.28	0.42	0.54	0.54	4.20	190	+6.30
10	-	1.25	0.29	0.46	0.46	6.90	290	-0.30
11	0.070	1.25	0.30	0.51	0.51	7.50	320	+0.90
12	0.070	1.25	0.56	0.61	0.61	2.40	104	+1.90
13	0.155	1.25	0.61	0.68	0.68	2.50	110	+3.80
14	0.145	1.25	0.62	0.67	0.67	2.10	95	+6.30
15	0.065	1.25	0.61	0.65	-	1.80	-	-
16	0.100	1.25	0.61	0.66	-	2.00	-	-
17	0.130	1.25	0.61	0.67	-	2.20	-	-
18	0.155	1.25	0.61	0.68	-	2.50	-	-
19	0.075	1.10	0.46	0.55	0.55	3.40	150	+4.00
20	0.075	1.10	0.54	0.61	0.62	2.50	107	+1.90

Table 2. Experimental results of NH_3-air-water system. (Sphere column and packed column, 25°C and 1 atm)

Run no.	Flowrates at inlet			Concentrations in out-going liquid		$(\text{Moles of } CO_2)/(\text{Moles of air})$ in out-going gas
	NH_3	CO_2	Air	$CO_2 \times 10^3$	Total NH_3	
	cc/min			gmole/litre		%
(Sphere column: liquid flowrate = 0.8 cc/s, height = 49 cm)						
1	-		1830	1.63	-	5.0
2	52		1770	5.72	0.050	4.8
3	100	94	1730	7.76	0.085	4.9
4	152		1670	9.40	0.133	4.8
5	180		1650	11.45	0.164	4.8
6	-		1730	3.27	-	10.1
7	52		1670	11.45	0.051	10.0
8	100	195	1630	14.71	0.086	9.9
9	152		1580	17.60	0.132	9.8
10	180		1550	20.05	0.163	9.8
11	-		1625	4.91	-	15.1
12	52		1580	12.70	0.048	15.1
13	100	295	1520	19.23	0.090	15.1
14	152		1470	26.60	0.142	15.3
15	180		1450	29.45	0.168	15.2
(Packed column: superficial liquid velocity = 0.4 cm/s, height = 158 cm.						
16	-		74000	1.63	-	4.9
17	2100		72000	5.93	0.049	4.8
18	4100	3800	70000	8.00	0.087	4.8
19	6200		68000	9.50	0.130	4.8
20	7300		67000	11.60	0.161	4.7
21	-		70000	3.27	-	10.0
22	2100		68000	12.00	0.048	9.8
23	4100	7900	66000	15.00	0.089	9.8
24	6200		64000	17.60	0.131	9.8
25	7300		62500	20.40	0.165	9.8
26	-		66000	4.90	-	15.1
27	2100		63900	13.00	0.047	15.2
28	4100	12000	61900	19.50	0.089	15.1
29	6200		59800	27.00	0.140	15.2
30	7300		58700	29.70	0.168	15.1

Table 3. Experimental results of CO_2-SO_2-air-aq. monoethanolamine system. (Sphere column and packed column, 25°C and 1 atm)

Run no.	Concentration in inlet gas		Air flow-rate	Concentration of amine at inlet	SO_3^{--} in outgoing liquid	CO_2 in outgoing liquid	CO_2 in outgoing gas
	SO_2	CO_2					
	%		litre/min		gmole/litre		%
(Sphere column: liquid flowrate = 0.8 cc/s, height = 49 cm)							
1	7.9	10.3	1.57	0.53	0.125	0.106	4.4
2	7.9	16.7	1.44	0.53	0.117	0.127	9.3
3	4.5	10.3	1.64	0.52	0.072	0.110	4.3
4	4.6	10.4	1.64	0.82	0.073	0.147	2.4
5	4.5	13.6	1.57	0.82	0.071	0.200	4.0
(Packed column: superficial liquid velocity = 0.4 cm/s, height = 158 cm)							
6	4.45	10.30	66.5	0.52	0.071	0.106	4.4
7	4.55	10.35	66.5	0.82	0.073	0.145	2.5
8	4.50	13.60	63.9	0.82	0.071	0.198	4.0

Table 4. Experimental results of CO_2-air-2,6 dimethylmorpholine system. (Sphere column and packed column, 25°C and 1 atm)

Run no.	Concentration of amine at inlet	Air flowrate	Concentration of CO_2 in gas stream		CO_2 in outgoing liquid
			Inlet	Outlet	
	gmole/litre	litre/min	%		gmole/litre
(Sphere column: liquid flowrate = 0.8 cc/s, height = 49 cm)					
1	0.4	1.73	10.3	3.8	0.110
2	0.4	1.60	17.1	8.9	0.135
3	0.4	1.50	22.5	14.6	0.143
(Packed column: superficial liquid velocity = 0.4 cm/s, height = 158 cm)					
4	0.4	70.0	10.4	3.9	0.108

the following ranges: (i) both reactions were effectively instantaneous, (ii) one reaction was effectively instantaneous, the other was in an intermediate range between pseudo first order and instantaneous reactions.

Table 5 shows the experimental results in the sphere and packed columns. In these experiments, pure CO_2 was absorbed into a mixture of amine solutions and the total absorption rates and the concentrations of CO_2 (as carbamate etc.) were measured. In the sphere and packed columns k_L and (aH/L) were the same. The total absorption rate in the packed column can be predicted from the measured total absorption rates in the sphere column, $(\overline{RAN})$, using eqn (23), and the last column of Table 5 shows the percentage difference between the predicted $(\bar{R}aH_xH)_p$ and the measured $(\bar{R}aA_xH)$ values and the difference between the two is always less than ±4%.

DISCUSSION AND CONCLUSIONS

Comparison of the sphere column experiments with the respective packed column experiments under the proper "scaling" conditions indicates that if the inlet compositions of the liquid and gas are the same, the outlet compositions are likely to be closely the same (see Tables 1–5). When a gas is absorbed in the sphere and packed columns, the total absorption rate in the latter can be predicted from the measurement of total absorption rate in the sphere column using eqn (23). The last columns of Tables 1, 2 and 5 show the percentage difference between the predicted and measured total absorption rates, $(\bar{R}aA_xH)_p$ and $(\bar{R}aA_xH)$ and the maximum difference is always less than ±7%. It appears therefore (from the results of experiments with five essentially chemically different systems) that the criteria of simulation as given here are probably sound.

For the first system, the experiments correspond to a situation where a fast reaction near the interface may be followed by another reaction in the bulk. Danckwerts and McNeil[10] give some simplified approximate methods to predict the absorption rate for carbonation ratios either less than 0·5 or greater than 0·5. When the carbonation ratio is in the region of 0·5–0·6 (e.g. runs 3, 7, 19, etc. in Table 1) neither method will predict the absorption rate sufficiently accurately. Thus the model experiments can take the place of calculation.

For the second and the third systems, the results correspond to two different cases of simultaneous absorption of two gases. For the second system, two gases CO_2 and NH_3, are absorbed into water simultaneously and a reaction takes place in the liquid. Conse-

Table 5. Experimental results of CO_2–aq. Monoethanolamine (or Diethanolamine) + another amine system. (Sphere column and Packed column, 25°C and 1 atm) (Sphere column: Liquid flowrate = 0.8 $cm^3/s.$, Height = 49 cm) (Packed column: Superficial liquid velocity = 0.4 cm/s., Height = 158 cm)

Run no.	Concn. of mono-ethanolamine	Concn. of amine	CO_2 in outgoing liq: Sphere column	CO_2 in outgoing liq: Packed column	Total abs. rates: $(\overline{RAN})$	Total abs. rates: $(\overline{RaA_xH})$	$\frac{(\overline{RaA_xH})-(RaA_xH)}{(\overline{RaA_xH})} \times 100$
	g/mole/litre		gmole/litre		$cm^3/s.$		%
MEA + Di-isopropanolamine							
1	0.300	—	0.178	—	3.52	—	—
2	0.300	0.040	0.192	—	3.91	—	—
3	0.300	0.080	0.212	0.206	4.35	170.0	−3.5
4	0.142	0.038	0.125	0.124	2.45	98.0	−4.5
MEA + 2-Methylaminoethanol							
5	0.335	—	0.195	—	3.92	—	—
6	0.335	0.115	0.256	0.256	5.15	213.0	+2.1
7	0.335	0.225	0.308	0.388	6.33	260.0	+1.1
MEA + Monoisopropanolamine							
8	0.296	—	0.173	—	3.54	—	—
9	0.296	0.080	0.210	0.213	4.44	185.0	+2.4
10	0.296	0.264	0.300	0.300	6.17	250.0	+0.1
11	0.132	0.118	0.158	0.155	3.03	124.0	+0.7
MEA + Diethanolamine							
12	0.396	—	0.222	—	4.80	—	—
13	0.396	0.074	0.255	0.258	5.20	213.0	+0.9
14	0.396	0.130	0.284	0.283	5.90	238.0	−0.7
Diethanolamine + Di-isopropanolamine							
15	0.243	—	0.138	—	2.85	—	—
16	0.243	0.045	0.147	0.147	3.02	125.0	+2.4
17	0.243	0.135	0.172	0.176	3.55	145.0	+0.7

quently, the effect of absorption of the second gas should generally be to increase the absorption rate of the other gas. On the other hand, for the third system, the two gases, CO_2 and SO_2, compete for the same reactant. Thus in general, the effect of the presence of one of the gases is to reduce the rate of absorption of the other. The difficulty of the task of predicting the absorption rate of each gas for a given boundary condition (i.e. given bulk concentration of reactant and concentrations of each gas at the gas-liquid interface) is well known. In practice, the gases will probably be diluted (with air for instance) and firstly it is necessary to estimate the concentration of each gas at the interface from their concentrations in the bulk of the gas. Furthermore, in a packed column, the compositions of liquid and gases are changing along the column and the bulk concentrations of the reactant and the gases at the chosen level cannot be calculated using a simple material balance equation.

For the fifth system, the experiments correspond to absorption of a gas into a solution containing two reactants. In this case, theoretical predictions would have been also hard to make, as the bulk concentration of the reactants at a given level could not be calculated using a material balance equation.

According to the above discussion, it appears that in several cases the theoretical predictions would have been either too complicated—if indeed possible—or would lead to large errors in several instances. On the other hand, it appears (from the experimental results of sphere and packed columns) that if one builds up a model of the packed column based on the criteria which are given here, the behaviour of the packed column (for instance total absorption rate or the composition of gas leaving the column etc.) can be obtained from the results of the sphere column sufficiently accurately from laboratory experiments.

NOTATION

a interfacial area per unit volume of packed column, cm^{-1}

a' interfacial area per unit height in the string-of-spheres, cm^{-1}

A interfacial area per sphere, cm^2
A_x cross-sectional area of the packed column, cm^2
B^0 concentration of reactant B in the bulk of liquid, $gmole/cm^3$
C^0 concentration of dissolved gas in the bulk of liquid, $gmole/cm^3$
C^* concentration of dissolved gas at interface, in equilibrium with liquid at interface, $gmole/cm^3$
d diameter of spheres, cm
D internal diameter of sphere column, cm
D_B diffusivity of dissolved reactant B, cm^2/s
D_C diffusivity of dissolved gas C, cm^2/s
E enhancement factor, i.e. factor by which rate of absorption is increased by reaction
g gravitational acceleration, cm/s^2; or flowrate of gas in string-of-spheres, gmole/s
G superficial gas velocity in packed column, cm/s
$\bar{G}$ mean superficial gas velocity, cm/s
H height of packed column, cm
H_1 height of laboratory model, cm
He Henry's law constant, atm cm^3/gmole
k_L physical liquid-side mass transfer coefficient, cm/s
k_G gas-side mass transfer coefficient, $gmole/cm^2 s$ atm
L superficial liquid velocity, cm/s
N total number of spheres in a string
p partial pressure of soluble gas in bulk of gas, atm
P pressure in absorber (generally atmospheric), atm
Q liquid flowrate in string-of-spheres or wetted wall column, cm^3/s
r rate of reaction of C per unit volume, $gmole/cm^3 s$
$\bar{R}$ average rate of absorption per unit area, $gmole/cm^2 s$
$(\bar{R}AN)$ total rate of absorption in a string-of-spheres, gmole/s
$(\bar{R}aA_xH)$ total rate of absorption in packed column, gmole/s
$(\bar{R}aA_xH)_p$ predicted total absorption rate (from total rate of absorption in sphere column) in packed column, gmole/s
$\bar{t}_r$ mean residence time of liquid in packed column, defined in eqn (6)
u_r relative velocity of gas and liquid surface, cm/s
u_s velocity of liquid surface, cm/s
u_0 superficial gas velocity in string-of-spheres, cm/s
v volume of liquid per unit volume of packed space, liquid hold-up
V volume of liquid per sphere, cm^3
z number of moles of reactant reacting with each mole of A
θ angle as shown in Fig. 5
ν_L kinematic viscosity of liquid, cm^2/s.

REFERENCES

[1] Danckwerts P. V. and Alper E., *Trans. Instn Chem Engrs.* 1975 **53** 34.
[2] Danckwerts P. V., *Gas-Liquid Reactions.* p. 221. McGraw-Hill, New York 1970.
[3] Danckwerts P. V. and Gillham A. J., *Trans. Instn Chem. Engrs.* 1966 **44** T42.
[4] Alper E., Ph.D. Thesis, Cambridge University, Cambridge, England 1972.
[5] Davidson J. F., Cullen E. J., Hanson D. and Roberts D., *Trans. Instn Chem. Engrs.* 1959 **37** 122.
[6] Danckwerts P. V. and Rizvi S. F., *Trans. Instn Chem. Engrs.* 1971 **49** 124.
[7] Seidell A. and Linke W. F., *Solubilities of Inorganic and Organic Compounds.* Van Nostrand, New York 1952.
[8] Danckwerts P. V., *Gas-Liquid Reactions.* p. 149. McGraw-Hill, New York 1970.
[9] Stephens E. J. and Morris G. A., *Chem. Engng Prog.* 1951 **47** 232.
[10] Danckwerts P. V. and McNeil K. M., *Trans. Instn Chem. Engrs.* 1967 **45** T32.
[11] Astarita G., *Mass Transfer with Chemical Reactions.* p. 116. Elsevier, Amsterdam 1967.
[12] Danckwerts P. V., *Gas-liquid Reactions.* p. 213. McGraw-Hill, New York 1970.
[13] Mannford Doble M., Ph.D. Thesis, Cambridge University, Cambridge, England 1972.
[14] Sharma M. M., Ph.D. Thesis, Cambridge University, Cambridge, England 1972.
[15] Danckwerts P. V., *Gas-Liquid Reactions.* p. 200. McGraw-Hill, New York 1970.

APPENDIX

(i) *Simultaneous absorption of various gases into a liquid containing a single reactant*

When n gases are absorbed simultaneously and each reacts with the same dissolved reactant B, the equivalent of eqns (1), (3) and (4) are

$$\frac{G_{\rm in}\left(\sum_1^n z_n p_{{\rm in},n}\right) - G_{\rm out}\left(\sum_1^n z_n p_{{\rm out},n}\right)}{v_m P} = -L\left(\sum_1^n z_n C^\circ_{{\rm in},n} - \sum_1^n z_n C_{{\rm out},n} + L(B^\circ_{\rm in} - B^\circ_{\rm out}\right) \quad (24)$$

$$\frac{aH}{L} = \int_{\rm in}^{\rm out} \frac{z_1 dC_1^0}{\sum_1^n z_n \bar{R}_n} + \int_{\rm in}^{\rm out} \frac{z_2 dC_2^0}{\sum_1^n z_n \bar{R}_n} + \ldots + \int_{\rm in}^{\rm out} \frac{z_n dC_n^0}{\sum_1^n z_n \bar{R}_n} - \int_{\rm in}^{\rm out} \frac{dB^0}{\sum_1^n z_n \bar{R}_n} \quad (25)$$

$$\frac{vH}{L} = \int_{\rm in}^{\rm out} \frac{dB^0}{\sum_1^n z_n r_n} \quad (26)$$

where $\sum_{1}^{n} z_n p_{in,n} = (z_1 p_{in,1} + z_2 p_{in,2} + \ldots + z_n p_{in,n})$ etc., $p_{in,1}$ and $p_{out,1}$ are the partial pressures of solute gas C_1, $C^0_{in,1}$ and $C^0_{out,1}$ are the concentrations of dissolved but unreacted gas. If in the laboratory absorber and the packed column the values of $k_{L,1}$ and $k_{G,1}$, the physical liquid-side and gas-side coefficients for the solute gas C_1, are the same and if we make the inlet compositions of gas and liquid and the values of (vH/L) or (a/v), (aH/L) and $(L/\bar{G})$ the same in both absorbers we may expect the right hand sides of the eqns (25) and (26) to be the same. This means that the necessary conditions for simulating the packed column in a model are the same as when there is one solute gas and one reactant (namely $k_{L,1}$, $k_{G,1}$, (aH/L), (vH/L) or (a/V) and $(L/\bar{G})$).

(ii) *Absorption of a gas into a liquid which contains various reactants*

In this case the equivalent of eqns (1), (3) and (4) gives

$$\frac{(G_{in}p_{in} - G_{out}p_{out})}{v_m P} = -L(C^0_{in} - C^0_{out}) + L\sum_{1}^{n}\left(\frac{B^0_{in,n} - B^0_{out,n}}{z_n}\right. \qquad (27)$$

$$\frac{aH}{L} = \int_{in}^{out}\frac{dC^0}{\bar{R}} - \int_{in}^{out}\frac{dB_1{}^0}{z_1\bar{R}} + \int_{in}^{out}\frac{dB_2{}^0}{z_2\bar{R}} + \ldots + \int_{in}^{out}\frac{dB_n{}^0}{z_n\bar{R}} \qquad (28)$$

$$\frac{vH}{L} = -\int_{in}^{out}\frac{dB_1{}^0}{z_1 r_1} = -\int_{in}^{out}\frac{dB_2{}^0}{z_2 r_2} = \ldots = -\int_{in}^{out}\frac{dB_n{}^0}{z_n r_n} \qquad (29)$$

We may therefore expect that the necessary conditions for simulation to be the same as before (i.e. the same k_L, k_G, (aH/L), (vH/L) or (v/a) and $(L/\bar{G})$ according to eqns (27)–(29).

B11

Surface instability during the absorption of CO_2 by monoethanolamine solutions

BRIAN *et al.* [1] have discussed some anomalous effects observed when CO_2 is absorbed by solutions of monoethanolamine (MEA). It appears that under certain conditions convection currents may be set up in the liquid as a result of the absorption process and that these currents enhance the rate of absorption.

We have observed such convection currents directly. A drop of 1*M* MEA solution, containing fine suspended particles, was placed on a microscope slide, illuminated from the side and viewed against a dark background. A stream of pure CO_2 was allowed to flow over the solution. Convection currents became observable under the microscope within one second, leading to convection cells of the familiar "Bérnard cell" type with a scale of about 0·25 mm. Convection ceased after some seconds, presumably when the amine had been completely carbonated. This happened much more rapidly than it would if molecular diffusion alone had been responsible for the transport of CO_2 and amine in the solution.

It should be pointed out that in order to account for some of the observations cited by Brian *et al.*, convection currents must be assumed to arise quickly enough to affect the rate of absorption when the gas and liquid have been in contact for 10^{-2} sec or less. Our observations were not sufficiently refined to throw any light on the early history and growth of the convection pattern; it seems likely that the scale of the pattern may initially be small and may increase with time of contact [2, 3].

The surface-tensions of 1*M* MEA solutions, both uncarbonated and freshly saturated with CO_2, were measured at 19°C by the drop-weight method. The surface-tension of the carbonated solution was found to be about 2 per cent higher than that of the uncarbonated solution. Qualitatively, an increase in surface-tension on carbonation could be the reason for the instability of the surface, the formation of convection currents being an example of Marangoni instability [4, 5]. The convection currents would be driven by the difference in surface-tension between parts of the surface which had been exposed for some time and those composed of liquid which had newly arrived from the bulk. However, the observed increase in surface-tension is small and we had some doubt as to whether it was sufficient to account for the vigorous effects observed.

The carbonation reaction is

$$CO_2 + 2RNH_2 \rightarrow RNHCOO^- + RNH_3^+$$

where $-R$ is $-C_2H_4OH$. Thus the products of the reaction are ions and this is probably why the surface-tension rises. However, at equilibrium the concentration of a solute which increases the surface-tension of the solvent is less in the neighbourhood of the surface than in the bulk of the liquid. The total deficit of the solute in the surface region can be calculated from the Gibbs equation, if the relationship between surface-tension and concentration of solute is known, but it is not possible to calculate the actual concentration at the surface.

However, when CO_2 is being absorbed by the amine solution, the system is not at equilibrium and the concentration of the ionic products is actually higher at the surface than in the bulk of the solution, into which they are diffusing. In the common case in which the reaction between CO_2 and amine is pseudo-first-order in the CO_2 concentration, the concentration of the products at the surface can be calculated [6]. In the case of 1*M* MEA and CO_2 at 1 atm, the concentration of each of the ions at the surface will be about $4\sqrt{t}$ gmole/l. at 25°C, where t is the contact-time in seconds (for contact times greater than 10^{-3} sec). Thus the absorption process ensures a substantial concentration of ions at the surface for contact times of the order of 10^{-2} sec or more and it seems possible that this may lead to a substantially higher surface-tension than that of the depleted *equilibrium* surface of a carbonated solution.

In order to test this hypothesis, we tried to measure the surface-tension of an amine solution into which CO_2 was being absorbed, using Rayleigh's "oscillating jet" method [7]. A jet of the liquid issues from an elliptical hole. Surface-tension forces tend to reduce the jet to a circular shape, but inertia causes "overshooting" and the jet becomes elliptical once more, with its major axis at right-angles to the original direction. The frequency of this reversal of the shape of the cross section of the jet is related to the surface-tension of the liquid. Observations on jets of 1*M* and 2*M* MEA solutions in air and in CO_2 at 1 atm revealed no detectable difference in surface-tension; the method was not very accurate, but an increase in surface-tension of 15 per cent would have been detected. Times of contact up to $3 \cdot 3 \times 10^{-2}$ sec were used.

The experiment therefore failed to demonstrate that a solution which is absorbing CO_2 has a markedly increased surface-tension. However, this may not be conclusive. If convection currents, driven by differences in surface-tension, are bringing streams of liquid of low surface-tension to the surface, the latter will be a mosaic of regions of high and low surface-tension. It is not unreasonable to suppose that in the oscillating-jet experiment such a patchy surface would behave as though its tension were that of the regions of lowest surface-tension; if such a surface is stretched, it responds by increasing the area of the patches of low surface-tension, and it is their resistance to extension which will be measured.

P. V. DANCKWERTS
A. TAVARES DA SILVA

Department of Chemical Engineering
Pembroke Street
Cambridge

REFERENCES

[1] BRIAN P. L. T., VIVIAN J. E. and MATIATOS D. C., *A.I.Ch.E.Jl* 1967 **13** 28.
[2] ORELL A. and WESTWATER J. W., *Chem. Engng Sci.* 1961 **16** 127.
[3] ORELL A. and WESTWATER J. W., *A.I.Ch.E.Jl* 1962 **8** 350.
[4] STERNLING C. V. and SCRIVEN L. E., *A.I.Ch.E.Jl* 1959 **5** 514.
[5] SCRIVEN L. E. and STERNLING C. V., *Nature* 1960 **187** 186.
[6] DANCKWERTS P. V., *Chem. Engng Sci.* 1967 **22** 472.
[7] RAYLEIGH J.W.S. *Proc. R. Soc.* 1879 **29** 71.

B12

Gas absorption with instantaneous reaction

Abstract – When an absorbed gas undergoes an instantaneous reaction, reversible or otherwise, in solution the rate of absorption is

$$\bar{R} = k_L \Delta C$$

where ΔC is the amount of gas required to saturate unit volume of the bulk liquid to a partial pressure equal to that at the interface, and k_L is the mass-transfer coefficient for physical absorption. This expression is valid when the diffusivities of all species are equal.

ABSORPTION INTO QUIESCENT LIQUID

IN CERTAIN circumstances the absorption of a gas in a liquid is followed by instantaneous reaction of the dissolved gas with a reactant in solution in the liquid, or with the liquid itself. The reaction may be irreversible (e.g. absorption of NH_3 in HCl solution) or reversible (e.g. absorption of H_2S in amine or alkali solution; absorption and partial hydrolysis of SO_2 in water). The reactions may occur virtually instantaneously (as do those referred to above); or they may be regarded as instantaneous under appropriate circumstances. In the following it is assumed that the reaction is fast enough for equilibrium to reign at every point and instant. The kinetics are then irrelevant.

Olander[1] has given extensive consideration to instantaneous, reversible reactions. Gioia and Astarita[2] have recently given a general method for calculating rates of absorption of H_2S into various alkaline solutions, if the diffusivities of all species are assumed to be equal. The present communication presents a general solution to the problem of gas absorption followed by instantaneous reversible reactions of any type, in which the diffusivities of all species are taken to have the same value, D. The reaction in solution is

$$A + z_1 B_1 + \ldots \rightleftharpoons y_1 P_1 + y_2 P_2 + \ldots \quad (1)$$

A being the dissolved gas, B_1 etc. the reactants, P_1 etc. the products. The composition is initially uniform, the concentrations being A^0, B_1^0, P_1^0, etc. At time $t = 0$ the partial pressure of A at the interface is raised from π^0 to π^*, thus raising the concentration at the surface ($x = 0$) from A^0 to A^*, and subsequently maintained at this value.

The concentrations of the various species at any point in the liquid are related at all times by the equilibrium expression:

$$Ka = \frac{(p_1)^{y_1}(p_2)^{y_2}\ldots}{(b_1)^{z_1}(b_2)^{z_2}\ldots} \tag{2}$$

where a, p_1, etc., b_1, etc. are the concentrations of dissolved gas, products and reactants respectively, and K is the equilibrium constant.

The rate of absorption of the gas under these circumstances is

$$R = \Delta C\sqrt{(D/\pi t)} \tag{3}$$

where t is the time elapsed since the concentration at the surface was raised to A^* and ΔC is the amount of gas required to saturate unit volume of absorbent when the ambient partial pressure is raised from π^0 to π^*. In other words, ΔC is the amount of gas which must be added to unit volume of absorbent in order to raise the concentration of unreacted dissolved gas from its initial value A^0 (in equilibrium with a partial pressure π^0) to a final value of A^* (in equilibrium with π^*) and simultaneously to raise the equilibrium concentration of A in the reacted form by the amount required by the equilibrium condition. The validity of Eq.(3) is demonstrated in the Appendix.

The initial concentrations are related by

$$KA^0 = \frac{(P_1^0)^{y_1}\ldots}{(B_1^0)^{z_1}\ldots}. \tag{4}$$

Sufficient information must be given to determine the initial combined concentration of A in the reacted and unreacted forms. If ΔA is the increase in the concentration of A in the *reacted* form when the concentration of *unreacted A* rises from A^0 to A^*, then:

$$KA^* = \frac{(P_1^0 + y_1\Delta A)^{y_1}\ldots}{(B_1^0 - z_1\Delta A)^{z_1}\ldots} \tag{5}$$

$$\Delta C = A^* - A^0 + \Delta A\,. \tag{6}$$

The value of ΔA may be found from the initial composition and Eq (5); ΔC is then found from Eq. (6). The quantity ΔC may also be found by other methods of calculation, such as those used in the examples given below.

ABSORPTION INTO AGITATED LIQUIDS

Applying Eq. (3) to any form of the surface-renewal model, we get

$$\bar{R} = k_L\Delta C \tag{7}$$

where $\bar{R}$ is the average rate of absorption per unit area of interface and k_L is the liquid-side transfer coefficient for absorption without reaction. It is easily shown, using arguments similar to those in the Appendix, that the same conclusion is reached if the film model of gas-absorption is used. Thus the correct driving force to use in the case of absorption with instantaneous reaction is the amount of gas required to re-saturate unit volume of the bulk liquid when the partial pressure of the gas is raised from π^0 to π^*. In the case of *irreversible* reaction with a single reactant B, this leads to the familiar expression

$$\bar{R} = k_L(A^* + B^0/z) \tag{8}$$

(since $\Delta A = B^0/z$, $A^0 = 0$, and from Eq. (6) $\Delta C = A^* + B^0/z$).

EXAMPLES

The method may be demonstrated by considering the absorption of SO_2 at 1 atm into a solution of $NaHSO_3$ (0·5 M). At 20°C the solubility of SO_2 at 1 atm is 1·82 gmole/l. The SO_2 is partially hydrolysed:

$$SO_2 + H_2O \rightleftharpoons H^+ + HSO_3^-$$

$$A \rightleftharpoons P_1 + P_2. \tag{9}$$

The concentration of H_2O is constant, and so the equilibrium constant is of the form:

$$K_1 = \frac{[H^+][HSO_3^-]}{[SO_2]} = 1{\cdot}7 \times 10^{-2}\ \text{gmole/l}. \tag{10}$$

When the solution is saturated with SO_2 gas at 1 atm:

$$[SO_2] = 1{\cdot}82\ \text{gmole/l}$$

$$[Na^+] = 0{\cdot}5\ \text{g ion/l}$$

$$\frac{[H^+][HSO_3^-]}{[SO_3]} = 1{\cdot}7 \times 10^{-2}\ \text{g ion/l}$$

and for electrical neutrality

$$[H^+] + [Na^+] = [HSO_3^-]$$

whence

$$[H^+] = 0{\cdot}055\ \text{g ion/l}$$

$$[HSO_3^-] = 0{\cdot}555\ \text{g ion/l}$$

and

$$\Delta C = [SO_2] + [HSO_3^-] - 0{\cdot}5 = 1{\cdot}875.$$

Thus

$$R = 1{\cdot}875 \times 10^{-3}\sqrt{(D/\pi t)}\ \text{gmole/cm}^2\text{sec}$$

in the case of absorption into a quiescent liquid, and

$$\bar{R} = 1{\cdot}875\times 10^{-3}k_L$$

when the gas is absorbed into an agitated liquid. The rate of absorption in either case is some 3 per cent greater than if the hydrolysis of SO_2 were suppressed completely. If the absorbent were pure water, the SO_2 present in reacted form at equilibrium with 1 atm SO_2 would be

$$[H^+] = [HSO_3^-] = \sqrt{(1{\cdot}7\times 10^{-2}\times 1{\cdot}8)} = 0{\cdot}175 \text{ gmole/l}$$

and since the initial concentration of SO_2 in the absorbent would be zero:

$$\Delta C = 1{\cdot}82 + 0{\cdot}175 = 1{\cdot}995 \text{ gmole/l}$$

so that the absorption-rate would be nearly 10 per cent greater than if the hydrolysis were suppressed. Lynn *et al.*[3] have studied the effects of hydrolysis and its suppression on the absorption of SO_2.

A second example is the absorption of CO_2 into a partly carbonated solution of an amine, under such conditions that the reaction may be treated as instantaneous:

$$CO_2 + 2R_2NH \rightleftharpoons R_2NCOO^- + R_2NH_2^+ \quad (11)$$

$$K_2 = \frac{[R_2NCOO^-][R_2NH_2^+]}{[CO_2][R_2NH]^2}. \quad (12)$$

Let the total concentration of amine in reacted and unreacted form be M: i.e.

$$M = [R_2NH] + [R_2NCOO^-] + [R_2NH_2^+]. \quad (13)$$

(It is shown in the Appendix that for this type of reaction the quantity defined by Eq. (13) is constant and uniform.) Let the carbonation ratio, α, be defined as the ratio in the bulk liquid of CO_2 absorbed to amine originally present; thus

$$\alpha = \frac{[R_2NCOO^-]^0 + A^0}{M} \quad (14)$$

where A^0 and $[R_2NCOO^-]^0$ are the bulk concentrations of unreacted CO_2 and of R_2NCOO^- respectively. Initially, then, the combined concentration of reacted and unreacted CO_2 in the bulk is

$$A^0 + [R_2NCOO^-]^0 = M\alpha. \quad (15)$$

Now let the liquid be saturated at a sufficient partial pressure of CO_2 to raise its concentration to A^*; electrical neutrality requires that $[R_2NHCOO^-] = [R_2NH_2^+]$, and thus, from Eq. (13),

$$[R_2NH] = M - 2[R_2NCOO^-] \quad (16)$$

and

$$K_2 = \frac{[R_2NCOO^-]^2}{A^*(M - 2[R_2NCOO^-])^2} \quad (17)$$

whence

$$[RNCOO^-] = \frac{M\delta}{1+\delta} \quad (18)$$

where

$$\delta = \sqrt{(K_2 A^*)}. \quad (19)$$

Thus,

$$\Delta C = A^* + \frac{M\delta}{1+\delta} - M\alpha \quad (20)$$

whence

$$\bar{R} = k_L\left\{A^* + M\left(\frac{\delta}{1+\delta} - \alpha\right)\right\}. \quad (21)$$

This expression is given incorrectly (also with some differences in nomenclature) by Danckwerts and Sharma[4] (their Eq. (14d)).

CONDITIONS FOR 'INSTANTANEOUS' REACTION

Whether we are considering transient diffusion into a quiescent liquid, or absorption into an agitated liquid, there are circumstances in which any reaction which follows the absorption can be treated as 'instantaneous', even though it may have a finite rate-constant. For transient absorption the condition in the case of irreversible reaction is[4]

$$\frac{Q_0}{2A^*}\sqrt{\left(\frac{\pi}{Dt}\right)} \gg 1 + \frac{B_1}{z_1 A^*}. \quad (22)$$

Here Q_0 is the amount of gas, per unit area of surface, which would be absorbed in time t if the reactant in the neighbourhood of the surface retained the bulk composition, instead of becoming depleted. If condition (22) is fulfilled, depletion has proceeded to the point at which equilibrium is reached in the neighbourhood of the surface and the rate of reaction is governed entirely by the diffusion of the various species. For absorption into agitated liquids the analogous condition is:

$$\frac{\bar{R}_0}{k_L A^*} \gg 1 + \frac{B_1^0}{z_1 A^*} \quad (23)$$

where $\bar{R}_0$ is the rate at which gas would be absorbed if there were no depletion. In the case of an irreversible reaction in which A and B react at the local rate

$$r = k a^m b^n \quad (24)$$

conditions (22) and (23) take the form[5–7]

$$\sqrt{(k(A^*)^{m-1}(B^0)^n t)} \gg 1 + \frac{B_1^0}{z_1 A^*} \quad (25)$$

and

$$\frac{\sqrt{(kD(A^*)^{m-1}(B^0)^n)}}{k_L} \gg 1 + \frac{B_1^0}{z_1 A^*}. \quad (26)$$

Danckwerts and Sharma[4] indicate the conditions under which certain types of reversible reaction can be treated as instantaneous: e.g. CO_2 absorbed in potassium carbonate-bicarbonate solution (converse of their condition (13)); the same, catalysed by arsenite (their condition (13c)); CO_2 absorbed in partly-carbonated amine solutions (their condition (14c)).

EQUAL DIFFUSIVITIES

Equations (3) and (7) are only strictly valid when the diffusivities of all the species participating in the reaction are equal. The diffusivities of small molecules in water do not vary very widely, so that Eqs. (3) and (7) often serve as approximations which are useful for engineering design purposes. However, special considerations arise when some of the species are ions. When the products of the reaction are two species of ion of opposite sign, and there are no other species of ion present in solution (as, for instance, when SO_2 is absorbed by water), the condition of electrical neutrality requires that the two species of ion diffuse at equivalent rates. The effective diffusivity of the two ions is given by [8]

$$D_{\pm} = (z_+ + z_-)\frac{D_+ D_-}{z_+ D_+ + z_- D_-} \tag{27}$$

where D_+, D_- are the self-diffusivities, and z_+, z_- the valencies of the two ions. The self-diffusivities are related to the equivalent ionic conductivities, λ, by

$$D_+ = \frac{RT}{F^2 z_+}\lambda_+ \tag{28}$$

$$D_- = \frac{RT}{F^2 z_-}\lambda_- \tag{29}$$

(where R is the gas constant, T the absolute temperature and F the Faraday) so that

$$D_{\pm} = \frac{RT}{F^2}\frac{(z_+ + z_-)}{z_+ z_-}\frac{\lambda_+ \lambda_-}{(\lambda_+ + \lambda_-)}. \tag{30}$$

The values of λ and D are influenced by the concentrations of electrolytes in the solution, and therefore vary to some extent in the diffusion zone if the difference in concentration between the surface and the bulk is large. For present purposes this effect will be ignored, however.

Eriksen[9] has measured the following self-diffusivities in 1 M NH_4NSO_4 solution at 22°C:

$$SO_2 \; 1{\cdot}76 \times 10^{-5} \text{ cm}^2/\text{sec}$$

$$HSO_3^- \; 1{\cdot}10 \times 10^{-5} \text{ cm}^2/\text{sec}.$$

At 25°C the ionic conductivity of H^+ (at infinite dilution) is 350 cm^2/ohm g. equivalent[5], giving

$$D_+ = 9{\cdot}35 \times 10^{-5} \text{ cm}^2/\text{sec}$$

whence

$$D_{\pm} \approx 1{\cdot}97 \times 10^{-5} \text{ cm}^2/\text{sec}$$

at 25°C. Thus, the effective diffusivity of $(HSO_3^- + H^+)$ is not very different from that of SO_2, and the expressions referring to equal diffusivity can be applied without much error.

When extraneous electrolytes are present in the absorbent the situation is a good deal more complicated[9]. For instance, when SO_2 is absorbed by $NaHSO_3$ solution, H^+ and HSO_3^- diffuse into the solution at different rates and the effective diffusivity of H^+ is likely to be several times that of HSO_3^-. In general whenever H^+ or OH^- is one of the diffusing species, and extraneous electrolyte is present, the effective diffusivity of H^+ or OH^- will be several times as large as those of other species. In these circumstances the equations developed above for the rate of absorption give a rough guide rather than an accurate prediction.

Robinson and Stokes[8] give the diffusivities and ionic mobilities of a number of electrolytes. Information on the diffusivities of non-electrolytes is scattered, but some of it is referred to by Danckwerts and Sharma[4].

NOTATION

A	dissolved gas
ΔA	increase in concentration of A in reacted form when bulk solution is saturated at partial pressure corresponding to A^*
a	local concentration of unreacted A
A^*, A^0	concentration of unreacted A at interface and in bulk
B	reactant
B^0	concentration of unreacted B in bulk
b	local concentration of unreacted B
C_m^*, C_m^0	values of c_m at interface and in bulk
c_m	defined by Eq. (7a)
ΔC	$= C_m^* - C_m^0$; amount of gas required to saturate unit volume of bulk solution when partial pressure rises from π^0 to π^*
D	diffusivity in solution
D_+, D_-	self-diffusivities of cations and anions in solution
$D_{\pm}$	effective diffusivity of single electrolyte
F	Faraday
K	equilibrium constant
K_1	equilibrium constant for reaction (9)
K_2	equilibrium constant for reaction (11)
k	rate-constant of reaction
k_L	liquid-film mass-transfer coefficient
M	m^{th} species of reactant; or concentration of amine in reacted and unreacted form
P	product
P^0	concentration of P in bulk

p local concentration of P
Q_0 amount of gas absorbed in time t if no depletion of reactant were to occur
R transient rate of gas absorption, per unit area of surface; or gas constant
$\bar{R}$ rate of gas absorption into agitated liquid, per unit area of surface
$\bar{R}_0$ rate of gas absorption if no depletion of reactant were to occur
t time since first contact of gas and liquid
T absolute temperature
u_m local concentration of M
U_m^*, U_m^0 concentrations of M at surface and in bulk
v_m stoichiometric number of M
x distance from surface
y stoichiometric number of P
z stoichiometric number of B
z_+, z_- valencies of cation and anion.
α parameter in Eq. (15A); or carbonation ratio, moles of CO_2 absorbed per mole of amine originally present
β parameter in Eq. (18A)
δ $\sqrt{(K_2A^*)}$
λ_+, λ_- equivalent conductivities of cation and anion
π^0, π^* initial and final partial pressures of gas.

REFERENCES

[1] OLANDER D. R., *A.I.Ch.E. Jl* 1960 **6** 233.
[2] GIOIA F. and ASTARITA G., *Ind. Engng Chem. Fundls* 1967 **6** 370.
[3] LYNN S., STRAATEMEIER J. R. and KRAMERS H., *Chem. Engng Sci.* 1955 **4** 49.
[4] DANCKWERTS P. V. and SHARMA M. M., *The Chemical Engineer*, October 1966, CE 224.
[5] BRIAN P. L. T., HURLEY J. F. and HASSELTINE E. H., *A.I.Ch.E. Jl* 1961 **7** 226.
[6] BRIAN P. L. T. and BEAVERSTOCK M. C., *Chem. Engng Sci.* 1965 **20** 47.
[7] HIKITA H. and ASAI S., *Int. Chem. Engng* 1964 **4** 332.
[8] ROBINSON R. A. and STOKES R. H., *Electrolyte Solutions*. Butterworth 1955.
[9] ERIKSEN T., *Chem. Engng Sci.* 1967 **22** 727.
[10] BRIAN P. L. T., BADDOUR R. F. and MATIATOS D. C., *A.I.Ch.E. Jl* 1964 **10** 727.

APPENDIX

The reaction can be written in generalised form:

$$A \rightleftharpoons \Sigma v_m M \tag{1A}$$

where M is the mth species and v_m its stoichiometric number (negative for B_1 etc. positive for P_1 etc). The sum is taken for all species except A. The equilibrium relationship can be written

$$Ka = \Pi(u_m)^{v_m} \tag{2A}$$

where u_m is the concentration of the mth species. Let the rate of disappearance of A at any point be r (mole/vol. time). Then the equations for one-dimensional diffusion accompanied by reaction are

$$D\frac{\partial^2 a}{\partial x^2} = \frac{\partial a}{\partial t} + r \tag{3A}$$

and

$$D\frac{\partial^2 u_m}{\partial x^2} = \frac{\partial u_m}{\partial t} - v_m r \tag{4A}$$

whence

$$D\frac{\partial^2}{\partial x^2}\left(a + \frac{u_m}{v_m}\right) = \frac{\partial}{\partial t}\left(a + \frac{u_m}{v_m}\right) \tag{5A}$$

or

$$D\frac{\partial^2 c_m}{\partial x^2} = \frac{\partial c_m}{\partial t} \tag{6A}$$

where

$$c_m = a + \frac{u_m}{v_m}. \tag{7A}$$

The above equation has to be satisfied subject to the following boundary conditions:

$$c_m \to C_m^*, \quad x = 0, \quad t \to 0 \tag{8A}$$

$$c_m = C_m^0, \quad x > 0, \quad t = 0 \tag{9A}$$

$$c_m = C_m^0, \quad x = \infty, \quad t > 0 \tag{10A}$$

where C_m^0 is the initial value of c_m in the absorbent, in equilibrium with a partial pressure Π^0 of gas and C_m^* is the composition obtained when the absorbent, initially of composition C_m^0, is saturated with A at a partial pressure Π^*. The value of c_m at the surface rises instantaneously to C_m^* when the pressure of gaseous A is raised, because the liquid at the surface ($x = 0$) becomes immediately saturated with A at the higher pressure and chemical equilibrium follows instantly.

In addition, the solution of Eq. (6A) must be compatible with two conditions on u_m, namely that the equilibrium relationship Eq. (2A) is fulfilled at all points and times, and that

$$\frac{1}{v_m}\left(\frac{\partial u_m}{\partial x}\right)_{x=0} \text{ is the same for every } m. \tag{11A}$$

Condition (11A) expresses the fact that at the surface ($x = 0$) the rate of arrival of each of the reactants (except A) and departure of each of the products is the same when all are expressed in equivalents.

The required solution for c_m is

$$\frac{c_m - C_m^0}{C_m^* - C_m^0} = erfc\left(\frac{x}{2\sqrt{(Dt)}}\right). \tag{12A}$$

It will be noted that the term $(C_m^* - C_m^0)$ is the same for every m, since

$$\Delta C \equiv C_m^* - C_m^0 = A^* - A^0 + \frac{U_m^* - U_m^0}{v_m}$$
$$\equiv A^* - A^0 + \Delta A. \tag{13A}$$

Here $(U_m^* - U_m^0)/v_m \equiv \Delta A$ represents the increase in the concentration of A in the reacted form, and $(C_m^* - C_m^0) \equiv \Delta C$ the increase in the concentration of reacted and unreacted A combined, when the solution is saturated with A at the higher pressure.

Since $erfc(\infty) = 0$, conditions (9A) and (10A) are fulfilled by Eq. (12A). Since $erfc(0) = 1$, $c_m = C_m^*$ at $x = 0$ for all values of $t > 0$, however, small, so condition (8A) is fulfilled. Equation (12A) gives

$$c_m = C_m{}^0 + \alpha\Delta C \tag{14A}$$

where $\alpha = erfc(x/2\surd(Dt))$. Thus

$$u_m = v_m(c_m - a) = v_m[C_m{}^0 + \alpha\Delta C - a].$$

Thus, condition (2A) will be fulfilled if a is a solution of

$$Ka = \Pi v_m[C_m{}^0 + \alpha\Delta C - a]^{v_m}. \tag{16A}$$

It is sufficient to show that there are real, positive solutions for a in the range $0 < \alpha < 1$, which make each term positive in the product on the right-hand side of Eq. (16A). Now consider the addition of A to an absorbent which initially has the composition $C_m{}^0 = A^0 + U_m{}^0/v_m$. When sufficient A is added to the absorbent to saturate it at the partial pressure Π^*, so that the concentration of unreacted A rises to A^*, the composition will be $C_m{}^* = A^* + U_m{}^*/v_m$. If less than this quantity of A is added, so that

$$\frac{c_m - C_m{}^0}{\Delta C} = \beta \tag{17A}$$

(with $0 < \beta < 1$), we have

$$u_m = v_m(c_m - a) = v_m[C^0 + \beta\Delta C - a] \tag{18A}$$

and since the solution is at equilibrium

$$Ka = \Pi v_m[C_m{}^0 + \beta\Delta C - a]^{v_m}. \tag{19A}$$

There are therefore real, positive solutions to Eq. (19A) for $0 < \beta < 1$, which make each term in Eq. (19A) positive; thus there are such solutions to Eq. (16A) when $0 < \alpha < 1$, and Eq. (12A) is compatible with the condition for equilibrium. (Eq. (16A) must be solved to find the concentration of A and of species M at any point; these are not required for present purposes.)

By differentiating Eq. (12A) we find

$$\left(\frac{\partial c_m}{\partial x}\right)_{x=0} = \left(\frac{\partial a}{\partial x}\right)_{x=0} + \frac{1}{v_m}\left(\frac{\partial u_m}{\partial x}\right)_{x=0} = -\Delta C/\surd(\pi Dt). \tag{20A}$$

It follows that

$$\frac{1}{v_m}\left(\frac{\partial u_m}{\partial x}\right)_{x=0}$$

is the same for every m, as required.

The rate of absorption is the sum of the rates at which A and equivalents of M leave the surface; thus

$$R = -D\frac{\partial}{\partial x}\left(a + \frac{u_m}{v_m}\right)_{x=0} = -D\left(\frac{\partial c_m}{\partial x}\right)_{x=0} \tag{21A}$$

$$= \Delta C\sqrt{\left(\frac{D}{\pi t}\right)} \tag{22A}$$

which is Eq. (3).

From Eq. (7A) we find

$$\Sigma u_m = -a\Sigma v_m + \Sigma v_m c_m \tag{23A}$$

and from Eq. (14A)

$$\Sigma v_m c_m = \Sigma v_m C_m{}^0 + \alpha\Delta C \Sigma v_m \tag{24A}$$

$$= \Sigma U_m{}^0 + (A^0 + \alpha\Delta C)\Sigma v_m. \tag{25A}$$

Thus,

$$\Sigma u_m = \Sigma U_m{}^0 + (-a + A^0 + \alpha\Delta C)\Sigma v_m \tag{26A}$$

so that in cases where $\Sigma v_m = 0$ (i.e. number of reactant molecules – exclusive of A – reacting is equal to number of product molecules arising) the combined concentration of reactant and product molecules is constant and uniformly equal to that in the bulk. This is the case for the reaction of CO_2 with amines, discussed above.

Résumé – Quand un gaz absorbé subit une réaction instantanée, réversible ou autre, le taux d'absorption en solution est de

$$\bar{R} = k_L\Delta C$$

ΔC représentant la quantité de gaz nécessaire pour saturer l'unité de volume du liquide en vrac à une pression des particules égale à celle à l'interface, et k_L le coefficient de transfert de la masse pour l'absorption physique. Cette expression est valable quand les diffusivités de toutes les espèces sont égales.

Zusammenfassung – Wenn ein absorbiertes Gas in Lösung einer augenblicklichen, reversiblen irreversiblen Reaktion unterworden wird, ist die Absorptionsgeschwindigkeit

$$\bar{R} = k_L\Delta C$$

wobie ΔC die zur Sättigung der Volumeneinheit der Flüssigkeitsmenge erforderliche Gasmenge bedeutet, bis zu einem Partaldruck, der dem an der Grenzfläche gleich ist und k_L der Massenübergangskoeffizient für physikalische Absorption ist. Dieser Ausdruck gilt, wenndas Diffusionsvermögen aller Spezies gleich ist.

B13

THE EFFECTS OF DIFFUSIVITY IN THE LIQUID ON RATES OF GAS ABSORPTION

By A. TAVARES da SILVA, Ph.D.†‡ and Professor P. V. DANCKWERTS, G.C., M.B.E., M.A., S.M., C.Eng., M.I.Chem.E.†

SYNOPSIS

Hydrogen sulphide was absorbed into aqueous solutions of amines in order to see whether the results could best be represented by the film model or the surface-renewal models. This system provides a good test because of the widely varying diffusivities of the amines. The results unequivocally support surface renewal. The diffusivities of H_2S and various amines in water were measured in the course of the work.

Introduction

The film model and the various forms of the surface-renewal model for the processes occurring on the liquid side of the interface during gas-absorption differ principally in the predicted effect of diffusivity. For instance, if we consider physical absorption with no reaction, the film model predicts that k_L will be proportional to D and the surface-renewal models that it will be proportional to $\sqrt{D}$ where D is the diffusivity of the gas in solution.§ Various more complex models predict dependence on other powers of D.[1] Vivian and King[2] have measured the rates of absorption in a packed column of a number of gases with different values of D and have found that k_L appears to vary with $\sqrt{D}$. However, an independent demonstration would strengthen the conclusions. Kozinski and King[3] measured rates of desorption of a number of gases from water in a stirred vessel. They found that at low stirring speeds k_L appeared to vary with the 0·5 or 0·6 power of D (at higher stirring speeds bubbles were entrained). Again, an independent test is desirable.

Although neither the film model nor the surface-renewal models may actually be a very faithful representation of absorption into agitated liquids, it is of some interest to discover which model yields predictions which fit more closely the observed effects of diffusivity on the rate of absorption. The choice of model may be a matter of practical importance; for instance when a gas is absorbed into a solution of a reactant which has a diffusivity very different from that of the dissolved gas.

The Use of Instantaneous Reaction

The fact that the diffusivities of dissolved gases are difficult to measure accurately and do not vary very widely suggests that a more conclusive test would be to use a series of reactants with different diffusivities. If a gas is absorbed into a solution of a reactant with which it reacts instantaneously and irreversibly, the rate of absorption depends on the diffusivity of the reactant as well as on that of the dissolved gas. The nature of the dependence predicted by the film model is different from that predicted by the surface-renewal models. Since the diffusivities of various reactants may vary over a wide range and can be fairly easily measured, the use of a system of this kind offers a prospect of more conclusive discrimination between the two types of model.

If the concentration of the dissolved gas at the interface is C^* and that of the reactant in the bulk of the solution is C' while the diffusivity of the reactant is D' we have the following expressions for absorption into an agitated liquid, followed by instantaneous, irreversible reaction:

Film model

$$R = k_L\left(C^* + \frac{D'C'}{Dz}\right) \quad . \quad . \quad (1)$$

(where R is the rate of absorption per unit area of interface, and z moles of reactant react with each mole of dissolved gas).

Surface-renewal models[4]

$$R = k_L \alpha C^* = \alpha C^* \sqrt{(Ds)} \quad . \quad . \quad (2)$$

where s is the rate of surface renewal or some similar quantity, and:

$$\alpha = 1/\mathrm{erf}\,(\beta/\sqrt{D}) \quad . \quad . \quad . \quad (3)$$

β being defined by:

$$zC^*\sqrt{D}\exp(\beta^2/D')\,\mathrm{erfc}(\beta/\sqrt{D'}) = C'\sqrt{D'}\exp(\beta^2/D)\,\mathrm{erf}\,(\beta/\sqrt{D}) \quad . \quad (4)$$

When α is much greater than unity, equation (3) can be replaced by the approximation:

$$\alpha \approx \sqrt{\left(\frac{D}{D'}\right)} + \frac{C'}{zC^*}\sqrt{\left(\frac{D'}{D}\right)} \quad . \quad . \quad (5)$$

The error is of order $1/2\alpha$.

If C'/zC^* is sufficiently large equation (1) becomes:

$$R \approx k_L D'C'/Dz \quad . \quad . \quad . \quad (6)$$

and since, according to the film model:

$$k_L = D/\delta \quad . \quad . \quad . \quad (7)$$

we can write equation (6) as:

$$R = \frac{D'}{\delta}\frac{C'}{z} \quad . \quad . \quad . \quad (8)$$

Equation (5) on the other hand becomes:

$$\alpha \approx \frac{C'}{zC^*}\sqrt{\left(\frac{D'}{D}\right)} \quad . \quad . \quad (9)$$

† Department of Chemical Engineering, Pembroke Street, Cambridge, England.

‡ *Present address:* Instituto Superior Tecnico, Lisbon, Portugal.

§ *Symbols have the meanings given them on p. 55.*

whence equation (2) becomes:

$$R = \frac{C'}{z}\sqrt{(D's)} \quad . \quad . \quad . \quad (10)$$

Thus, if we carry out absorption measurements with a sufficiently large value of C'/zC^*, keeping the hydrodynamics unchanged but using a series of reactants of different D', we can discriminate between the film and the surface-renewal models by seeing whether R varies as D' or as $\sqrt{D'}$. This method, originally suggested by Sharma,[5] has the advantage not only that D' can be varied over a wide range and can be accurately measured but that the values of C^* and D need not be known. The latter quantities cannot be measured directly; considerable uncertainty about their values may arise if the solution is sufficiently concentrated to make C^* and D depart markedly from their values in pure water. Unfortunately, for reasons given in the Appendix, it was not possible to operate with a sufficiently large value of C'/zC^* to use equations (8) and (10) and more complicated expressions involving C^* and D had to be used.

The Use of Hydrogen Sulphide and Amines

The procedure adopted in the experiments described in this paper was to determine the diffusivities of hydrogen sulphide and various amines, and then to measure the rate of absorption of hydrogen sulphide into water and solutions of various amines in a packed column and a stirred vessel. Conditions were such that the approximate equation (5) could be used with sufficient accuracy for the predictions based on surface-renewal theory. The solutions of amines were kept sufficiently dilute to justify the assumption that the diffusivity and solubility of hydrogen sulphide were the same as in pure water.

In what follows it is assumed that hydrogen sulphide reacts instantaneously and irreversibly with the amine, according to reaction (11), and that each molecule of hydrogen sulphide reacts stoichiometrically with one molecule of amine. The validity of these assumptions is discussed in the Appendix.

Figures 1 and 2 show the concentration profiles for a truly irreversible instantaneous reaction between hydrogen sulphide and amine, according to:

$$R_2NH + H_2S \rightarrow R_2NH_2^+ + HS^- \quad . \quad . \quad (11)$$

The diffusivities of all species have been taken as equal in drawing the figures; Fig. 1 refers to the film model and Fig. 2 to any one of the surface-renewal models.

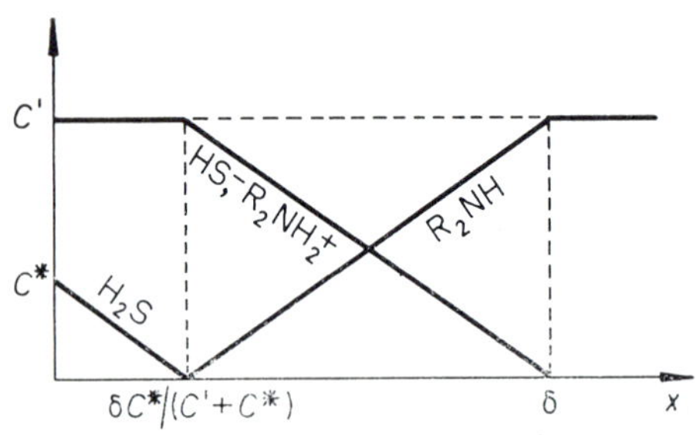

Fig. 1.—Absorption of hydrogen sulphide into amine solution; concentration profiles according to film model

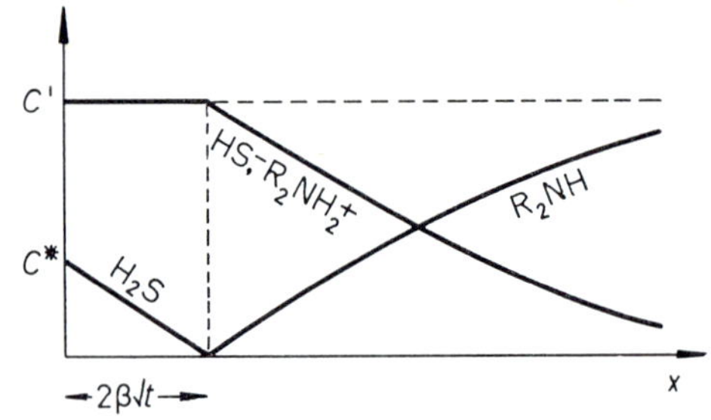

Fig. 2.—Absorption of hydrogen sulphide into amine solution; concentration profiles according to surface-renewal models

Measurements of Diffusivity by Absorption into a Jet

When hydrogen sulphide is absorbed into a laminar jet of water or a solution of an amine we have:

$$R = 2\alpha C^* \sqrt{\left(\frac{D}{\pi t}\right)} \quad . \quad . \quad (12)$$

or:

$$M = 4\alpha C^* \sqrt{(DLl)} \quad . \quad . \quad . \quad (13)$$

where R is the average rate of absorption per unit area of surface, M is the total rate of absorption, L the liquid flow-rate, l the length of the jet, and t the time for which the jet is exposed to the gas.

In the case of hydrogen sulphide and water $\alpha = 1$. Thus the measurements give D for hydrogen sulphide in water if C^*, the solubility of hydrogen sulphide, is known. In the case of amine solutions the value of α can be found from the experimental results by means of equation (13). The resulting values of α can be plotted against C'/C^* and the value of D'/D which gave the best fit to equation (3) and (4) found by trial and error. Hence D' is determined if D is known. The values of C^* and D for pure water were used throughout. The reduction in solubility due to the presence of the products of reaction was estimated by the method of van Krevelen and Hoftyzer[6] and found to be less than 1%. The viscosity of the solutions used was less than 2% higher than that of water, so little error was probably introduced in assuming the diffusivities to be the same as in pure water.

The rise in temperature at the surface due to the absorption and reaction of hydrogen sulphide in the amine solutions was calculated and found to be negligible (see Appendix). The individual runs with the laminar jet followed equation (13) closely, and there was no sign of an effect which might be attributed to convection currents near the surface, as apparently arise when carbon dioxide is absorbed by amine solutions.[7] The matter is discussed further by Danckwerts and Tavares da Silva.[8]

Absorption in a packed column

If we assume that the liquid is in " piston flow " (a postulate considered further in the Appendix) then in the case of absorption in a packed column of pure hydrogen sulphide (no diluent in the gas) without reaction:

$$2{\cdot}303\frac{d}{dH}\log_{10}\left(1 - \frac{M}{LC^*}\right) = -\frac{k_L a}{L} \quad . \quad (14)$$

where H is the height of packing exposed to the gas, M is the total rate of absorption, L is the total flow-rate of the liquid, and a is the effective interfacial area per unit packed volume.[9]

Thus $k_L a$ can be found by varying the height of packing and plotting $\log_{10}(1-M/LC^*)$ against H. When the absorbed hydrogen sulphide undergoes an instantaneous irreversible reaction with an amine the effect on the rate of absorption is different for the film and the surface-renewal models. Consider a column with a packed height, H, and let h be a height less than H, measured from the top of the packing. Then:

$$Ra\,dh = -\frac{L\,dC'}{A} \quad . \quad . \quad . \quad (15)$$

and:

$$M = \int_0^H ARa\,dh = L(C_T' - C_B') \quad . \quad (16)$$

where A is the cross-sectional area of the column and C_T' and C_B' are the concentrations of free amine at $h = 0$ and $h = H$ respectively. For the film model:

$$R = k_L C^* \left(1 + \frac{D'C'}{DC^*}\right) \quad . \quad . \quad (17)$$

Thus:

$$M = \int_0^H ARa\,dh = k_L aAC^* \int_0^H \left(1 + \frac{D'C'}{DC^*}\right) dh \quad . \quad (18)$$

$$\frac{dM}{dH} = k_L aAC^* \left(1 + \frac{D'C_B'}{DC^*}\right) \quad . \quad (19)$$

$$= k_L aAC^* \left[1 + \frac{D'}{DC^*}\left(C_T' - \frac{M}{L}\right)\right] \quad . \quad (20)$$

whence:

$$\frac{d}{dH} \ln\left(\frac{D}{D'} + \frac{C_T'}{C^*} - \frac{M}{LC^*}\right) = -\frac{k_L aA}{L}\left(\frac{D'}{D}\right) \quad . \quad (21)$$

With a given L and C_T' and a given C^* (fixed by the partial pressure of hydrogen sulphide) one can measure the total rate of absorption in the column, M, as a function of the height of packing, H. A plot of $-\ln\left(\frac{D}{D'} + \frac{C_T'}{C^*} - \frac{M}{LC^*}\right)$ against H would then give a straight line of slope $\frac{k_L aA}{L}\left(\frac{D'}{D}\right)$ if the film model were representative.

For the surface-renewal models we have:

$$R = k_L C^* \alpha \quad . \quad . \quad . \quad (22)$$

and in our conditions it was possible, with negligible error, to use equation (5) and put:

$$\alpha = \sqrt{\left(\frac{D}{D'}\right)} + \frac{C'}{C^*}\sqrt{\left(\frac{D'}{D}\right)} \quad . \quad (23)$$

With equations (15) and (16) this leads to:

$$\frac{d}{dH} \ln\left(\frac{D}{D'} + \frac{C_T'}{C^*} - \frac{M}{LC^*}\right) = -\frac{k_L aA}{L}\sqrt{\left(\frac{D'}{D}\right)} \quad . \quad (24)$$

In this case the plot would give a straight line of slope $\frac{k_L aA}{L}\sqrt{\left(\frac{D'}{D}\right)}$.

The first step, therefore, is to see whether plots of this type give straight lines; and then to see whether the slopes of these lines for various amines vary as $\left(\frac{D'}{D}\right)$ or $\sqrt{\left(\frac{D'}{D}\right)}$. The latter is done by making a logarithmic plot of the slopes of the lines against the corresponding values of D/D' and seeing whether this line itself has a slope of 1 or $\frac{1}{2}$ [the point: slope $= k_L aA/L$, $D/D' = 1$ should lie on the line, as can be seen by comparing equations (14), (21), and (24)].

Absorption in a stirred cell

If a stirred cell is operated at steady state with a continuous flow of liquid, and the liquid is well stirred so as to be uniform in concentration while the gas contains no diluent, we can obtain the following relationships.

For physical absorption:

$$M = LC_0 = k_L S(C^* - C_0) \quad . \quad . \quad (25)$$

where S is the area of the liquid surface, C_0 is the concentration of dissolved hydrogen sulphide in the vessel and in the outlet stream, L is the liquid flow-rate, and M is the total rate of absorption (the inflowing stream is pure water). Thus:

$$k_L S = \frac{M}{\left(C^* - \frac{M}{L}\right)} \quad . \quad . \quad (26)$$

Thus $k_L S$ can be calculated if M is measured. The value of k_L will of course depend on the stirring speed.

In the case of absorption followed by irreversible reaction with amine we have:

$$M = L(C_i' - C_0') \quad . \quad . \quad . \quad (27)$$

where C_i' and C_0' are the concentrations of amine in the inlet and outlet streams respectively.

For the film model:

$$M = k_L SC^* \left(1 + \frac{D'C_0'}{DC^*}\right) \quad . \quad . \quad (28)$$

and thus:

$$\frac{M/C^*}{\frac{D}{D'} + \frac{C_i'}{C^*} - \frac{M}{LC^*}} = k_L S\left(\frac{D'}{D}\right) \quad . \quad (29)$$

For the surface-renewal models:

$$M = k_L SC^* \left[\sqrt{\left(\frac{D}{D'}\right)} + \frac{C_0'}{C^*}\sqrt{\left(\frac{D'}{D}\right)}\right] \quad . \quad (30)$$

which leads to:

$$\frac{M/C^*}{\frac{D}{D'} + \frac{C_i'}{C^*} - \frac{M}{LC^*}} = k_L S\sqrt{\left(\frac{D'}{D}\right)} \quad . \quad (31)$$

Thus if either type of model is valid a plot of: $\log\left[\frac{M}{C^*}\Big/\left(\frac{D}{D'} + \frac{C_i'}{C^*} - \frac{M}{LC^*}\right)\right]$ against $\log(D'/D)$ at a given stirring speed should give a straight line passing through $\log(k_L S)$ when D'/D equals one; the slope would be one if the film model were valid and $\frac{1}{2}$ if the surface-renewal model were valid.

Experimental

Laminar jet

The jet apparatus was the same as that used by Sharma and Danckwerts.[10] Undiluted gas was used in all the experiments to eliminate gas-side resistance. The arrangement of the whole absorption system and the experimental procedure for operation at reduced pressure have been described by Tavares da Silva.[11] A steady flow of gas into the system was obtained by means of a fine adjustment valve and the required pressure was maintained constant by employing a Cartesian manostat with a water vacuum pump. The gas was saturated with water at the desired temperature. The absorption rate was measured with a soap-film meter. An air-ballast rotary pump was used to evacuate and purge the system. During the purging of the system effluent hydrogen sulphide was absorbed by a trap of iron oxide pellets between the system and the pump. The hydrogen sulphide cylinder was kept outside the building. The whole apparatus was kept in the open laboratory; however, it was possible to keep the liquid at 20°C (or 25°C) during a run and changes in temperature in other parts of the system were not big enough to cause any appreciable errors.

Packed column

The technique used for measuring absorption rates in the packed column with different heights of packing has been described by Danckwerts and Kennedy.[9] The column was packed with $\frac{1}{2}$ in (12·7 mm) ceramic Raschig rings. The internal diameter of the column was 4 in (10·2 cm) and a

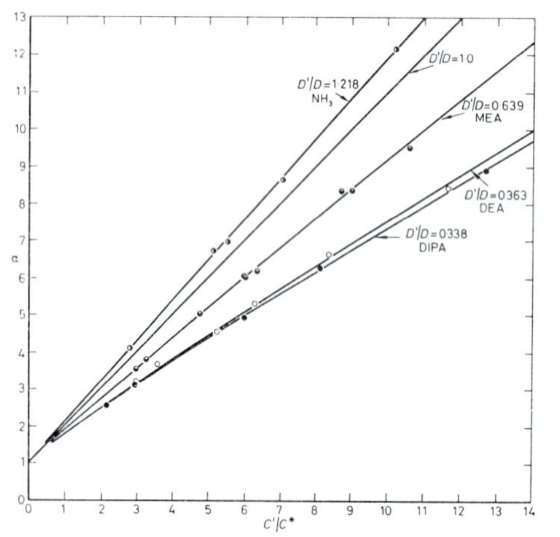

Fig. 3.—*Enhancement of absorption rate in jet as a function of concentration for various amines*

packing density of $3{\cdot}9 \times 10^5$ rings per cubic metre was used in all the experiments. The arrangement of the remainder of the apparatus was essentially the same as that used for the experiments at reduced pressure with the laminar jet. Soap-film meters with volumes of 0·6, 1·2, and 4 l were used for measuring the gas flow-rate.

Stirred vessel

The stirred vessel used in this work was similar to that used by Danckwerts and Gillham[12] and was operated with a continuous flow of liquid. It was constructed from a Pyrex glass column section of 10·5 cm i.d. and each end was closed with a Perspex plate sealed with rubber O-rings to allow the vessel to be operated at reduced pressure. The stirrer shaft passed through a vacuum seal made of Teflon and a small O-ring. The stirrer consisted of four Perspex strips, of rectangular cross-section, which were supported in the form of a cross at right-angles to the stirrer rod. The level of the liquid flowing through the cell was adjusted so that the bottom edge of the stirrer blades just cut the plane of the surface of the liquid. The performance of this cell is very sensitive to the liquid level and this is a disadvantage of this

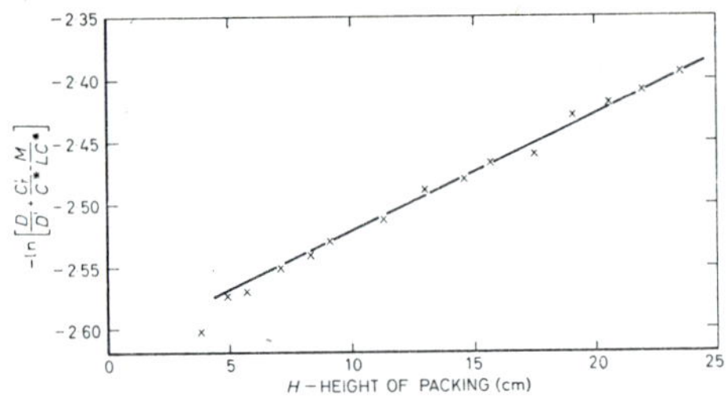

Fig. 4.—*Absorption in a packed column as a function of height. Absorbent: 0·052 M DIPA. $u = 0{\cdot}430$ cm/s*

type of apparatus. Under the conditions of operation used in this work the liquid surface remained reasonably flat with no vortex formation and with no entrainment of gas bubbles in the bulk of the solution. The general arrangement of the ancillary apparatus and the experimental procedure were similar to that used for the absorption experiments at low pressure in the laminar jet and the packed column.

The solutions

All the solutions used were prepared with technical grade substances and de-ionized water. The solutions were carefully and completely de-aerated by spraying them several times into an overhead reservoir under as low a vacuum as could be obtained. The concentrations of all the solutions were in the range 0·02–0·05 g mol/l and their physical properties were virtually indistinguishable from those of water.

Results

Laminar jet

In the experiments with the laminar jet, hydrogen sulphide at 1 atm and at reduced pressure (40–50 mm Hg) was absorbed into water at 20°C and at 25°C. The value of $C^*\sqrt{D}/p$ (where p is the pressure of hydrogen sulphide) was found to be independent of p. Values are reported in Table I, together with values of C^*/p selected from published data (Linke and

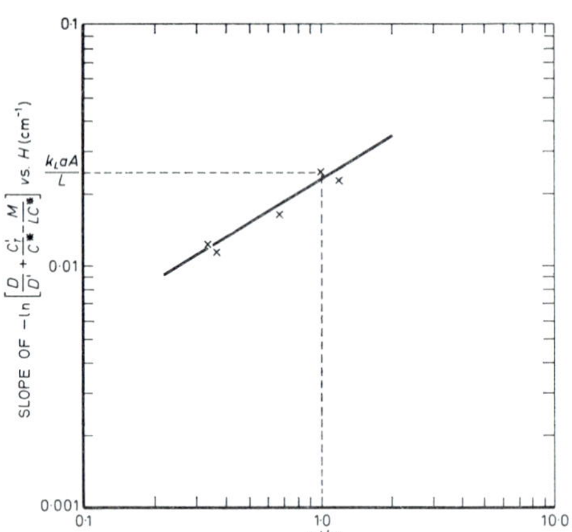

Fig. 5.—*Absorption in packed column—relationship of absorption rates to diffusivity ratio. $u = 0{\cdot}146$ cm/s*

TABLE I.—*Solubility and Diffusivity of Hydrogen Sulphide in Water*

Temperature (°C)	$10^3C^*/p$ $\left(\frac{\text{g mol}}{\text{cm}^3\text{atm}}\right)$	$10^6C^*\sqrt{D}/p$ $\left(\frac{\text{g mol}}{\text{cm}^2\text{s}^{\frac{1}{2}}\text{atm}}\right)$	10^5D (cm²/s)
20	0·115	0·421	1·34
25	0·102	0·392	1·48

Seidell;[13] Wright and Maass[14]) and of D for hydrogen sulphide in water calculated from the experimental results. Arnold[15] gives $10^5D = 1{\cdot}635$ at 20°C.

Hydrogen sulphide at reduced pressure (29–68 mm Hg) was absorbed into solutions of ammonia, monoethanolamine (MEA), diethanolamine (DEA), and di-isopropanolamine (DIPA). Various combinations of flow-rate, partial pressure and amine concentration were used to check the validity of equation (13). The solutions were sufficiently dilute (0·02–0·05 g mol/l) so that C^*/p, D, and D' could be taken to be the same as in pure water. All the experiments were conducted at 20°C. Values of the total absorption rate, M, were plotted against $\sqrt{(Ll)}$ for each concentration and partial pressure; the slope of the best straight line through the origin fitted by least squares to the experimental points was taken to be $4\alpha C^*\sqrt{D}$, according to equation (11). The final results are shown in Fig. 3 as a plot of α against C'/C^*. For each value of D'/D—that is for each amine—the points should lie on a line given by equations (3) and (4). The value of D'/D giving the line that best fitted the points for a given amine was found by minimising the sum of the squares of the deviations, a computer program being written for the purpose. The values of D'/D thus determined for the amines are given in Table II, together with their diffusivities, calculated as $D(D'/D)$. The values of D' cannot be very different from the values for infinite dilution.

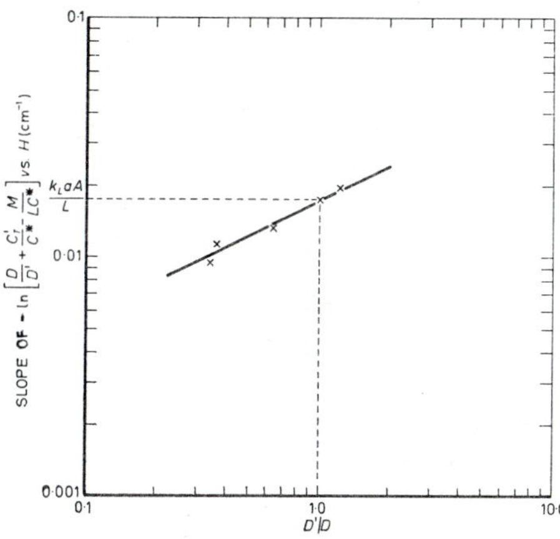

Fig. 6.—*Absorption in a packed column—relationship of absorption rates to diffusivity ratio.* $u = 0{\cdot}430$ *cm/s*

TABLE II.—*Diffusivities of Amines in Water at* 20°C

	D'/D	$10^5D'$(cm²/s)
NH_3	1·218	1·632
MEA	0·639	0·856
DEA	0·363	0·486
DIPA	0·338	0·453

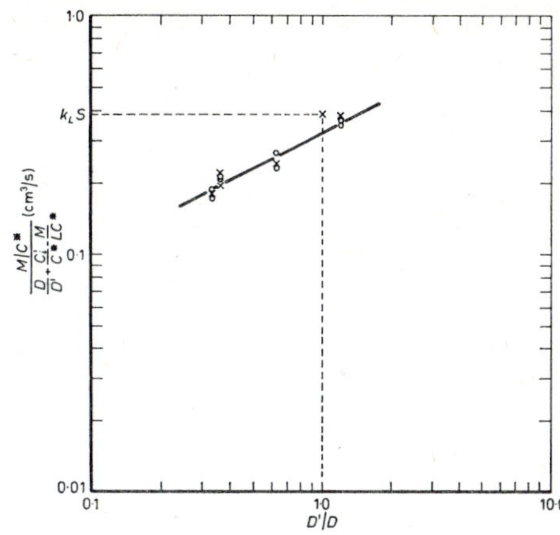

×: $L = 5$ cm³/s
○: $L = 10$ cm³/s

Fig. 7.—*Absorption in a stirred cell—relationship of absorption rates to diffusivity ratio.* 1·5 *rev/s*

Thomas and Furzer[16] report $10^5D = 1{\cdot}08$ for 0·12 M MEA at 25°C and $10^5D = 0{\cdot}567$ for 0·42 M DEA at 25°C. "*International Critical Tables*"[17] report $10^5D = 1{\cdot}77$ for 1 M NH_3 at 15°C.

Packed column

The measurements on the absorption of hydrogen sulphide in water in a packed column were made at superficial liquid velocities of 0·146 and 0·430 cm/s. The value of $k_L a$ at each flow-rate was determined by plotting $\log_{10}(1-M/LC^*)$ against H, as explained above. The points were found to lie on a straight line except over the top few centimetres of the packing; in this region, as found by Danckwerts and Ken-

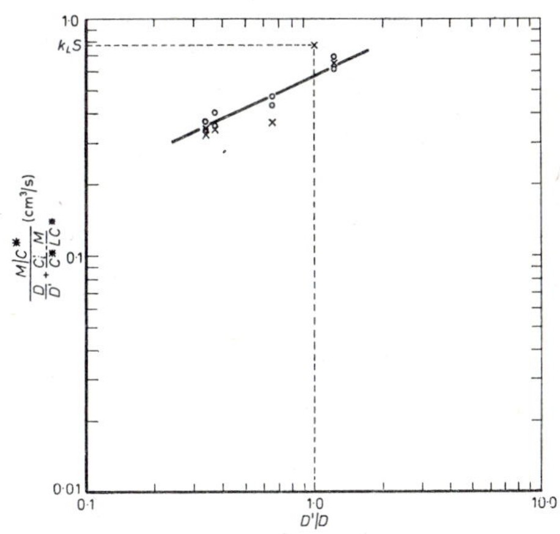

×: $L = 5$ cm/s
○: $L = 10$ cm/s

Fig. 8.—*Absorption in a stirred cell—relationship of absorption rates to diffusivity ratio.* 2 *rev/s*

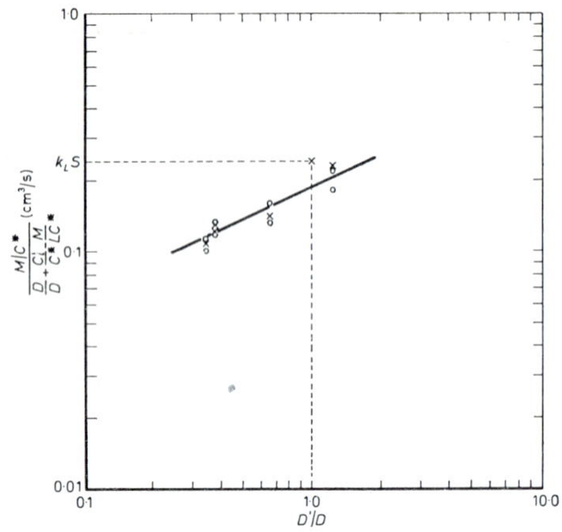

×: $L = 5$ cm³/s
○: $L = 10$ cm³/s

Fig. 9.—*Absorption in a stirred cell—relationship of absorption rates to diffusivity ratio.* 1 rev/s

nedy,[9] the value of $k_L a$ is anomalously high. The values of $k_L a$ so determined are given in Table III.

TABLE III.—*Values of $k_L a$ for Hydrogen Sulphide in Water at* 20°C *on* ½ in. *Raschig Rings* (3·9 × 10⁵ rings/m³)

u(cm/s)	$10^3 k_L a$(s⁻¹)
0·146	3·62
0·430	7·48

Hydrogen sulphide was also absorbed into amine solutions and the total rate of absorption, M, was measured with various heights of packing, H.

The results were plotted as $-\ln\left(\frac{D}{D'} + \frac{C_T'}{C^*} - \frac{M}{LC^*}\right)$ against H.

Figure 4 shows a typical plot for 0·0520 M DIPA. Except for the first few centimetres of packing the points fall on a straight line as expected. For each of the two superficial flow-rates the slopes of the lines for the four amines were plotted against D'/D; the points corresponding to $k_L a A/L$ at $D'/D = 1$ were also included. Figures 5 and 6 show the resulting logarithmic plots and the best straight lines fitted to the experimental results by least squares. The slopes of these lines were 0·60 at 0·146 cm/s and 0·49 at 0·430 cm/s. These results show that the surface-renewal model and not the film model is preferable for predicting absorption rates in packed columns, at any rate under the conditions of our experiments.

Stirred vessel

The physical mass-transfer coefficient in a stirred vessel, $k_L S$, was found at three stirring speeds by absorbing undiluted hydrogen sulphide into water. No significant change of $k_L S$ with the flow-rate of water was found. The rates of absorption of hydrogen sulphide into amine solutions were measured for the same three stirring speeds. Two different concentrations of each absorbent were used. For each stirring speed the results were plotted as

$$\log\left[\frac{M}{C^*}\Big/\left(\frac{D}{D'} + \frac{C_i'}{C^*} - \frac{M}{LC^*}\right)\right]$$

against log (D'/D) (see Figs 7, 8, and 9). The points representing $k_L S$ when $D'/D = 1$ are also shown; since these points, which referred to physical absorption, appeared to be systematically higher than the points referring to absorption into amines it was concluded that there might be a genuine physical difference between the two. The best straight lines, found by least squares, were drawn through the points representing absorption into amines; the slopes were found to be 0·45, 0·49, and 0·45 for 60, 90 and 120 rev/min respectively. None of the slopes was significantly different from 0·5 at the 95% confidence level. The chance that the slopes were equal to unity was in each case less than one in 2000. Thus it seems that in the stirred vessel also the penetration model rather than the film model should be used for predicting rates of gas absorption.

APPENDIX

Rate of Reaction

The reaction is:

$$H_2S + R_2NH \rightleftharpoons HS^- + R_2NH_2^+ \quad . \quad (29)$$

It involves the transfer of a proton from the hydrogen sulphide to the amine molecule. It can be considered instantaneous for present purposes,[18] as can the various other reactions considered below.

Equilibrium Composition

When hydrogen sulphide dissolves in water it dissociates according to:

$$H_2S \rightleftharpoons HS^- + H^+;\ K_1 = \frac{(HS^-)(H^+)}{(H_2S)} = 10^{-7 \cdot 0}\ \text{g mol}/l \quad (32)$$

$$HS^- \rightleftharpoons S^= + H^+;\ K_2 = \frac{(S^=)(H^+)}{(HS^-)} = 10^{-13 \cdot 90}\ \text{g mol}/l \quad (33)$$

(the values of the two equilibrium constants are those for 25°C).[19] The solubility of hydrogen sulphide in water is about 0·1 g mole/l atm at 25°C (see Table I); it can thus be calculated that for a solution of hydrogen sulphide in water:

$$\frac{(HS^-)}{(H_2S)} = \frac{10^{-3}}{\sqrt{p}} \quad . \quad . \quad . \quad (34)$$

where p is the partial pressure in atmospheres of the hydrogen sulphide corresponding to its concentration. The concentration of $S^=$ is very much less than that of HS^-. Thus, less than 1% of the hydrogen sulphide is dissociated when the concentration corresponds to more than 10^{-2} atm. Danckwerts[20] has shown that if the diffusivities of all species are equal the correct driving-force for calculating the rate of absorption in a case of this kind is the total solubility of hydrogen sulphide in the ionised and un-ionised forms combined. In this particular case, however, the fraction ionised is so small that we can neglect its variation with concentration and simply use the solubility of hydrogen sulphide at one atmosphere multiplied by the partial pressure of hydrogen sulphide as the driving force.

In an aqueous solution of an amine, the equilibrium:

$$R_2NH + H_2O \rightleftharpoons R_2NH_2^+ + OH^-;\ K_b = \frac{(R_2NH_2^+)(OH^-)}{(R_2NH)} \quad (35)$$

is set up. A quantity quoted more often than K_b is K_a:

$$K_a = \frac{(R_2NH)(H^+)}{(R_2NH_2^+)} = \frac{K_w}{K_b} \qquad (36)$$

where $K_w = (H^+)(OH^-)$ is the ionic product of water. The values of K_a for some amines are given in Table IV.

TABLE IV.—*Dissociation Constants of Amines at Infinite Dilution* (25°C)

Species	K_a (g mol/l)	Reference
ammonia	$5{\cdot}37\times10^{-10}$	Hall & Sprinkle[21]
monoethanolamine MEA	$3{\cdot}63\times10^{-10}$	Hall & Sprinkle[21]
diethanolamine DEA	$1{\cdot}318\times10^{-9}$	Hall & Sprinkle[21]
triethanolamine TEA	$1{\cdot}698\times10^{-8}$	Hall & Sprinkle[21]
di-isopropanolamine DIPA	$3{\cdot}16\times10^{-10}$	quoted by Sharma[5]
hexamethyleneimine HMI	$9{\cdot}55\times10^{-12}$	Searles *et al.*[22]

When hydrogen sulphide reacts with a solution of an amine the reaction involves the hydroxyl ions present in the solution as well as the un-ionised amine molecules; thus the reaction:

$$H_2S+OH^- \rightarrow HS^- + H_2O \qquad (37)$$

takes place as well as reaction (29).

The combined concentration of undissociated amine plus hydroxyl ion in the bulk of the solution is the same as the concentration of amine before dissociation, but the hydroxyl ions diffuse faster to the reaction zone than do the amine ions. Thus the rate of reaction is higher than if the amine were undissociated. For present purposes it is necessary that the amount of hydrogen sulphide reacting with the hydroxyl ions should be small compared to that reacting with amine—say less than 5%. This requires that in the bulk solution:

$$\frac{(OH^-)}{(R_2NH)} < 0{\cdot}05 \qquad (38)$$

or:

$$\sqrt{\left(\frac{K_w}{K_a C'}\right)} < 0{\cdot}05 \qquad (39)$$

where C' is the total concentration of amine. Thus the condition:

$$K_a C' > 4\times10^{-12} \qquad (40)$$

must be fulfilled at 25°C (when $K_w = 10^{-14}$ g ion/*l*). The lowest concentration of amine used in this work was 0·02 g mol/l; the condition will then be fulfilled if K_a is greater than 2×10^{-10} g ion/l. As may be seen from Table IV this is the case for all the amines listed except hexamethyleneimine.

In writing reaction (29) above it is assumed that no appreciable amount of $S^=$ ions is formed at the point where the hydrogen sulphide and amine molecules meet; this would alter the stoichiometry of the reaction. Considering a point to the right of $x = \delta C^*/(C^*+C')$ in Fig. 1 or $x = 2\beta\sqrt{t}$ in Fig. 2, we see that:

$$(HS^-) \approx (R_2NH_2^+) \quad \text{and} \quad (R_2NH) \approx C' - (HS^-)$$

[assuming that $(S^=) \ll (HS^-)$].

Considering the equilibrium:

$$R_2NH + HS^- \rightleftharpoons R_2NH_2^+ + S^= \qquad (41)$$

we see that:

$$\frac{K_2}{K_a} = \frac{(R_2NH_2^+)(S^=)}{(R_2NH)(HS^-)} \approx \frac{(S^=)}{C'-(HS^-)} \qquad (42)$$

Thus in the worst case cited in Table I (except hexamethylene imine), namely di-isopropanolamine:

$$\frac{(S^=)}{(HS^-)} = \left(\frac{C'}{(HS^-)} - 1\right) \times 10^{-4{\cdot}4} \qquad (43)$$

Thus, moving to the right on Fig. 1 or 2, the proportion of the anions in the form of $S^=$ remains very small until the total concentration of anions has itself fallen to a very small value so that the actual concentration of $S^=$ at any point is much less than C'. In these circumstances the number of molecules of amine reacting with each molecule of hydrogen sulphide absorbed is very nearly equal to one.

The Effect of Reversibility on the Rate of Absorption

The reaction is not truly irreversible, as suggested by reaction scheme (11). The equilibrium constant of the actual reaction (29) is:

$$\frac{(HS^-)(R_2NH_2^+)}{(H_2S)(R_2NH)} = \frac{K_1}{K_a} \qquad (44)$$

The greater the value of K_1/K_a, the more closely the rate of absorption will approach that predicted on the basis of irreversible reaction. The deviation due to reversibility can be estimated in two ways—namely by using Olander's solution for the film model[23] and by solving the equations for transient diffusion accompanied by reversible instantaneous reaction (20). In both cases the diffusivities of all species have been assumed equal; it seems unlikely that order-of-magnitude arguments would lead to markedly different conclusions if unequal diffusivities were taken into account. (The full solution of the film model case shows that the effect of the reverse reaction increases as the ratio of the diffusivity of the amine to that of the products increases. This effect is independent of the diffusivity of hydrogen sulphide.)

In the case of both models, the result is:

$$R = k_L C^* \left\{1 + \frac{K_1}{2K_a}\left[\sqrt{\left(1+\frac{4K_a C'}{K_1 C^*}\right)} - 1\right]\right\} \qquad (45)$$

whereas for the irreversible reaction:

$$R = k_L C^* \left(1 + \frac{C'}{C^*}\right) \qquad (46)$$

Thus, when $K_a C'/K_1 C^* \ll 1$, the relative error introduced by ignoring the reversibility of the reaction is less than $K_a C'/K_1 C^*$. It is found that for values of C' and C^* such as those employed in our experiments the error was no greater than the experimental error in the cases of ammonia, MEA, DEA, and DIPA. In the case of triethanolamine the large value of K_a made the error too large to ignore. These conclusions apply to absorption in the jet as well as to absorption in the packed column and stirred vessel.

The condition $K_a C'/K_1 C^* \ll 1$ imposes a lower limit on the value of C^*/C'. As mentioned in the text, if this ratio could have been made small enough a simplified procedure could have been used in analysing the experimental results; but in practice this procedure could not be used as the required low value of C^*/C' was incompatible with the low value of $K_a C'/K_1 C^*$ required to avoid reverse reaction.

The requirement that little hydroxyl ion must be formed by the hydrolysis of the amine implies that the value of K_a for the amine must not be too small; while the requirement that $K_a C'/K_1 C^* \ll 1$ implies that K_a must not be too large. Thus, only amines having values of K_a within a limited range could be used.

Temperature Effects

It can be shown that in the case of an instantaneous reaction the rise in temperature at the surface is independent of the time of exposure, and given approximately by:

$$\Delta T \approx \frac{\left(C^* + \dfrac{C'}{z}\right)}{\rho\sigma}(H_A + H_R)\sqrt{\frac{D}{\gamma}}.$$

Here ΔT is the temperature rise, ρ and σ are the density and the specific heat of the liquid ($\rho\sigma = 1{\cdot}0$ cal/cm^3 °C for water), γ the thermal diffusivity of the liquid ($\gamma \approx 1{\cdot}5 \times 10^{-3}$ cm^2/s for water), H_A the heat of absorption of hydrogen sulphide, and H_R the heat of reaction of dissolved hydrogen sulphide with amine (both positive if exothermic). Bichowsky and Rossini[24] give $H_A = 4750$ cal/g mol and also the following heats of reaction:

$$H_2S \rightarrow H^+ + HS^- \qquad -6000 \text{ cal/g mol}$$

$$H^+ + NH_3 \rightarrow NH_4^+ \qquad +12300 \text{ cal/g mol}$$

whence: $NH_3 + H_2S \rightarrow HS^- + NH_4 \qquad +6300$ cal/g mol

Thus, for ammonia with $C' = 0{\cdot}05$ and $C^* = 0{\cdot}005$ g mol/l:

$$\Delta T = \frac{10^{-3}(0{\cdot}005 + 0{\cdot}050)}{1{\cdot}0}\sqrt{\left(\frac{1{\cdot}34 \times 10^{-5}}{1{\cdot}5 \times 10^{-3}}\right)}(4750 + 6300)$$

$$= 0{\cdot}055°\text{C}.$$

Thus in this case, and by extension in the cases of the other systems used, the rise in temperature was small enough to be neglected.

Maldistribution of Liquid in Packed Column

It is known that under some circumstances the liquid flowing over a packing tends to segregate at the walls. However, Mannford Doble[25] has shown that this segregation is unimportant if the height of the packing is less than three times the diameter of the column. In our case the height of packing never exceeded 2·5 times the diameter of the column.

Longitudinal Dispersion of Liquid in a Packed Column

Equations (14), (21), and (24) are derived on the assumption that the liquid moves through the packing in piston flow. In fact there is some longitudinal dispersion. Stemerding[26] showed that for 13 mm Raschig rings the apparent axial diffusion coefficient, D_e, was about 6 cm^2/s at all velocities. The effect of longitudinal dispersion can be estimated from figures given by Levenspiel[27] and by Kramers and Westerterp.[28] It depends on the quantity $D_e f/uH$, where f is the fraction of the volume of the column occupied by liquid, and can be taken from the results of Mannford Doble.[25] To use the figures referred to we must assume the rate of absorption to be proportional to the amine concentration which is roughly true. It is then found that the effect of longitudinal dispersion on the rate of absorption is quite negligible.

Acknowledgments

This work has been carried out when one of the authors (A.T.S.) was on leave of absence from Comissão de Estudos de Energia Nuclear, Lisbon, Portugal, which he wishes to thank for having rendered possible this research. Financial support from Comissão Coordenadora da Investigação para a OTAN, Lisbon, and from the Calouste Gulbenkian Foundation, Lisbon, is gratefully acknowledged.

Symbols Used

A = cross-sectional area of column.
a = interfacial area per unit volume of packing.
C^* = saturated concentration of hydrogen sulphide.
C' = concentration of amine in bulk liquid.
C_T' = concentration of amine at top of packing.
C_B' = concentration of amine at bottom of packing.
C_0 = concentration of hydrogen sulphide at the outlet of stirred cell.
D = diffusivity of dissolved hydrogen sulphide.
D' = diffusivity of amine.
H = height of packing.
H_A = heat of absorption.
H_R = heat of reaction.
h = heat measured downwards from top of packing.
k_L = liquid-film mass transfer coefficient.
K_a = defined by equation (34).
K_b = defined by equation (33).
K_w = ionic product of water.
K_1 = first dissociation constant of hydrogen sulphide.
K_2 = second dissociation constant of hydrogen sulphide.
L = total flow-rate of liquid.
l = length of jet.
M = total rate of absorption of hydrogen sulphide.
p = partial pressure of hydrogen sulphide.
R = rate of absorption per unit interfacial area.
S = area of surface in stirred cell.
s = fractional rate of surface renewal.
ΔT = rise of temperature at surface of liquid.
u = superficial velocity of liquid in packed column.
z = stoichiometric factor (= 1 for hydrogen sulphide–amine system).
α = enhancement factor, defined by equations (2) and (3).
β = defined by equation (4).
γ = thermal diffusivity of liquid.
δ = thickness of diffusion film.
ρ = density of liquid.
σ = specific heat of liquid.

The above quantities may be expressed in any set of consistent units in which force and mass are not defined independently.

References

1 King, C. J. *Industr. Engng. Chem.*, (*Fund.*), 1966, **5**, 1.
2 Vivian, J. E. and King, C. J. *A.I.Ch.E. Jl*, 1964, **10**, 221.
3 Kozinsky, A. H. and King, C. J. *A.I.Ch.E. Jl*, 1966, **12**, 109.
4 Danckwerts, P. V. *Industr. Engng Chem.*, 1951, **43**, 1460.
5 Sharma, M. M. *Ph.D. Thesis*, 1964. University of Cambridge.

6 Van Krevelen, D. W. and Hoftyzer, P. J. in " *Chimie Industrielle: XXIme Congrés International de Chimie Industrielle* ", 1948, p. 168. (Paris: Societé de Productions Documentaires).

7 Brian, P. L. T., Vivian, J. E., and Matiatos, D. C. *A.I.Ch.E. Jl*, 1967, **13**, 28.

8 Danckwerts, P. V. and Tavares de Silva, A. *Chem. Engng Sci.*, 1967, **22**, 1513.

9 Danckwerts, P. V. and Kennedy, A. M. *Trans. Instn chem. Engrs*, 1954, **32**, S53.

10 Sharma, M. M. and Danckwerts, P. V. *Chem. Engng Sci.*, 1963, **18**, 729.

11 Tavares de Silva, A. *Ph.D. Thesis*, 1967. University of Cambridge.

12 Danckwerts, P. V. and Gillham, A. J. *Trans. Instn chem. Engrs*, 1966, **44**, T42.

13 Linke, W. F. and Seidell, A. " *Solubilities of Inorganic and Metal-organic Compounds* ", 1958, Vol. I. (New York: D. van Nostrand Co. Inc.).

14 Wright, R. W. and Maas, O. *Canad. J. Res.*, 1932, **6**, 94.

15 Arnold, J. H. *J. Am. chem. Soc.*, 1930, **52**, 3937.

16 Thomas, W. J. and Furzer, I. A. *Chem. Engng Sci.*, 1962, **17**, 115.

17 " *International Critical Tables* ", 1929, Vol. V, p. 64. (New York: McGraw-Hill Book Co. Inc.).

18 Eigen, M. and Kustin, K. *J. Am. chem. Soc.*, 1960, **82**, 9852.

19 Marconny, G. *Electrochim. Acta*, 1959, **1**, 58.

20 Danckwerts, P. V. *Chem. Engng Sci.*, 1968, **23**, T.B.P.

21 Hall, N. F. and Sprinkle, M. R. *J. Am. chem. Soc.*, 1932, **54**, 3469.

22 Searles, S., Tamres, M., Block, T., and Quaternam, L. A. *J. Am. chem. Soc.*, 1956, **78**, 4917.

23 Olander, D. R. *A.I.Ch.E. Jl*, 1960, **6**, 233.

24 Bichowsky, F. R. and Rossini, F. D. " *The Thermochemistry of Chemical Substances* ", 1963. (New York: Reinhold Publications Inc.).

25 Mannford Doble, M. G. *Ph.D. Thesis*, 1966. University of Cambridge.

26 Stemerding, S. *Chem. Engng Sci.*, 1961, **14**, 209.

27 Levenspiel, O. " *Chemical Reaction Engineering* ", 1962. (New York: John Wiley & Sons Inc.).

28 Kramers, H. and Westerterp, K. R. " *Elements of Chemical Reactor Design and Operation* ", 1963. (London: Chapman and Hall Ltd.).

The manuscript of this paper was received on 1 *April*, 1968.

B14

Chemical reaction and effective interfacial areas in gas absorption

G. E. H. JOOSTEN† and P. V. DANCKWERTS

Department of Chemical Engineering, Pembroke Street, Cambridge, England

(*Received* 1 *June* 1972)

Abstract—Experiments have been carried out in a packed column to test the hypothesis that the interfacial area effective for gas absorption with chemical reaction depends on γ, where

$$\gamma = \frac{\text{factor by which reaction increases capacity of absorbent}}{\text{factor by which reaction increases rate of absorption}}.$$

For physical absorption, $\gamma = 1$. For absorption into certain buffer solutions, $\gamma \gg 1$, and the effective interfacial area is much larger than for physical absorption. For absorption with "instantaneous reaction", $\gamma \sim 1$ and the effective interfacial area is equal to that for physical absorption. Provided γ exceeds a certain value, the effective interfacial area does not depend on γ, so that values of the effective interfacial area determined by reactions with a sufficiently high γ have a general significance. The experiments show that it is not simply the occurrence of a chemical reaction but the value of γ to which it leads which determines the effective interfacial area. Values of the interfacial area determined with high γ should not be used for the design of physical absorbers.

THERE are numerous references in the literature to the fact that the *effective* area of interface between gas and liquid in an irrigated packed column depends on the transfer process occurring between gas and liquid. The subject is summarised by Danckwerts[1] (p. 217). In particular, investigators have sought to show that the area effective for the evaporation of the irrigating liquid is larger than the area effective for the physical absorption of a gas. It seems probable that the liquid running over the packing forms at some points very thin films and at other

†Present address: Koninklijke/Shell-Laboratorium, Amsterdam, The Netherlands.

points stagnant pools in which the rate of turnover will be low. Such regions are as effective for evaporation as those covered by fast-moving liquid of adequate depth. However, in the case of physical absorption the thin regions will tend to become saturated and cease to contribute to the rate of absorption, and the exposed surfaces of the slow-moving pools will likewise become saturated. We may therefore suppose that less of the surface will be effective for physical gas absorption than for evaporation, and there is a considerable body of evidence which suggests that this is the case.

The present paper is concerned with the effective interfacial area when a gas is absorbed and reacts with a substance dissolved in the liquid. This is particularly important because a considerable number of determinations of interfacial areas have been made with the help of chemical reactions (e.g. Danckwerts and Rizvi [3]), and it is important to know whether the values obtained have any general significance or whether they are characteristic only of the particular reactions employed. Onda *et al.*[6] have shown that the area effective for evaporation is equal to that effective for the absorption of CO_2 into solutions of NaOH, but it seemed to us that the problem needed to be more clearly set out and more precisely investigated.

When a reactant is added to an absorbing liquid, there are in general two effects. In the first place the quantity of gas which can be absorbed by unit volume of the liquid is increased by some factor C by virtue of the reaction, and thus it will take longer for thin and slow-moving patches of the liquid surface to become saturated; this effect therefore tends to increase the area which is effective for gas absorption. On the other hand, the reaction will increase the rate of absorption by some factor E (the enhancement factor), and this tends to have the opposite effect. The net effect of the reactant on the effective interfacial area may therefore be expected to depend on the quantity γ, which is

$$\gamma = \frac{C}{E} = \frac{\text{factor by which reactant increases capacity of solution}}{\text{factor by which reactant increases rate of absorption}}. \quad (1)$$

We may expect the effective interfacial area to be greater than that for physical absorption of a non-reacting solute gas in the same solution if $\gamma > 1$, smaller than that for the non-reacting gas if $\gamma < 1$, and the same if $\gamma = 1$.

In the experiments described below we have used three types of solution for the absorption of CO_2: (a) non-reacting solutions; (b) buffer solutions in which the absorbed CO_2 underwent a pseudo-first order reaction and in which $\gamma \gg 1$; (c) solutions in which the CO_2 underwent an "instantaneous" reaction, and $\gamma \sim 1$. In view of the difficulty of separating a and $k_L{}^0$ we have compared the values of $(k_L{}^0a)$ in the various solutions, the values being referred to as $(k_L{}^0a)_{\text{phys}}$, $(k_L{}^0a)_{\text{chem}}$ and $(k_L{}^0a)_{\text{inst}}$ respectively. Since the physical properties of the solutions were similar, $k_L{}^0$ was to a first approximation the same in each case (this is discussed further later on), and the ratio of the $(k_L{}^0a)$ values depended principally on the variation of a with γ. Here, $k_L{}^0$ is the liquid-film transfer coefficient for absorption without reaction and a the effective gas–liquid interfacial area per unit volume of packed space.

EXPERIMENTAL

The packed column. This was 11 cm i.d. and packed with ½ in. ceramic Raschig rings or Intalox saddles. The liquid was introduced by 64 separate nozzles fixed in a plate immediately above the packing. The column was operated by the so-called differential height technique, first described by Danckwerts and Kennedy [2]. This technique consists of operating with the bottom of the packing flooded and with different heights of packing exposed, thus allowing end-effects to be virtually eliminated. The packing was initially cleaned with chromic acid solution, and was kept wet between experiments. The gas was pure CO_2 at atmospheric pressure (thus there was no gas-side resistance). The whole apparatus, and the other apparatus described below, were in a constant temperature room at 25°C.

The laminar jet apparatus (Fig. 1). This was similar to that used by Kramers *et al.*[4], with certain modifications. It was surrounded by a water-jacket through which water at 25°C was circulated; this was found to be desirable in order to prevent fluctuations in temperature of the gas space (Schröder [8]). The chamber was designed to minimise the effect of spilled liquid on the absorption rate (liquid spillage always occurred at the beginning of an experiment, when the

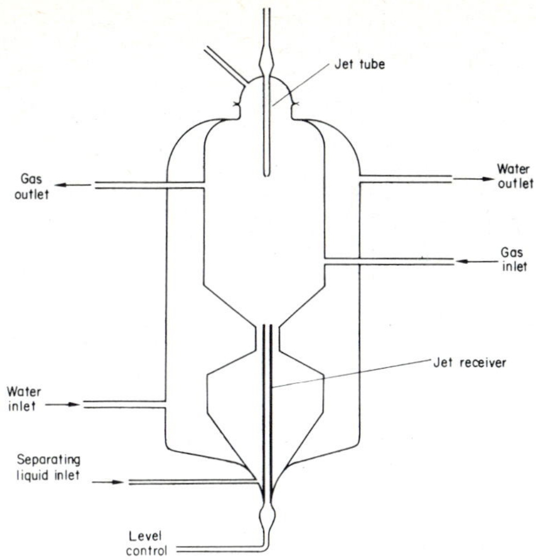

Fig. 1.

liquid flow was started). Spilled liquid sank to the bottom of a pool of 1,3 di-isopropyl benzene (having low density and vapour pressure). The surface of the organic liquid was at the neck of the vessel. The organic liquid was saturated with CO_2 before each experiment. The length of the jet could be varied stepwise by placing spacer-rings between the lid and the main body of the absorption chamber. The ancillary apparatus was as described by Danckwerts [1] (p. 84).

The wetted-wall column. This was the same as that described by Roberts and Danckwerts [7] (see Danckwerts [1] p. 73).

Solutions used. Two different buffer solutions were used to determine $(k_L{}^0a)_{\text{chem}}$; one consisted of 1 g mol K_2CO_3 + 1 g mol $KHCO_3$ per l., the other of 1·69 g mol K_2HPO + 0·26 g mol KH_2PO_4 per l. In order to preserve dynamical similarity in the packed column, the solutions used to determine $(k_L{}^0a)_{\text{phys}}$ and $(k_L{}^0a)_{\text{inst}}$ were matched with the two buffer solutions with respect to viscosity and density. The compositions of the solutions used are shown in Table 1.

CHARACTERISTICS OF THE SOLUTIONS USED

The carbonate buffer. The overall reaction in this case is

$$CO_2 + CO_3^{=} + H_2O \rightarrow 2HCO_3^{-}. \qquad (2)$$

Conditions were such that dissolved CO_2 underwent a pseudo-first order reaction in solution (for criteria for this see Danckwerts [1] p. 246) and the concentration of CO_2 in the bulk was negligible. The reaction of CO_2 in the buffer can be catalysed by the addition of potassium arsenite (Danckwerts [1] p. 244). The pseudo-first order rate constant, k_1, depends on the concentration of arsenite. Values of k_1 with different concentrations of added arsenite were determined with the laminar jet apparatus, using a well-established technique (Danckwerts [1] p. 89). It was found that

$$k_1 = 1{\cdot}2 + 538(\text{KAsO}_2)\ \text{sec}^{-1} \qquad (3)$$

where (KAsO_2) is in g mol/l. The highest concentration of arsenite used was 0·04 g mol/l.

Considering the packed column, and using the Danckwerts surface-renewal model, we have

$$Ra = (k_L{}^0a)_{\text{chem}} A^* \sqrt{\left(1 + \frac{Dk_1}{k_L{}^0}\right)} \qquad (4)$$

whence

$$(Ra)^2 = ((k_L{}^0a)_{\text{chem}} A^*)^2 \left(1 + \frac{Dk_1}{k_L{}^0}\right) \qquad (5)$$

Table 1.

	Matching the carbonate buffer		Matching the phosphate buffer	
Density (g/cm^3)	1·16		1·24	
Viscosity/η_{H_2O}	1·57		2·29	
	Inert solution	"Instantaneously" reacting (0·20M NaOH)	Inert solution	"Instantaneously" reacting (0·18M NaOH)
Sodium nitrate, g	272	268	405	400
Sugar, g	92	89	169	166
Sodium hydroxide, g	—	9·4	—	9·1
Water, g	1000	1000	1000	1000

where Ra is the observed rate of absorption per unit packed volume, A^* the physical solubility of CO_2 in the solution, and D the diffusivity of dissolved CO_2. Thus, if absorption rates (Ra) are measured at various values of k_1, and $(Ra)^2$ plotted against k_1, a straight line should result with an intercept of $((k_L{}^0a)_{chem}A^*)^2$ on the ordinate. Thus, if the value of A^* is known, that of $(k_L{}^0a)_{chem}$ can be deduced. Methods for estimating A^* are discussed in Appendix 1. (This method of determining $(k_L{}^0a)_{chem}$ assumes that a_{chem} is independent of R and hence of γ. This question is discussed later.)

The values of γ for the carbonate buffer solutions ranged from 30 to 60, depending on the arsenite concentration.

The phosphate buffer. The overall reaction in this case is

$$H_2O + CO_2 + HPO_4^{=} \rightarrow HCO_3^{-} + H_2PO_4^{-}. \qquad (6)$$

The value of k_1 in this solution is so small (0·2 sec^{-1}) that the specific absorption rate is given by the limiting form of Eq. (3):

$$Ra = (k_L{}^0a)_{chem}A^* \qquad (7)$$

and it is only necessary to know A^* to deduce $(k_L{}^0a)_{chem}$ from the absorption rate. The value of γ for this solution was about 16.

In the case of both types of buffer it was verified by calculation that the rate of reaction was fast enough to maintain equilibrium in the bulk of the reactant and that the equilibrium concentration of CO_2 was in all circumstances much less than A^* and hence could be neglected (Danckwerts [1] p. 244).

The solutions containing NaOH. These solutions were designed to give conditions approaching "instantaneous reaction", i.e. the case where the rate of reaction is determined entirely by the diffusion of the CO_2 and the reactant towards the reaction zone, and is independent of the kinetics of the reaction.

The overall reaction in this case is

$$CO_2 + 2OH^- \rightarrow CO_3^{=} + H_2O. \qquad (8)$$

The reaction is second-order, the rate per unit volume being given by

$$k_{OH}(OH^-)(CO_2). \qquad (9)$$

The reaction can be considered "instantaneous" in the above sense when

$$\frac{\sqrt{(Dk_{OH}(OH^-))}}{k_L{}^0} \gg E_i. \qquad (10)$$

Here (OH^-) is the bulk concentration of sodium hydroxide and E_i is the value which the enhancement factor E would have if the reaction were truly instantaneous. When

$$\frac{(OH^-)}{2A^*} > 3 \qquad (11)$$

(which was the case in our experiments), E_i is given to a close approximation by

$$E_i = \sqrt{\frac{D}{D_{OH}}} + \frac{(OH^-)}{2A^*}\sqrt{\frac{D_{OH}}{D}} \qquad (12)$$

where D_{OH} is the effective diffusivity of OH^-. Using the following estimated or actual values: $D_{OH}/D = 1{\cdot}67$, $k_{OH} = 10^4$ l./g mol sec, $(OH^-) = 0{\cdot}2$ g mol/l., $k_L = 5\times10^{-3}$ cm/sec, $A^* = 2\times10^{-2}$ g mol/l., $D = 1{\cdot}4\times10^{-5}$ cm^2/sec, condition (10) combined with Eq. (12) takes the form

$$34 \gg 7{\cdot}5. \qquad (13)$$

This indicates an error of only about 4 per cent in assuming E for the NaOH-containing solution to be equal to E_i.

The factor C by which the reaction enhances the capacity of the solution for the absorption of CO_2 is

$$C = 1 + \frac{(OH)}{2A^*}. \qquad (14)$$

The enhancement factor, E, is given by Eq. (12) (since $E = E_i$). Thus, using the values of D_{OH}/D and A^* given above, we can calculate that the factor γ has a value of about 0·8 for 0·2 M NaOH, and more for a solution depleted by absorption of CO_2 during its passage through the column.

In order to interpret the results of absorption experiments in packed columns it was necessary to know E_iA^* as a function of the degree of conversion of NaOH to Na_2CO_3, which increased progressively as the liquid passed through the column. Although this could be calculated, as above, in view of the uncertain values of some of the quantities involved it was decided to determine values of E_iA^* experimentally, using the wetted-wall column, and solutions representing various amounts of absorbed CO_2. The method is described in Appendix 2.

RESULTS AND DISCUSSION

Determination of $(k_L{}^0a)_{phys}$. The absorption rate M gmol/sec in the column is determined with various heights, h, of packing exposed. The cross-sectional area of the column is X. Then

$$dM = (k_L{}^0a)_{phys}(A^* - A)Xdh \quad (15)$$

where A is the bulk concentration of CO_2 at height h. But the liquid entering the column is free from CO_2; thus

$$A = M/L \quad (16)$$

where L is the liquid flow-rate. Thus Eq. (15) becomes

$$\frac{d\log_{10}\left(1-\frac{M}{LA^*}\right)}{dh} = -\frac{X}{2{\cdot}303L}(k_L{}^0a)_{phys}. \quad (17)$$

The value of $(k_L{}^0a)_{phys}$ at a particular flow-rate is determined from the slope of a plot of $\log_{10}\times(1-M/LA^*)$ against h. Values of $(k_L{}^0a)_{phys}$ for the two inert solutions at various flow-rates are shown in Figs. 2–4.

Determination of $(k_L{}^0a)_{chem}$. As regards the carbonate buffer,

$$\frac{1}{X}\frac{dM}{dh} = Ra \quad (18)$$

so that (Ra) can be determined from a plot of M against h. $(k_L{}^0a)_{chem}$ is determined as described above, by altering the arsenite concentration and plotting $(Ra)^2$ against k_1. Some plots of this kind are shown in Fig. 5.

As regards the phosphate buffer, $(k_L{}^0a)_{chem}$ is found simply by determining Ra and dividing by the estimated value of A^*, according to Eq. (7).

Values of $(k_L{}^0a)_{chem}$ are shown in Figs. 2–4 for various liquid flow-rates.

Determination of $(k_L{}^0a)_{inst}$. The differential equation relating the absorption rate to the height of packing exposed is

$$dM = (k_L{}^0a)_{inst}A^*E_iXdh \quad (19)$$

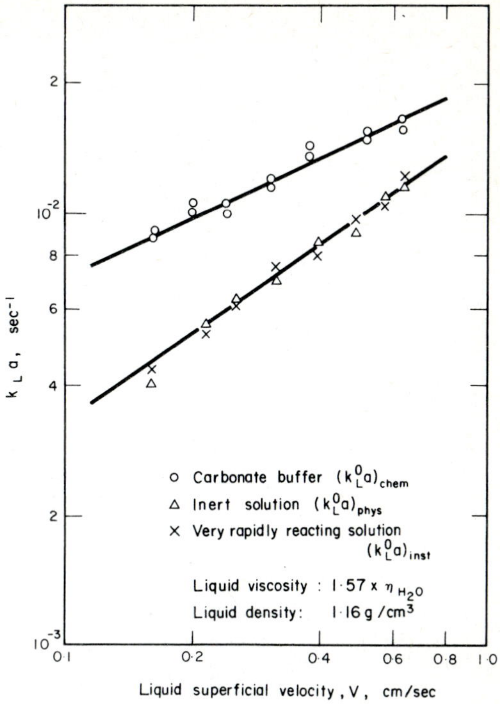

Fig. 2. ½ in. Raschig rings. Comparison of $(k_L{}^0a)_{phys}$, $(k_L{}^0a)_{chem}$ and $(k_L{}^0a)_{inst}$.

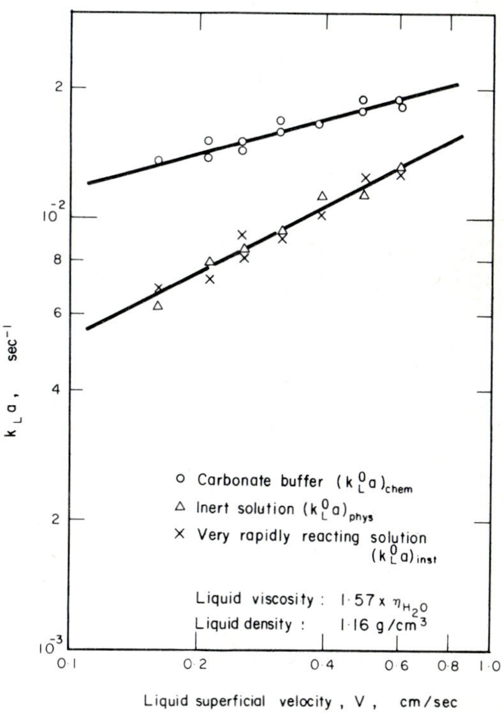

Fig. 3. ½ in. Intalox saddles. Comparison of $(k_L{}^0a)_{phys}$, $(k_L{}^0a)_{chem}$ and $(k_L{}^0a)_{inst}$.

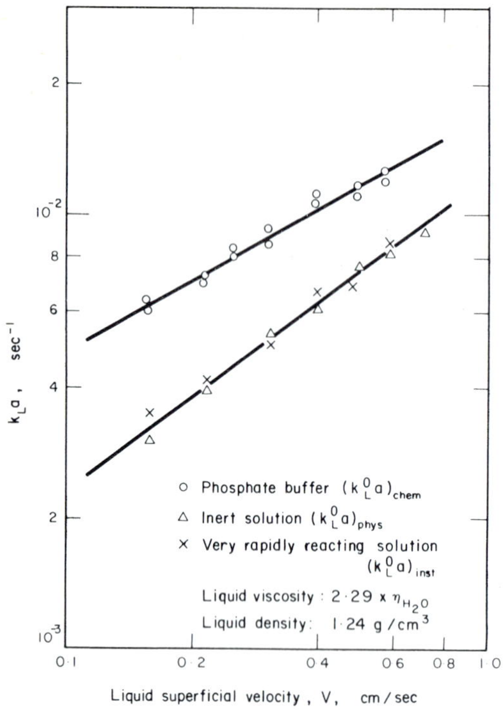

Fig. 4. ½ in. Raschig rings. Comparison of $(k_L{}^0a)_{phys}$, $(k_L{}^0a)_{chem}$ and $(k_L{}^0a)_{inst}$.

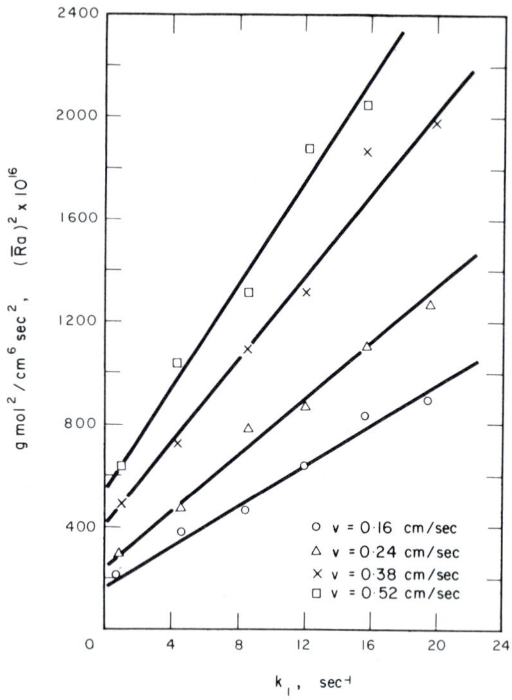

Fig. 5. The square of the specific absorption rate plotted against the pseudo-first order rate constant k_1 in the solution. Packing: ½ in. Raschig rings; carbonate buffer solution.

and from a material balance (for the solution matching the carbonate buffer):

$$(OH^-) = 0{\cdot}20\times 10^{-3} - 2\frac{M}{L}\text{ gmol/cm}^3 \quad (20)$$

where (OH^-) is the concentration (gmol/cm^3) of NaOH unconverted to carbonate. A^*E_i is related to (OH^-) (for the solution matching the carbonate buffer) by Eq. (26) (Appendix 2). Hence

$$-d\ln(3{\cdot}4\times 10^{-5} + (OH^-)) = \frac{1{\cdot}24}{L}(k_L{}^0a)_{inst}X dh \quad (21)$$

where (OH^-) is determined from Eq. (20) and is in gmol/cm^3. Thus, for a given value of L, M is determined as a function of h, and $\ln(3{\cdot}4\times 10^{-5}+(OH^-))$ plotted against h. From the slope of the plot, $(k_L{}^0a)_{inst}$ is determined. The solutions matching the phosphate buffer are treated similarly. Values of $(k_L{}^0a)_{inst}$ for the two solutions are shown in Figs. 2–4 for various liquid flow-rates.

DISCUSSION

As can be seen from Figs. 2–4, $(k_L{}^0a)_{chem}$ is larger than $(k_L{}^0a)_{phys}$ and $(k_L{}^0a)_{inst}$, the latter two quantities being almost equal at any given flow-rate. The fact that these latter two are so nearly equal is a matter of some surprise, since $\gamma < 1$ for the case of $(k_L{}^0a)_{inst}$. However, this fact may be compensated for by the fact that the reaction was not quite within the "instantaneous" regime, a circumstance which tends to increase γ somewhat. However, the results show clearly that $(k_L{}^0a)$ is larger when $\gamma \gg 1$ than it is when $\gamma \simeq 1$, and that it is the value of γ and not simply the presence or absence of a chemical reaction which determines $(k_L{}^0a)$.

Figure 6 shows the ratio $(k_L{}^0a)_{chem}/(k_L{}^0a)_{phys}$. The ratio is almost the same for the carbonate buffer as it is for the phosphate buffer, despite the very different values of γ. The results suggest that even in the case of the phosphate buffer, the value of γ is large enough to render effective the whole of the available wetted surface (for the concept of availability, see later). In practice this

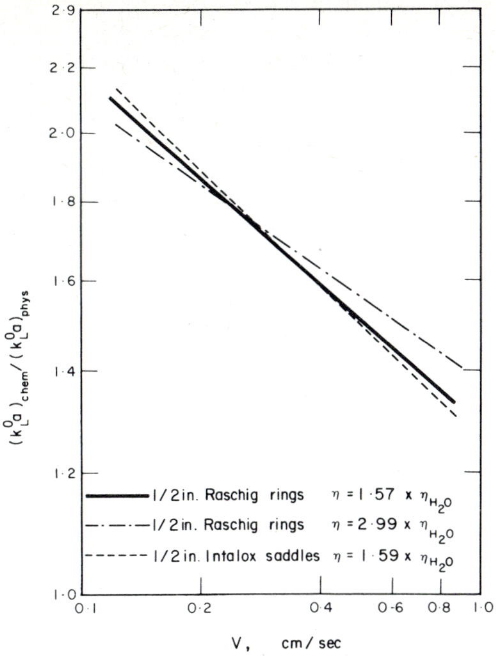

Fig. 6. Ratio $(k_L{}^0a)_{chem}/(k_L{}^0a)_{phys}$ as a function of the superficial liquid velocity, v.

means that when interfacial areas are determined by the chemical method the values of a found will be independent of the value of γ, providing this exceeds some minimum value (which depends, presumably, on the packing and flow-rate). Thus, these values of $(a)_{chem}$ have a general significance and are not specific to the particular reaction used. Danckwerts and Rizvi [3] have shown that values of a determined by dissimilar chemical reactions are in good agreement.

It is possible, from the plots of $(Ra)^2$ against k_1 in the case of the carbonate buffers, to determine the actual value of $(a)_{chem}$, the effective specific interfacial area (Danckwerts [1] p. 210). The value of $(a)_{chem}$ depends on the flow-rate, but even at the highest flow-rate used (0·62 cm/sec) its value was somewhat less than half the total geometrical area in the case of both Raschig rings and Intalox saddles. On the other hand, visual observation suggested that the entire surface of the packing was wetted by the liquid. In an irrigated packing, therefore, the situation appears to be as follows: only a fraction of the surface is "available" for mass transfer, and the fraction of the "available" surface which is effective for mass transfer depends on the value of γ. The "unavailable" part of the surface is covered by an absolutely stationary film of liquid which is never renewed and thus does not contribute towards mass transfer when the stationary state is reached, even when $\gamma \gg 1$. In experiments in which liquid is evaporated from the packing, the "unavailable" parts of the surface will be dried, and thus in the stationary state the "available" surface is the same as the wetted surface. This accounts for the findings of Onda *et al.* [6] that the area effective for evaporation of the liquid was equal to $(a)_{chem}$. It may be noted that Danckwerts and Rizvi [3] showed that with larger packings and high liquid flow-rates $(a)_{chem}$ may become equal to the geometrical surface area of the packing (and in fact may slightly exceed it, perhaps because of the formation of drops).

The fact that $(k_L{}^0a)_{chem}$ is independent of γ provided this exceeds a certain value validates the method described above for determining $(k_L{}^0a)_{chem}$ for the carbonate buffer, in which $(k_L{}^0a)_{chem}$ was assumed to be independent of k_1, and explains why there is no systematic curvature in the plots shown in Fig. 5.

The ratio $(k_L{}^0a)_{chem}/(k_L{}^0a)_{phys}$ is probably not simply equal to the ratio of the effective interfacial areas, $(a)_{chem}/(a)_{phys}$. Although the physical properties of the liquids were the same, it is probable that the mass transfer coefficients $(k_L{}^0)_{chem}$ and $(k_L{}^0)_{phys}$ were different. The part of the surface which is effective in the physical case consists of liquid in brisk motion, whereas in the chemical case comparatively sluggish regimes are also included. Thus we should expect $(k_L{}^0)_{chem} < (k_L{}^0)_{phys}$, nd the ratio of the effective interfacial areas would then be even larger than the ratio of the $(k_L{}^0a)$ values.

Clearly, values of a or $(k_L{}^0a)$ determined by the chemical method with $\gamma \gg 1$ should not be used for the design of physical absorbers.

Acknowledgement—G. E. H. J. wishes to thank the managing directors of the Koninklijke/Shell-Laboratorium Amsterdam for their support during his stay in Cambridge.

NOTATION

- A^* solubility of CO_2 at 1 atm, 25°C gmol/cm^3
- a effective interfacial area for gas absorption per unit volume of packed space
- C factor by which chemical reaction increases capacity of liquid for absorbing gas
- D diffusivity of CO_2 in solution

d diameter of wetted-wall column
E factor by which chemical reaction enhances rate of absorption of gas
E_i value of E when chemical reaction is instantaneous
h height of packing exposed to gas; or height of wetted-wall column
I ionic strength
k_1 first-order reaction-rate constant
k_L^0 mass transfer coefficient in absence of reaction
k_{OH} second-order rate constant for reactions of CO_2 with OH^-
L liquid flow-rate to packed column
M absorption rate in packed column
q absorption rate on wetted-wall column
R rate of absorption per unit effective interfacial area in packed column
t time of exposure of liquid to gas in wetted-wall column
X cross-sectional area of packed column

Subscripts
chem referring to buffer solutions
phys referring to inert solutions
inst referring to solutions containing NaOH

REFERENCES

[1] DANCKWERTS P. V., *Gas–Liquid Reactions* McGraw-Hill, New York 1970.
[2] DANCKWERTS P. V. and KENNEDY A. M., *Chem. Engng Sci.* 1958 **8** 201.
[3] DANCKWERTS P. V. and RIZVI S. F., *Trans. Instn Chem. Engrs* 1971 **49** 124.
[4] KRAMERS H., BLIND M. P. P. and SNOECK E., *Chem. Engng Sci.* 1961 **14** 115.
[5] MARKHAM A. E. and KOBE K. A., *J. Am. Chem. Soc.* 1941 **63** 449.
[6] ONDA K. E., SADA E. and TAKEUCHI H., *J. Chem. Engng Japan* 1968 **1** 62.
[7] ROBERTS D. and DANCKWERTS P. V., *Chem. Engng Sci.* 1962 **17** 961.
[8] SCHRÖDER W., *Compt. Rend.* 36^e Congres Int. Chimie Industrielle, p. 248, Brussels 1967.

APPENDIX 1

The solubility of CO_2 *in the buffer solutions*

As explained above, it is necessary to know A^* (the solubility of CO_2 at 1 atm and 25°C) for the two buffer solutions in order to determine $(k_L^0 a)_{chem}$. Three methods were used to determine A^*:

(a) The van Krevelen–Hoftijzer method extended to mixed electrolytes (Danckwerts [1] p. 20). This is not applicable to the phosphate buffer as the required parameters have not been determined.

(b) The N_2O method. According to the principles of the van Krevelen–Hoftijzer method and the parameters for the two gases (Danckwerts [1] p. 20), the solubilities of CO_2 and N_2O in a given electrolyte solution are related by

$$\log_{10}\frac{A^*_{CO_2}}{A^*_{N_2O}} = 0{\cdot}019 \times I \tag{22}$$

where I is the ionic strength of the solution (g ion/l.). The solubility of N_2O in the two solutions was measured by the method of Markham and Kobe [5] and the solubility of CO_2 deduced.

Table 2. Estimates of the solubility of carbon dioxide in the buffer solutions at 25°C, 760 mm Hg. (gmol/cm³)

	Carbonate buffer	Phosphate buffer
Predicted by method (a)	$1{\cdot}68 \times 10^{-5}$	—
Predicted by method (b)	$1{\cdot}51 \times 10^{-5}$	$0{\cdot}95 \times 10^{-5}$
Predicted by method (c)	$1{\cdot}50 \times 10^{-5}$	$0{\cdot}92 \times 10^{-5}$
Used in this work	$1{\cdot}51 \times 10^{-5}$	$0{\cdot}94 \times 10^{-5}$

(c) The laminar jet method. By absorbing CO_2 into a laminar jet of the solution, the value of $A^*\sqrt{D}$ can be determined (Danckwerts [1] p. 89). The value of D was determined by the N_2O method described in Appendix 2 (Eq. 25), and hence A^* deduced.

The solubilities predicted by the various methods are shown in Table 2.

APPENDIX 2

Determination of E_iA^* *as function of absorbed* CO_2

Provided certain conditions are fulfilled (Danckwerts [1] p. 44) the absorption of CO_2 into the sodium hydroxide solutions on the wetted-wall column proceeds in the "instantaneous reaction" regime. These conditions were fulfilled in our experiments. Then

$$q = 2\pi h d A^* E_i \sqrt{\frac{D}{\pi t}} \tag{23}$$

where q is the total absorption rate into the wetted column, d is the diameter and h the height of the column, and t the time of exposure of the falling liquid to the gas; the latter is easily calculated if the flow-rate, density and viscosity of the liquid are known (Danckwerts [1] p. 73), and thus $a^*E_i\sqrt{D}$ is found. This was done for solutions representing different degrees of carbonation of the original OH^- solution.

Two methods were used to determine D in the solutions. Firstly the presence of NaOH and Na_2CO_3 in the solutions was ignored, in view of the large excess of other solutes, and the diffusivity of CO_2 in a similar solution free of NaOH was determined, using the wetted-wall column. The rate of physical absorption is

$$q = 2\pi h d A^* \sqrt{\frac{D}{\pi t}}; \tag{24}$$

hence $A^*\sqrt{D}$ is determined. A^* can be found by the method of Markham and Kobe [5] for determining solubilities, and so

D is found. The second method was to assume that the solutes have the same effect on the diffusivity of N_2O as they do on that of CO_2. Using the wetted-wall column, D_{N_2O} was determined as above for pure water and the solution. D for CO_2 in pure water is well-known. We then put

$$\frac{D_{soln}}{D_{water}} = \frac{D_{N_2O,\,soln}}{D_{N_2O,\,water}} \tag{25}$$

and hence find D_{soln}. The results of the two methods agreed within 5 per cent, and the average value was used.

The values of E_iA^* found for the solutions matching the carbonate buffer were found to be given by

$$E_iA^* = 2{\cdot}12\times 10^{-5}+0{\cdot}62[OH^-]\ \text{g mole/cm}^3 \tag{26}$$

and those for the solutions matching the phosphate by

$$E_iA^* = 1{\cdot}10\times 10^{-5}+0{\cdot}64[OH^-]\ \text{g mole/cm}^3 \tag{27}$$

where $[OH^-]$ is the concentration (g mol/cm^3) of NaOH unconverted to carbonate. The value of γ in these solutions varies with the degree of carbonation, having a minimum value of about 0·8 in the uncarbonated solutions and a maximum value of about 0·9 at the bottom of the absorption column.

B15

The effects of catalysis on rates of absorption of CO_2 into aqueous amine–potash solutions

P. V. DANCKWERTS and K. M. MCNEIL†

Abstract—The effect of CO_2-hydration catalysts on the rate of absorption of CO_2 into mixed solutions of amine and potash is considered theoretically. The conclusions are supported by measurements of absorption rates in a stirred cell, using sodium arsenite as a catalyst. In general, the catalyst decreases the absorption rate at low carbonation ratios and increases it at high carbonation ratios.

† Present address: Amoco Chemicals Corporation, Whiting, Indiana, U.S.A.

B16

THE ABSORPTION OF CARBON DIOXIDE INTO AQUEOUS AMINE SOLUTIONS AND THE EFFECTS OF CATALYSIS

By Professor P. V. DANCKWERTS, G.C., M.B.E., M.A., S.M., C.Eng. (MEMBER),* and K. M. McNEIL, B.Sc., Ph.D. (GRADUATE)*†

SYNOPSIS

When carbon dioxide is absorbed by aqueous amine solutions, a fast reaction leads to the formation of amine carbamate. This is followed by a slow reaction in the bulk of the liquid, leading to partial hydrolysis of carbamate to bicarbonate. When the second reaction is complete the pressure of carbon dioxide is lower and the concentration of free amine higher. The addition of a catalyst, such as sodium arsenite, speeds up the second reaction and may thereby increase the rate of absorption and the capacity of the absorbent.

Calculations are compared with the results of absorption measurements in a stirred cell. The agreement is good and indicates that the mechanism assumed for the absorption process is correct. A catalyst can substantially increase the rate of absorption and the capacity of the solution when the ratio (moles carbon dioxide absorbed/moles amine initially present) is greater than ½. This might form the basis of an economic industrial process for carbon dioxide absorption.

* Department of Chemical Engineering, Pembroke Street, Cambridge.

† *Present address*: Amoco Chemicals Corporation, P.O. Box 431, Whiting, Indiana, U.S.A.

B17

DESORPTION OF CARBON DIOXIDE FROM AQUEOUS POTASH SOLUTIONS WITH AND WITHOUT THE ADDITION OF ARSENITE AS A CATALYST

By C. N. S. McLACHLAN, Ph.D.†‡ and Professor P. V. DANCKWERTS, G.C., M.B.E., C.Eng. (FELLOW)†

SYNOPSIS

The rate of desorption of carbon dioxide at elevated temperatures from carbonated potash solutions has been measured in a 15·4 cm dia. bubble-cap column containing a single plate and a single bubble-cap. The gas-phase and liquid-phase mass transfer coefficients, k_G and k_L, and the interfacial area per unit volume of liquid hold-up, a, have been measured as a function of the superficial gas velocity and the submergence. Values of k_G and k_L determined from these measurements have been used together with estimated values of the relevant physical and chemical constants to predict the specific rate of desorption.

For the solutions containing no catalyst agreement is reasonable. However, in solutions containing arsenite as a catalyst there is a considerable discrepancy between the predicted specific rates of desorption and the corresponding values obtained from the measured rates of desorption. Reasons for the deviation are discussed and a possible explanation is proposed.

† Department of Chemical Engineering, University of Cambridge, Pembroke Street, Cambridge CB2 3RA.

‡ *Present address:* D.S.I.R., Chemistry Division, Private Bag, Petone, N.Z.

SECTION C

Reactions of Carbon Dioxide with Bases and Catalysts in Aqueous Solution

The papers of SECTION C are concerned with the homogeneous reaction kinetics of carbon dioxide in various aqueous solutions (although the homogeneous rates were generally inferred from rates of absorption). Papers C1-3 represent the work of M.M. Sharma (now Professor at Bombay University) the most diligent research student I ever supervised. My contribution was confined mostly to suggest-

-ions; the experimental work and the chemical theory were provided principally by Sharma. Paper C1 is concerned with an exhaustive test of the power of various weak acid anions to catalyse the reaction of carbon dioxide with water. One particular type of anion - "Group A", including arsenite - forms a family which conforms to the Brønsted relationship between basic strength of anion and catalytic constant, while anions of other types do not fall into this category. The test did not show up any alternative to arsenite (already used) as a catalyst for industrial use. However, hypochlorite (OCl^-) and formaldehyde hydrate had enough catalytic power to be worth investigating for use in experimental work related to gas absorption.

It should be mentioned that the catalytic effect of arsenite is a good deal more complicated than appears in Paper C2 (for instance). This is brought out in Paper B17 and by C.N.S. McLachlan, N.Z. J.Sci. 1971 14 634.

Paper C2 is concerned to widen the investigation of the performance of arsenite, hypochlorite and formaldehyde (the use of arsenite and hypochlorite in experimental investigations is described in several papers in SECTION B). Paper C1 also reports on the kinetics of the reaction of carbon dioxide in monoisopropanolamine solutions, since this is an absorbent with industrial possibilities. Sharma (Trans. Faraday Soc. 1965 61 681) also reported on the rates of reaction of many other amines. Paper C3 reports on the rates of reaction of some of these amines with carbonyl sulphide, which often has to be removed from synthesis gases. The rates are in general about 1% of those with carbon dioxide.

Paper C4 is a highly speculative attempt to explain why Himmelblau and Babb, using radiocarbon to measure the rate of interconversion of carbon dioxide and bicarbonate in solution, found an ostensible value for the rate constant of the reaction between carbon dioxide and hydroxyl ions which is about 10 times as great as the actual value. The paper created considerable interest (judging from the number of requests for reprints) but I have not seen any comments on our explanation from competent chemists.

Papers C5 and C6 are concerned with the kinetics of the reaction of carbon dioxide with mono- and diethanolamine. Paper C5 suggests an explanation for the experimental results of Hikita (cited there) which showed that the kinetics for the two amines appeared to be fundamentally different. This explanation is strongly supported by the experiments described in Paper C6. However, the effect of alkali carbonate on the diethanolamine reaction is surprising as will duly be reported in "Chemical Engineering Science".

The papers in SECTION C lead to the boundary between the descriptive and mechanistic aspects of some reactions of interest to chemical engineers and emphasise that the latter too often fail to consult their chemist colleagues. I leave it to the reader to decide which of these papers is likely to excite the derision of pure chemists.

C1

Catalysis by Brönsted bases of the reaction between CO_2 and water

BY M. M. SHARMA AND P. V. DANCKWERTS

Dept. of Chemical Engineering, Pembroke Street, Cambridge

Received 30*th July*, 1962

The catalysis of the reaction $CO_2+H_2O \rightarrow HCO_3^- + H^+$ by a number of anions has been studied at 0°C. Anions which have a negatively-charged oxygen atom and at least one hydroxyl group attached to the same central atom, and in which the negative charge is not delocalized by resonance effects, form a single group which obeys the Brönsted relationship. Anions of other types are in general weaker catalysts for a given pK.

The removal of CO_2 from mixtures of gases by liquid absorbents is an important industrial process in several contexts. For economic reasons the solution must (*a*) have a large capacity for CO_2, (*b*) be capable of regeneration by driving off the CO_2 (usually by heating or stripping with steam and/or air), and (*c*) have a high specific absorption rate. Solutions of potassium carbonate (which is more soluble than sodium carbonate) are satisfactory in respect of (*a*) and (*b*), and are in common use. However, the dissolved CO_2 must react with the solution before the absorptive capacity of the latter can be realized, and this reaction is relatively slow. Normally the dissolved CO_2 reacts with the OH^- in the solution:

$$CO_2+OH^- \rightarrow HCO_3^-. \tag{1}$$

The second order rate-constant is about 6000 l./mole sec at 20°C; thus, in an equimolal mixture of carbonate and bicarbonate, with a pH of about 10, the CO_2 will undergo a pseudo-first order reaction with a rate constant of about 0·6 sec^{-1}. The CO_2 will also react with water:

$$CO_2+H_2O \rightarrow HCO_3^- + H^+, \tag{2}$$
$$H^+ + OH^- \rightarrow H_2O.$$

The rate-determining step, reaction (2), has a rate constant of about 0·02 sec^{-1} at 20°C, and thus plays a minor part in the absorption process. However, reaction (2) can be catalyzed by a variety of agents: arsenite ion is used as a catalyst in one industrial process,[1] thus speeding up the absorption and desorption of CO_2. Roberts and Danckwerts[2] have shown that the catalytic constant for arsenite ions (defined below) is about 160 l./mole sec at 25°C.

Roughton and Booth [3,4] and Kiese and Hastings [5] have studied the catalysis of reaction (2) and its reverse extensively because of the physiological importance of the latter. It was found that a number of anions of weak acids, such as selenite, arsenite, tellurate, hypochlorite, hypobromite, etc., possess a high catalytic power, but no general correlation between ionization constant, structure and catalytic power has been put forward. A critical examination of the data suggested to us that the various anions might be grouped according to the distribution of the negative charge, and that the Brönsted relationship might be obeyed within each group.

The work reported here was undertaken to throw more light on the catalytic effect of various types of anions, and at the same time as a search for catalysts of possible industrial value. The catalysts tried include the anions of (A) telluric, germanic, arsenious and silicic acids, chloral hydrate, chloral alcoholate, and the hydrates of butyl chloral, glyoxal, formaldehyde, acetaldehyde, and diacetyl: (B) trichloroethanol and trifluoroethanol; hydrogen peroxide; (C) some sugars; and (D) sulphurous, selenous, tellurous, phosphoric and phenyl arsonic acids and phenol.

EXPERIMENTAL

The rate of reaction was measured by a manometric method, similar to that employed by Roughton and Booth.[6] The reaction flask, compensating flask and CO_2 flask were all kept in a water-bath maintained at the desired temperature. The solution in the reaction flask (10 ml) was vigorously agitated by a magnetic stirrer operated through the bottom of the water-bath, which was made of a sheet of Perspex. All the experiments were carried out at a total pressure of 152 mm Hg (0·2 atm) and a maximum partial pressure of CO_2 of about 25 mm Hg. The constants of the apparatus were evaluated by the method given by Dixon.[7]

The value of the reaction-rate constant calculated from the observations is based on the assumption that the solubility of CO_2 in the buffers employed was substantially the same as in water. The solubility of a gas in a solution with which it reacts cannot be determined by equilibrium measurements. The available data on the solubility of CO_2 in a number of electrolytes akin to the constituents of the buffers employed, as also the general thermodynamic considerations pertaining to the solubility of electrolytes,[8] indicate that this assumption is unlikely to lead to an error greater than 5%.

The manometer readings were recorded every 15 sec, the first reading being taken after 30 sec. Normally, readings were continued for 5-10 min. The first 15 to 60 sec, depending upon the pH and the catalyst employed, represented the period during which physical saturation was taking place. At least 5 or 6 points beyond this time were used to obtain the reaction rate. All the experiments were carried out in triplicate or quadruplicate, the maximum variation in the rate-constants being about ±5%. A series of experiments with phosphate and veronal buffers gave results comparable with those obtained by Roughton and co-workers.[9, 10] Spot checks were made at frequent intervals with these buffers. The catalytic activity of the desired material was obtained by carrying out experiments with 2 or 3 different concentrations of the catalyst in a given buffer. The pH values of the buffers employed were measured with a pH meter (Radiometer, Copenhagen). Four buffer solutions were employed, namely, phosphate pH 7·52 (0°C) ($M = 0{\cdot}132$, [ion]/[acid] $= 5{\cdot}2$, $\mu = 0{\cdot}35$), veronal pH 8·75 (0°C) ($M = 0{\cdot}12$, [ion]/[acid] $= 3$, $\mu = 0{\cdot}09$), carbonate buffer—I pH 9·5 (0°C) ($M = 0{\cdot}2$, $[HCO_3^-]/[CO_3^{2-}] = 4$, $\mu = 0{\cdot}28$), and carbonate buffer-II pH 10·1 (0°C) ($M = 0{\cdot}14$, $[HCO_3^-]/[CO_3^{2-}] = 1$, $\mu = 0{\cdot}28$).

MATERIALS

All the materials employed were either A.R. or technical grade. The acetaldehyde and formaldehyde solutions were analyzed by the hydroxylamine hydrochloride method. The aldehydes tend to undergo reactions in alkaline solution so all the solutions were prepared and stored at 0°C. All experiments with aldehydes were carried out on the day on which the solutions were prepared. Consistent results were obtained, indicating that degradation of the materials employed was insignificant during the period of storage.

RESULTS AND DISCUSSION

In general, the rate constant of the reaction in any particular buffer was found to be a linear function of the concentration of the total catalyst added (ionized + unionized forms), in conformity with the expressions

$$R = (k_u + k_B[B^-])[CO_2], \tag{3}$$

$$[B^-] = ([B^-]+[HB])/(1+[H^+]/K_A), \tag{4}$$

where R = rate of reaction of CO_2, moles/l. sec, $[CO_2]$ = concentration of CO_2 mole/l., $[B^-]$ = concentration of anion, moles/l., [HB] = concentration of conjugate acid, moles/l., k_u = rate constant for hydration of CO_2 in absence of catalyst, sec^{-1}, k_B = catalytic rate constant corresponding to base B^-, l./mole sec, K_A = stoichiometric dissociation constant of HB, g ion/l.

The velocity constant k_B can be calculated from the slope of the plot of reaction-rate constant against catalyst concentration (ionized + unionized), together with the pH of the buffer solution and the pK of the acid conjugate to the catalytic anion. The effect of ionic strength on the dissociation of the acid should be known, but such data are not always available and an assumption had to be made in some cases. On the basis of the thermodynamic considerations and the data available on a number of substances of the type under consideration, a correction factor of $-0{\cdot}2$ was applied to the value of pK at infinite dilution to get its value at an ionic strength of 0·28.

Experimental results are presented in table 1. Values of the quantity k, defined by

$$R = (k_u + k\{[B^-]+[HB]\})[CO_2], \tag{5}$$

are also reported, so that k_B can be recalculated if necessary in the light of further information about pK values.

All the aldehydes investigated exist in the solution in the hydrated form. For acetaldehyde and diacetyl hydration is incomplete and it is therefore necessary to consider the hydration equilibria when calculating the catalytic power of the hydrate anion. The relevant references are given in table 1.

The anions of group A have a negative charge localized on a single oxygen atom, and there is no possibility of this being spread over the molecule by resonance. The atom to which the ionized oxygen atom is attached also carries at least one unionized hydroxyl group, except with chloral alcoholate. The large catalytic activity of ions of this type may be due to the ability of the unionized hydroxyl group to attach itself to and orient favourably one of the reacting molecules. A similar role for the hydroxyl group in the alkaline hydrolysis of α- hydroxy cyclopentyl acetate has been suggested by Bruice and Fife.[34a]

The effect of pH on the catalytic power of the anions of group A is not great except with germanite. It is known that variations of pH change the nature of the species present in germanite solutions[14,15]. The plot of log k_B against pK_a for this group is shown in fig. 1, which indicates that the results can be fitted reasonably well to a straight line of slope 0·56, although the catalytic activities of acetaldehyde hydrate and diacetyl hydrate are rather low. The Brönsted relation thus holds among members of the group. It is interesting that the catalytic activity of choral alcoholate appears to be nearly the same as that of chloral hydrate, although its anion carries an $-OC_2H_5$ rather than an $-OH$ group. It is likely that this substance was hydrolyzed to chloral hydrate.

Trichloroethanol, trifluoroethanol (group B) and the sugars (group C) display much lower apparent catalytic powers than substances in group A having comparable pK values. Indeed, it seems probable that the observed rates of reaction of CO_2 did not correspond to catalysis of reaction (2), but could be accounted for by the formation of mono-alkyl carbonates:

$$RO^- + CO_2 \rightarrow RCO_3^- . \tag{6}$$

Reactions of this type have been investigated extensively by Faurholt and co-workers [33,34] and their data on the kinetics indicate that with sucrose, lactose and glucose the whole of the observed rate of reaction in our experiments can be accounted for in this way. In our experiments on substances in this group, the quantity of "catalyst" present in the solution was always much larger than the amount of CO_2 absorbed. The carbonates, when formed, hydrolyze slowly. It is therefore not possible to assign true values of k_B to substances in this group. They are almost certainly much less than the apparent values reported in table 1.

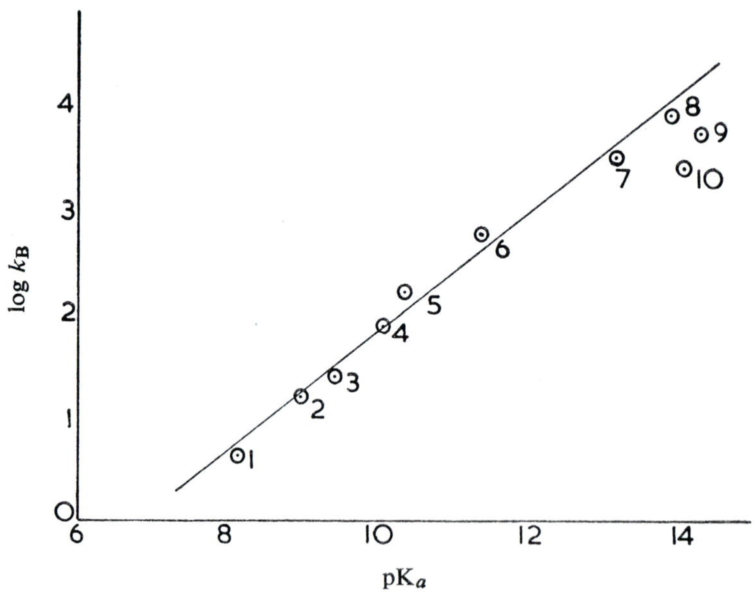

FIG. 1.—Brönsted plot for catalysts of group A at 0°C

The substances in group A do not, as far as is known, form stable carbonates, and in any case the amounts of catalyst were in some cases much less than the amounts of CO_2 absorbed. Hydrogen peroxide and acetophenoxime, which do not form carbonates, exhibit very low catalytic powers as compared to substances in group A with comparable pK values.

The data of Kiese and Hastings [5] (table 2B) indicate that ClO^- and BrO^- are powerful catalysts; the effect of pH on the catalytic power of ClO^- is not significant but with BrO^- it is very pronounced. The latter ion is rather unstable in alkaline solution and this might partly account for the observed behaviour.

Table 1.—Catalytic power of various substances, temp. 0°C.

k, rate constant in terms of total concentration of catalyst, l./mole sec; k_B, rate constant for Brönsted base, l./mole sec.

no.	catalyst	pk_1 $\mu = 0$	pk_1 $\mu = 0.28$	maximum concentration of catalyst (ion + acid) M	phosphate buffer pH 7·52 (0° C) k	phosphate buffer pH 7·52 (0° C) k_B	veronal buffer pH 8·75 (0° C) k	veronal buffer pH 8·75 (0° C) k_B	carbonate buffer, I pH 9·5 (0° C) k	carbonate buffer, I pH 9·5 (0° C) k_B	carbonate buffer, II pH 10·1 (0° C) k	carbonate buffer, II pH 10·1 (0° C) k_B	remarks
	GROUP A												
	Anions of:												
1	telluric acid, $Te(OH)_6$ [11-13a]	8·1	7·76						4·6	4·7	4·75	4·75	
2	germanic acid, $Ge(OH)_4$ [14-15]	9·0	8·85	0·005	0·572	12·8	8·8	21·2	25·5	31·2			
3	arsenious acid, $As(OH)_3$ [14, 16-18]	9·45	9·35	0·01	0·345	23·7	5·2	25·9	19·0	32·4	26	30	
4	silicic acid $Si(OH)_4$ [19]	10·1	11·5 *	0·005					0·9	91	3·1	81·5	
5	chloral hydrate, $CCl_3CH(OH)_2$ [20-22]	10·4	10·2 *	0·0005					30	180	130	294	
5*a*	chloral alcoholate, $CCl_3—CH(OEt)(OH)$		10·2 *	0·0005					32	192			
6	butyl chloral hydrate, $CH_3—CH(Cl)—C(Cl_2)—CH(OH)_2$	11·4 *	11·2 *	0·001					12·2	620	35	735	
7	glyoxal hydrate, $CH(OH)_2—CH(OH)_2$ or $CH(OH)_2—CHO$	13·2 *	13·0 *	0·0075					1·28	4050	5·1	4050	
8	formaldehyde hydrate: (methylene glycol) $CH_2(OH)_2$ [21, 23-25]	13·9	13·7 *	0·02					0·52	8250	2	7980	
9	acetaldehyde hydrate $CH_3—CH(OH)_2$ [21, 24-26]	14·0	13·8 *	0·03					0·1235	2470	0·615	3070	hydration equilibria considered
10	diacetyl hydrate, $CH_3—C(=O)—C(OH)_2—CH_3$ [26]	14·3 *	14·1*	0·025							0·515	5150	hydration equilibria considered

*estimated

TABLE 1.—CATALYTIC POWER OF VARIOUS SUBSTANCES, TEMP. O°C.

k, rate constant in terms of total concentration of catalyst, 1./mole sec; k_B, rate constant for Brönsted base, 1./mole sec.

no.	catalyst	pk_1 $\mu = 0$	pk_1 $\mu = 0.28$	maximum concentration of catalyst (ion + acid) M	phosphate buffer pH 7·52 (O°C) k	phosphate buffer pH 7·52 (O°C) k_B	veronal buffer pH 8·75 (O°C) k	veronal buffer pH 8·75 (O°C) k_B	carbonate buffer, I pH 9·5 (O°C) k	carbonate buffer, I pH 9·5 (O°C) k_B	carbonate buffer, II pH 10·1 (O°C) k	carbonate buffer, II pH 10·1 (O°C) k_B	remarks
	GROUP B												
	Anions of:												
1	trifluoroethanol CF_3CH_2OH [27]		12·8	0.02							0·26	130	
2	trichloroethanol CCl_3CH_2OH [28]		12·7	0·02							0·4	160	
3	acetophenoxime [29]		11·7	0·01							<0·25	<10	
4	H_2O_2		12·2	0·05	0·03				<0·03				
	GROUP C												
	Anions of:	pk_1											
1	glucose [33]	12·9		0·1							0·04	25	
2	fructose [24]	12·5		0·1							0·12	30	
3	sucrose [34]	13·36		0·05							0·39	710	
4	lactose [34]	12·75		0·05							0·36	160	
	GROUP D		pK_2										
1	sulphite, SO_3^{2-} [5]		7·0	0·005	1·0	1·3	0·64	0·65					
2	hydrogen phosphate, HPO_4^{2-}		7·1	0·15		0·016							
3	selenite* SeO_3^{2-} [5]		8·2	0·005 (5°C)	2·08 3·1	12·0 17·9	2·17	2·78					
4	tellurite, TeO_3^{2-} [30–31]		8·0	0·002	4·4	17·7	8·0	20·8	30·0	31·0			
5	phenylarsonate,[32] $C_6H_5AsO_3^{2-}$		8·7							<0·15			
6	phenolate $C_6H_5O^-$		10·2							<0·25			
7	carbonate, CO_3^{2-}	10·5	10·1							<0·1	<0·1		

†experiments with O·2M phosphate buffer, pH 6·83 (O°C) gave $k = 0·51$ and $k_B = 12·5$

TABLE 2A.—DATA OBTAINED BY ROUGHTON AND BOOTH[3] USING THE BUFFERS OF THE APPROPRIATE SPECIES; TEMP. = 0°C

no.	catalyst	pK		k_B, l./mole sec
1	phosphite	pK_2	6·7	0·013
2	arsenate	pK_2	6·8	0·013
3	cacodylate	pK_1	6·3	0·018
4	phosphate	pK_2	7·1	0·017
5	sulphite		7·0	1·89
6	selenite		8·0	3·57
7	borate	pK_1	9·5	0·315
8	tellurate	pK_1	7·8	1·26
	KIESE AND HASTINGS, 5°C[5]			
9	phosphate	pK_2	7·1	≈0·03
10	pyrophosphate	pK_3	6·7	
		pK_4	9·4	≈0·08-0·3
11	borate	pK_1	9·5	≈1·5

TABLE 2B.—DATA OBTAINED BY KIESE AND HASTINGS;[5] TEMP. 5°C. CALCULATIONS HAVE BEEN DONE ON THE BASIS OF k_u = 0·0036 sec^{-1} AT 5°C

no.	catalyst	pK	phosphate buffer pH = 6·9 k	k_B	phosphate buffer 7·55 k	k_B	borate buffer 9·0 k	k_B	carbonate buffer 10·0 k	k_B
		pK_2	pH=7·0							
1	sulphite	7·0	1·09	2·18	1·8	2·3	2·57	2·57	0·41	0·41
2	selenite	8·2	0·68	11·5	2·78 *	15·1	2·25	2·63	0·515	0·52
		pK_1								
3	ClO^-	7·55	1·35	7·45	5·14	10·3	13·5	13·5	13·5	13·5
4	BrO^-	9·0	21·2	2680	72	2020	360	720	180	190

* agrees well with our data.

The data of Roughton and Booth and of Kiese and Hastings indicate that the catalytic power of borate is very low in view of the pK value of boric acid. This material apparently falls in group A and might be expected to give rise to $B(OH)_2O^-$ ions. However, it is known that in aqueous solutions boric acid gives rise to $B(OH)_4^-$[36] which can accept a proton only with simultaneous expulsion of a water molecule.

In fig. 2 are plotted the results for anions of group D together with some other ions for which the catalytic constants have been evaluated at or close to 0°C. These anions are all apparently of the types in which the residual negative charge is spread over the ion by resonance rather than localized on the oxygen atom from which the hydrogen ion has been removed. All these materials except sulphite, selenite and tellurite, which are discussed below, exhibit much lower catalytic powers than substances in group A with comparable pK values. Bell and Higginson[29] in their extensive experiments on the acid-base catalysis of the hydration of acetaldehyde have

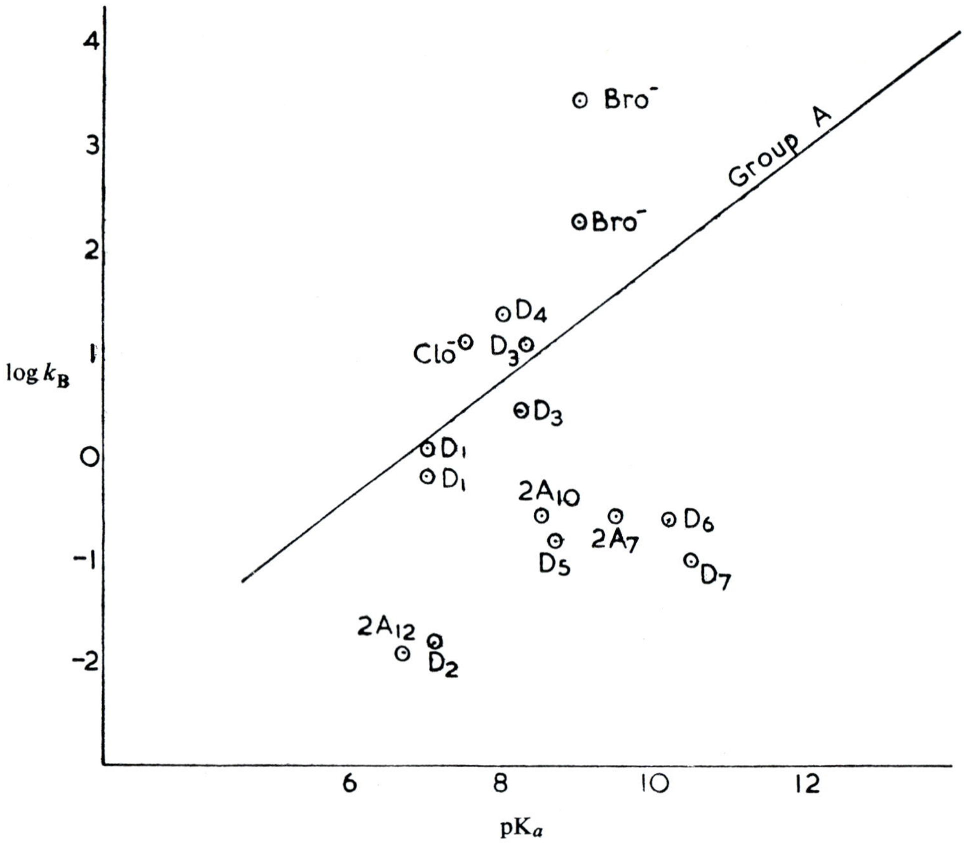

FIG 2.—Catalytic constants of various species at or about 0°C.

observed that charge distribution in the anion plays a very significant role in base catalysis.

Sulphite, $O = S(O^-)_2$, selenite, $O = Se(O^-)_2$, and tellurite, $O = Te(O^-)_2$ have apparently delocalized charges and yet possess catalytic powers comparable to the materials in group A. It appears that there may be a number of special circumstances involved. Both our measurements and those of Kiese and Hastings [5] showed that pH had a pronounced effect on the catalytic activity of sulphite, selenite and tellurite. It has been reported [35] that sulphite has a tendency to acquire a structure of the type $S^+(O^-)_3$ which might account for its comparatively high catalytic power. It is conceivable that selenite ions exist in solution in the hydrated form $(HO)_2Se(O^-)_2$ which would have localized charges. [30a] The extent of hydration may well vary with pH. There is a considerable controversy [30-31,30a] regarding the nature of tellurous acid and it has been suggested that it may well have the structure $Te(OH)_4$ which would give rise to $(HO)_3TeO^-$ and $(HO)_2Te(O^-)_2$.

TABLE 3.—EFFECT OF TEMPERATURE ON CATALYTIC POWER OF ANIONS OF ARSENIOUS AND GERMANIC ACID

pH range 8·5-8·8 ; veronal buffer

no.	temp. °C	$k_{As(OH)_2O^-}$	$k_{Ge(OH)_3O^-}$
1	0	25·9	21·2
2	5	41·4	41·7
3	10	71·5	67·8
4	15	95·5	102

The effect of temperature on the catalytic power of arsenious and germanic acid is recorded in table 3. The data have been correlated by the method of least squares and the following results obtained.

$$As(OH)_2O^-: \log k_B = 9{\cdot}837 + \log T - 2967/T, \ (T \text{ in } °K)$$

$$\Delta H^{\neq} = 13{,}570 \text{ cal/mole}; \ \Delta S^{\neq} = -2{\cdot}2 \text{ cal/deg. mole.}$$

$$Ge(OH)_3O^-: \log k_B = 11{\cdot}617 + \log T - 3470/T,$$

$$\Delta H^{\neq} = 15{,}880 \text{ cal/mole}; \ \Delta S^{\neq} = +6{\cdot}0 \text{ cal/deg. mole.}$$

The relevant data for the hydration and hydroxylation reactions of CO_2 are as follows:

	$\Delta H^{\neq}$	$\Delta S^{\neq}$
(*a*) hydration	14,600	−16·8
(*b*) hydroxylation	12,620	+ 1·76.

It appears, therefore, that the essential effect of the catalyst is to change the entropy of activation favourably. A number of base-catalysed reactions are reported to exhibit this characteristic feature. The above figures also suggest that the hydroxylation reaction may itself be regarded as an example of catalysis of reaction (2) by an anion. The pK of water is normally taken as 16·69 (i.e., 14·94+log 55·5) at 0°C. On the basis of fig. 1 it would appear that if it is regarded as a member of group A the pK_a value corresponding to its catalytic power would be about 12. Bell [37] has pointed out that invariably the catalytic effect of OH^- falls well below that corresponding to the apparent pK value of water and the " true " value appears to be in the range 12-13.

From a practical point of view it seems likely that the most powerful catalysts are to be sought among anions of the type of those in group A which do not form stable carbonates. There are not many inorganic substances which belong to group A and which have reasonable solubilities. The hypobromite ion possesses a particularly high catalytic power but suffers from the serious drawback of instability under industrial conditions in addition to being extremely corrosive. The pH values of the carbonate +bicarbonate solutions encountered in industrial practice range from 8·5 to 10 and therefore materials having their pK values in the range 8-11 would be the most attractive, as they would yield sufficient quantities of ions. Arsenite occupies a key position in this group by virtue of its high catalytic power, favourable ionization constant, stability, reasonably high solubility in both acid and salt form, availability and low cost. Inorganic substances like germanate and tellurite have comparable catalytic power and ionization constant but are relatively unstable and have low solubility. Tellurite, like sulphite and selenite, is highly susceptible to oxidation. Technical and economic considerations rule out most of the materials which we have investigated. It is unlikely that there is an alternative inorganic substance which is as effective a catalyst as arsenite and yet possesses its other attractive features. Most of the organic substances are ruled out because of their instability. However, the use of formaldehyde merits further consideration. In spite of its relatively high volatility and tendency to undergo reaction in alkaline solutions it might in certain circumstances find industrial applications. In an equimolal carbonate+bicarbonate solution at 20°C 1 % by weight of formaldehyde would be expected to increase the total rate of reaction of CO_2 eight-fold, an effect comparable to that obtainable with about 0·5 % by weight of As_2O_3, The effect of formaldehyde on the rate of absorption of carbon dioxide in alkaline solutions has already been noted by Riou and co-workers,[38] Killeffer [39] and Fujikawa.[40] In future work, attention will be directed towards higher specific reaction rates.

We wish to acknowledge valuable discussions with Dr. K.-A. Melkersson, Dr. D. A. Ratkowsky, Mr. R. P. Bell and Dr. R. J. P. Williams.

1 Reisenfeld and Mullowney, *Pet. Ref.*, 1959, **38,** no. 5, 161.
2 Roberts and Danckwerts, *Chem. Eng. Sci.*, in press.
3 Roughton and Booth, *Biochem. J.*, 1938, **32,** 2049.
4 Roughton, *Harvey Lectures*, 1943, **39,** 96.
5 Kiese and Hastings, *J. Biol. Chem.*, 1940, **132,** 267.
6 Roughton and Booth, *Biochem. J.*, 1946, **40,** 309.
7 Dixon, *Manometric Methods* (Cambridge University Press, 1952).
8 van Krevelen and Hoftyzer, *Chim. Ind., XXIeme Congrés Int. de Chim. Ind.*, 1948, p. 168.
9 Pinsent and Roughton, *Trans. Faraday Soc.*, 1951, **47,** 263.
10 Pinsent, Pearson and Roughton, *Trans. Faraday Soc.*, 1956, **52,** 1512.
11 Antikainen, *Suomen kemist. B*, 1955, **28,** 135.
12 Antikainen, *Suomen kemist. B*, 1957, **30,** 201.
13 Earley, *J. Amer. Chem. Soc.*, 1959, **81,** 1295.
13a Ellison *et al.*, *J. Amer. Chem. Soc.*, 1962, **84,** 1820.
14 Lourijsen-Teysseore, *Bull. soc. Chim.*, 1955, 1118.
15 Antikainen, *Suomen kemist. B*, 1957, **30,** 123.
16 Jander and Hofmann, *Z. anorg. Chem.*, 1958, **296,** 134.
17 Antikainen and Rossi, *Suomen kemist. B*, 1959, **32,** 185.
18 Antikainen and Trevanen, *Suomen kemist. B*, 1961, **34,** 3.
19 Greenberg, *J. Amer. Chem. Soc.*, 1958, **80,** 6509, and personal communication.
20 Gustafsson and Johanson, *Acta. Chem. Scand.*, 1948, **2,** 42.
21 Bell and Onwood, paper to be published (personal communication).
22 Federlin, *Bull. Soc. Chim.*, 1953, no. 10, C-11.
23 Walker, *Formaldehyde* (Reinhold Publ. Corpn., N.Y., 2nd ed., 1953).
24 Rumpf, *Bull. Soc. Chim.*, 1946, 546.
25 Bell and McTigue, *J. Chem. Soc.*, 1960, 2983.
26 Bell and McDougall, *Trans. Faraday Soc.*, 1960, **56,** 1281.
27 Henne and Kelly, *J. Amer. Chem. Soc.*, 1952, **74,** 1426.
28 Ballinger and Long, *J. Amer. Chem. Soc.*, 1960, **82,** 795
29 Bell and Higginson, *Proc. Roy. Soc. A*, 1949, **197,** 141.
30 Issa and Awad, *J. Physic. Chem.*, 1954, **58,** 918.

[30a] Brintzinger and Ratanarat, *Z. anorg. Chem.*, 1935, **222**, 317.
[31] Ricketts and Tresselt, *J. Amer. Chem. Soc.*, 1959, **81**, 2305.
[32] Pressman and Brown, *J. Amer. Chem. Soc.*, 1943, **65**, 540.
[33] Thamsen, *Acta Chem. Scand.*, 1956, **10**, 1165.
[34] Smidt and Thamsen, *Acta Chem. Scand.*, 1956, **10**, 1172.
[34a] Bruice and Fife, *J. Amer. Chem. Soc.*, 1962, **84**, 1973.
[35] Sidgwick, *The Chemical Elements and their Compounds* (Oxford University Press, 1950).
[36] Ingri *et al.*, *Acta Chem. Scand.*, 1957, **11**, 1034.
[37] Bell, *Acid-Base Catalysis* (Oxford University Press, 1941); *The Proton in Chemistry* (Methuen, 1961).
[38] Riou *et al.*, *Compt. rend.*, 1927, **184**, 325; 1928, **186**, 1543, 1727.
[39] Killeffer, *Ind. Eng. Chem.*, 1937, **29**, 1293.
[40] Fujikawa, *J. Chem. Soc., Japan, Ind. Chem. Sect.*, 1949, **52**, 39

C2

Fast reactions of CO_2 in alkaline solutions—
(a) Carbonate buffers with arsenite, formaldehyde and hypochlorite as catalysts
(b) Aqueous monoisopropanolamine (1-amino-2-propanol) solutions

M. M. Sharma and P. V. Danckwerts

Department of Chemical Engineering, Pembroke Street, Cambridge

Abstract—(a) The anions of certain weak acids are known to catalyse the reaction between CO_2 and water. The catalytic powers of arsenite, formaldehyde hydrate and hypochlorite ions have been determined under various conditions by measuring rates of absorption of CO_2 into carbonate buffer solutions containing these anions. The energies of activation have been estimated.

(b) Carbon dioxide reacts rapidly with monoisopropanolamine according to second-order kinetics to give the amine carbamate. The rate constant of the reaction has been determined over a range of temperatures. Its magnitude is similar to that for other alkanolamines which have been investigated. The energy of activation is 10,500 cal/g mole.

(a) Carbonate Buffers with Arsenite, Formaldehyde and Hypochlorite as Catalysts

In a previous paper [1] the authors have reported on the catalysis of the hydration of CO_2 by the anions of weak acids, such as arsenite, which is used in the industrial absorption of CO_2. It was found that anions which have a negatively charged oxygen atom and at least one hydroxyl group attached to the same central atom, and in which the negative charge is not delocalized by resonance effects, form a single group which obeys the Brönsted relationship. The work reported here is an extension of the previous work. The transient rates of absorption of CO_2 into carbonate buffers containing arsenite, formaldehyde and hypochlorite as catalysts have been measured, enabling the catalytic constants of the various species to be determined. The measurements were made over the temperature range of 15 to 45°C. In the case of arsenite measurements have also been made up to 60°C using a potassium carbonate solution of concentration comparable to that employed in industrial practice. The information on formaldehyde enables one to assess its possible industrial significance. The data on arsenite solutions may contribute to the rational design of industrial absorption equipment.

The reaction-rate data were derived from rates of absorption of the gas into a carefully controlled cylindrical jet of the solution, as described by NIJSING *et al.* [2], KRAMERS *et al.* [3] and SCRIVEN and PIGFORD [4]. Some work was also carried out on the rates of absorption of CO_2 into carbonate buffers containing arsenite and formaldehyde in a wetted-sphere column as described by DAVIDSON and CULLEN [5].

EXPERIMENTAL

The principal design features of the jet apparatus were akin to those employed by NIJSING *et al.* [2], KRAMERS *et al.* [3] and SCRIVEN and PIGFORD [4]. The diameters of the jets ranged from 0·067 to 0·179 cm (eight in all), and their lengths from 1·48 to 7·07 cm. The jet lengths were measured with a cathetometer. The diameters of the jets were separately measured by a travelling microscope. Under the relevant conditions the variation in jet diameter along the length was of the order of $\pm$ 1 per cent. Most of the jets were formed at the sharply contracted tips of glass tubes. Two of them were obtained by joining capillary tubes of relevant size to a wider bore tube and subsequently grinding down the capillary tube so that a disc-shaped portion less than 1 mm long was left. The receivers were similar to that described by SCRIVEN and PIGFORD (4). With the combination of different jets and liquor rates it was possible to vary the contact time between gas and liquid in the range of 0·004 to 0·06 sec. The whole apparatus was kept in a constant temperature room where the temperature could be maintained within about 2°C at temperatures from 15 to 30°C. The gas was saturated with water at the desired temperature and stored in a balloon at atmospheric pressure. The absorption rates were measured with soap film meters (constructed from 5, 10 and 50 cm^3 burettes). The absorption chamber was frequently purged with CO_2 saturated with water. For experiments at temperatures higher than 30°C the jet apparatus was put into a Perspex chamber (75 × 75 × 16 cm) and hot air at the controlled temperature was circulated through the chamber. It was possible to maintain the desired temperature within $\pm$ 1°C.

Solutions used

All the substances were technical grade and deionized water was used for preparation of solutions. Hypochlorite solution was analysed by the standard iodometric techniques and formaldehyde solution was analysed by the hydroxylamine hydrochloride method.

RESULTS

The hydrodynamics of the various jets were checked by studying the physical absorption of

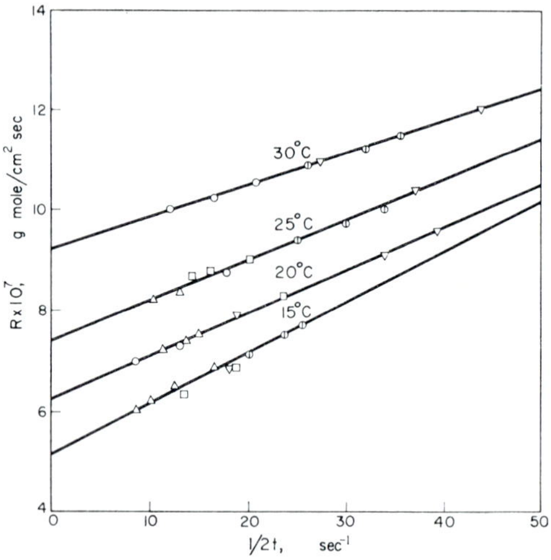

FIG. 1. Rate of absorption of CO_2 in carbonate buffer containing 0·4M arsenite when $kt > 1$ (cf. equation 1).

CO_2 in water at 26·5°C. For a "perfect" jet the average rate of absorption is given by $4C^*\sqrt{(D)}\sqrt{(Ll)}$, where C^* is the solubility of CO_2 in water, D the diffusivity of dissolved CO_2, L the volumetric liquid flow rate and l the length of jet. The value of $C^*\sqrt{(D)}$ found for CO_2 in water was within 3 per cent of the interpolated value, obtained from the data of NIJSING *et al.* [2], of $1{\cdot}48 \times 10^{-7}$ g mole/-$cm^2 sec^{1/2}$. The discrepancy is of the order of magnitude of the correction proposed by SCRIVEN and PIGFORD [4] and BEEK [6] for the effect of non-uniform velocity distribution in the jet immediately after it issues from the nozzle. Some jets were found to exhibit a considerable deviation (>5 per cent) from the ideal behaviour and were, therefore, discarded. Values of $C^*\sqrt{(D)}$ for CO_2 have also been obtained at 37·5 and 45°C. This information is required for interpreting the data on absorption of CO_2 in aqueous monoisopropanolamine solutions at the corresponding temperatures. The values obtained are as follows:

Temperature (°C)	$C^*\sqrt{(D)} \times 10^7$ (g mole/$cm^2 sec^{1/2}$)
37·5	1·25
45	1·15

It may be added that slight imperfections in the jet would have a very much smaller effect on rates of absorption into reacting solutions studied than into water.

It has been shown [7, 8] that the following equations hold for absorption accompanied with first-order reaction:

A $$R = C^* \sqrt{(Dk)} \left(1 + \frac{1}{2kt}\right) \qquad (1)$$

when $kt > 1$

B $$R\sqrt{(t)} = \frac{2}{\sqrt{(\pi)}} C^* \sqrt{(D)} \left(1 + \frac{kt}{3}\right) \qquad (2)$$

when $kt < 1$

where R is the average rate of absorption per unit area of liquid surface, k the total reaction rate constant and t the contact time between gas and liquid. (The last term in equation (2) is incorrectly given as $kt/2$ in Ref. [7].) Thus by working under suitable conditions it is possible to evaluate $C^*\sqrt{(D)}$ and k separately. When kt exceeds 25 then equation (1) may be approximated as

$$R = C^* \sqrt{(Dk)} \qquad (1a)$$

These conditions were fulfilled in the case of absorption in the wetted-sphere column.

The experimental results obtained with a carbonate buffer containing 0·4M arsenite are shown in Figs. 1 and 2. The pertinent data obtained for this and other solutions are reported in Tables 1 to 4. Within the limits specified below, the reaction of dissolved CO_2 in the jet could be regarded as pseudo-first order, with

$$k = k_1 + k_2[C]$$

k_1 being the rate of the reaction due to the buffer employed (estimated on the basis of work of ROBERTS and DANCKWERTS [9] and PINSENT *et al.* [10]), and k_2 the catalytic rate constant based on the total concentration, [C], of catalyst in both ionized and unionized forms.

The rate constant k_B for the anion which is the actual catalyst is given by

$$k_2[C] = k_B[B]$$

where [B], the concentration of the anion, is given by

$$[B] = \frac{[C]}{1 + [H^+]/K_A}$$

where K_A is the stoichiometric dissociation constant of the conjugate acid. Thus in order to evaluate k_B it is necessary to know K_A and the effect on it of ionic strength. The relevant references for the ionization constant of the various species were reported by the authors in a previous paper [1]. The effect of ionic strength on K_A can only be approximated for the solutions employed. In the case of arsenious acid the error in the estimated extent of ionization is unlikely to be greater than 2 per cent

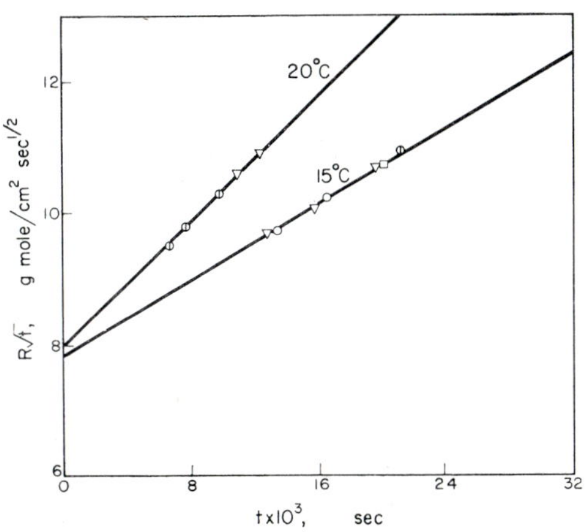

FIG. 2. Rate of absorption of CO_2 in carbonate buffer containing 0·4M arsenite when $kt < 1$ (cf. equation 2).

Table 1. Absorption of CO_2 *in sodium carbonate buffer containing arsenite*

$$\frac{[CO_3^{2-}]}{[HCO_3^-]} = 3;\ Na_2CO_3 = 0{\cdot}6M,\ NaCl = 0{\cdot}1M,$$

$$[\text{Sodium arsenite}] = 0{\cdot}4M,\ I = 2{\cdot}5,\ pH\ (25°C) \approx 10{\cdot}0.$$

Temperature °C	$C^*\sqrt{(D)} \times 10^8$ (g mole/cm² sec$^{1/2}$) equation (1)	equation (2)	k (sec^{-1}) equation (1)	equation (2)	k_1 (sec^{-1})	k_2(C) (sec^{-1})	k_B (l/g mole sec)
15	7·2	6·9	52	55	0·6	51·4	140
20	7·3	7·1	73·5	85	1·2	72·3	198
25	7·7		92		2·2	89·8	246
30	7·7		143		4·0	139	380

as the pH of the solution is considerably higher than the pK. However, in the case of formaldehyde the error in true value of pK_A will have a direct effect on the calculated values of k_B. Hypochlorous acid has a pK of about 7·6 at 20°C and hence at about pH 9·3 almost all of the material exists in the form of ions. The values of k_B reported in Tables 1 to 4 are based on the values of $k_2[C]$ evaluated from equation (1). The rate constants derived from equation (2) are also reported for comparison. It may be added that the nature of equations (1) and (2) suggest that they are best used to yield estimates of $C^*\sqrt{(Dk)}$ and $C^*\sqrt{(D)}$ respectively.

The data on $C^*\sqrt{D}$ reported in Tables 2–5 indicate that, in general, the variation of $C^*\sqrt{D}$ with temperature is not very significant, the decrease in C^* with increasing temperature being in part compensated by the increase in D. Also, the fractional reduction in C^* due to the presence of electrolytes may decrease with increase in temperature. Values

Table 2. Absorption of CO_2 *in potassium carbonate buffers containing arsenite*

$$\frac{[CO_3^{2-}]}{[HCO_3^-]} = 3;\ K_2CO_3 = 1{\cdot}5M,$$

[Potassium arsenite] = 0·2M; I = 5·2 (=5·0 for buffer only)

Temperature (°C)	$C^*\sqrt{(D)} \times 10^8$ (g mole/cm² sec$^{\frac{1}{2}}$) equation (1) *Buffer*	*Buffer with arsenite*	$C^*\sqrt{(Dk)} \times 10^7$† (g mole/cm²sec) *Buffer with arsenite*	k (sec^{-1}) equation (1) *Buffer* k_1	*Buffer with arsenite*	k_2(C) (sec^{-1})	k_B (l/g mole sec)
30		6·2		6‡	132	126	640
40		5·5		15‡	302	287	1460
50	5·27		12·5	36	560	524	2700
60	5·0		14·5	110	840	730	3700

† Under conditions wherein equation (1a) is valid.
‡ Estimated.

Table 3. Absorption of CO_2 *in sodium carbonate buffer containing formaldehyde*

$$\frac{[CO_3^{2-}]}{[HCO_3^-]} = 10;\ Na_2CO_3 = 0{\cdot}5M,\ [HCHO] = 0{\cdot}4M,\ I = 1{\cdot}55$$

Temperature (°C)	$C^*\sqrt{(D)} \times 10^8$ (g mole/cm² sec$^{\frac{1}{2}}$) equation (1)	equation (2)	k (sec^{-1}) equation (1)	equation (2)	k_1 (sec^{-1})	k_2(C) (sec^{-1})	pH − pK	k_B (l/g mole sec)
20	9·4	9·2	34·2	28	6	28·4	2·82	18700
25	9·9	9·55	47·3	58	9	38·3	2·78	23000
30	9·05		74·5		12	62·5	2·72	32800

Table 4. Absorption of CO_2 *in sodium carbonate buffers containing hypochlorite*

$$\frac{[CO_3^{2-}]}{[HCO_3^-]} = 1;\ Na_2CO_3 = 0{\cdot}5M,\ [HClO] = 0{\cdot}05M,\ I = 2{\cdot}05$$

Temperature (°C)	$C^*\sqrt{(D)} \times 10^8$ (g mole/cm² sec$^{\frac{1}{2}}$) equation (1)	equation (2)	k (sec^{-1}) equation (1)	equation (2)	k_B† (l/g mole sec)
20	7·6	7·6	71·5	68	1430
25	7·1		108·5		2170
35	7·1		152		3040

† The contribution of $k_1 < 1$ per cent.

Table 5. Rate constants for arsenite and formaldehyde obtained by different methods

	k_B (l/g mole sec)						
	Arsenite					*Formaldehyde*	
Temperature (°C)	*Veronal buffer* I = 0·09 *Manometric* [1]	*Wetted wall column. Carbonate buffer* [9]	*Wetted sphere column. Carbonate buffer*	*Liquid jet* *Carbonate buffer Sodium*	*Liquid jet* *Potassium*	*Wetted sphere column Carbonate buffer*	*Liquid jet. Carbonate buffer*
		I = 1	I = 2·5	I = 2·5	I = 5·2	I = 2·5	I = 1·55
15	95·5			140			
25	185†	164	220	246		30000	23000
30				380	640		

† Unpublished data of authors.

for the rate constants of arsenite and formaldehyde obtained by different methods under related conditions are compared in Table 5.

In evaluating the reaction rate constant from the wetted sphere column data, $C^*\sqrt{D}$ values reported in Table 1 were employed. Taking into consideration the variation in the ionic strength and pH the agreement among the data found by the various techniques may be considered as satisfactory. Ionic strength and the nature of the ionic species present are known to have a considerable effect on rate constants of such reactions [10].

The values of the rate constant for hypochlorite are 30 times higher than values calculated from the data of Kiese and Hastings [11] for 5°C, using a temperature coefficient based on an activation energy of 10,000 cal/mole.

The energies of activation for two catalysts, namely arsenite and formaldehyde, have been found to be 12,600 (13,000 for potassium carbonate solutions) and 10,200 cal/mole respectively. These figures appear to be in reasonable agreement with our earlier work [1]. The plot of log k_B against $1/T$ for hypochlorite does not give a straight line, perhaps due to some decomposition of hypochlorite at 35°C. However, an approximate value of 10,000 cal/mole may be assigned to the activation energy.

It is thought that otherwise the error in the value of the rate constants reported in this paper is unlikely to be greater than $\pm$ 10 per cent.

The results for formaldehyde suggest that it is unlikely to be useful industrially as, apart from its comparatively high volatility and tendency to undergo Cannizaro's reaction, the proportion of formaldehyde existing in the form of ions in (for instance) partly-carbonated carbonate solutions would be quite low and thus its net catalytic effect would not be great. Unfortunately hypochlorite is also unlikely to be of industrial value because of its poor stability and corrosiveness. The data on arsenite in potassium carbonate solutions suggest that in industrial conditions the rate of absorption is likely to be augmented by a factor of about 10 to 20 as compared to solutions containing potassium carbonate alone.

(b) Aqueous Monoisopropanolamine Solutions

Monoethanolamine and diethanolamine are familiar industrial reagents for the absorption of CO_2. With the growth of the petrochemical industry increasing quantities of propylene and its derivatives have been made available in recent years and monoisopropanolamine (MIPA) may prove to have some technical or economic advantages as an absorbent. The work reported here was undertaken to investigate transient rates of absorption of CO_2 and the kinetics of its reaction in aqueous MIPA solutions as a prelude to further experiments which would be required to assess its industrial possibilities.

The following reactions would be expected to contribute to the overall rate of reaction

(i) $$OH^- + CO_2 \rightarrow HCO_3^- \quad (4)$$
$$HCO_3^- + OH^- \rightarrow CO_3^{2-} + H_2O \text{ (Instantaneous)}$$

(ii) $$CH_3\text{—}\underset{\displaystyle OH}{\underset{|}{CH}}\text{—}CH_2\text{—}NH_2 + CO_2 \rightarrow RNHCOO^- + H^+ \quad (5)$$
$$H^+ + RNH_2 \rightarrow RNH_3^+ \text{ (Instantaneous)}$$

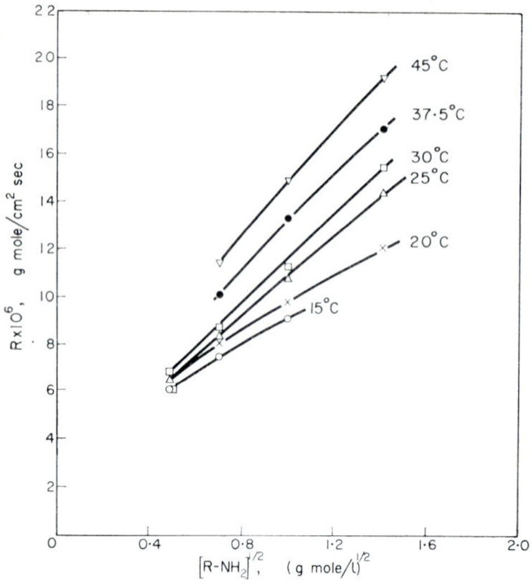

FIG. 3. Rate of absorption of CO_2 in aqueous mono-isopropanolamine solutions.

(iii) CH_3—CH—CH_2—NH_2 + CO_2 → R—CH—CH_2NH_2 (6) [with O^- on the CH of the amine and CO_3 on the CH of the product]

However, the available data [12] indicate that the contribution of steps (i) and (iii) are insignificant. Thus the rate data obtained in this work are those for reaction (ii), the formation of carbamate. The rate of the reaction will presumably be proportional to the concentrations of CO_2 and of amine, as in the case of all amines previously investigated [13].

At sufficiently short contact times the reaction of dissolved CO_2 will display pseudo first-order kinetics and the following equation will hold [12, 14, 15].

$$R = C^* \sqrt{(Dk_P[\text{MIPA}])} \qquad (7)$$

provided $\sqrt{(k_P[\text{MIPA}]\,t)} \ll \left[1 + \frac{[\text{MIPA}]}{2C^*}\right]$ and $k_P[\text{MIPA}]t \gg 1$. These conditions were fulfilled in the experiments reported here.

EXPERIMENTAL

The jet apparatus described in Part (a) was used.

RESULTS

Experimental values of the absorption rate obtained under conditions wherein equation (7) holds are reported in Fig. (3) for six different temperatures and four different concentrations of MIPA absorbing CO_2 at 1 atm pressure. (Temperatures of 37·5 and 45°C are commonly employed in industrial practice). There is some departure from a straight line relationship between R and $\sqrt{([\text{MIPA}])}$, which is probably due mainly to the decrease in the diffusivity with increasing concentrations of MIPA. In order to evaluate the reaction rate constant the solubility, C^*, of CO_2 has been assumed to be the same as in water. The values for the diffusivities of dissolved CO_2 were calculated by assuming the relationship $D\mu/T$ = constant, where μ is the viscosity of the solution in g/cm sec; the relevant values in water are known (see Part a). The viscosities of the solutions were taken to be equal to those of monoethanolamine solution [16] (on wt. per cent basis) for want of data on the viscosity of MIPA solutions.

The reaction rate constants evaluated from the measurements on 1M MIPA solutions are plotted in Fig. (4) to obtain the value of the activation energy, which was found to be 10,500 cal/mole. This is close to the value of 11,600 cal/mole obtained for the reaction of CO_2 with aqueous ammonia [17]. At 18°C the rate constant for monoethanolamine is 3200 l/g mole sec [18] compared to the interpolated value of 4100 for MIPA.

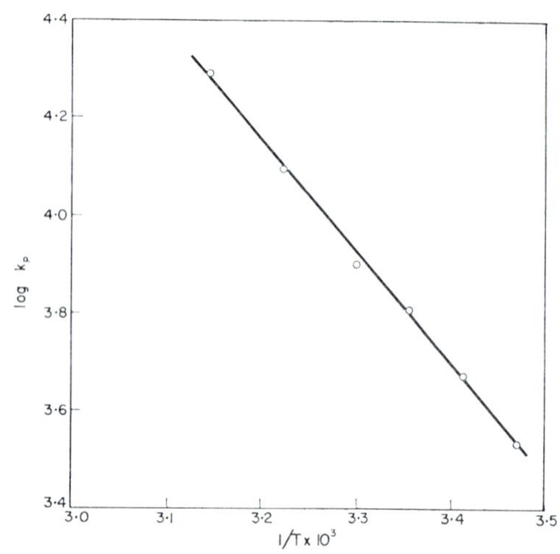

FIG. 4. Arrhenius plot for the rate of reaction between CO_2 and monoisopropanolamine.

NOTATION

C^* Saturated concentration of CO_2 in solution g mole/cm^3
D Diffusivity of dissolved CO_2 cm^2/sec
E Energy of activation cal/mole
I Ionic strength g ion/l.
k Total reaction rate constant sec^{-1}
k_1 Rate constant due to buffer sec^{-1}
k_2 Catalytic rate constant based on total concentration of catalyst l/g mole sec
k_B Catalytic rate constant based on concentration of anion l/g mole sec
k_P Rate constant for reaction between CO_2 and monoisopropanolamine l/g mole sec
K_A Stoichiometric dissociation constant of the acid g ion/l
l Length of liquid jet cm
L Liquid flow rate cm^3/sec
R Average rate of absorption per unit area of interface g mole/cm^{2}sec
t Contact time sec
T Temperature °K
μ Viscosity of solution g/cm sec

Acknowledgement—One of the authors (M.M.S.) wishes to express his gratitude to the Shell Fund for Chemical Engineering of the University of Cambridge, an award from which enabled this work to be carried out.

REFERENCES

[1] SHARMA M. M. and DANCKWERTS P. V., *Trans. Faraday Soc.* 1963 **59**, 386.
[2] NIJSING R. A. T. O., HENDRIKSZ R. H. and KRAMERS H., *Chem. Engng. Sci.* 1959 **10**, 88.
[3] KRAMERS H., BLIND M. P. P. and SNOECK E., *Chem. Engng. Sci.* 1961 **14**, 115.
[4] SCRIVEN L. E. and PIGFORD R. L., *Amer. Inst. Chem. Engrs. J.* 1958 **4**, 439.
[5] DAVIDSON J. F. and CULLEN E. J., *Trans. Inst. Chem. Engrs.* 1957 **35**, 51.
[6] BEEK W. J. Ph.D. Thesis, Delft 1963.
[7] DANCKWERTS P. V., *Trans. Faraday Soc.* 1950 **46**, 300.
[8] DANCKWERTS P. V. and KENNEDY A. M., *Chem. Engng. Sci.* 1958 **8**, 201.
[9] ROBERTS D. and DANCKWERTS P. V., *Chem. Engng. Sci.* 1962 **17**, 961.
[10] PINSENT B. R. W., PEARSON L. and ROUGHTON F. J. W., *Trans. Faraday Soc.* 1956 **52**, 1512.
[11] KIESE M. and HASTINGS A. B., *J. Biol. Chem.* 1940 **132**, 267.
[12] EMMERT R. E. and PIGFORD R. L., *Amer. Inst. Chem. Engrs. J.* 1962 **6** 171, 702.
[13] JENSEN A. and FAURHOLT C., *Acta. Chem. Scand.* 1952 **6**, 385.
[14] ASTARITA G., *R. C. Accad. Napoli* 1960 **54**, 27.
[15] ASTARITA G., *Chem. Engng. Sci.* 1961 **16**, 202.
[16] DOW CHEMICAL CO., Bulletin on monoethanolamine; Union Carbide Co., Bulletin on alkanolamines and morpholines.
[17] PINSENT B. R. W., PEARSON L., ROUGHTON F. J. W., *Trans. Faraday Soc.* 1956 **52**, 1594.
[18] JENSEN M. B., JORGENSEN E. and FAURHOLT C., *Acta. Chem. Scand.* 1954 **8**, 1137.

C3

Absorption of carbonyl sulphide in amines and alkalis

(*Received* 3 *August* 1964)

THE removal of carbonyl sulphide (COS) from mixtures of gases by liquid absorbents is an important industrial operation in several contexts. Refinery and synthetic gases, derived from either petroleum fractions or coal, often contain significant amounts of COS. Recently, a number of new natural gas supplies have been reported to contain more than 500 ppm of COS [1–3]. The manufacture of olefins, notably C_2H_4 and C_3H_6, from petroleum fractions also entails absorption of COS because of the close boiling points of COS and C_3H_6 [4]. It is necessary to remove COS down to a few ppm for several reasons, such as catalyst sensitivity to COS in subsequent operations, statutory regulations regarding the sulphur content of domestic gas supplies, etc. (CO_2 is generally removed simultaneously).

The common absorbents used in practice are aqueous monoethanolamine (MEA), diethanolamine (DEA), di-isopropanolamine (DIPA) and sodium hydroxide solutions, the latter being non-regenerable. The amines can be (at least in part) regenerated by stripping with steam or hot air. It is known, however, that when MEA is used to absorb COS a considerable amount of the amine which reacts with COS is lost in the form of non-regenerable products and it is therefore considered desirable in practice to use DEA [6, 7], which does not form appreciable amounts of non-regenerable products. The use of DEA, however, calls for a considerably larger plant, because of the lower rate of reaction of CO_2 with DEA as compared to MEA, as reported elsewhere. There is no quantitative information available in the published literature on rates of reaction of COS with the above substances. The relevant reactions are:

$$RNH_2 + COS \rightarrow RNHCOS^- + H^+ \quad (1)$$

$$RNH_2 + H^+ \rightarrow RNH_3^+ \text{ (Instantaneous)} \quad (1a)$$

$$COS + 4OH^- \rightarrow CO_3^{=} + S^{=} + 2H_2O \quad (2)$$

There is a close analogy with the reactions of CO_2:

$$RNH_2 + CO_2 \rightarrow RNHCOO^- + H^+ \quad (3)$$

$$RNH_2 + H^+ \rightarrow RNH_3^+$$

$$CO_2 + 2OH^- \rightarrow CO_3^{=} + H_2O. \quad (4)$$

Both COS and CO_2 are linear molecules and because of the similarity in their modes of reaction with amines and alkalis it was thought that there might be a close relationship between the relative rates of reaction of CO_2 and COS with various amines. It also seemed likely that the hydrolysis of COS,

$$COS + H_2O \rightarrow CO_2 + H_2S$$

might be prone to base catalysis similar to that found for the hydration of CO_2 [5]. It will be clear from what follows that the rate of reaction of COS with MEA, DEA, DIPA and OH^- is not more than about 1 per cent of the corresponding rate of reaction of CO_2. The rate of absorption of COS into solutions of these substances is determined by the rate of reaction and is consequently very low.

It is perhaps not generally realized that when the partial pressure of CO_2 in the gas is low (e.g. reduction of the mole fraction of CO_2 from 2×10^{-2} to 5×10^{-6} at a total pressure of about 20 atm), the rate of absorption of the CO_2 is also determined by its rate of reaction with the amine. (The controlling mechanism is simultaneous diffusion of CO_2 or COS and its pseudo first-order reaction with the amine.) Thus it appears worthwhile to investigate the use of new amines which have higher rates of reaction with CO_2 and COS and will thus absorb them more rapidly, and which also possess other characteristics desirable from the industrial point of view. The more fundamental aspects of this work will be communicated in a separate paper.

Experimental

The following types of apparatus were used for measuring the rate data according to circumstances; all have been described elsewhere. (a) Manometric [5] (b) Wetted wall column [8] (c) Stirred cell [9] and (d) Jet apparatus [10]. Over twenty amines and a few typical hydrolysis catalysts were investigated, at a temperature of 25°C.

Results

The intrinsic rate constant for the hydrolysis of COS is $0{\cdot}001\ sec^{-1}$ at 25°C [10] (for CO_2 the corresponding rate constant is $0{\cdot}025\ sec^{-1}$ [11]). The hydrolysis of COS, like that of CO_2, proves to be catalysed by bases. However, the catalysis of hydrolysis of COS is very unlikely to have any practical significance, as even the catalysed reaction is very slow. Thus, in the case of catalysis by arsenite ions the rate constant is only about 0·2 l/g mole sec at 25°C (for CO_2 the corresponding rate constant is about 200 l/g mole sec [12]). The second-order rate constant for the reaction of COS with OH^- is about 12 l/g mole sec at 25°C (for CO_2 the corresponding rate constant is about 10,000 l/g mole sec [11]). The effects on the rate of absorption of COS of varying the concentration of NaOH, and of substituting KOH for NaOH solution, are very similar to those found in the case of CO_2

The rate of reaction of COS with most of the amines investigated is about 1 per cent of the corresponding rate for CO_2 at 25°C. Unfortunately MEA reacts relatively even more slowly with COS, the rate being only about 0·25 per cent of the corresponding rate for CO_2. The values of the second order rate constants for reaction between COS and some of the more important amines are reported in Table 1.

Table 1. *Second-order rate constants for reaction between* COS *and amines. Temperature* = 25°C.

Amine	*Rate constant* (l/g mole sec)
Monoethanolamine	16
Diethanolamine	11
Di-isopropanolamine	6
Morpholine	220
2, 6, Dimethyl morpholine	400
Piperidine	750
Hexamethylene-imine	1800
2-Methylaminoethanol (N-Methyl ethanolamine)	220
2-Ethylaminoethanol	250

From an industrial point of view it is desirable to use an amine which not only exhibits a relatively high reaction rate constant but also possesses a low mol. wt. (so that for a given molar concentration a lower weight per cent of the material is required) and which is non-volatile, readily available and inexpensive. The most promising material appears to be 2-Methylamino ethanol (MAE), $CH_3NHCH_2CH_2OH$ (b.p.160°C; can be readily manufactured by the reaction of methylamine with ethylene oxide and is available from some companies for development purposes). The rate-constant for the reaction of this amine with CO_2 is about 30,000 l/g mole sec at 25°C. Solutions of this amine might absorb COS about five times faster than DEA solutions under similar conditions. In the partial pressure range 0·5 to 10^{-4} atm the rate of absorption of CO_2 in MAE solutions is likely to be more than twice as fast as into MEA solutions of the same concentration. The volatility and viscosity of MAE solutions are comparable with those of corresponding MEA solutions. It appears, therefore, that this amine may have considerable advantages as an absorbent for both CO_2 and COS. However, it will be necessary to examine the various aspects of the regeneration of carbonated amine solution (particularly the amount of non-regenerable by-products formed) before a final recommendation can be made.

Ethylaminoethanol is just as good as MAE but it has a higher molecular weight (89) than the latter (75). Morpholine, piperidine and hexamethylene-imine are probably too volatile; dimethyl morpholine (b.p. 147°C) might be useful under special circumstances.

Acknowledgement—Some of the information in this communication forms a part of the provisional specification for a British Patent (Application No. 30019/64).

One of the authors (M.M.S.) wishes to express his gratitude to the Shell Fund for Chemical Engineering of the University of Cambridge for an award which enabled this work to be carried out.

M. M. Sharma
P. V. Danckwerts

Department of Chemical Engineering
University of Cambridge

References

[1] Negrerie H. *Rev. Inst. franç. Pétrole, Ann. Combus. Liquides.* 1962 **17** 164.
[2] Bailleul M. *J. Inst. Petrol.* 1963 **49** 99.
[3] Berlie E. M. *Hydrocarbon Processing and Pet. Refin.*, 1962 **41** 117.
[4] Evans E. B. (Edited), *Modern Petroleum Technology*, The Inst. of Pet., London, 3rd Ed. (1962) p. 420.
[5] Sharma M. M. and Danckwerts P. V. *Trans. Faraday Soc.* 1963 **59** 386.
[6] Pearce R. L. Arnold J. L. and Hall H. K. *Pet. Refin.* 1961 **40** 8, 121.
[7] Kohl A. L. and Riesenfeld F. C. *Gas Purification*, McGraw-Hill Book Co. (N.Y.) 1960.
[8] Roberts D. and Danckwerts P. V. *Chem. Engng. Sci.* 1962 **17** 961.
[9] Gillham A. J. and Danckwerts P. V. unpublished work.
[10] Thomson H. W. Kearton C. F. and Lamb S. A. *J. Chem. Soc.* 1935 1003.
[11] Pinsent B. R. W. Pearson L. and Roughton F. J. W. *Trans. Faraday Soc.* 1956 **53** 1512.
[12] Sharma M. M. and Danckwerts P. V. *Chem. Engng Sci.* 1963 **18** 729.

C4

Kinetics of the Conversion of Bicarbonate to Carbon Dioxide

By P. V. Danckwerts and K.-A. Melkersson *

Dept. of Chemical Engineering, Cambridge University

Received 7th February, 1962

Himmelblau and Babb [1] added bicarbonate labelled with C-14 to solutions containing unlabelled bicarbonate and CO_2. They measured the rate at which C-14 appeared in the CO_2 and deduced values of the rate-constant for the reaction between CO_2 and OH^-. These values were up to 130 times as high as those found by other workers. A new reaction-mechanism is proposed to account for their experimental results, involving an exchange reaction between carbon-containing compounds.

Himmelblau and Babb [1] (hereafter referred to as HB) have reported experiments in which carbon-14, in the form of bicarbonate ions, was introduced into solutions of sodium bicarbonate containing CO_2. They measured the rate at which carbon-14 appeared in the CO_2, and ascribed the observed rate of the process to the following two reactions:

$$H^+ + H\overset{*}{C}O_3^- \underset{k_{-1}^{\circ}}{\overset{k_1^{\circ}}{\rightleftharpoons}} H_2O + \overset{*}{C}O_2; \tag{1}$$

$$H\overset{*}{C}O_3^- \underset{k_{-2}}{\overset{k_2}{\rightleftharpoons}} OH^- + \overset{*}{C}O_2. \tag{2}$$

* now at Boliden Mining Company, Skelleftehamn, Sweden.

By varying the $[CO_2]/[HCO_3^-]$ ratio in the solutions, they were able to separate the effect of reaction (1) from the supposed effect of reaction (2). The rate-constants for reaction (1) found at various temperatures agreed with those found by other workers. However, they calculated that the rate which had to be ascribed to reaction (2) to account for the observations was some 50-130 times as fast (depending on the temperature) as the actual rate of reaction (2), which has been well-established by a number of workers. In fact, reaction (2) is too slow to contribute appreciably to the observed reaction rate in solutions of the pH (5·5-7·5) used by HB.

Reaction (1) is known to be catalyzed by the anions of weak acids, including the carbonate ion. The rate-constant for the catalyzed reaction is approximately

$$k_1 \approx k_1^\circ(1 + l_A[A]), \tag{3}$$

where [A] is the concentration of the anion, k_1° the forward rate constant in the absence of catalysts and l_A the catalytic multiplication constant characteristic of the anion (at this stage of the argument we shall not distinguish between concentrations and activities). The catalytic effect of the rate-constant of the reverse reaction, k_{-1}°, is, of course, identical, i.e., $k_{-1} \approx k_{-1}^\circ (1 + l_A[A])$.

Roughton and Booth [2] report that carbonate ion in 1 M concentration increases k_{-1}° by a factor lying between 1 and 25 at 0°C, suggesting a value of $l_{CO_3^{2-}}$ lying between 0 and 24 l./mole. Some measurements by Roberts [3] indicated that the value of $l_{CO_3^{2-}}$ at 25°C might be about 22 l./mole.

It might be expected that bicarbonate ions would also catalyze reaction (1), although both the Brönsted rule and a scrutiny of the catalytic effects of other ions seem to suggest that since bicarbonate is a much weaker base than carbonate it will be a correspondingly weaker catalyst. Roughton and Booth [2] report that it is a very weak catalyst. However, the concentration of bicarbonate ions was very much higher than that of carbonate ions in the solutions used by HB, and this would enhance the relative importance of the catalytic effect of the former. We have accordingly analyzed the experimental results reported by HB to see whether catalysis by carbonate and bicarbonate ions might account for their observations.

We have followed much the same method as used by HB, but we do not agree with the way in which they have introduced thermodynamic quantities into their rate equations. For instance, they would write the net forward rate of reaction (1) as

$$\frac{d}{dt}(\overset{*}{C}O_2) = k_1^\circ(H^+)(H\overset{*}{C}O_3^-) - k'_{-1}(\overset{*}{C}O_2)(H_2O), \tag{4}$$

where $k'_{-1} = k_{-1}^\circ/(H_2O)$, k_{-1}° being the pseudo-first order constant. (We use symbols in round brackets to denote activities and in square brackets to denote concentrations.) However, if the concept of the transition-state is accepted (and nothing else would justify the introduction of thermodynamic quantities), the rate equation should be written

$$\frac{d}{dt}[\overset{*}{C}O_2] = \frac{k_1^\circ}{\gamma_1^\ddagger}(H^+)(H\overset{*}{C}O_3^-) - \frac{k'_{-1}}{\gamma_1^\ddagger}(\overset{*}{C}O_2)(H_2O), \tag{5}$$

where $\gamma_1^\ddagger$ is the activity coefficient of the activated complex. With this correction we can use the general line of argument followed by HB to get expressions for the net rate of reaction either for a combination of reactions (1) (uncatalyzed) and (2), as did HB, or for reaction (1) catalyzed by carbonate and bicarbonate ions. The solutions used by HB were virtually at chemical equilibrium throughout the reaction (the amount of labelled bicarbonate added being much less than that of the unlabelled species). Since the net rate of interconversion of the various unlabelled species was virtually zero during the exchange reactions, it follows from eqn. (5) that

$$\frac{(CO_2)(H_2O)}{(H^+)(HCO_3^-)} = \frac{k_1^\circ}{k'_{-1}}, \tag{6}$$

if the dependence of the rate-constants on isotopic constitution is neglected.

Combining this with eqn. (5) and rearranging, we get

$$\frac{d\,[\overset{*}{C}O_2]}{dt\,[CO_2]} = k'_{-1}\frac{\gamma_{CO_2}}{\gamma_1^\ddagger}(H_2O)\left\{\frac{[H\overset{*}{C}O_3^-]}{[HCO_3^-]} - \frac{[\overset{*}{C}O_2]}{[CO_2]}\right\}, \tag{7}$$

because $[CO_2]$ can be regarded as virtually constant, and the activity coefficients of labelled and unlabelled compounds will be almost equal.

The contribution of reaction (2) to the rate is

$$\frac{d}{dt}[\overset{*}{C}O_2] = \frac{k_2}{\gamma_2^\ddagger}(H\overset{*}{C}O_3^-) - \frac{k_{-2}}{\gamma_2^\ddagger}(OH^-)(\overset{*}{C}O_2). \tag{8}$$

In view of the equilibrium among the unlabelled species we can put

$$\frac{k_2}{k_{-2}} = \frac{(OH^-)(CO_2)}{(HCO_3^-)} = \frac{K_W}{K_1}, \tag{9}$$

where $K_1 = (HCO_3^-)(H^+)/(CO_2)$ is the apparent first dissociation constant of " carbonic acid ", as defined by Harned and Owen.[5] Substituting eqn. (9) into eqn. (8) and rearranging we find

$$\frac{d\,[\overset{*}{C}O_2]}{dt\,[CO_2]} = k_{-2}\frac{K_W}{K_1}\frac{\gamma_{CO_2}}{\gamma_2^\ddagger}\frac{(HCO_3^-)}{(CO_2)}\left\{\frac{[H\overset{*}{C}O_3]}{[HCO_3^-]} - \frac{[\overset{*}{C}O_2]}{[CO_2]}\right\}. \tag{10}$$

Adding eqn. (7) and (10), and transforming further according to the method of HB, we get

$$-\frac{R}{R^*}\frac{d}{dt}\ln(R_\infty - R) = \gamma_{CO_2}\left\{\frac{k'_{-1}}{\gamma_1^\ddagger}(H_2O) + \frac{k_{-2}}{\gamma_2^\ddagger}\frac{K_W}{K_1}\frac{(HCO_3^-)}{(CO_2)}\right\}, \tag{11}$$

where R = ratio of C-14 to C-12 in gas at time t, R_∞ = ratio at isotopic equilibrium, R^* = ratio of C-14 bicarbonate injected to C-12 bicarbonate.

We shall, like HB, ignore the small changes in (H_2O) from one solution to another, and write $k'_{-1}(H_2O) = k^\circ_{-1}$. If the expression on the left-hand side of eqn. (11) is denoted by S, it then becomes

$$S = k^\circ_{-1}\frac{\gamma_{CO_2}}{\gamma_1^\ddagger} + k_{-2}\frac{K_W}{K_1}\frac{\gamma_{HCO_3^-}}{\gamma_2^\ddagger}\frac{[HCO_3^-]}{[CO_2]}. \tag{12}$$

This is the combined rate-equation for uncatalyzed reaction (1) + reaction (2). It will be seen that each term of the rate-equation contains a thermodynamic factor which is a ratio of activity coefficients. Since precise values cannot be assigned to the activity coefficients of the activated complexes (tacitly assumed equal to unity by HB) there is no point in trying to assign precise values to the others. As a rough approximation we can say that the activity coefficient of an ionic species is uniquely determined by its charge. The thermodynamic factors in eqn. (12) would then be equal to unity. (In fact, these factors will vary somewhat with the composition and ionic strength of the solution.)

The rate equation for reaction (1) catalyzed by carbonate ion is

$$\frac{d}{dt}[CO_2^*] = \frac{l_{CO_3^{2-}}(CO_3^{2-})}{\gamma_3^\ddagger}\{k_1^\circ(H^+)(H\overset{*}{C}O_3^-) - k_{-1}^\circ(CO_2^*)\}, \tag{13}$$

and that for the reaction catalyzed by bicarbonate ion,

$$\frac{d}{dt}[CO_2^*] = \frac{l_{HCO_3^-}(HCO_3^-)}{\gamma_4^\ddagger}\{k_1^\circ(H^+)(H\overset{*}{C}O_3^-) - k_{-1}^\circ(CO_2^*)\}. \tag{14}$$

The combined rate equations for reaction (1), catalyzed and uncatalyzed, and reaction (2) can be written:

$$S = -\frac{R_\infty}{R^*}\frac{d}{dt}\ln(R_\infty - R) = k_{-2}\frac{K_W}{K_1}\frac{\gamma_{HCO_3^-}}{\gamma_2^\ddagger}\frac{[HCO_3^-]}{[CO_2]} + k_{-1}^\circ\gamma_{CO_2}\left\{\frac{1}{\gamma_1^\ddagger} + \frac{l_{CO_3^{2-}}(CO_3^{2-})}{\gamma_3^\ddagger} + \frac{l_{HCO_3^-}(HCO_3^-)}{\gamma_4^\ddagger}\right\}. \tag{15}$$

Since the solutions were virtually at chemical equilibrium throughout the reaction we can put

$$(CO_3^{2-}) = K_2(HCO_3^-)^2/K_1(CO_2), \tag{16}$$

where

$$K_2 = (H^+)(CO_3^{2-})/(HCO_3^-).$$

Eqn. (15) then becomes

$$S = k_{-2}\frac{K_W}{K_1}\frac{\gamma_{HCO_3^-}}{\gamma_2^\ddagger}\frac{[HCO_3^-]}{[CO_2]} + k_{-1}^\circ\frac{\gamma_{CO_2}}{\gamma_1^\ddagger} + k_{-1}^\circ\, l_{CO_3^{2-}}\frac{K_2}{K_1}\frac{\gamma_{HCO_3^-}^2}{\gamma_3^\ddagger} \times \frac{[HCO_3^-]^2}{[CO_2]} + k_{-1}^\circ\, l_{HCO_3^-}\frac{\gamma_{CO_2}\gamma_{HCO_3^-}}{\gamma_4^\ddagger}[HCO_3^-]. \tag{17}$$

Here again, the thermodynamic factors can be taken as approximately equal to unity.

If, like HB, we ignore the catalysis of reaction (1), eqn. (12) leads us to expect that a plot of S against $[HCO_3^-]/[CO_2]$ would give a straight line of slope approximately $k_{-2}K_W/K_1$ and intercept k_{-1}°. The results of HB at 0°, 10° and 20° are plotted in this way in fig. 1. The plots of the results at each temperature approximate fairly closely to straight lines. The values deduced for k_{-1}° are: 0°, 0·0020; 10°, 0·0066; 20°, 0·0151 sec^{-1}, in reasonable agreement with values obtained by other workers, such as Pinsent *et al.*[4] However, the slopes of the lines drawn through the experimental points give the following ostensible values for k_{-2}: 0°, $0{\cdot}79\times10^5$ (72); 10°, $1{\cdot}54\times10^5$ (68); 20°, $1{\cdot}71\times10^5$ (29) l./mole sec. (The ostensible values for k_{-2} are different from those calculated by HB because of our different treatment

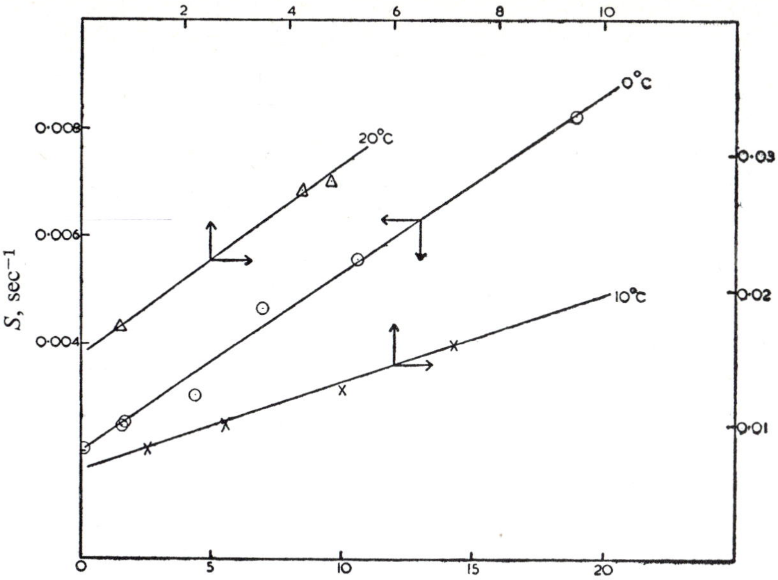

$[HCO_3^-]/[CO_2]$

FIG. 1.

of the thermodynamic factors. Values for K_1, K_2 and K_W were taken from Harned and Owen,[5] and refer to infinite dilution.) The figures in brackets are the ratios of the ostensible values to those found by Pinsent *et al.*[4] The discrepancy is far greater than can be accounted for by ionic strength effects. It may also be seen that the temperature-dependence of the ostensible values of k_{-2} is suspiciously unusual. Reactions (1) and (2) alone, therefore, do not account for the observed results, and there is no doubt that some important reaction path has been neglected in arriving at eqn. (12).

If the values of k_{-2} found by Pinsent *et al.*[4] and the value of $l_{CO_3^{2-}}$ estimated above from the measurements of Roberts are inserted into eqn. (17), it appears that neither the first term on the right-hand side of eqn. (17) (representing reaction (2)) nor the third (representing the catalysis of reaction (1) by carbonate ions) amount to more than $2\frac{1}{2}$ % of the second term (which represents uncatalyzed reaction (1)). We may, therefore, write for eqn. (17),

$$S \approx k^{\circ}_{-1}\{1 + l_{HCO_3^-}[HCO_3^-]\}. \tag{18}$$

A plot of S against $[HCO_3^-]$ at one temperature should, if no other reactions occur, give a straight line of slope $k^{\circ}_{-1}l_{HCO_3^-}$ and intercept k°_{-1}. The results of HB are plotted in this way in fig. 2, together with the best lines drawn through the points by the least-squares method. The values of $l_{HCO_3^-}$ deduced from the slopes are: 5·7, 7·7 and 8·6 l./mole at 0, 10 and 20°C respectively. These values are much too large to be compatible with the experimental findings of Roughton and Booth,[2] and the deviation of some of the individual points from the lines is much greater than the probable experimental error assessed by HB, and cannot be explained by ionic

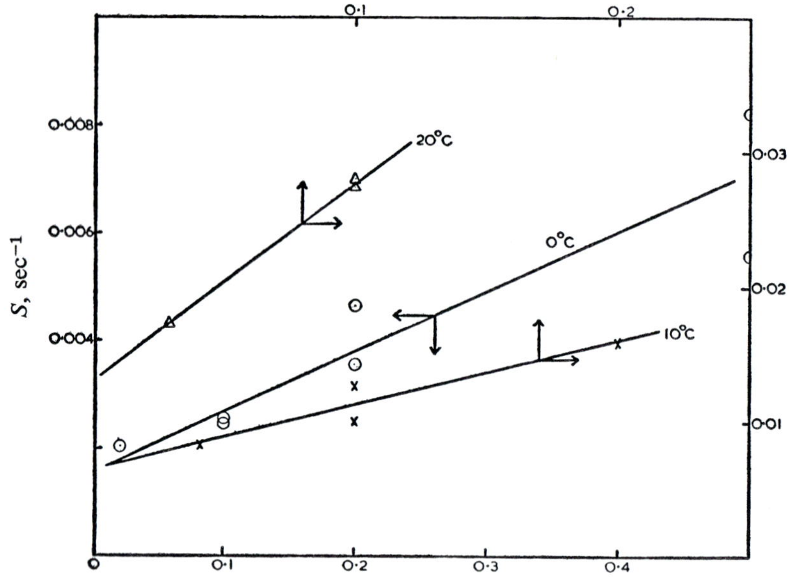

$[HCO_3^-]$, g ion/l.

FIG. 2.

strength or specific ion effects (the two points for 0°C with $[HCO_3^-] = 0{\cdot}5$ g ion/l. refer to solutions with virtually the same ionic composition, as do the two with $[HCO_3^-] = 0{\cdot}2$). It does not seem, therefore, that the catalysis of eqn. (1) by bicarbonate ions fully accounts for the observations.

In case the catalytic power of carbonate ions should in fact be much greater than the estimate given above, which has been made from rather sketchy experimental evidence, we have replotted the results in the form $(S-k_1^\circ)/[HCO_3^-]$ against $[HCO_3^-]/[CO_2]$. As can be seen from eqn. (17), this would give a straight line of slope $l_{CO_3^{2-}}\ K_2/K_1$. In fact, the points on the plot gave no sign of conforming to a linear relationship, showing that catalysis by carbonate and bicarbonate ions cannot account for the results.

One is forced to seek another reaction-path which will give rise to a rate-equation similar to eqn. (12), but which will involve an exchange reaction between two carbon-containing molecules, which would not be observable except by the isotopic labelling technique. Such a reaction path is shown in fig. 3.

$$
\overbrace{}^{A}
$$

$$
B\left\{
\begin{array}{lcl}
\overset{*}{C}O_2 . OH^- & \underset{k_{-a}}{\overset{k_a}{\rightleftharpoons}} & H\overset{*}{C}O_3^- \\
+\ CO_2 & & \\
k_b \downarrow\uparrow k_b & & \\
CO_2^* + & & \\
CO_2 . OH^- & \underset{k_{-a}}{\overset{k_a}{\rightleftharpoons}} & HCO_3^-
\end{array}
\right.
$$

FIG. 3.

The C-12 compounds are all at equilibrium in the buffer-solution. We have assumed the existence of a species written $\overset{*}{C}O_2 . OH^-$. It is further assumed that this species can react with CO_2, with the exchange of the OH^- group (reaction B), and that this process is much faster than reaction A (i.e., $k_b(CO_2) \gg k_a$), so that the species participating in it are constantly at equilibrium. Thus

$$(CO_2^* . OH^-)(CO_2) = (\overset{*}{C}O_2)(CO_2 . OH^-). \tag{19}$$

It can easily be shown that this reaction mechanism leads to the rate-equation

$$\frac{d}{dt}\frac{[CO_2^*]}{[CO_2]} = k_{-a}\frac{\gamma_{HCO_3^-}}{\gamma_a^+}\frac{[HCO_3^-]}{[CO_2]}\left\{\frac{[H\overset{*}{C}O_3]}{[HCO_3^-]} - \frac{[CO_2^*]}{[CO_2]}\right\}. \tag{20}$$

Comparison with eqn. (10) and (12) shows that the quantity there denoted by $k_{-2}K_W/K_1$ can be identified with k_{-a} (if the thermodynamic factors are put equal to unity) and therefore that the slopes of the lines in fig. (1) are equal to the values of k_{-a}, namely, at 0°, $0{\cdot}33 \times 10^{-3}$; 10°, $1{\cdot}3 \times 10^{-3}$; 20°, $2{\cdot}8 \times 10^{-3}$ sec^{-1}. The temperature dependence of k_{-a} is reasonable and indicates an activation energy of about 17,000 cal/mole.

We have also considered the following reaction schemes:

$$H\overset{*}{C}O_3^- + HCO_3^- \rightleftharpoons \overset{*}{C}O_2 + CO_3^{2-} + H_2O; \tag{21}$$

$$H\overset{*}{C}O_3^- + CO_2 \rightleftharpoons CO_2^* + HCO_3^-. \tag{22}$$

However, eqn. (21) leads to a rate-equation indistinguishable from that for catalysis of reaction (1) by carbonate ions, and eqn. (22) to a rate-equation indistinguishable from that for catalysis by bicarbonate ions, so neither is in accord with the experimental results.

Koefoed and Engel,[6] in order to explain some kinetic measurements in CO_2+ bicarbonate solutions, have recently proposed that in addition to reactions (1) and (2) the following reaction may under certain circumstances contribute appreciably to the overall rate:

$$5HCO_3^- + 6H^+ \rightleftharpoons H_{11}C_5O_{15}^+ \overset{X}{\rightleftharpoons} CO_2 + H_{11}C_4O_{13}^+ \rightleftharpoons CO_2 + H_2O + 4HCO_3^- + 5H^+. \tag{23}$$

The rate-determining step is the reaction labelled X. The others are assumed to proceed instantaneously to equilibrium.

It can be shown by methods similar to those used above that the contribution of eqn. (23) to the observed rate in the experiments of HB would be

$$S \approx k_{-5}K_1^5[CO_2]^5/[HCO_3^-], \tag{24}$$

where k_{-5} is the overall rate-constant for reaction (23) proceeding from right to left. Koefoed and Engel estimated the value of k_{-5} (k_F in their nomenclature) to be $1{\cdot}7 \times 10^{33}$ (g ion, l. sec) at 0°C. This means that the contribution of reaction (23) to the observed rate at 0°C would have been negligibly small except in the most acid of the solutions used by HB. With $[HCO_3^-] = 0{\cdot}01$ g ion/l., $[CO_2]/[HCO_3^-] = 9{\cdot}73$, the contribution to S is calculated from eqn. (24) to be $0{\cdot}0019\ sec^{-1}$. This is almost equal to the observed total value of S under these conditions, namely, $0{\cdot}0020\ sec^{-1}$, which is already completely accounted for by reaction (1). This throws doubt on the conclusions of Koefoed and Engel, which would not, in any case, serve qualitatively to explain the results of HB.

It therefore seems reasonable to suggest that the results of HB are to be accounted for by the reaction scheme shown in fig. 3. The species $CO_2 . OH^-$ may be a normal intermediate in reaction (2):

$$HCO_3^- \rightleftharpoons CO_2 . OH^- \rightleftharpoons CO_2 + OH^-.$$

The species is stable in the sense that it is separated by an activation barrier from both $(CO_2 + OH^-)$ and HCO_3^-. It may comprise a linear arrangement of the carbon and oxygen atoms: O . . . C . . . O . . . O—H.

[1] Himmelblau and Babb, *Amer. Inst. Chem. Eng. J.*, 1958, **4**, 143.

[2] Roughton and Booth, *Biochem. J.*, 1938, **32**, 2049.

[3] Roberts, D., *Thesis* (Imperial College of Science and Technology, London, 1961), p. 81.

[4] Pinsent, Pearson and Roughton, *Trans. Faraday Soc.*, 1956, **52**, 1512.

[5] Harned and Owen, *The Physical Chemistry of Electrolytic Solutions* (Reinhold Publishing Corp., New York, 3rd edn., 1958), pp. 754, 755, 760.

[6] Koefoed and Engel, *Acta Chem. Scand.*, 1961, **15**, 1319.

C5

THE REACTION OF CO_2 WITH ETHANOLAMINES

P. V. DANCKWERTS

Department of Chemical Engineering, Pembroke St., Cambridge CB2 3RA, England

(*Accepted* 14 *November* 1978)

Abstract—Measurements of the rates of homogeneous reaction of mono-, di- and tri-ethanolamine by various workers are critically compared. Some discrepancies remain unexplained but it seems probable that a zwitterion is the intermediate in the formation of carbamate and that the reaction of DEA (but not of MEA) is catalysed by bases.

The absorption of CO_2 by solutions of alkanolamines is an operation of such industrial importance that the kinetics of the reactions between CO_2 and these amines are of considerable interest. I shall confine myself to aqueous solutions of monoethanolamine (MEA) and diethanolamine (DEA), with a glance at triethanolamine (TEA), although other ethanolamines and other solvents are sometimes used. Only the forward reaction will be considered here, although the reverse reaction is equally important.

In the recent chemical engineering literature it has been usual to write the reaction as

$$CO_2 + R_2NH \rightarrow R_2NCOOH \qquad (1)$$

$$R_2NCOOH + R_2NH \rightarrow R_2NCOO^- + R_2NH_2^+ \qquad (2)$$

the first step being bimolecular, second-order and rate determining, while the second step was supposed to take place instantaneously. The carbamic acid can be regarded as fairly strong, with $pK_a < 7$, and the zwitterion $R_2\overset{+}{N}HCOO^-$ is insufficiently stable to be formed in appreciable concentration; these suppositions are borne out in the case of $(H_2NCOO^- + NH_4^+)$ by the electrical conductivity measurements of Faurholt[1] which showed that the equivalent conductivity of ammonium carbamate is not much less than that of ammonium chloride.

However, recent rapid-mixing investigations by Hikita *et al.*[2] show that while the reaction of CO_2 with MEA is first-order, that with DEA is second-order in the concentration of amine (in the amine concentration ranges 0.0152–0.177 and 0.174–0.719 M respectively, between temperatures of about 6°C and 40°C).

Reference to earlier publications shows that the kinetics of the reactions between CO_2 and amines in general and MEA and DEA in particular are more complicated than had previously been assumed by chemical engineers. The results to be considered are:

(a) *Jensen et al.*[3]—MEA *and* DEA. The rates were determined by the "competitive" method; 10% CO_2 in air was bubbled through solutions containing amine and NaOH (each at 0.1 and 0.2 M) at 18°C. The solutions were then analysed by adding $BaCl_2$ and filtering-off the barium carbonate precipitated; the carbamate (and any alkyl carbonate formed from the hydroxyl groups of the amines—assumed in [3] to be of negligible amount) remain in solution. The reaction and analysis have to be performed quickly because the carbamate is subject to hydrolysis. This is more marked in the case of DEA, and a calculated correction had to be applied with consequent doubts as to their accuracy. Depletion of the reagents by preferential reaction near the interface was probably negligible and had been the subject of some investigations by Jensen *et al.*[4]. The initial concentrations of amine and NaOH, the amount of CO_2 absorbed (mol/l) and the percentage of the CO_2 absorbed which was not precipitated as carbonate were recorded. The rate of formation of carbamate was assumed to be $k_{am}[CO_2]$ [Am] and k_{am} was calculated from

$$\frac{\%\ \text{non-carbonate}}{\%\ \text{carbonate}} = \frac{k_{am}[\text{Am}]}{k_{OH}[OH^-]}. \tag{3}$$

The results have been recalculated using the value of k_{OH} given by Pinsent *et al.*[5]

$$\log_{10} k_{OH} = 13.65 - 2895/T. \tag{4}$$

(T°K) and the mean values of the concentrations of amine and OH^- between the beginning and end of the reaction calculated by means of the stoichiometric relations

$$CO_2 + \text{Am} + OH^- \rightarrow \text{AmCOO}^- + H_2O \tag{5}$$

$$CO_2 + 2OH^- \rightarrow CO_3^{2-} + H_2O. \tag{6}$$

The resulting value of k_{am} for MEA at 18°C is $4065 \pm 5\%$ l/mols (based on 3 experiments with $[OH^-]$ and [MEA] both initially at 0.1 and 0.2 M), compared with 3934 interpolated from the results of Hikita *et al.*[2]. There is possibly a slight dependence of k_{am} on $[OH^-]$, but doubling the latter only increased the former by 7–8%. The results for DEA were discarded in favour of those below.

(b) *Jørgenson*[6]—DEA. Similar experiments at 0°C, carried out more rapidly, thus reducing the error due to hydrolysis. The concentrations of amine and OH^- were permutated at the levels 0.1, 0.2 and 0.3 M. Jørgenson interpreted the results as follows:

(i) a second-order reaction of CO_2 with DEA to form carbamate;

(ii) a reaction system

$$C-OH + OH^- \rightleftharpoons C-O^- + H_2O \tag{7}$$

$$C-O^- + CO_2 \rightarrow C-OCO_2^- \tag{8}$$

where (8) is the rate-determining step. Thus (if the fraction of the –OH groups ionised were small) the rate of formation of alkyl carbonate would be proportional to $[CO_2][DEA][OH^-]$, giving

$$\frac{\%\ \text{non-carbonate}}{\%\ \text{carbonate}} = \frac{[\text{Am}](a_0 + a_1[OH^-])}{k_{OH}[OH^-]}. \tag{9}$$

However, a glance at the results indicates the presence of a second-order term in [DEA]. Linear regression analysis along the lines of (9) gives a fit with a coefficient of confirmation 0.86; inclusion of a term a_2[DEA] gives

$$R = [CO_2][DEA](692 + 3380[OH^-] + 1056[DEA]) \tag{10}$$

with a coefficient of confirmation 0.93 (R being the rate of reaction of amine, mol/l. s).

Jørgensen also carried out a set of experiments in which CO_2 reacted with NaOH solutions containing diethylamine (which forms no alkyl carbonate), TEA (which forms no carbamate) and a mixture of NaOH and the two amines. The % non-carbonate observed in the last experiment did not agree well with that predicted from the results of the first two, no doubt because of a misinterpretation of the kinetics as suggested above.

Comparing (10) with an extrapolation of Hikita's results (in which the concentration of OH^- formed by the hydrolysis of DEA was small enough for the second term in (10) to be ignored) to 0°C by

$$\log_{10}(R/[CO_2][DEA]^2) = 12.41 - 2775/T \quad (11)$$

we find $R/[CO_2][Am]s^{-1}$

[DEA] = 0.174 M	Hikita	31
	(10)	876
[DEA] = 0.719 M	Hikita	126
	(10)	1451.

There is a gross discrepancy in the rates and (10) is far from leading to values which are second-order in [DEA], because of the dominance of the term 692. Even if this term were ignored, (10) would yield a third-order rate constant $R/[CO_2][DEA]^2$ with a value of 1056 l^2/mol^2s, comapred to the value 176 from (11). This 6-fold discrepancy remains inexplicable.

DISCUSSION

Caplow[7] investigated the rates of reaction of CO_2 with a number of amines (not alkanolamines) with CO_2 at varying pH, using a sophisticated (but apparently not very accurate) version of the "competitive" method. He fitted the kinetics to the expression

$$R/[CO_2][Am] = a_0 + a_1[OH]. \quad (12)$$

The reaction was definitely catalysed by OH^- but he was unable to determine whether or not there was a third term, $a_2[Am]$, indicating catalysis by the amine. He considered the mechnism to be the formation of a zwitterion followed by the removal of a proton by a base. B:

$$CO_2 + R_2NH \underset{k_{-1}}{\overset{k_1}{\rightleftharpoons}} R_2\overset{+}{N}HCO_2^- \quad (13)$$

$$R_2\overset{+}{N}HCO_2^- + B \xrightarrow{k_B} R_2NCO_2^- + BH^+ \quad (14)$$

the second step being rate-determining. The base may be water ($\rightarrow H_2O^+$), leading to the term a_0 or OH^- ($\rightarrow H_2O$), leading to the term a_1. One might expect a third term $a_2[Am]$ corresponding to specific catalysis by the amine ($\rightarrow AmH^+$).

There is a close analogy between the formation of carbamates by CO_2 and the formation of substituted ureas by cyanic acid:

$$OCNH + R_2NH \rightarrow OC\begin{matrix}\diagup NR_2\\ \diagdown NH_2\end{matrix}. \quad (15)$$

Jensen[8] points out that the rate constants (for an assumed simple second-order reaction) for CO_2 are in general about 12 times those for cyanic acid. Williams and Jencks[9] have investigated reaction (15) thoroughly. They show that the reaction of weakly-basic amines is subject to general acid-base catalysis, with a term such as $a_2[Am]$, but that strong bases (like MEA) react according to simple second-order kinetics. The results are consistent with a mechanism analogous to (13) and (14) above, but it is concluded that in the case of strong bases the proton-removal step (14) is so fast that the forward rate constant, k_1, in (13) becomes controlling.

In general, then, it seems reasonable to suppose that the mechanism of formation of carbamates is as shown in (13) and (14). If [Z] is the concentration of zwitterion (at quasi-steady state) then

$$R = k_1[CO_2][Am] - k_1[Z] = [Z]\Sigma k_B[B] \quad (17)$$

the term $\Sigma k_B[B]$ indicating the contribution of the various bases present to the rate of removal of protons. Thus

$$\frac{R}{[CO_2][Am]} = \frac{k_1}{1 + \dfrac{k_{-1}}{\Sigma k_B[B]}}. \quad (17)$$

If the second term in the denominator is $\ll 1$ we have simple second-order kinetics, as in the case of MEA:

$$R = k_1[CO_2][Am]. \quad (18)$$

Almost every zwitterion is deprotonated before it can revert to CO_2 + amine, and the rate-controlling step is the formation of the zwitterion which is not subject to catalysis. Conversely, if the second term is $\gg 1$ we get

$$R/[CO_2][Am] = \frac{k_1}{k_{-1}}\Sigma k_B[B]. \quad (19)$$

This reaction scheme seems to fit all the phenomena. Trial shows that Jørgensen's results[6] with DEA can only be fitted to (17) if the second term in the denominator dominates, leading to (10). The term 692 could be attributed to deprotonation by water.

If we consider solely the results of Hikita *et al.* with MEA and DEA we can explain them by assuming that k_{-1} is very much larger for DEA than for MEA—that is, the zwitterion is much less stable. This could plausibly be attributed to seteric effects; the $-C_2H_2OH$ chains may be enlarged by hydrogen-bonding to water molecules (there does not seem to be a similar difference in behaviour between mono-, and diethylamine). The difference in stability of the zwitterions may be parallel by the difference in stability of the carbamates[10]: The equilibrium constant

$$\frac{[Am][CO_2]}{[H^+][AmCO_2^-]}$$

is (at 20°C) 5.8×10^4 for MEA and 3.8×10^5 for DEA. Steric effects may also lead to k_B being smaller for DEA than for MEA, because it is more difficult for the nitrogen atom to reach and remove the proton. Moreover, the lower basic strength (by a factor of 0.26) of DEA is likely to have a "Brønsted" effect on the magnitude of k_B, which may be expected to vary as a fractional power of this ratio.

Thus it is assumed that catalysis by the amine dominates in the case of DEA and that (19) becomes

$$R = \frac{k_1 k_B}{k_{-1}} [CO_2][DEA]^2$$

when CO_2 reacts with a solution containing only DEA. Possibly, if the concentration of DEA were much greater than 0.7 M (concentrations up to 3 M may be used in practice) the second term in the denominator of (17) might become comparable with unity, or even less, and the order of reaction with respect to DEA might become less than 2 (similarly at concentrations of MEA less than 0.015 M the reaction might tend to a higher order than unity in [MEA]).

It may be supposed that the term 3380 $[OH^-]$ in (10) is at least in part due to the catalysis of the formation of carbamate by OH^- and not solely to the formation of alkyl carbonate as assumed by Jørgensen[6]. A more important consideration, from the industrial point of view, is that MEA and DEA are often mixed with potassium carbonate solution (up to 3 M original carbonate) as absorbents for CO_2. Carbonate ion, which is a stronger base than MEA or DEA (pK_a at 20°C 10.38, 9.81 and 9.22 respectively) and which is less liable to steric hindrance, might be expected to catalyse the carbamatation reaction in the case of DEA. It would be of great interest to find out whether this is the case as DEA is though to enhance the rate of absorption by carbonate solutions more than can be accounted for by an additive effect. The analysis of rate measurements would be complicated by the fact that amine added to carbonate/bicarbonate solution reacts in part to form carbamate.

The formation of alkyl carbonate

It is generally assumed that in the reaction of CO_2 with MEA and DEA the rate of formation of alkyl carbonate can be ignored compared to the formation of carbamate. Since Jørgensen's[6] attempt to determine the relative rates in the case of MEA does not seem to be valid one may turn to the reaction of CO_2 with TEA, which forms no carbamate. One might assume that the rate of reaction of the alkyl group (Jørgensen asserts that not more than one alkyl group in any molecule of TEA is carbonated) in DEA is of the same order as in TEA. Three sets of determinations of the rate of reaction of TEA with CO_2 are available. Jørgensen and Faurholt[11] and Jørgensen[6] measured the rate by the competitive method at 0°C and 18°C respectively (the barium alkyl carbonate is soluble in water). In each set the concentration of OH^- was kept constant while that of TEA was varied, so it is not possible to verify that the reaction is first-order with respect to $[OH^-]$ as assumed by the authors, who tool the mechanism to be (7), (8), the latter being the rate-determining step, leading to

$$R = k[CO_2][TEA][OH^-]. \tag{20}$$

Another possibility, which would be kinetically indistinguishable, is that the oxygen atom of the alkyl group attacks the carbon atom of CO_2 and that OH^- removes a proton. The values of $R/[CO_2][Am][OH^-]$ are found (after recalculation) to be 5930 at 0°C and 12,730 $l^2/mol^2 s$ at 18°C. Hikita *et al.*[2], using the rapid-mixing method, found rates of reaction which were much larger than could be accounted for by (20) in view of the low values of the pH in solutions containing no NaOH (the pK_a for TEA is 8.22 at 0°C). The concentrations of TEA in Hikita's experiments ranged from 0.335 to 1.06 M and the reaction was found to be first-order with respect to both amine and CO_2.

Using Hikita's results, one finds the ratio of the rates or reaction of TEA to that of MEA to be 1% at 0°C and 2% at 100°C, while the ratio for TEA to DEA is almost independent of temperature, being 6% when [DEA] = 0.174 M and 1% when [DEA] = 0.719 M. Under practical conditions it is probably justifiable to ignore the formation of alkyl carbonate from MEA and DEA in relation to the formation of carbamate in determining the total rate of reaction and hence of absorption.

This discussion does nothing to reconcile the rate-constants determined by Hikita *et al.*[2] with those deduced from rates of CO_2 absorption. This topic will be duscussed separately.

Acknowledgements—I am grateful for suggestions from Dr. I. Fleming and Dr. A. Kirby.

REFERENCES

[1] Faurholt C., *J. Chim. Phys.* 1925 **22** 1.
[2] Hikita H., Asai S., Ishikawa H. and Honda M., *Chem. Engng J.* 1977 **13** 7.
[3] Jensen M. B., Jørgensen E. and Faurholt C., *Acta Chem. Scand.* 1954 **8** 1137.
[4] Jensen A., Jensen M. B. and Faurholt C., *Acta Chem. Scand.* 1954 **8** 1129.
[5] Pinsent B. R. W., Pearson L. and Roughton F. J. W., *Trans. Faraday Soc.* 1956 **52** 1512.

[6] Jørgensen E., *Acta Chem. Scand.* 1956 **10** 1612.
[7] Caplow M., *J. Am. Chem. Soc.* 1968 **90** 6795.
[8] Jensen M. B., *Acta Chem. Scand.* 1959 **13** 289.
[9] Williams A. W. and Jencks W. P., *J. Chem. Soc.* Perkin II 1974 1753, 1760.
[10] Danckwerts P. V. and Sharma M. M., *The Chemical Engineer* Oct. 1966 CE 244.
[11] Jørgensen E. and Faurholt C., *Acta Chem. Scand.* 1954 **8** 1141.

C6

REACTION OF CO_2 WITH ETHANOLAMINES: KINETICS FROM GAS-ABSORPTION

S. S. LADDHA and P. V. DANCKWERTS*
Department of Chemical Engineering, University of Cambridge, Pembroke St., Cambridge CB2 3RA, England

(*Accepted* 2 *October* 1980)

Abstract—CO_2 at very low pressure was absorbed in a stirred cell into aqueous solutions of MEA and DEA. The rates of absorption were determined from the fall in pressure. The results were analysed on the assumption (for which experimental support is given) that the solubilities and diffusivities of CO_2 in these solutions bear a constant proportionality to those of N_2O, which have been measured. The results for MEA at 25°C agree very closely with those of Hikita *et al.*[1], determined by a rapid-mixing thermal method. The results for DEA appear to support a reaction mechanism proposed by Danckwerts[13], the reaction tending to be second-order in amine at low concentrations and first-order at high concentrations.

INTRODUCTION

Hikita *et al.*[1], using a rapid-mixing thermal method, showed that the reaction between CO_2 and monoethanolamine (MEA) in aqueous solution

$$CO_2 + 2RNH_2 \rightarrow RNCO_2^- + RNH_3^+ \qquad (1)$$

appeared to be first-order in the amine concentration (Am):

$$r_M = k_2(CO_2)(Am) \qquad (2)$$

(r_M being the homogeneous reaction-rate of CO_2) while the corresponding reaction of diethanolamine (DEA) appeared to be second-order:

$$r_D = k_3(CO_2)(Am)^2. \qquad (3)$$

Hikita *et al.*[2] also absorbed CO_2 at 1 atm into laminar jets of MEA and DEA solutions and in order to reconcile the homogeneous kinetics with the observed rates of absorption had to make assumptions about the solubilities of CO_2 in solutions of MEA and DEA which appear to be anomalous[3].

In the experiments discussed below we absorbed CO_2 at an initial partial pressure of about 10^{-2} atm into a pool of amine solution in a stirred cell similar to that described by Danckwerts and Alper[4], in which a four-bladed stirrer "wiped" the surface of the liquid. The rate of rotation was 70 rpm. The low partial pressure of CO_2 resulted in a low absorption rate, so that the depletion of amine at the surface was negligible. The condition for the concentration to be uniform is

$$R \ll \frac{1}{2} k_L(Am)\sqrt{(D_B/D_A)} \qquad (4)$$

where D_A is the diffusivity of CO_2 and D_B that of the amine ($D_B/D_A \sim 1/2$) and was comfortably satisfied in all our experiments. Under these circumstances the rate of absorption of CO_2 per unit area, R, is given by

$$R = pS\sqrt{(D_A[k_a(Am) + k_{OH}(OH^-)])} \qquad (5)$$

where p is the partial pressure of CO_2, S its solubility,

*Author to whom correspondence should be addressed.

(Am) the concentration of amine, k_a a rate-coefficient which may be a function of (Am)—e.g. $k_3(Am)$ for DEA according to Hikita *et al.*[1]—and $k_{OH}(OH^-)$ the contribution made to the reaction rate by the hydroxyl ions, whcih is small and can be calculated with sufficient precision (pK_a is 9.50 for MEA and 8.92 for DEA at 25°C[5]; k_{OH} is 8322 l/mol s[6]). It is easily shown that in a sealed absorber with a gas space of volume V, CO_2 being absorbed into a liquid surface of area A

$$\beta = k_a(Am) = \frac{1}{D_A}\left[\frac{V}{SR_GTA}\left(\frac{-\mathrm{d}\ln p}{\mathrm{d}t}\right)\right]^2 - k_{OH}(OH^-) \tag{6}$$

where R_G is the gas-constant and T the absolute temperature.

In the case of a gas such as N_2O which does not react with the amine the quantity k_LS, where k_L and S are the mass-transfer coefficient and the solubility, can be determined in a similar way:

$$k_LS = \frac{V}{AR_GT}\left(\frac{-\mathrm{d}\ln p}{\mathrm{d}t}\right). \tag{7}$$

In order to calculate β from (6) it is necessary to estimate D_AS^2 for CO_2. Laddha *et al.*[3] have shown that the solubilities of CO_2 and N_2O bear a constant ratio (1.37) to one another in water and in aqueous solutions of a number of alcohols (mol fraction up to 0.06) and thus there is every reason to suppose this ratio to hold in solutions of MEA and DEA. For similar reasons one might expect the diffusivities of the two gases in solution to exhibit a constant ratio.

Now, if the stirrer in the stirred cell brings about some kind of "surface-renewal" the ratios of the values of k_LS for CO_2 and N_2O at a given stirring-speed will be the same as the ratios of $S\sqrt{D}$[7]. The values of $S\sqrt{D}$ for CO_2 and N_2O in water at 25°C are well-established[8–11] and the ratio is 1.43. Our measurements in the stirred cell showed that the ratio of k_LS in water was 1.43 and furthermore that the ratio was within 3% of 1.43 in a variety of solutions of alcohols (see Table 1). We also did some experiments in a laminar jet at 25°C which showed that the ratio of $S\sqrt{D}$ was equal to 1.43 to within 4% for aqueous solutions of glycerol and glycol (m.f. 0.06).

Since S and D have been measured for N_2O in solutions of MEA and DEA[9–11] it seems highly probable that values of $S\sqrt{D}$ for CO_2 can be assigned by multiplying those for N_2O by the factor 1.43. Prof. Sada has been kind enough to communicate the actual values of $S\sqrt{D}$ for N_2O measured in a laminar-jet apparatus. If those for DEA are plotted against the concentration a point of inflection occurs in the range 2.5–3.5 M. There is no corresponding feature in the curves of solubility or viscosity. We therefore have some reservation about applying the analogy with CO_2 in our experiment at 2.88 M. This point is discussed below.

EXPERIMENTAL

The stirred vessel (9.2 cm i.d.) was similar to that described by Danckwerts and Alper[4] except that the stirrer was driven by a small electric motor enclosed in the gas-space so that a shaft passing through a gland was not necessary and the system could be positively sealed. Another internal motor drove a gas stirrer and it was found that above a certain speed gas-side resistance ceased to decrease and was presumably negligible.

The surface area, A, of the liquid was determined by geometrical considerations. The volume, V, of the gas in the vessel and its associated connections was measured by evacuating it, connecting it to another vessel of known volume containing air at atmospheric pressure and measuring the resultant pressure.

The apparatus consisted of the absorber and a reference vessel containing a little of the solution under investigation. The pressure difference between these two vessels could be recorded on a chart by a micromanometer. A fixed volume (500 cm³) of the solution was put into the stirred cell and the whole system made vacuum-tight, evacuated and held under vacuum until bubbles of air ceased to be evolved by the liquid. The two vessels were then isolated from the pump and one another. At this stage the pressure on each side of the micromanometer was the saturated water–vapour pres-

Table 1. Comparative values of k_LS for CO_2 and N_2O in various solutions at 25°C

$V = 637\ cm^3$ $A = 64\ cm^2$

Partial pressure of gas, $p \simeq 1$ atm $\alpha = 1.43\ (k_LS)_{N_2O} / (k_LS)_{CO_2}$

Solute	Mole Fraction	$10^4(1/p)(\Delta p/\Delta t), s^{-1}$ CO_2	N_2O	$10^8\ k_LS, mol/cm^2.atm.s$ CO_2	N_2O	α
Water	1.00	2.31	1.62	9.44	6.60	1.00
Glycol	0.06	1.70	1.16	6.92	4.72	0.98
Glycerol	0.06	1.12	0.76	4.56	3.09	0.97
Diethyleneglycol	0.06	1.27	0.90	5.17	3.65	1.01

sure and the deflection was zero. A small amount of CO_2 (~ 0.02 atm) was admitted to the stirred cell and the stirrers were started. The chart was started when the pressure of the CO_2 was about 0.01 atm. The pressure was recorded over 10–85 s. The experiment could be repeated by allowing the deflection to fall to zero and admitting more CO_2 (the CO_2 absorbed was very much less than the amine in the liquid).

The record could conveniently be analysed by reading from the chart the time θ required for the pressure of CO_2 to fall by a factor ϕ: $-\mathrm{d}\ln p/\mathrm{d}t = \ln \phi/\theta$.

In order to measure k_L the stirred vessel was purged with the appropriate gas at atmospheric pressure and then isolated. The difference, Δp, between the pressure in the vessel and the atmosphere at time θ was recorded. Since Δp was much less than the pressure of the gas, p, $-\mathrm{d}\ln p/\mathrm{d}t = \Delta p/p\theta$. The back-pressure of the dissolved gas was negligible.

All experiments were done at 25°C with a stirrer speed of 70 rpm.

The experimental results are shown in Tables 1 and 2.

DISCUSSION OF RESULTS

The values of β (eqn 6) as inferred from the pressure measurements are given in Table 2 and in Fig. 1 $\log\beta = \log k_a(Am)$ is plotted against $\log(Am)$. As regards MEA, the extrapolation of the homogeneous results of Hikita *et al.*[1] agrees extradordinarily well with our findings, establishing that the reaction is indeed second-order in the amine concentration and that the method for estimating S^2D_A for CO_2 in the amine solutions is correct (and that therefore the figures given by Hikita *et al.*[2] for the solubilities are incorrect). Furthermore, it has been shown[11] that the diffusivities of N_2O in solutions of amine cannot be correlated by expressions of the form $D\mu^{\kappa}$ = const. The same is presumably true for CO_2.

As regards DEA, Danckwerts [13] suggested that the reaction mechanism of CO_2 with amine might be

$$CO_2 + R_2NH \underset{k_{-2}}{\overset{k_2}{\rightleftharpoons}} R_2NH^+CO_2^- \quad (8)$$

$$R_2NH^+CO_2^- + R_2NH \xrightarrow[k_b]{} R_2NCO_2^- + R_2NH_2^+ \quad (9)$$

(the nomenclature is changed here) leading to a homo-

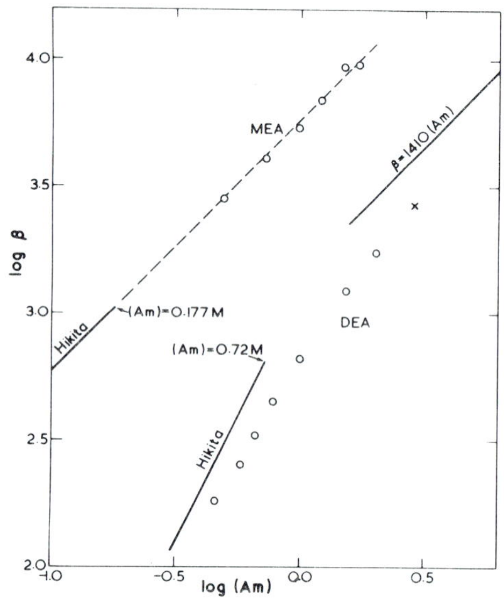

Fig. 1.

Table 2. Reaction rates of CO_2

Amine	(Am) mol/l	10^7 S √D mol/cm^{2}atm.s$^{1/2}$	10^3(-dln p/dt) s^{-1}	$k_a(Am)+k_{OH}(OH^-)$ s^{-1}	$k_{OH}(OH^-)$ s^{-1}	$\beta = k_a(Am)$ s^{-1}
MEA	0.49	1.47	19.29	2853	33	2820
	0.73	1.46	22.99	4109	40	4061
	0.98	1.44	25.96	5385	46	5339
	1.21	1.43	29.35	6980	51	6929
	1.48	1.39	32.93	9300	57	9243
	1.71	1.36	32.66	9556	61	9495
DEA	0.46	1.42	4.92	199	16	183
	0.57	1.41	5.70	271	18	253
	0.66	1.39	6.40	351	19	332
	0.785	1.38	7.34	469	21	448
	0.99	1.35	8.71	690	23	667
	1.50	1.29	11.22	1254	29	1225
	1.98	1.24	12.83	1774	33	1741
	2.88	(1.11)	14.20	(2711)	40	(2671)

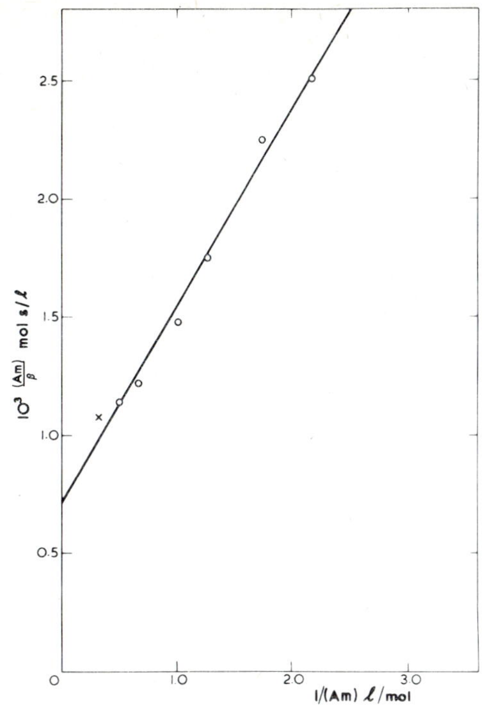

Fig. 2.

geneous rate of reaction

$$r_D = k_a(CO_2)(Am) = \frac{(CO_2)(Am)}{\frac{1}{k_2}+\frac{1}{k_3(Am)}} \tag{10}$$

where $k_3 = k_2 k_b / k_{-2}$.

If this were so a plot of $(Am)/k_a(Am) = (Am)/\beta$ against $1/(Am)$ should be a straight line of slope $1/k_3$ and intercept $1/k_2$. Such a plot is shown for DEA in Fig. 2. Linear regression analysis leads to

$$\frac{(Am)}{\beta} = \frac{(Am)}{k_a(Am)} = \frac{1}{1410} + \frac{1}{1200(Am)} \tag{11}$$

((Am) in mol/1) with a coefficient of confidence of 0.99. As explained above, we have some reservations about the value of $s\sqrt{D}$ in 2.88 M DEA (inferred rate-constants shown by crosses in Figs. 1 and 2). This value of β does not correlate very well with the others shown in Fig. 2 and has not been taken into account in the analysis leading to eqn (11). If it were it would lead to a more pronounced falling-off to the second-order reaction at higher concentrations, without appreciably affecting the inferred value of k_3. Since the two terms on the right-hand side of eqn (11) are of comparable magnitude the results lend very strong support to eqn (10). The third-order coefficient k_3 has a value of $1200\,(mol/l)^{-2}s^{-1}$, compared with the value of 1253 found by Hikita *et al.*[1], a difference of only 5%. The value of k_2 is found to be 1410 l/mol s, much less than that for MEA (5868), as might be expected from the lower basic strength of DEA and the fact that the approach of a CO_2 molecule to the nitrogen atom is hindered, in the case of DEA, by two rather than one—$(CH_2)OH$ groups (no doubt with hydrogen-bonded water molecules).

There is, however, one important discrepancy between our results and those of Hikita *et al.*[1], namely that they found the reaction to be third-order when the amine concentration was as high as 0.72 M whereas we found that third-order kinetics were not attained (although fairly closely approached) at concentrations as low as 0.5 M. Very recently Alvarez-Fuster *et al.*[12] have reported experiments on the rate of absorption of CO_2 into amine solutions in a wetted-wall column which suggest that the kinetics of the reaction reported by Hikita *et al.*[1] are correct, both as to order of reaction and rate-constants, for MEA and DEA, third-order kinetics being maintained in the latter case up to 0.8 M. We have information from the British Gas Corporation (private communication) that experiments with a stopped-flow apparatus indicate that the reaction between CO_2 and DEA is first-order in the concentration of amine in the range 0.01–0.03 M. The kinetics of the CO_2–DEA reaction still admit room for speculation.

Acknowledgement—This work was supported by the Science Research Council.

NOTATION

A	surface area of absorbing liquid, cm^2
(Am)	concentration of amine, mol/l
D	diffusivity of dissolved gas, cm^2/s
D_A	diffusivity of dissolved CO_2, cm^2/s
D_B	diffusivity of amine in solution, cm^2/s
k_L	mass-transfer coefficient for dissolved gas, cm/s
$k_a, k_2, k_3, k_{-2}, k_b$	rate-constants defined by eqns (8)–(10)
p	partial pressure of CO_2, atm
Δp	differential pressure at time t
R	specific rate of absorption of CO_2, $mol/cm^2\,s$
R_G	gas constant, cm^3 atm/mol °K
r_M, r_D	rates of reaction of CO_2 in MEA and DEA solutions, mol/l s
S	solubility of gas, mol/cm^3 atm (pp. of gas)
T	temperature, °K
V	volume of gas in absorber, cm^3
β	$k_a(Am)$ (eqn 6), s^{-1}

θ time elapsed for decrease in préssure by factor ϕ, s

REFERENCES

[1] Hikita H., Asai S., Ishikawa I. and Honda M., *Chem. Engng J.* 1977 **13** 7.

[2] Hikita H., Asai S., Katsu Y. and Ikuno S., *A.I.Ch.E.J.* 1979 **25** 3.

[3] Laddha S. S., Diaz M. and Danckwerts P. V., *Chem. Engng Sci.* 1981 **36** 226.

[4] Danckwerts P. V. and Alper E., *Trans. Inst. Chem. Engrs* 1975 **53** 74.

[5] Perrin D. D., *Dissociation Constants of Organic Bases in Aqueous Solution.* Butterworths, London 1965.

[6] Pinsent B. R. W., Pearson L. and Roughton F. J. W., *Trans. Faraday Soc.* 1956 **52** 1512.

[7] Danckwerts P. V., *Ind. Engng Chem.* 1951 **43** 1460.

[8] Yasunishi A. and Yushida F., *J. Chem. Engng Data* 1979 **24** 11.

[9] Sada E. and Kito S., *Kagaku Kōgaku* 1972 **36** 218.

[10] Sada E., Kumazawa H. and Butt M. A., *J. Chem. Engng Data* 1977 **22** 11.

[11] Sada E., Kumazawa H. and Butt M. A., *J. Chem. Engng Data* 1978 **23** 161.

[12] Alvarez-Fuster C., Midoux N., Laurent A. and Charpentier J. C., *Chem. Engng Sci.* 1980 **35** 1717.

[13] Danckwerts P.V., *Chem. Engng. Sci. 1979* **34** 443.

SECTION D

Residence - Time Distributions and Related Topics

Paper D1 is possibly my most influential contribution to chemical engineering. It may seem unsophisticated now but at the time it fulfilled a major need. Like most conceptual advances, it represents a crystallisation of ideas which at the time were scattered and ill-defined. Kekulé's concept of the benzene ring came to him in a flash on the top of a tram. The central idea of Paper D1 came to me during a tea-break at Cambridge, supporting my appraisal of the value of "academic indolence". I would press on my colleagues in the U.S.A. the value of having morning and afternoon coffee-breaks, although they may seem a "waste of time". Paper D1 has now reached the status of being a primary reference which is seldom cited (it would be presumptious of me to recall Newton's "Principia"). It has been overtaken by a snowfall of more sophisticated concepts, many derived from control theory. However, I think it has served and continues to serve working chemical engineers because it does not blind them with mathematics.

Paper D2 presents an extension of these ideas and indeed I have developed a number of similar extensions, obvious enough, upon which I have written and lectured but not published in English.

Two features of Paper D1 are worth comment. One is the treatment of dispersion in pipe-flow; this made no allowance for lateral dispersion at right-angles to the mean flow, which was the subject of two famous papers by G.I. Taylor concerning molecular diffusion in capillaries (Proc. Roy. Soc. 1953 A219 186) and turbulent diffusion in a pipeline (Proc. Roy. Soc. 1954 A223 446). The second feature is the introduction of the "Danckwerts boundary conditions" for the flow of fluid through a bed of packing of finite depth, which gave rise to a lengthy and almost metaphysical discussion in the literature although I thought that I had fully justified them in this first publication.

Papers D3 and D4 refer to experiments on the heroic scale to which I have always aspired but which is so difficult to attain in practice - namely, tracer tests in full-scale operating plant. Both of these were successful only because of the energy and executive ability of G. Place. For some years I had the ambition to do a similar experiment on a blast-furnace; but the sensitivity of blast-furnace managers to meddling from outside is enormous - quite understandably in view of the economic consequences of any interference in production and the fact that blast-furnace management is as much an art as a science. No doubt in 20 years' time the blast-furnace will be as obsolete as the Bessemer converter.

Paper D5 describes a particularly elegant chemical method of flow visualisation by continuous-flow systems. It was applied to a model of the spray-drier discussed in Paper D4, although it was not possible to attain dynamic similarity. The idea behind this method was first floated by Denbigh (cited) who used a chemical system which is considerably less corrosive than that specified in Paper D4 and which might in practice be preferable.

Finally, Paper D6 is one of several (not reproduced here) which apply the concept of the autocorrelogram in discussing periodicity in the performance of a chemical plant (related, for instance, to the eight-hour shift) and the effect of hold-up and mixing (e.g. in a gasometer) on the variability of a stream of material. This paper was not published in a journal in which it made much impact on contemporary ideas (the same might be said of the Abbe Mendel's experiments on sweet peas, which had considerably more potential than Paper D6).

An example of the lack of communication between different disciplines is given by the book "Basic Principles of the Tracer Method" C.W. Sheppard (Wiley, 1962) which contains no reference to chemical engineering sources and which was not cited by chemical engineers for some years after its publication.

D1

Continuous flow systems

Distribution of Residence Times

P. V. Danckwerts

Department of Chemical Engineering, Tennis Court Road, Cambridge, England

(*Received* 24 *October* 1952)

Summary—When a fluid flows through a vessel at a constant rate, either " piston-flow " or perfect mixing is usually assumed. In practice many systems do not conform to either of these assumptions, so that calculations based on them may be inaccurate. It is explained how distribution-functions for residence-times can be defined and measured for actual systems. Open and packed tubes are discussed as systems about which predictions can be made. The use of the distribution-functions is illustrated by showing how they can be used to calculate the efficiencies of reactors and blenders. It is shown how models may be used to predict the distribution of residence-times in large systems.

Résumé—Quand, dans un récipient, on introduit, à vitesse constante, un fluide donné, on suppose généralement soit un mélange parfait, soit un " écoulement frontal parfait." En pratique, de nombreux systèmes s'écartent de l'une ou l'autre de ces hypothèses simplificatrices et les calculs qui en résultent sont plus ou moins inexacts. L'auteur expose, pour des systèmes réels, comment l'on peut définir et mesurer des fonctions de distribution pour la " durée de séjour ": ceci peut s'appliquer à des tubes vides ou munis de garnissages. Par emploi de ces fonctions de distribution, l'auteur montre comment on peut calculer l'efficacité des réacteurs ou des mélangeurs. Des modèles peuvent être utilisés pour prévoir la répartition des " durées de séjour " dans des systèmes de grandes dimensions.

When a stream of material flows steadily through a vessel such as a pipe or a tank, in which it takes part in some process such as chemical reaction, heat- or mass-transfer, or simple mixing, it is usual to make use of one of the following assumptions for the purposes of calculation :

(*a*) The fluid in the vessel is completely mixed, so that its properties are uniform and identical with those of the outgoing stream. This assumption is frequently made the basis of calculations on stirred reactors or blenders.

(*b*) Elements of fluid which enter the vessel at the same moment move through it with constant and equal velocity on parallel paths, and leave at the same moment. This type of behaviour will be referred to as " piston flow," and is normally assumed when considering flow through heat-exchangers, catalytic reactors, packed towers, chromatographic columns, etc.

It is clear that there are many cases in which neither type of flow corresponds exactly to the facts – for instance, fluid in laminar flow in a pipe, or gas flowing through a fluidised catalytic reactor or blast furnace. It is of some importance to investigate the discrepancies between the assumed and actual behaviour of such systems, and where necessary to allow for them in making calculations. This paper is intended to clarify some of the problems presented by steady-flow systems, and to show how their behaviour can be investigated and quantitatively specified. Some of the concepts and mathematical expressions which appear below have already been used by Gilliland and Mason [8], [9] in the course of a study of flow through fluidised beds.

F-diagrams and Age-distribution Functions

The volume of the vessel occupied by the fluid is V, and the volumetric rate of inflow and outflow of fluid is assumed constant and equal to v. Suppose some property of the inflowing fluid undergoes a sudden change from one steady value to another; for instance, let the colour change from white to red. Call the fraction of red material in the outflow at time θ later be $F(\theta)$. The plot of $F(\theta)$ vs. $v\,\theta/V$ will be called an " F-diagram." Figs. 1 (a)-(d) show F-diagrams for some representative types of system.

Perfect piston flow (1(a)) will never occur with Newtonian fluids; there will always be some longitudinal mixing, due to viscous effects and molecular or eddy-diffusion. 1 (b) illustrates the departure from piston flow caused by restricted longitudinal mixing. 1 (c) is the diagram for perfect mixing ; the equation of the curve is easily shown to be :

$$F(\theta) = 1 - e^{-v\theta/V}. \tag{1}$$

It cuts the ordinate $v\theta/V = 1$ at $(1 - \frac{1}{e})$, and its initial slope is unity. 1 (d) shows a diagram of the type to be expected when there is a good deal of " dead water " in the system ; a considerable fraction of the fluid is trapped in eddies, and spends much more than the average length of time in the vessel, while most of the flow takes place through a restricted channel.

The F-diagram of a system is easy to obtain (for instance, by injecting tracer materials into the entering stream), and its shape will clearly give a good deal of information about the behaviour of the fluid flowing through the vessel. As will be shown later,

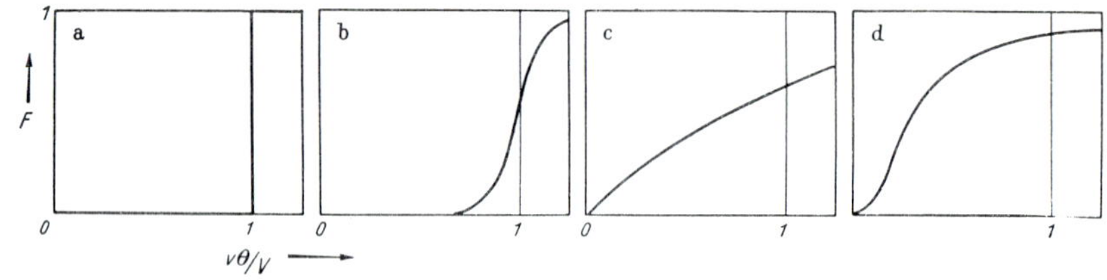

Fig. 1. F-diagrams.

(*a*) Piston flow; (*b*) Piston flow with some longitudinal mixing; (*c*) Complete mixing; (*d*) Dead water.

much of this information can be summarised by two numbers which can be derived from the F-diagram. The diagram also enables certain calculations to be made concerning the performance of the system when used, for instance, as a blender or reactor.

The shape of the F-diagram depends on the relative times taken by various portions of the fluid to flow through the vessel, or in other words, on the distribution of residence-times. Those elements of the material in the vessel which have been in it for a time θ are said to have an "age" θ, and the fraction of the material in the system having at any instant ages between θ and $(\theta + d\theta)$ is $I(\theta)\,d\theta$. The fraction of the material having ages between θ and $(\theta + d\theta)$ at the moment of leaving the system is $E(\theta)\,d\theta$. I and E may be called the internal and exit age-distribution functions respectively. The relationship between E, I and F can be shown by imagining the ingoing stream, as before, to change from white to red at time $\theta = 0$. At a time θ later the balance sheet for the red material is :

Entered : $v\theta$.

Still in system : $V \int_0^{\theta} I(\theta')\,d\theta'$.

Left system : $v \int_{\theta''=0}^{\theta} \int_{\theta'=0}^{\theta''} E(\theta')\,d\theta'\,d\theta''$.

(The last follows because the rate of outflow of red material at any time θ'' after the change of colour is $v \int_0^{\theta''} E(\theta')\,d\theta'$). Thus the conservation equation for red material is

$$\frac{v\theta}{V} = \int_0^{\theta} I(\theta')\,d\theta' + \frac{v}{V} \int_{\theta''=0}^{\theta} \int_{\theta'=0}^{\theta''} E(\theta')\,d\theta'\,d\theta''. \tag{2}$$

Also, the total fraction $F(\theta)$ of red material in the outflow at time θ is

$$F(\theta) = \int_0^{\theta} E(\theta')\,d\theta'. \tag{3}$$

Differentiating (2) with respect to θ :

$$1 - \int_0^{\theta} E(\theta')\,d\theta' = \frac{V}{v} I(\theta), \tag{4}$$

and hence from equation (3) :

$$1 - F(\theta) = \frac{V}{v} I(\theta). \tag{5}$$

Equation (4) gives the relationship between $E(\theta)$ and $I(\theta)$, while equation (5) shows how the latter function can be determined from the F-diagram.

The same information can be obtained in a somewhat different way, which may prove more convenient under some circumstances. Suppose a quantity Q of some substance (a radioactive tracer might be chosen for ease of estimation) is injected into the entering stream virtually instantaneously – that is, within a period very short compared to V/v. As before, samples are taken at various times θ after the injection ; the concentration of the injected material in the exit stream at time θ is $C(\theta)$. Then it can be shown, by methods similar to those used above, that

$$C(\theta) = \frac{Q}{v} \frac{d}{d\theta} \left\{ F(\theta) \right\} = \frac{Q}{v} E(\theta). \tag{6}$$

Plots of $\frac{VC(\theta)}{Q}$ vs. $\frac{v\theta}{V}$, which may be called C-diagrams, are shown in Fig. 2 for the systems whose F-diagrams are given in Fig. 1. Note that

$$\frac{V}{Q} \int_0^{\infty} C(\theta)\,d\left(\frac{v\theta}{V}\right) = 1, \tag{7}$$

so that the area under every C-diagram is equal to unity.

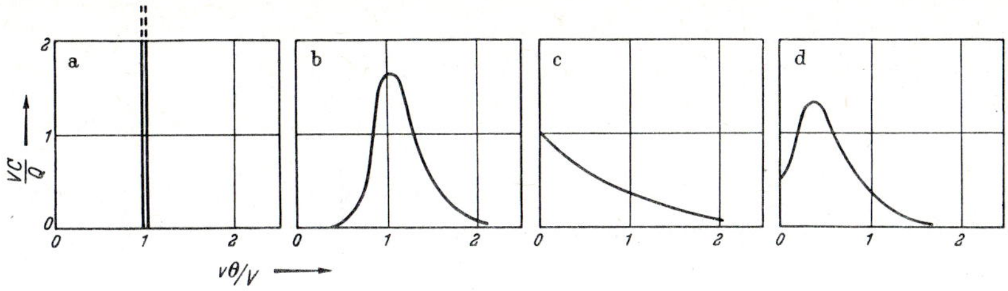

Fig. 2. *C*-diagrams.

(*a*) Piston flow (*b*) Piston flow with some longitudinal mixing (*c*) Complete mixing (*d*) Dead water.

It has been assumed in the foregoing that the forms of the functions I and E do not vary with time. In fact, however, instability of flow may lead to fluctuations in the internal and exit age-distributions, and cause observed values of $F(\theta)$ to oscillate about some smooth curve of the kind shown in Fig. 1. These fluctuations will generally be random in nature, and by making a number of determinations of the F-diagram and averaging, the mean form of the diagram can be obtained. The problems raised by such fluctuations will be considered on another occasion; for the moment attention will be confined to cases where they may be ignored.

It will be noted that

$$\int_0^\infty I(\theta)\,d\theta \equiv \frac{v}{V}\int_0^\infty \left\{1 - F(\theta)\right\} d\theta = 1 \tag{8}$$

and

$$\int_0^\infty E(\theta)\,d\theta = 1 \tag{9}$$

(because the integrals in equations (8) and (9) represent the total fractions of material, in the vessel and at the exit respectively, having ages between 0 and ∞ — namely, *all* the material in each case). Equation (8) implies that the area between the curve $F(\theta)$ vs. $v\theta/V$ and the line $F(\theta) = 1$ is equal to unity. It can be seen immediately that the two shaded areas are equal in Fig. 3 whatever the shape of the curve.

The average age $\bar{\theta}_E$ of the material leaving the vessel is always V/v. This can be shown as follows:

$$\bar{\theta}_E = \int_0^\infty \theta\,.\,E(\theta)\,d\theta. \tag{10}$$

Hence, from equation (3)

$$\frac{v\bar{\theta}_E}{V} = \int_0^\infty \frac{v\theta}{V}\frac{d\,F(\theta)}{d\,\theta}\,d\theta = \int_{F=0}^{1} \frac{v\theta}{V}\,d\,F(\theta). \tag{11}$$

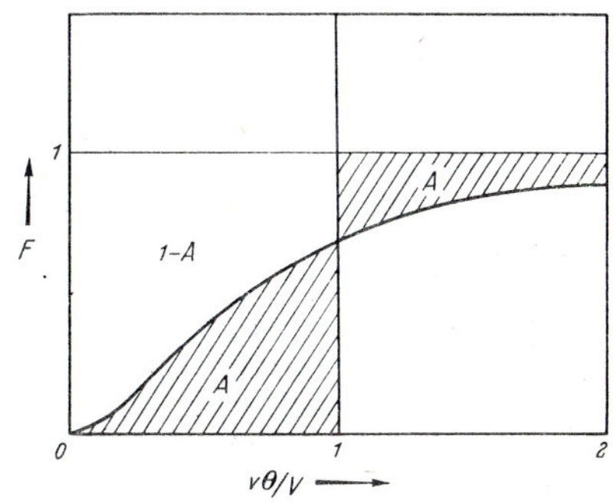

Fig. 3.

It can be seen ~~that~~ from Fig. 3 that the last integral is equal to 1.

The average age of the material in the system at any time is $\bar{\theta}_I$, where

$$\bar{\theta}_I = \int_0^\infty \theta\,.\,I(\theta)\,d\theta = \frac{v}{V}\int_0^\infty \theta\{1 - F(\theta)\}\,d\theta. \tag{12}$$

Integrating by parts we have

$$\bar{\theta}_I = \frac{v}{2V}\int_{F=0}^{1} \theta^2\,d\,F(\theta). \tag{13}$$

The integral in (12) or (13) can be evaluated graphically with the help of the F-diagram.

Hold-Back and Segregation

The systems illustrated in Figs. 1 (a)-(d) show in progressively greater degree a quality which will be called "hold-back." In a system which displays hold-back, some elements of fluid spend more, others less, than the average time, V/v, in the vessel. In a system with piston-flow (1 (a)), there is no hold-back. Quantitatively, the magnitude of the hold-back will be denoted by H, and can be conveniently defined as the area under the F-diagram between $v\theta/V = 0$ and $v\theta/V = 1$ (area A in Fig. 3):

$$H = \frac{v}{V} \int_{\theta = 0}^{V/v} F(\theta)\, d\theta. \tag{14}$$

H varies from 0 for piston-flow to values approaching 1 when most of the space in the vessel is dead water. For the completely mixed vessel (Fig. 1 (c)), $H = 1/e$. The magnitude of H is a measure of the deviation from piston-flow. Its significance is most easily visualised as follows: if the colour of the inflowing stream changes suddenly from white to red, H is equal to the fraction of the vessel which will still be occupied by white fluid after a volume of red fluid equal to the volume of the vessel has flowed in.

The hold-back is of importance in connection with chemical reactors and the successive flow of fluids through pipes and other vessels. In the former case it will usually be true to say that if the reactants are adequately mixed before or shortly after entering the vessel, the output of the reactor will be greatest when the hold-back is least (although the reverse may be true for autocatalytic reactions). A high hold-back means that much of the volume of the reactor is occupied by material which has already undergone reaction, while much reactant passes rapidly through the vessel by a " short-circuit " route. The magnitude of the hold-back gives only a general idea of the behaviour of the system, however; the whole F-diagram (and possibly other information) is needed if the actual performance of the reactor is to be calculated. This subject is discussed in a later section.

An indication of the efficiency of mixing in a system can be given by a single quantity, S, which will be called the " segregation," and can be derived from the F-diagram as follows. If one superimposes the F-diagrams for a perfectly-mixed system (equation (1)) and for an imperfectly mixed system such as that represented in Fig. 1 (b), the result will be similar to Fig. 4.

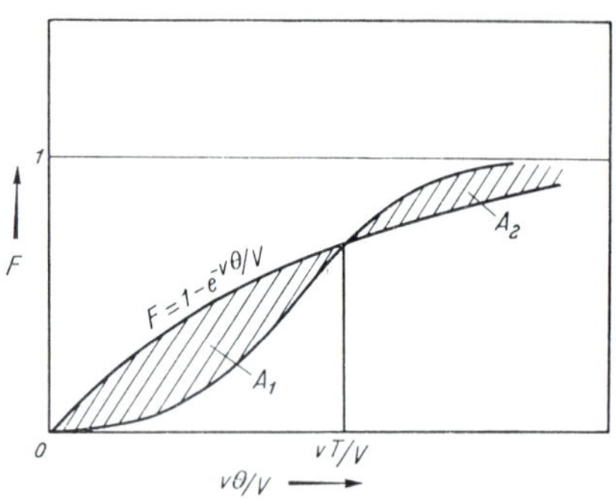

Fig. 4.

The degree of departure of the system from perfect mixing will be indicated by the size of the shaded area ($A_1 + A_2$). However, since the area between each curve, and the line $F(\theta) = 1$ is the same, the two shaded regions have the same area ($A_1 = A_2$). It is thus convenient to define the segregation S quantitatively as the area A_1 between the F-diagram of the system and the curve $F(\theta) = 1 - e^{-v\theta/V}$, up to the point (at $\theta = T$) where the curves cross; it is thus equal to half the total shaded area.

When there is dead water in the system the positions of the curves may be inverted, as in Fig. 5.

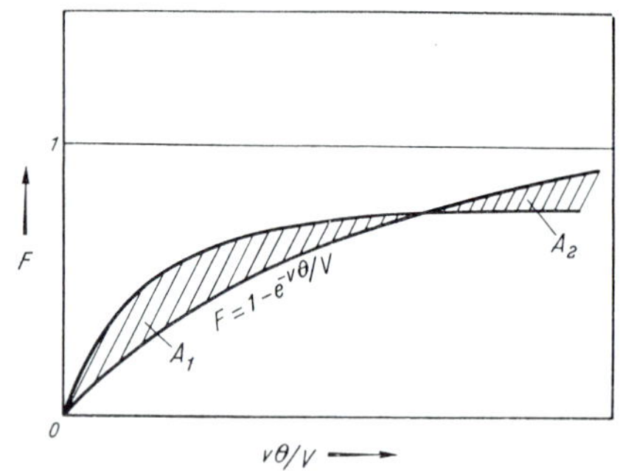

Fig. 5.

The segregation is then given the negative value $-A_1$ to indicate the nature of the departure from perfect mixing. In a case such as that illustrated in Fig. 1 (*b*), the two curves may cut twice, as indicated in Fig. 6.

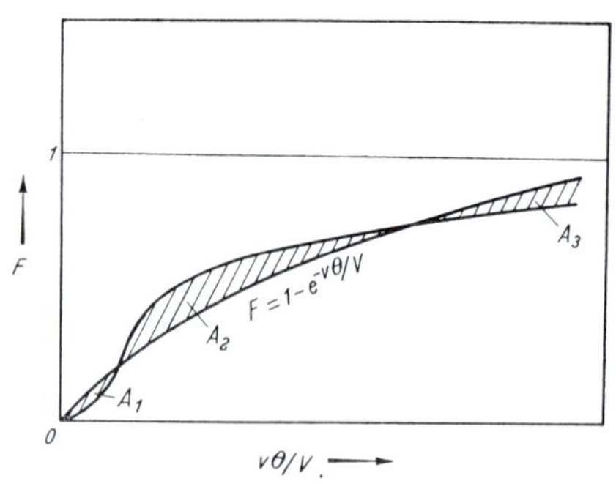

Fig. 6.

In such cases $A_1 + A_3 = A_2$, and the segregation is equal to $-A_2$. S varies from $+1/e$ for piston-flow to values approaching -1 when most of the space in the system is dead water.

The sense in which the word " mixing " has been used here should be clearly understood. We are not concerned with homogeneity, but with the age-distribution of material in the vessel and in the outgoing stream. If these distributions are the same, the system is said to be perfectly mixed, although there might, in the experiment described, be a perceptible non-uniformity of colour. On the other hand, when a fluid flows through a pipe with a high degree of turbulence, the segregation as defined above may approach that for piston-flow, although mixing in directions at right-angles to the axis of flow may be very efficient, ensuring that at any given moment the composition will be uniform across each cross-section and in the outflowing fluid.

As pointed out by GILLILAND and MASON [9], although the F-diagram for a fluidised bed may be quite close to that for perfect mixing, other tests show that longitudinal mixing is in fact not very efficient, and moreover a considerable proportion of the gas passes through the bed in the form of bubbles.

Under some circumstances, when the fluid concerned is a gas, molecular diffusion between neighbouring elements of fluid may play a material part in determining the shape of the F-diagram. In liquids, however, diffusion coefficients are so small that only hydrodynamic effects will usually be of significance.

FLOW THROUGH BEDS OF SOLIDS

A very large number of chemical engineering operations involve the flow of a fluid through beds of stationary solid particles. Calculations are usually based on the assumption that piston-type flow occurs in such systems. This is certainly not exactly true, since longitudinal mixing must take place. Many cases, however, are probably amenable to the mathematical treatment which follows.

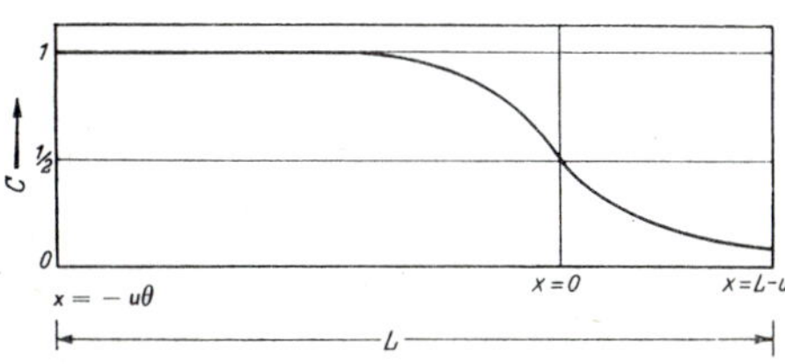

Fig. 7.

Consider a packed tubular vessel of length L, through which fluid flows with a mean axial velocity u. At time $\theta = 0$ the colour of the fluid changes from white to red. Now if the flow were of piston-type, the plane boundary between red and white would move down the tube with velocity u. We shall denote this imaginary plane by $x = 0$ and use it as the origin of a frame of reference moving down the tube with uniform velocity u, so that at time θ the plane $x = 0$ is distant $u\theta$ from the entry, and the x-coordinates of the ends of the tube are $(-u\theta)$ and $(L - u\theta)$ respectively. (See Fig. 7).

The longitudinal velocity of any element of fluid relative to the plane $x = 0$ will fluctuate irregularly. At times the element will be close to a solid surface, and viscous forces will slow it down, while at other times it will be near the centre of a channel and moving at a velocity greater than the mean; part of the time may be spent in regions of turbulence where the element will undergo rapid and irregular fluctuations in velocity. If the packing is quite randomly arranged, without any channelling, each element of fluid will travel at the same average velocity, and will experience fluctuations of the same average magnitude and frequency. It is clear that the ordinary "random walk" theory [10] can be applied to such behaviour; the result will be redistribution of red and white material according to the laws of diffusion. That is, if c is the mean concentration (volume fraction) of red material at a plane x at time θ,

$$\frac{\partial c}{\partial \theta} = D \frac{\partial^2 c}{\partial x^2}. \tag{15}$$

D is a "diffusivity" which must be determined empirically; it will presumably depend on the viscosity, density and velocity of the liquid, and on the size and shape of the packing. The following boundary conditions will represent the facts with sufficient precision under many circumstances:

$$\left.\begin{array}{ll} c = 0,\ x > 0, & \theta = 0 \\ c = 1,\ x < 0, & \theta = 0 \\ c = 0,\ x = \infty, & \theta > 0 \\ c = 1,\ x = -\infty, & \theta > 0 \end{array}\right\}. \tag{16}$$

(The above conditions hold if the concentration of white material at the entry falls virtually to zero at a time very much less than L/u. It can be shown that this will occur if $4D/Lu \ll 1$ – that is, provided mixing is not too effective or the tube too short). Under these conditions the solution to (15) is

$$c = \tfrac{1}{2}\left[1 - \operatorname{erf}\left(\frac{x}{2\sqrt{D\theta}}\right)\right] \tag{17}$$

(where $\operatorname{erf}\left(\dfrac{x}{2\sqrt{D\theta}}\right) = \dfrac{2}{\sqrt{\pi}}\displaystyle\int^{x/2\sqrt{D\theta}} e^{-y^2} dy$; numerical values may be found in tables and are shown in Fig. 8). Hence

$$2\,F(\theta) = 1 - \operatorname{erf}\left(\frac{L - u\theta}{2\sqrt{D\theta}}\right), \tag{18}$$

and since $v/V = u/L$ (V being the void volume of the vessel):

$$2\,F(\theta) = 1 - \operatorname{erf}\left(\frac{1 - v\theta/V}{2\sqrt{\dfrac{v\theta}{V}\cdot\dfrac{D}{Lu}}}\right). \tag{19}$$

The F-diagram will thus be determined entirely by the value of D/Lu. It will have the general shape shown in Figs. 1 (b) and 9.

If we let

$$\operatorname{erf}(z) = 1 - 2F \tag{20}$$

equation (19) becomes, on rearrangement,

$$\frac{1 - v\theta/V}{z\sqrt{v\theta/V}} = 2\sqrt{\frac{D}{Lu}}\,. \qquad (21)$$

(Values of z corresponding to given values of F can be found with the help of Fig. 8).

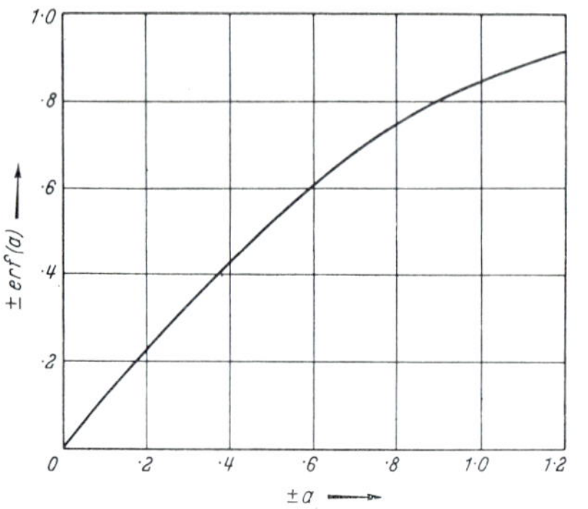

Fig. 8. $\text{Erf}\,(a) = \frac{2}{\sqrt{\pi}} \int_0^a e^{-y^2} dy$

For a given flow-rate and depth of packing, the left-hand side of equation (21) should be constant if the diffusion equation (15) is obeyed. However equation (21) is not very suitable for testing the diffusion hypothesis or for determining the value of D since the expression on the left-hand side is very sensitive to small errors in F over the range of greatest importance. The value of D is probably more easily determined from the slope of the F-diagram at $v\theta/V = 1$. Differentiating (19) with respect to $v\theta/V$ and putting $v\theta/V = 1$, we find

$$\left[\frac{dF}{d\,(v\theta/V)}\right]_{v\theta/V=1} = \tfrac{1}{2}\sqrt{\frac{Lu}{\pi D}}\,. \qquad (22)$$

D/Lu can be calculated from this expression, the F-diagram calculated from equation (19), and the results compared with the experimental curve. D should, of course, be independent of L for a given u.

The hold-back in the system is equal to $\frac{1}{L}\sqrt{DV/\pi v}$, or $\frac{1}{L}\sqrt{DL/\pi u}$.

Fig. 9 shows the F-diagram obtained when water flowed through a bed of $\frac{3}{8}$-in. Raschig rings (tube diameter 4·8 cm ; $L = 140$ cm ; fractional free volume 0·62 ; volumetric flow-rate 4·5 cc/sec). The curve calculated from equation (19) with $D/Lu = 0{\cdot}013$ is compared with the experimental points and shown to agree well.

The liquid running over the packing in an absorption tower is presumably subject to the same kind of mixing, although channelling may prevent equation (15) from being very closely obeyed. The quantity V in this case is equal to the total liquid hold-up.

The effect of longitudinal mixing on the performance of tubular reactors is discussed later.

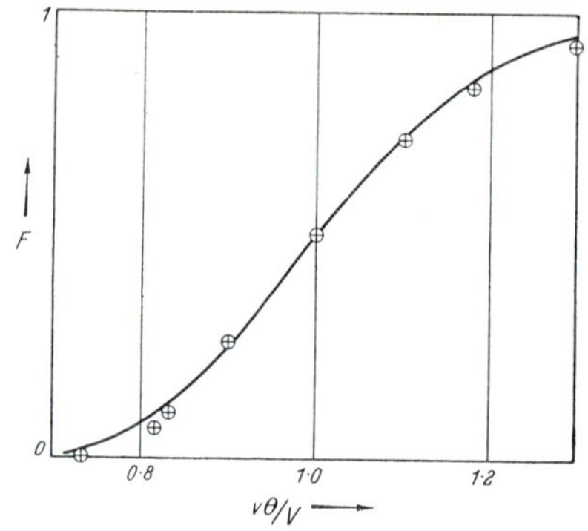

Fig. 9. F—diagram calculated for flow through packed column compared with experimental points.

Flow in Pipes

Hold-back will always arise when a true fluid flows through a pipe. In the first place, there is a variation in velocity from the axis to the wall of the pipe, so that the central "core" of fluid moves with a velocity greater than the mean, while the fluid near the wall lags behind. This effect will be most marked when flow is laminar ; in turbulent flow the velocity is more uniform across the pipe, and the hold-back is less, although eddy-diffusion contributes to the longitudinal mixing. In some circumstances molecular diffusion may also contribute appreciably to the hold-back. It will be convenient to discuss laminar and turbulent flow separately.

When viscous liquids such as heavy petroleum oil or rayon dope are pumped through pipes the flow may be laminar. If the flow is changed, say, from one type of petroleum to another, or from normal to pigmented dope, the mean composition of the liquid leaving the pipe will change only gradually from one pure component to the other, and a large volume of mixed liquid will be discharged. If certain simplifying assumptions are made, the form of the F-diagram is easily calculated. (Similar calculations have been made by Bosworth [2], and by Fowler and Brown [7].)

In the first place, entrance-effects will be ignored (as in the derivation of Poiseuille's formula), and the liquid will be assumed to be everywhere in unaccelerated laminar flow with a parabolic velocity-distribution ; provided the ratio of the length to the diameter of the pipe is sufficiently large ($\gg 0{\cdot}06\ Re$) entrance effects will in fact be unimportant. Secondly, it will be assumed that the two liquids have the same density and viscosity, and behave as Newtonian fluids. Thirdly, molecular diffusion will be ignored.

If $\bar{u}$ is the mean velocity and u the velocity at a distance r from the axis of a pipe of radius R, we have the well-known relationship

$$u = 2\bar{u}\left(1 - \frac{r^2}{R^2}\right). \qquad (23)$$

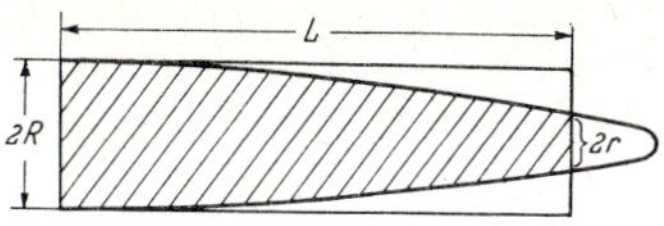

Fig. 10.

The follow-up liquid will thus first appear at the exit of a pipe of length L after a time $L/2\bar{u}$; after a greater time, θ, the situation will be as shown in Fig. 10, with a core of follow-up liquid of diameter $2R\sqrt{1 - L/2\bar{u}\theta}$ at the exit. Allowing for the variation of velocity from the centre to the edge of the core, the fraction of follow-up liquid in the stream leaving the pipe is seen to be

$$F(\theta) = \frac{1}{\pi R^2 \bar{u}} \int_0^{R\sqrt{1-L/2\bar{u}\theta}} 2\bar{u}\left(1 - \frac{r^2}{R^2}\right) . 2\pi r . dr, \quad (24)$$

$$\left.\begin{aligned} &= 1 - L^2/4\bar{u}^2\theta^2, \ \bar{u}\theta/L > \tfrac{1}{2} \\ &= 1 - V^2/4v^2\theta^2, \ v\theta/V > \tfrac{1}{2} \end{aligned}\right\} . \quad (25)$$

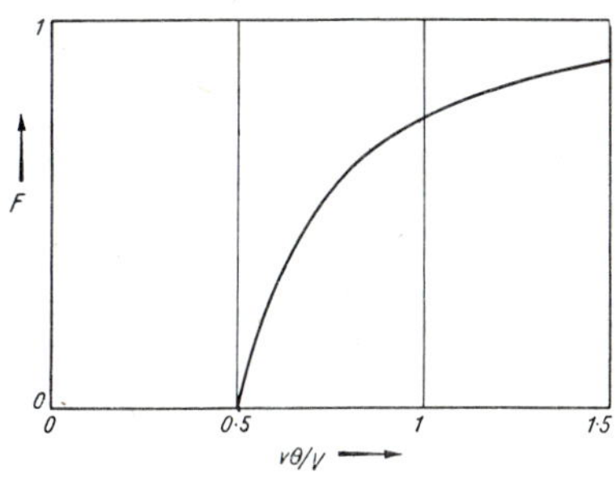

Fig. 11. F—diagram for viscous flow in pipe.

The resulting F-diagram is shown in Fig. 11. Its shape is independent of the length and diameter of the pipe, and of the viscosity and velocity of the fluid, (provided of course that flow is laminar). The hold-back is $\frac{1}{4}$. Bosworth [2] has calculated the effect of molecular diffusion on systems of this kind (his F_τ is equivalent to the present author's $E(\theta)$). He shows that it can be ignored providing the following conditions are fulfilled :

$$\left.\begin{aligned} R &> 13\sqrt{D_m L/\bar{u}} \\ L &> 6{\cdot}5 \times 10^4 \, D_m/\bar{u} \end{aligned}\right\},$$

(where D_m is the molecular diffusivity), and points out that in the case of gases turbulence usually occurs while R is too small for the first condition to be satisfied. With liquids, however, which have much higher Schmidt numbers ($\mu/\rho D_m$), there may be a considerable range of pipe-diameters for which flow is laminar and the first condition is also satisfied.

Bosworth [3] has also derived expressions for $E(\theta)$ (F_τ in his nomenclature) for turbulent flow in pipes, starting from simplified expressions for the values of the axial velocity and eddy-diffusivity at various distances from the pipe wall. The following expression for $F(\theta)$ can be derived from Bosworth's formula :

$$\left.\begin{aligned} F(\theta) &= 1 - 2\alpha^n \times \\ &\times \left[1 - \beta n - \frac{1}{2}\beta - \alpha^n\left(\frac{1}{2} - \frac{5}{2}\beta n - \frac{5}{4}\beta\right)\right. \\ &\qquad \left. - \beta(2n+1)\left(\alpha^{2n} - \frac{1}{4}\alpha^{3n}\right)\right], \end{aligned}\right\} \quad (26)$$

where

$$\frac{1}{\alpha} = 2\frac{v\theta}{V} \times$$

$$\times \left[(1 - \beta n - \beta/2)\frac{n}{n-1} + \left(-\frac{1}{2} + \frac{5}{2}\beta n + \frac{5\beta}{4}\right) \times \left(\frac{2n}{2n-1}\right) - \beta(2n+1)\left(\frac{3n}{3n-1} + \frac{4n}{4n-1}\right)\right]$$

$$\beta = (n-1)^2 R/0{\cdot}32\, nL,$$

$$n = 7, \ 2{,}000 < Re < 100{,}000,$$

$$n = 8, \ Re > 100{,}000,$$

R being the radius and L the length of the pipe.

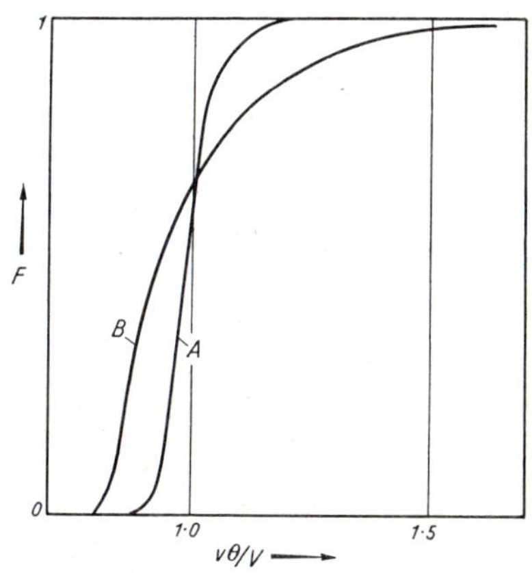

Fig. 12. Experimental (A) and calculated (B) F—diagrams. Water in circular tube, $Re = 6940$, $R/L = 6{\cdot}3 \times 10^{-4}$.

Comparison can be made with the experimental results of Fowler and Brown [7]. Fig. 12 shows an experimental F-diagram compared with that calculated from Bosworth's formula. In Fig. 13 a comparison is made between calculated and observed values of $\frac{v}{V}(\theta_{0\cdot8} - \theta_{0\cdot2})$ for various Reynolds numbers in a given pipe (i.e. F has the value 0·8 at $\theta_{0\cdot8}$, and the value 0·2 at $\theta_{0\cdot2}$). It will be seen that Bosworth's formula predicts (at least under the conditions chosen for the comparison) a much greater degree of longitudinal mixing than is actually observed, and also that it fails to predict the marked effect of velocity of flow on the F-diagram for a given pipe.

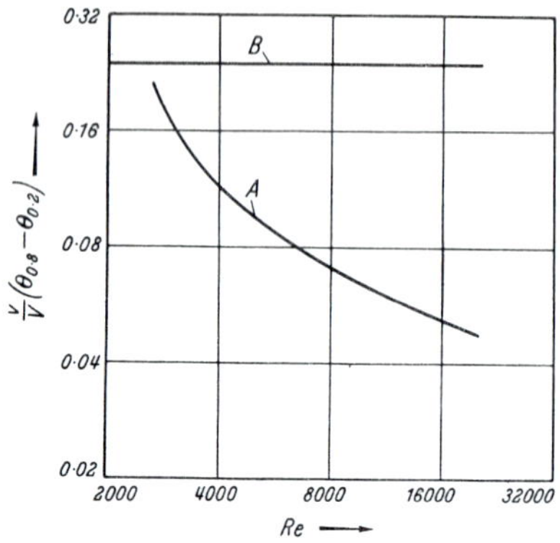

Fig. 13. Experimental (A) and calculated (B) values of $v/V\,(\theta_{0.8} - \theta_{0.2})$. $R/L = 6{\cdot}3 \times 10^{-4}$.

Reactors

The performance of a steady-flow reactor can be calculated from the F-diagram provided the reaction is first-order, or pseudo-first-order, with a velocity-constant which does not vary from place to place in the reactor. For instance, if an effluent containing a radioactive element is allowed to flow through a reservoir, the concentration of the element at the exit can be calculated if its radioactive decay-constant k and the F-diagram of the system are known.

Consider those elements of fluid which have ages between θ and $(\theta + d\theta)$ at the exit. A fraction $(1 - e^{-k\theta})$ of the reactant originally contained in these elements will have reacted during its passage through the vessel. Hence the total fraction $(1 - f)$ of reactant which reacts during its passage through the vessel is

$$1 - f = \int_0^\infty (1 - e^{-k\theta})\, E(\theta)\, d\theta. \tag{27}$$

From equation (3)

$$E(\theta)\, d\theta = d\,F(\theta). \tag{28}$$

Hence :

$$\left.\begin{aligned} 1 - f &= \int_{\theta=0}^\infty (1 - e^{-k\theta})\, dF(\theta) \\ &= 1 - \int_{\theta=0}^\infty e^{-k\theta}\, dF(\theta). \end{aligned}\right. \tag{29}$$

f can be evaluated by graphical integration if the F-diagram is available.

In the special case, previously discussed, of a tubular packed vessel, the expression for F given in equation (19) is unsuitable for use in equation (29) because of the approximations which have been made. Instead we set up the differential equation for a tubular reactor with longitudinal diffusion as well as flow (changes in volume are assumed not to occur, so that the mean longitudinal velocity, u, is the same at all cross-sections). The equation is easily shown, by making a balance on a section of differential length dy, to be

$$\frac{d^2c}{dy^2} - \frac{u}{D}\frac{dc}{dy} - \frac{kc}{D} = 0\,. \tag{30}$$

where c is the concentration of reactant at cross-section y. The concentration of reactant in the entering stream is c^*; owing to diffusion the concentration just within the entrance of the reactor, at $y = 0$, is less than c^*. The following boundary condition expresses the fact that the rate at which reactant is fed to the reactor is equal to the rate at which it crosses plane $y = 0$ by combined flow and diffusion.

$$uc^* = uc - D.\, dc/dy,\ y = 0 \tag{31}$$

At the outlet $(y = L)$

$$ufc^* = uc - D.\, dc/dy,\ y = L,$$

fc^* being the concentration of the exit stream. Now if dc/dy were negative, the concentration in the exit stream would be greater than that at the end of the packing. If dc/dy were positive the concentration would pass through a minimum somewhere in the reactor and then rise towards the downstream end. Intuition suggests that neither of these situations can arise, so that the boundary condition must be :

$$\frac{dc}{dy} = 0,\ y = L. \tag{32}$$

The solution to equation (30) with these two boundary conditions is

$$\left.\begin{aligned} \frac{c}{c^*} &= \exp\left(\frac{uy}{2D}\right) \times \\ &\times \left[\frac{2(1+a)\exp\frac{ua}{2D}(L-y) - 2(1-a)\exp\frac{ua}{2D}(y-L)}{(1+a)^2\exp\frac{uaL}{2D} - (1-a)^2\exp\frac{-uaL}{2D}}\right], \end{aligned}\right\} \tag{33}$$

where

$$a = \sqrt{1 + 4kD/u^2}.$$

The value of $(1 - c/c^*)$ at the exit $(y = L)$ is equal to $(1 - f)$, the fraction reacted. Hence

$$\left.\begin{aligned} 1 - f &= 1 - \\ &\frac{4a}{(1+a)^2\exp\frac{-uL}{2D}(1-a) - (1-a)^2\exp\frac{-uL}{2D}(1+a)}\,. \end{aligned}\right\} \tag{34}$$

As $D \to 0$ this expression tends to

$$1 - f = 1 - \exp \frac{-kL}{u} \qquad (35)$$

which is the well-known solution for "piston-flow." As $D \to \infty$, we find

$$1 - f = \frac{kL}{u + kL} \qquad (36)$$

which is the solution for complete mixing. For very small values of D

$$1 - f \approx 1 - (1 + k^2\, DL/u^3) \exp \frac{-kL}{u}. \qquad (37)$$

From this it can be seen that the effect of diffusion is to decrease the fractional conversion compared to that for piston-flow, but that the effect will be negligible provided $k^2\, DL/u^3 \ll 1$, or $[\ln f]^2\, D/Lu \ll 1$. Thus for specified operating conditions it is possible to decide, if an approximate value of D is available, whether neglect of longitudinal diffusion will lead to serious error in calculating the output of the reactor.

The foregoing treatment of reactors will seldom be applicable to practical problems except as a rough guide. In the first place, the heat of reaction often gives rise to temperature-gradients, and hence point-to-point variations in the value of the reaction-velocity constant, so that the chance of a molecule reacting depends on its path through the reactor, as well as its residence-time. Secondly, if the reaction is of order other than first, the chance of a given molecule reacting depends on the molecules which it encounters in its passage through the reactor; the nature of these encounters is largely determined by diffusional processes caused by point-to-point variations in composition in the fluid, which cannot be deduced from the F-diagram. Second-order reactions between imperfectly mixed fluids have been considered by the author [5]. Denbigh [6] has considered second-order reactions taking place in a fluid in laminar flow in a pipe, under conditions such that molecular diffusion may be ignored.

Blenders

A continuous-flow blender is a mixing vessel into which flows a stream of material of continuously-varying composition. In the vessel elements of material which have entered at different times are mixed, so that the outflowing stream shows less variation in composition than the input. The case of a perfectly-mixed blender has been considered in detail elsewhere [4], with special reference to the flow of town gas through a gas-holder, and consequent smoothing-out of fluctuations in the calorific value. Beaudry [1] has considered the special case of a perfectly-mixed blender with a feed consisting of batches of material of finite volume. A blender which displays segregation will be less effective in reducing fluctuations than one in which perfect mixing occurs.

Suppose the concentration of some component in the entering stream has a concentration c_i which fluctuates with time. Its value at any time is $(\bar{c} + \delta_i)$, where $\bar{c}$ is the mean value of c_i which is assumed to show no trend. A convenient measure of magnitude of the fluctuations in c_i is its standard deviation σ_i:

$$\sigma_i^2 = \overline{\delta_i^2}. \qquad (38)$$

σ_i is assumed to show no trend. The concentration $c_0 = (\bar{c} + \delta_0)$ of the outgoing stream has a standard deviation σ_0. The problem is to find the ratio σ_0/σ_i. The rates of inflow and outflow are assumed constant and equal to v. The concentration $c_0\,(t)$ in the outflow at time t is then given by

$$c_0\,(t) = \int_{\theta=0}^{\infty} c_i\{t - \theta\}\, E\,(\theta)\, d\theta, \qquad (39)$$

where $c_i\{t - \theta\}$ is the concentration in the inflow at time $(t - \theta)$. Hence

$$\delta_0\,(t) = \int_{\theta=0}^{\infty} \delta_i\{t - \theta\}\, E\,(\theta)\, d\theta \qquad (40)$$

and

$$[\delta_0\,(t)]^2 = \left[\int_{\theta=0}^{\infty} \delta_i\{t - \theta\}\, E\,(\theta)\, d\theta\right]^2. \qquad (41)$$

However, it is generally true that

$$\left[\int_0^{\infty} f(y)\, dy\right]^2 = 2 \int_{y=0}^{\infty} \int_{r=0}^{\infty} f(y) f(y + r)\, dy\, dr, \qquad (42)$$

where $f(y)$ is any function of y. Applying this to equation (42):

$$\left.\begin{aligned} [\delta_0\,(t)]^2 &= 2 \int_{\theta=0}^{\infty} \int_{r=0}^{\infty} \delta_i\{t - \theta\} \times \\ &\times \delta_i\{t - \theta - r\}\, E\,(\theta)\, E\,(\theta + r)\, d\theta\, dr. \end{aligned}\right\} \qquad (43)$$

Averaging this with respect to t

$$\left.\begin{aligned} \sigma_0^2 = \overline{\delta_0^2} &= 2 \iint \overline{\delta_i\{t - \theta\}\, \delta_i\{t - \theta - r\}} \times \\ &\times E\,(\theta)\, E\,(\theta + r)\, d\theta\, dr. \end{aligned}\right\} \qquad (44)$$

Now the quantity

$$R\,(r) = \frac{\overline{\delta_i\{t'\}\, \delta_i\{t' - r\}}}{\sigma_i^2} \qquad (45)$$

(the numerator being averaged with respect to t') is known as the autocorrelation coefficient or serial correlation coefficient of c_i for a time-interval r. Substituting (45) in (44) we have

$$\sigma_0^2/\sigma_i^2 = 2 \int_{\theta=0}^{\infty} \int_{r=0}^{\infty} R\,(r)\, E\,(\theta)\, E\,(\theta + r)\, d\theta\, dr. \qquad (46)$$

$R(r)$ may be found as a function of r from a representative record of c_i vs. t in various ways which have been discussed elsewhere (4). (The relationship between $R(r)$ and r indicates whether the fluctuations in c_i are on the whole rapid or slow, and whether they display any regular periodicity). Assuming that $R(r)$ is known for a sufficient range of values of r, in order to evaluate the integral in (46) it is necessary to know $E(\theta)$ as a function of θ. The necessary information is contained in the F-diagram of the system, and the following is perhaps the simplest way of doing the calculation. First fit some simple function of θ (such as a power-series) to the F-diagram up to a value θ_1, at which $F(\theta_1)$ is very close to 1. Differentiate this function with respect to θ. From equation (3) we see that

$$\frac{dF(\theta)}{d\theta} = E(\theta), \tag{47}$$

so that $E(\theta)$ is now a known function of θ. For a given value of θ, $E(\theta + r)$ is the value obtained by replacing θ by $(\theta + r)$. It will now be possible to evaluate the integral

$$I(r) = \int_{\theta=0}^{\theta_1} E(\theta)\, E(\theta + r)\, d\theta \approx \int_{\theta=0}^{\infty} E(\theta)\, E(\theta + r)\, d\theta \tag{48}$$

algebraically for a number of values of r. The integral

$$\sigma_0^2/\sigma_i^2 = 2\int_{r=0}^{\infty} R(r)\, I(r)\, dr \tag{49}$$

can now be evaluated graphically. (In practice, of course, the upper limit will be taken at some value of r for which the integrand has become vanishingly small).

The Use of Models for Predicting F-diagrams

Under certain circumstances the F-diagram of a large system can be predicted with confidence from that for a model. The conditions to be fulfilled are (*a*) The model must be geometrically similar to the system; (*b*) The Reynolds number must be the same in the model and the system; (*c*) Gravity-waves, density-differences, surface tension, and other influences apart from inertia and viscosity, must be unimportant in determining the behaviour of the fluid in both model and system.

In the simple case of a fluid flowing through a tank or a fixed bed of solids, the same fluid being used in the model as in the system, equality of the Reynolds numbers requires that

$$\frac{v_1}{v_2} = \frac{L_1}{L_2} \tag{50}$$

where v_1, v_2 are the volumetric flow-rates and L_1, L_2 the linear dimensions of the model and system respectively. That is, in a 1/10 scale model the volumetric flow-rate must be 1/10 that in the full-scale system. If the system incorporates a rotating stirrer, the angular velocities, ω, of the stirrers in model and system are related by

$$\frac{\omega_1}{\omega_2} = \left(\frac{L_2}{L_1}\right)^2 \tag{51}$$

In the 1/10 scale model the stirrer must rotate 100 times as fast as in the system.

If these conditions are fulfilled the flow-patterns of the fluid will be geometrically similar in model and system, hence both will have the same F-diagram.

The simple type of model experiment described here cannot be applied to systems in which the fluid is of non-uniform density or viscosity owing to variations in temperature or composition from point to point, or to fluidised beds.

References

[1] Beaudry, J. P.; Chem. Eng. 1948 **55** 112. [2] Bosworth, R. C. L.; Phil. Mag. 1948 **39** 847. [3] Bosworth, R. C. L.; Phil. Mag. 1949 **40** 314. [4] Danckwerts, P. V. and Sellers, E. S.; Ind. Chemist 1951 **27** 395; Coke and Gas 1952 **14** 247. [5] Danckwerts, P. V.; Appl. Sci. Res. A 1952 **3** 279. [6] Denbigh, K. G.; J. Appl. Chem. 1951 **1** 227. [7] Fowler, F. C. and Brown, G. G.; Trans. Amer. Inst. Chem. Eng. 1943 **39** 491. [8] Gilliland, E. R. and Mason, E. A.; Ind. Eng. Chem. 1949 **41** 1191. [9] Gilliland, E. R. and Mason, E. A.; Ind. Eng. Chem. 1952 **44** 218. [10] Kennard, E. H.; Kinetic Theory of Gases, 1938 New York, McGraw-Hill, p. 268. [11] Hull, D. E. and Kent, J. W.; Ind. Eng. Chem. 1952 **44** 2745. [12] Hull, D. E., Kent, J. W. and Lee, R. D.; World Oil 1949 **129** 188. [13] Smith, S. S. and Schulze, R. K.; Petroleum Eng. 1948 **19** 94; 1948 **20** 330. [14] Lapidus, L. and Amundson, N. R.; J. Phys. Chem. 1952 **56** 984. [15] Bernard, R. A. and Wilhelm, R. H.; Chem. Eng. Prog. 1950 **46** 233. [16] Arthur, J. R., Linett, J. W., Raynor, E. J. and Sington, E. P. C.; Trans Faraday Soc. 1950 **46** 270. [17] Morales, M., Spinn, C. W. and Smith, J. M.; Ind. Eng. Chem. 1951 **43** 225.

Note added in Proof

Flow in Pipelines—Hull and Kent [11] have recently published an account of experiments on petroleum flowing in a 182-mile long 10-inch diameter pipe line. A radioactive tracer was used to obtain (in effect) C-diagrams for various lengths of pipe. With a Reynolds Number of 20,000-30,000 the diagrams obtained were close to those which would be expected if the "diffusion" equation (15) were followed, D having a value of 1·3 sq. ft./sec. Further experimental work on mixing in pipe-lines is referred to by the authors [12], [13].

Flow in Packed Beds—Lapidus and Amundson [14], have analysed the consequences of longitudinal mixing in adsorption columns. Bernhard and Wilhelm [15] have measured the apparent *radial* diffusivity in fluids flowing through beds of solids. It seems likely that the longitudinal diffusivity D (equation [15]) will have a different value. Arthur *et al.* [16] and Morales *et al.* [17] have shown that the fluid velocity differs markedly from the mean in the neighbourhood of the wall of the column.

Notation.

A = area under curve
c = concentration
$C(\theta)$ = concentration at exit of vessel at time θ.

c^* = concentration of stream entering reactor
D = apparent diffusivity
D_m = molecular diffusivity
$E(\theta)$ = distribution-function for residence-times θ

$$\text{erf}(z) = \frac{2}{\sqrt{\pi}} \int_0^z e^{-y^2}\, dy$$

$F(\theta)$ = fraction of material in outflow which has been in system for a time less than θ
f = fraction of reactant unreacted
H = hold-back (equation (14))
$I(\theta)$ = distribution-function for "ages," θ, of material in system
k = first-order reaction-velocity constant
L = length of tube or reactor
n = parameter in equation (26)
Q = quantity of tracer material injected
R = radius of pipe
$R(r)$ = serial correlation-coefficient of c_i for time-interval r
r = radial distance from tube-axis; or time-interval
Re = Reynolds number
S = segregation (Figs. 4, 5, 6)
t = time
u = velocity of flow
$\bar{u}$ = mean velocity of flow in open pipe
V = volume of vessel
v = volumetric flow-rate
x = distance referred to coordinates moving with velocity u
y = distance referred to stationary coordinates
z = defined by equation (20)
α = parameter in equation (26)
β = parameter in equation (26)
δ_i, δ_0 = instantaneous deviations of c_i, c_0 from $\bar{c}$
θ = time-interval, residence-time or "age"
σ_i, σ_0 = standard deviations of c_i, c_0 from $\bar{c}$
ω = angular velocity

D2

Local residence-times in continuous-flow systems

P. V. DANCKWERTS
Imperial College of Science and Technology, London.

(*Received* 24 *January* 1958)

Summary—If a tracer is injected at the inlet of a continuous-flow system in the form of a step or pulse signal, the course of the subsequent change in concentration at a point within the system, can be used to determine quantitatively the average "age" as an indication of the degree of stagnation of the material at this point.

Résumé—L'injection d'un traceur, à l'alimentation d'un système en écoulement continu, sous la forme d'un signal discontinu ou d'un signal d'impulsion permet de déterminer quantitativement la vitesse de remplacement ou le degré de stagnation de la matière, en un point du système, par suite des fluctuations de concentration en ce point.

Zusammenfassung—Wenn ein Indikator stossweise in den Eintritt eines kontinuierlichen Strömungssystems injiziert wird, kann man den Verlauf der nachfolgenden Konzentrationsänderung an einem Punkt innerhalb des Systems zur quantitativen Bestimmung der Verdrängungsgeschwindigkeit oder der Aufenthaltsdauer des Stoffes in diesem Punkt verwenden.

ELSEWHERE in this issue (p. 74), SPALDING shows that if a pulse consisting of a quantity of tracer is injected at the inlet of a steady-state flow system at $t = 0$, then the quantity

$$\int_0^\infty c \,.\, dt$$

(where c is the local concentration of tracer at time t after injection) will have the same value at every point in the system. This suggests a method of determining the rate of replacement or degree of stagnation of the material in different parts of the system.

SPALDING's proposition is discussed further in the Appendix, where it is shown that the distribution of "ages" of the material in a volume-element around the point at which c is measured can be deduced from the variation of c with t. (The "age" of a particle of material is the time elapsed since it entered the system). The fraction of the material having ages between t, $(t + dt)$ is $(vc/Q) \, . \, dt$, where v is the rate of flow of material through the system, and Q the quantity of tracer injected.

It follows that the average age, $\bar{\theta}$, of the material at the point of measurement could be determined by evaluating graphically the integral

$$\bar{\theta} = \frac{v}{Q} \int_{0}^{\infty} ct \, . \, dt.$$

In practice this may be hard to evaluate because of uncertainties about the tail of the curve. For many purposes the median age, τ, will be just as useful as an index of the local rate of replacement. Half the material at the point has an age greater than τ; hence

$$\int_{0}^{\tau} c \cdot dt = \frac{1}{2} \frac{Q}{v}$$

The value of τ is easily found by plotting c against t until the area under the curve is equal to $Q/2v$.

The same information can be obtained by raising the concentration of tracer in the inlet stream at $t = 0$ from zero to a steady value C. The value of c/C at the point of measurement at time t is then equal to the fraction of the material having an age less than t, and the median age is equal to the value of t at which c/C reaches $\frac{1}{2}$.

These techniques offer a quantitative method of investigating rates of replacement of material in various parts of a continuous-flow plant (or of air in a room or water in a pond). For instance, a contour map of median ages throughout the plant might be made, and would show up clearly dead spots and by-passing effects. In general one would expect the median or average ages near the inlet to approach zero, and their values near the outlet to approach those for the material in the whole stream leaving the system. However, in stagnant pockets the age of the material might be considerably greater than that of the leaving stream, and channelling or by-passing effects would be shown up as regions where the age was relatively low.

Appendix

A quantity Q of tracer is injected into the inflowing stream. If Q is finite, the injection cannot be truly instantaneous; it will be assumed that it occurs at a constant rate over a short time-interval δt. Let the distribution function for the ages (θ) of the particles of material in the volume element, which contains a quantity δm of material be $f(\theta)$. At any time t after injection the age of the tracer will be between t, $(t + \delta t)$, and hence the fraction of the material in the volume-element having the same age as the tracer is $f(t) \, . \, \delta t$. Since a quantity $v \, . \, \delta t$ entered the system with the pulse of tracer, and therefore has the same age, the ratio of tracer to other material entering during the pulse was $Q/v \, . \, \delta t$. Provided therefore that tracer and other material migrate together, the quantity of tracer in the specified volume-element at time t is

$$\frac{Q}{v \, . \, \delta t} f(t) \, . \, \delta t \, . \, \delta m = \frac{Q f(t)}{v} \, . \, \delta m$$

and hence the ratio c, of tracer to the whole of the other material in the volume element is given by

$$c = Q \, . \, f(t)/v.$$

D3

The distribution of residence-times in an industrial fluidised reactor

P. V. Danckwerts, J. W. Jenkins and G. Place

Department of Chemical Engineering, Tennis Court Road, Cambridge, England

(*Received* 8 *February*, 1954)

Summary—The distribution of residence-times of the gas flowing through a fluidised catalyst regenerator has been determined by using helium as a tracer. Only about 5 % of the gas spends less than half, and about 5 % more than twice, the mean residence-time in the regenerator. It appears that the flow through the fluidised bed itself is much closer to piston-flow than to complete mixing, the hold-back being less than 0·1.

Résumé—En utilisant l'hélium comme indicateur, les auteurs ont étudié la distribution des " durées de séjour " d'un gaz qui s'écoule au travers d'un régénérateur de catalyseur fluidisé. Tandis que 5 % de la masse totale du gaz présente une " durée de séjour " inférieure à la moitié de la valeur moyenne, il y en a aussi 5 % dont la " durée de séjour " dépasse le double de cette valeur. Il semble qu'au travers d'un lit fluidisé, l'écoulement gazeux se rapproche beaucoup plus d'une veine cylindrique rigide que d'un régime turbulent, la " rétension " reste inférieure à 0·1.

When gas flows through a fluidised bed of solid particles, there is a considerable spread of residence-times ; different molecules of entering gas spend different lengths of time in passage, so that part of the gas spends less, and part more than the average length of time in the bed. This spread of residence-times, which is caused by back-mixing, or departure from piston-flow, is one of the most important characteristics of a fluidised reactor, as it will influence the nature and rate of the reaction. As yet, however, there is very little published information on the subject.

Gilliland and his co-workers [3] have measured the distribution of residence-times (and related characteristics) in model fluidised beds. However, there is no certainty that these reproduce the behaviour of the large-scale beds used industrially. For instance, in small beds the solid material can be seen to follow a circulatory path, with dimensions determined by those of the confining vessel, and this circulation of the solid helps to bring about back-mixing of the gas [3]. However, it is not possible to predict the way in which this motion will develop as the absolute dimensions of the bed are increased. It would be exceedingly difficult, perhaps impossible, to ensure rigorous dynamical similarity between fluidised beds whose linear dimensions are in the proportion of, say, 120 to 1 (40 ft to 4 in.). Such similarity was not attempted in Gilliland's work.

Some of the difficulties involved in the use of models can be avoided by carrying out experiments on industrial equipment. Other difficulties then arise, however ; for instance, the points at which measurements are made are often dictated by their accessibility, and it is not usually possible to cover a great range of variables. Nevertheless, information obtained in this way may provide a wholesome test of the value of data obtained in the laboratory, as well as being specifically applicable to the equipment on which the experiments are made.

The only published work which provides information on the distribution of residence-times in a large fluidised bed is that of Askins *et al* [1]. Samples of gas were taken from a number of points in a cracking-catalyst regenerator. The uniformity of composition of these samples suggested that the gas in the bed was completely mixed ; however, there was reason to suppose that the samples were not truly representative of the gas passing the sampling-points, since proportionately less gas was drawn from the " bubbles " which occur in the bed than from the interstitial gas flow. This phenomenon was also encountered by Gilliland and Mason [3].

The experiments described here were carried out on a catalyst regenerator virtually identical with that used by Askins *et al* [2] although the air-rate and quantity of catalyst may have differed in the two cases.

A study of the residence-times of gas in a blast-furnace, using radon as a tracer, was carried out by VOICE [6], and has some points of similarity with the work described here.

GENERAL PRINCIPLES

The problem of defining and measuring the distribution of residence-times in continuous-flow systems has been discussed elsewhere by DANCKWERTS and by other authors ([2], [3], [4], [5]). In outline, the method adopted here was to inject a dose of tracer (helium) at the entrance to the system, and to determine its time-distribution in the outflowing stream.

Fig. 1 is a diagrammatic view of the regenerator. Spent catalyst from the reactor is blown up the catalyst riser by a stream of air and enters the bottom cone of the regenerator. Between the bottom cone and the main body of the regenerator is a grate-like grid. The area of the slots in the grid which allows the passage of gas and catalyst is approximately 10% of the total cross-section of the cylindrical part of the regenerator. Catalyst drawn from above the grid passes down two pipes to the catalyst recycle coolers (only one is indicated on the diagram); the catalyst is blown up through the coolers into the bottom cone by a stream of air. Roughly half the air enters the regenerator via the main riser, the remainder entering with the recycled cooled catalyst and through the pipe A. The regenerated catalyst is withdrawn from the space above the grid, and the gases pass out via the cyclones and the stack.

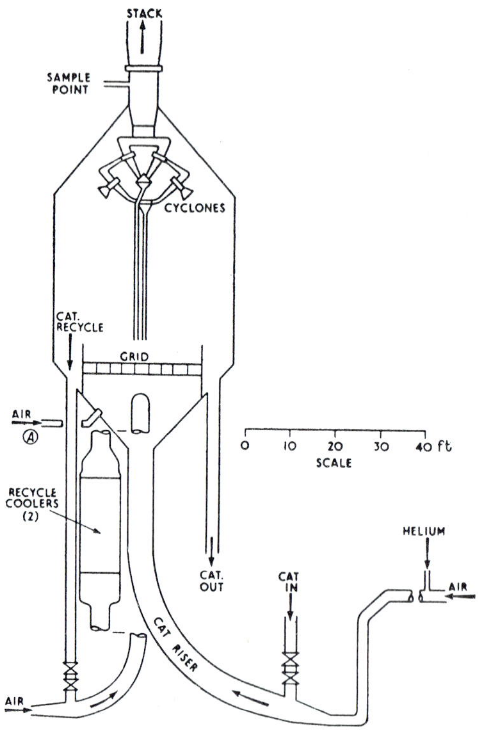

Fig. 1. Fluidised catalyst regenerator.

The procedure in these experiments was to inject rapidly a measured amount of helium into the air-line to the catalyst riser, and subsequently to take a series of samples of the stack gas at measured intervals. The ratio of helium to nitrogen was determined in these samples. Helium was used as a tracer (*a*) because of its unique molecular weight, which enables it to be estimated easily by mass-spectrometry; (b) because there is a negligible amount normally present in the stack gas; (*c*) because it would not be adsorbed by the catalyst.

The amount of information which can be derived from such an experiment depends on the validity of a cumulative series of assumptions.

(*a*) Assuming no gross experimental errors, and provided the injection is completed in a time much less than the mean residence-time in the regenerator (c. 30 secs) (a condition fulfilled in these experiments), the experiment gives unequivocally the distribution of residence-times of the helium atoms in their passage from the injection-point to the sampling-point.

(*b*) Assuming that molecular diffusion and gravitational convection play a negligible part in determining the paths of gas-molecules in the regenerator, in view of the high flow-velocities reigning in most regions; and that the injected helium is uniformly and rapidly spread over the cross-section of the air-pipe by turbulence ($Re \approx 1{\cdot}5 \times 10^6$); then the distribution of residence-times of the nitrogen which enters via the riser will be the same as that of the helium. The helium can then be regarded as "labelling" the nitrogen with which it enters.

(*c*) Assuming that the spread of residence-times between the injection-point and the bottom of the regenerator is very small (i.e. all the helium atoms enter the regenerator in a time much less than the mean residence-time in the regenerator) and that the time of passage between these points (c. 5 sec) can be calculated, it is possible to find the distribution of residence-times of the nitrogen in the regenerator itself.

(*d*) Virtually all the air entering the regenerator does so by roughly the same path, as it enters via the bottom cone below the grid. The stream of air and catalyst entering via the riser is split by a distributor into a number of non-vertical streams. The interaction of these with the streams entering from the catalyst coolers and by pipe A no doubt gives rise to a high degree of agitation in the space below the grid. If it is assumed that this brings about perfect mixing in this space (i.e. entering gas molecules are rapidly and uniformly dispersed throughout the space) then the distribution of residence-times will be the same for nitrogen entering by any of the various routes. The experiment can then be taken to give the distribution of residence times of the nitrogen flow as a whole in its passage through the regenerator.

The quantity of nitrogen leaving with the regenerated catalyst is known to be negligible in comparison with the total flow, so that virtually all nitrogen leaves through the stack. A small amount of fluidising air is put into the downcomers to the recycle coolers, and it is considered that this prevented any significant amount of helium being carried round the recycle cooling circuits ; the residence-times measured therefore refer to the regenerator exclusive of the coolers.

Experimental

Helium Injection—The helium was introduced from the storage cylinder into a pressure-vessel from which air had been displaced by filling it with water. The volume of gas in the vessel was about 1·3 cu. ft. and the pressure about 150 p.s.i.g. before and about 50 p.s.i.g. after helium had been released into the air-line. The pressure-vessel was connected to the air-line by a $\frac{3}{4}$ in. pipe about 3 ft. long, which was closed by a simple cock. The injection of helium was accomplished by opening and closing the cock as quickly as possible. The timing drum (see below) indicated that helium was flowing into the air-line for a period of less than 0·2 sec. in both runs. The quantity of helium injected could be calculated from the ambient temperature, the volume of the pressure-vessel, the initial pressure, and the pressure after the residual gas had returned to ambient temperature. The gauge was tested after the experiment and found to be accurate.

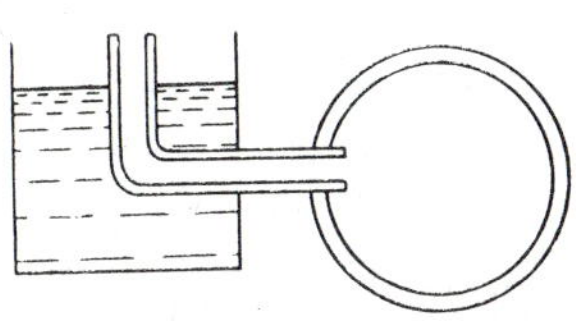

Fig. 2. Sampling pipe and branch, with cooling bath.

Sampling—Samples of the stack-gas were taken through a 4 ft. long 1 in. pipe projecting into the stack. This pipe was vented freely to the atmosphere, and the pressure-drop (9 p.s.i.) was sufficient to cause the gas to flow through the pipe at a high velocity. 12 side-branches of $\frac{1}{8}$ in. I.D. tube were brazed into the pipe, and passed through a water-bath so that their outer ends remained cool (Fig. 2). The sample vessels were attached to these branch tubes by rubber pressure-tubing. A sample vessel is shown in Fig. 3. A three-way tap C gave access either to the purge vessel or to the sample-vessel proper. Each vessel was connected via a header-pipe to a vacuum pump, and before sampling both vessels were evacuated

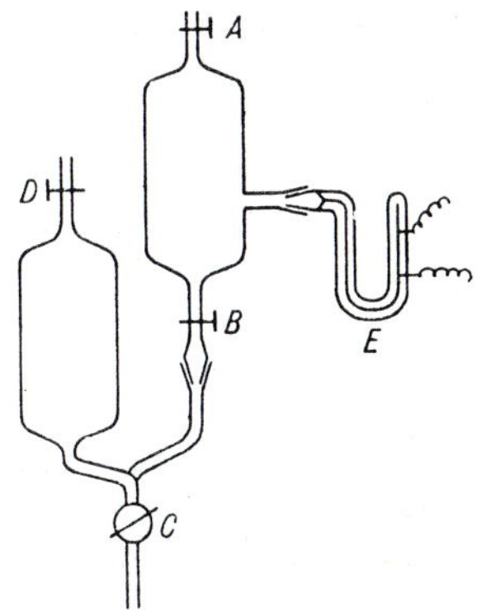

Fig. 3. Purge and sample vessels.

down to tap C, which was turned so as to isolate them from the stack-gas and from one another. Tap A was then closed (tap D could be left open if the vacuum pump was kept running). When it was desired to take a sample, a rapid turn of tap C through some 220° brought about the following sequence of events : (a) the purge vessel was opened to the branch-tube, and sucked through it several times its own volume of stack-gas ; (b) the purge vessel was isolated ; (c) the sampling vessel was opened to the branch-tube and sucked in a sample of stack-gas ; (d) the sampling vessel was isolated. Tap B was closed at leisure, and the sample removed for analysis.

By this method the branch-tube was flushed out with the gas to be sampled immediately before the sample was taken.

Time Measurements—The time of injection and the times at which the samples were taken were recorded by an electrically-operated stylus on a rotating drum. The stylus drew a horizontal line except when any one of 13 electrical circuits was closed, when it moved upwards and drew a line at a higher level so long as the circuit remained closed. The injection of helium caused one circuit to close and remain closed so long as the helium flow continued (see Fig. 4). The stylus thus recorded the time and duration of the injection. As each sample-vessel was filled, a thread of mercury rose into the evacuated arm of a small capillary-tube manometer (E, Fig. 3). In doing so it momentarily made a contact between two platinum wires, and the stylus recorded a " blip " on the drum. Action at the injection, sampling and timing-points was coordinated by field-telephone.

Run 1.

Nitrogen flow rate : $n = 1{\cdot}85$ *lb. mole/sec.*
Calculated time of passage from injection-point to base of regenerator = 5·4 *sec.*

Time after entry of He	t (*sec.*)	9·6	15·1	20·6	25·2	30·7	41·8	46·8	51·8
$(He/N_2) \times 10^6$	$10^6\,c'$	0	146	351	276	208	116	72·5	55·8
$(He/CO_2) \times 10^6$	$10^6\,a'$	0	926	2180	1850	1310	752	476	374
CO_2/N_2	c'/a'	—	0·158	0·161	0·149	0·159	0·155	0·151	0·149
$(He/N_2) \times 10^6$	$10^6.\ a'.\ (c'/a')_m$	0	143	338	286	202	116	73·5	57·7

$$\left(\frac{c'}{a'}\right)_m = 0{\cdot}155\ ; \qquad \frac{CO_2}{N_2}\ (\text{Orsat}) = 0{\cdot}148$$

$$\text{Helium injected : } Q' = 0{\cdot}0195 \text{ lb. mole ; } n\int_0^\infty c'\,dt = 0{\cdot}0149 \text{ lb. mole}$$

$$\text{Mean residence time : } \frac{N}{n} = 33{\cdot}7 \text{ sec ; } \quad \int_0^\infty c't\,dt \Big/ \int_0^\infty c'\,dt = 32{\cdot}0 \text{ sec.}$$

Run 2.

Nitrogen flow-rate : $n = 1{\cdot}87$ *lb. mole/sec.*
Calculated time of passage from injection-point to base of regenerator = 5·3 *sec.*

Time after entry of He	t (*sec.*)	4·7	11·0	14·2	16·2	18·6	21·2	23·1	31·7	49·7	60·1
$(He/N_2) \times 10^6$	$10^6\,c'$	0	16·6	67·0	147	198	244	256	165	46·5	23·1
$(He/CO_2) \times 10^6$	$10^6\,a'$	0	112	435	961	4350	1670	1730	1040	323	156
CO_2/N_2	c'/a'	–	0·148	0·154	0·152	0·146	0·146	0·148	0·145	0·146	0·146
$(He/N_2) \times 10^6$	$10^6.\ a'.\ (c'/a')_m$	0	16·6	64·5	143	200	248	256	154	47·9	23·1

$$\left(\frac{c'}{a'}\right)_m = 0.148\ ; \qquad \frac{CO_2}{N_2}\ (\text{Orsat}) = 0{\cdot}148$$

$$\text{Helium injected : } Q' = 0{\cdot}0191 \text{ lb. mole ; } n\int_0^\infty c'dt = 0{\cdot}0114 \text{ lb. mole.}$$

$$\text{Mean residence time : } \frac{N}{n} = 33{\cdot}5 \text{ sec. ; } \quad \int_0^\infty c'\,tdt \Big/ \int_0^\infty c'dt = 33{\cdot}3 \text{ sec.}$$

Analyses—The samples were analysed with a Metropolitan Vickers MS2 Mass Spectrometer. No modifications were made to the instrument, which was currently in use for routine hydrocarbon gas analysis. However, in order to achieve a high sensitivity for the detection of helium (0·0005 %) the sample pressure on the high pressure side of the introduction leak to the ionization chamber was increased to 20 mm.Hg.

Some of the samples were contaminated by air which leaked into the sample vessels ; a correction was made for this by determining the ratio of O_2 to N_2 in the samples (the stack gas itself contained not more than 0·2 mol % O_2, dry basis).

The accuracy of the reported ratios of helium to nitrogen was estimated as :

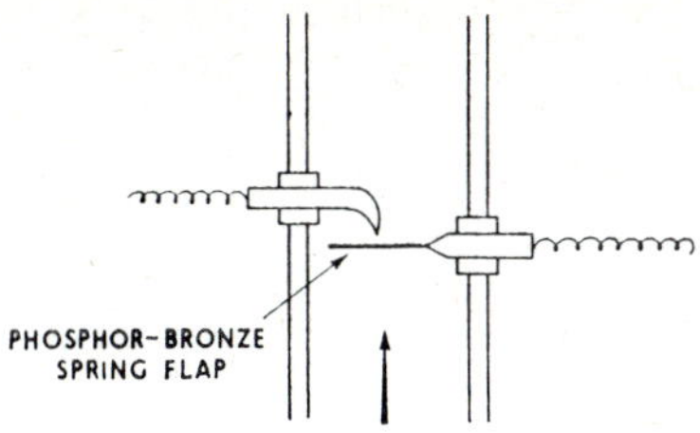

Fig. 4. Helium injection timing device.

$$\frac{\text{true (He/N}_2)}{\text{reported (He/N}_2)} = \frac{c}{c'} = \alpha (1 \pm \delta)$$

the error δ having a maximum value of 0·1 and the coefficient α (which was used only in the helium material balance) being a constant for the samples in any one run. The ratios of the true and reported He/CO_2 ratios were similarly related, α having the same value. It was found that at the abnormally high sample pressures, mass discriminatory effects made the absolute value of α difficult to determine. These discriminatory effects, which were absent for heavier molecules ($M \gg 20$) at 20 mm.Hg., and for helium atoms at lower sample pressures were thought to be due either to the onset of hydrodynamic flow, or more probably to space-charge effects within the ionisation chamber. On the basis of synthetic blends which were made up and analysed, α was at first believed to lie between 0·9 and 1·1. However, in view of the discrepancy in the helium balance (see " Interpretation of Results ") further calibrating measurements were made under conditions simulating more closely those prevailing during the analysis of the experimental samples. The calibrations indicated that α had the value of 1·33 for Run 1, and 1·49 for Run 2. The former gives a helium balance to about 1 %, but the discrepancy in the latter is still about 10 %.

Interpretation of Results

The tables show the reported ratios of He to $N_2(c')$ and He to $CO_2(a')$. The ratio (c'/a') should be equal to the ratio of CO_2 to N_2 in the sample, and hence should be virtually constant in each run and equal to the mean ratio obtained from the analyses of the stack-gas carried out before and after each run. This provides a useful check on the consistency of the analyses.

Multiplying the He/CO_2 ratios, a', by the mean value $(c'/a')_m$ for the run gives a second series of values of the He/N_2 ratios in each run. Since these contain independent information, they have been given equal weight with the directly-determined values of c' in plotting the distribution-curves.

From a knowledge of the dimensions of the air-pipe and catalyst-riser, and the volume of the catalyst present in the latter, it was possible to calculate the delay between the injection of the helium and its arrival at the bottom of the regenerator. This delay was subtracted from the time at which a sample was taken, to obtain the value of t, the time between the entry of the helium at the bottom of the regenerator and taking the sample. The calculation of the delay (c. 5 sec) was subject to the uncertainties discussed below in connection with the calculation of the mean residence-time in the regenerator, but the error involved is not thought to have been great enough to have any substantial effect on the subsequent interpretation of the results.

If c is the mole ratio of He to N_2 in the stack-gas at time t, and n is the total rate of feed of nitrogen (which virtually all leaves by the stack), the rate of escape of He up the stack at time t is nc. Thus the total quantity Q of helium injected is given by

$$Q = n \int_0^\infty c\mathrm{d}t. \tag{2}$$

A fraction $\frac{nc}{Q} \, . \, \mathrm{d}t$ of the helium escapes between t and $(t + \mathrm{d}t)$ – i.e. this fraction of the helium has residence-times between t and $(t + \mathrm{d}t)$. Thus the mean residence-time of the helium as a whole is

$$\theta = \frac{n}{Q} \int_0^\infty ct\mathrm{d}t$$

$$= \int_0^\infty ct\mathrm{d}t \Big/ \int_0^\infty c\mathrm{d}t \, . \tag{3}$$

In order to determine Q and θ from equations (2) and (3), c' and $c't$ can be plotted against t, and the curves integrated graphically. However, each curve has a " tail " extending beyond the time of the last observation. In order to estimate the area under the tail, $\ln c'$ was plotted against t for each run, and it was found that the later values of c' in each case fell on a straight line with slope

$$-\frac{\mathrm{d} \ln c'}{\mathrm{d}t} = 0{\cdot}065 \text{ sec}^{-1}.$$

Assuming that the tails of the c' vs. t curves continue to obey this relationship, the relevant integrals can be easily computed. (In fact, since some helium will find its way into the pores of the catalyst, from which it will diffuse out slowly, the tail probably becomes flatter than the above expression suggests. The above method thus probably under-estimates the mean residence-time. However the discrepancy is probably not large, as the mean residence-time θ computed from equation (3) agrees well with that estimated from plant operating data. The good helium balance obtained in Run 1 also tends to support this view.)

θ can also be found by dividing the quantity N of nitrogen present in the regenerator at any instant by the nitrogen feed-rate, n :

$$\theta = \frac{N}{n}. \tag{4}$$

N is computed from the void-volume of the regenerator and the temperature, pressure and composition of the gas. The computation is subject to various uncertainties :

(*a*) The quantity, and hence the volume, of the catalyst in the regenerator cannot be measured directly and must be inferred from the pressure drop in the catalyst bed. (The pore-volume of the catalyst was included in the total void-volume, of which it comprised about 5%.)

(*b*) The pressure fell from about 16 p.s.i.g. at the bottom to about 9 p.s.i.g. at the top of the regenerator. There was a drop of about 1 p.s.i. across the grid. Estimated average values were used for the spaces above and below the grid.

(*c*) The temperature above the grid was fairly uniform (600°C) ; however, there was a considerable rise in temperature between the bottom of the regenerator and the grid (from about 540 to 600°C) and an estimated average was used for this relatively small volume (5% of total void space).

(*d*) No information was available concerning the variation of composition of the gas with the level in the regenerator (although the work of ASKINS *et al* [1] suggests that it may be uniform above the grid). A uniform value of 71 mole % N_2 (wet basis) throughout the regenerator was used in the computation. This corresponds to the composition of the stack gas. Since the actual concentration of nitrogen could at no point have been greater than 79 % or less than 65%, the error introduced by this approximation was probably considerably less than 10%.

In the table of results the value of the mean residence-time has been calculated from N/n for each run, and can be compared with the value obtained by integration according to equation (3). The agreement is satisfactory considering the uncertainties involved. The values of θ obtained from equation (3) are used in subsequent calculations.

On the other hand, there is a considerable discrepancy in each run between Q', the quantity of helium calculated from the pressure and volume of the container, and the quantity Q calculated from equation (2) and the reported helium ratios. In view of the internal consistency of each run, the agreement between the residence-times calculated by each method, and the agreement (demonstrated later) between the reduced distribution-curves for the two runs, it seems clear either that the reported quantity, Q', of helium injected was larger by a factor α (different in each run) than the true quantity Q; or that the reported helium ratios, c' and a', were smaller by a factor $\frac{1}{\alpha}$ (different but *constant* in each run) than the true ratios c and a. For the purpose of interpreting the results it makes no difference which error has occurred. In either case, $cQ'/c'Q = \alpha$, and equation (2) takes the form

$$Q' = \alpha n \int_0^\infty c' \, dt, \tag{5}$$

while equation (3) remains valid (putting c' for c). α is then found to have the values 1·31 for Run 1, 1·67 for Run 2. As explained under "Analyses," further experiments with the mass spectrometer indicated that in fact the reported

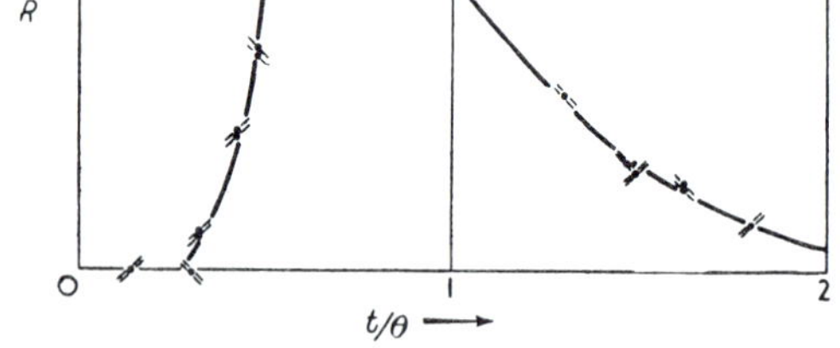

Fig. 5. *C*-diagram for Runs 1 and 2. The area under the curve between any two ordinates represents the fraction of material having residence-times between the corresponding values of *t*.

values c' should be multiplied by a factor of 1·33 in Run 1, and 1·49 in Run 2. The helium balance is thus very good for Run 1; the discrepancy in Run 2 remains unexplained.

It is desirable to plot the results in dimensionless form, as a reduced distribution-curve, or C-diagram [2]. This leads to a curve which is independent of the quantity of tracer used, and which directly represents the distribution of residence-times. The ordinate, $R\{t\}$, is the ratio of the concentration of helium in a sample taken at time t to that which would be obtained by dispersing the helium uniformly throughout the regenerator, and was calculated as follows:

$$R\{t\} = \frac{Nc}{Q} = \frac{N\alpha c'}{Q'} = \frac{N\alpha c'}{\alpha n \int_0^\infty c'\,dt}$$

$$= c'\frac{\theta}{\int_0^\infty c'\,dt} = c'\frac{\int_0^\infty c't\,dt}{\left[\int_0^\infty c'\,dt\right]^2} \qquad (6)$$

The abscissa is the ratio of t to the mean residence-time:

$$t/\theta = t\frac{\int_0^\infty c'\,dt}{\int_0^\infty c't\,dt}. \qquad (7)$$

It will be noticed that no figures derived from the plant operating data (e.g., n, N) are involved in calculating R and t/θ by this method. Only the reported values of the He/N_2 (and He/CO_2) ratios and the measurements of time are used. Equations (6) and (7) are exact as long as the error α is constant in each run.

R is plotted against t/θ in Fig. 5 to give the C-diagram for each run. Since operating conditions were almost identical during the two runs, the two curves should also be identical In fact the points appear to lie on a single curve. This is strong evidence for the internal consistency of the analyses for each run; it also suggests that the methods of evaluating $\int_0^\infty c'\,dt$ and $\int_0^\infty c't\,dt$ are not subject to any substantial errors. The area under the curve between any two values of t/θ is equal to the fraction of the material which has residence-times between the corresponding values of t.

The fraction, $F\{t\}$, of the helium (or nitrogen) spending less than a specified time, t, in the regenerator, is given by

$$F\{t\} = \int_0^t R\{t\}\,d\,(t/\theta). \qquad (8)$$

Values of F can therefore be obtained by graphical integration of the C-diagram, and are plotted against t/θ in Fig. 6, which is known as an F-diagram (2).

Discussion

The most striking feature of the diagrams (Figs. 5 and 6) is that none of the gas spends less than one quarter of the mean residence-time in the regenerator, and only some 5% of it spends less than half. This proves that under the conditions of

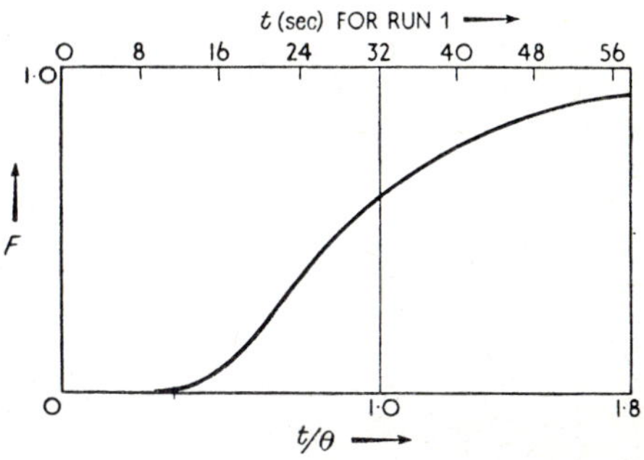

Fig. 6. F-diagram for Runs 1 and 2. The value of F for a given value of t is equal to the fraction of material spending times less than t in its passage through the system.

operation the gas in the regenerator (above the grid) is not completely mixed; in other words, there is no effective back-mixing between the top and bottom of the regenerator. The hold-back of the system is 0·17, whereas for a completely-mixed system it would be 0·37, and for piston-flow zero. (The hold-back is a measure of the departure from piston-flow, and is equal to the area under the F-diagram up to the ordinate $t/\theta = 1$; if the total quantity N of nitrogen present in the regenerator at a given instant were labelled, the hold-back would be the fraction of labelled nitrogen still present after a quantity N of unlabelled nitrogen had flowed in [2]). Only about 5% of the gas spends more than twice the mean residence-time in the system.

It is not possible to deduce from the experimental results the exact distribution of residence-times of the gas in its passage through the fluidized bed itself, since little is known *a priori* about the nature of the flow in the space above the bed, which has a profound influence on the overall distribution as measured. However, a surprising amount of information can be extracted by the method of analysis to be described.

It seems probable that the gas in the space below the grid is completely mixed, owing to the impingement of various streams at relatively high velocities. If this is so, the entering helium is instantaneously dispersed throughout the interstitial gas-space below the grid; its mean residence-time in this space is calculated (in Run 1) to have been 2·4 sec. The pressure-drop of 1 p.s.i. across the grid makes it probable that the gas flow-rate is uniform through all parts of the grid. The high velocity between the bars makes it highly unlikely that there is any back-mixing across the grid.

The height of the catalyst bed above the grid was estimated, from the pressure gradient and total pressure-drop in the bed, to be about 17 ft. (This implies a void-fraction of 0·73, excluding pores). Hence the total void-volume in the bed (including pores) was calculated to be 29% of the gas space above the grid, and the mean residence-time of the gas in the bed was taken to be 29% of that between grid and stack, or 8·5 sec.

If it is further assumed that there is no back-mixing between gas which has left the catalyst-bed and that within it, the flow through the regenerator may be represented as the flow through three compartments which discharge into one another in series (Fig. 7).

Gilliland and Mason [3] showed that the residence-time distribution in their model fluidised beds could be represented approximately by expressions of the form :

$$\left.\begin{aligned} F &= 0, & \frac{t}{\theta} &< \frac{S-1}{S} \\ F &= 1 - e^{\left(S-1-\frac{St}{\theta}\right)}, & \frac{t}{\theta} &\geqslant \frac{S-1}{S} \end{aligned}\right\} \quad (9)$$

where S is a parameter which depends mainly on the geometry of the bed, and θ the mean residence-time in the bed. When $S = \infty$ the above expression reduces to the F-diagram for piston-flow, and when $S = 1$ we have the expression for "complete mixing." Intermediate values of S represent systems with different degrees of hold-back between these limits; the hold-back in the system is $1/Se$. Some F-diagrams with different values of S are shown in Fig. 8.

If it is assumed that the distribution in the large fluidised bed in the catalyst regenerator also conformed to equation (9), it can be shown that the value of S must lie between certain limits. Regarding the regenerator as equivalent to three compartments in series (Fig. 7) with "complete mixing" in the first (A) and a distribution corresponding to equation (9) in the second (B), it is possible to relate the observed overall distribution to the unknown distribution in the third compartment (C) by the following expression :

$$F_3\{t\} = F\{\tau\} + \left[\frac{\theta_1}{\theta} + \frac{\theta_2}{S\theta}\right] R\{\tau\} + \frac{\theta_1\,\theta_2}{S\theta^2}\left[\frac{\mathrm{d}R\{t'\}}{\mathrm{d}(t'/\theta)}\right]_{t'=\tau}, \quad (10)$$

where $\tau = t + \theta_2\left(\frac{S-1}{S}\right)$.

In this expression $F_3\{t\}$ is the fraction of material which spends times less than t in its passage through the third compartment; θ_1, θ_2, θ are the mean residence times in A, B and the whole system respectively; and the number S, which determines the distribution in compartment B, is unknown. $F\{\tau\}$ and $R\{\tau\}$ are the values of F and R for the system as a whole (taken from Figs. 5 and 6 respectively in the present case) at a time $\tau = t + (S-1)\,\theta_2/S$; the last term (in square brackets) is the slope of the C-diagram (Fig. 5) for the whole system at time τ.

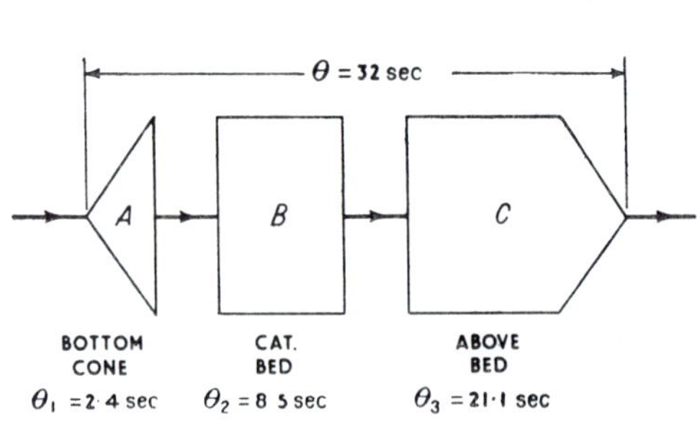

Fig. 7.

Thus, if a value is assigned to S, the corresponding F_3-diagram for the third compartment can be constructed from the data for the system as a whole. It is found by trial that if S is taken to be less than 2, the F_3-diagram has a region of negative slope, which represents a physical impossibility. If S is 5 or less the F_3-diagram displays a point of inflection (i.e. slope has a minimum value). This corresponds to a C_3-diagram with two peaks, and while such behaviour is possible for systems in which, for instance, the gas flows through two parallel and dissimilar channels, it can be confidently assumed that it does not occur in the space above the bed. The conclusion is that S must be greater than 5 for flow through the fluidised bed, the hold-back being less than 0·075. In other words (see Fig. 8) the flow of gas through the bed does not depart very far from piston-flow.

It should be emphasised that this conclusion, in its qualitative form, does not depend on the exact validity of the assumptions made, or on exact conformity with equation (9); these merely serve to simplify the calculations. Provided there is no substantial degree of back-mixing between the three "compartments," and the estimate of the void-volume of the bed is approximately correct, the experiment can be taken to show that the flow through the bed is much closer to piston-flow than to complete mixing.

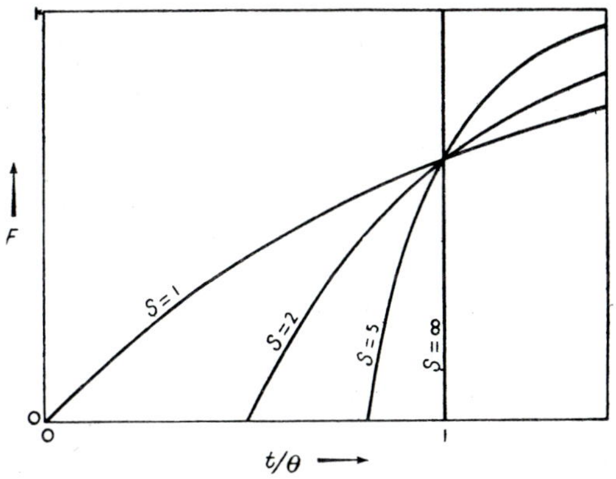

Fig. 8. $F = 1 - e^{\left(S-1-\frac{St}{\theta}\right)}$.

Conclusions

Although the distribution of residence-times for the gas flowing through the fluidised bed cannot be deduced exactly from the experiments, it can be concluded with some confidence that it is much closer to piston-flow than to complete mixing, and has a hold-back of less than 0·1.

Acknowledgment

The experimental work described was carried out during a summer vacation course at the Stanlow Refinery of the Shell Refining and Marketing Company. We are indebted to the Company for their generosity in enabling the work to be carried out and allowing the results to be published, and to the technological staff for their collaboration at all stages.

Notation

c = true mole ratio of He to N_2 in sample
c' = reported value of above
a = true ratio of He to CO_2 in sample
a' = reported value of above
Q = true quantity of He injected (lb. mole)
Q' = apparent value of above calculated from pressure change in container
$\alpha = cQ'/c'Q$
N = nitrogen contained in regenerator (lb. mole)
n = feed-rate of nitrogen to regenerator (lb. mole/sec.)
θ = mean residence time of nitrogen in regenerator (sec.)
t = time after injection of helium,less calculated time of passage from injection-point to bottom of regenerator (sec.)
$R = Nc/Q$ (see equation (4))
F = fraction of material spending less than specified time in passage

$$\tau = t - \theta_2 \left(\frac{S-1}{S}\right)$$

S = parameter in equation (9)

Subscripts :

1, 2, 3 refer to gas-spaces below grid, in catalyst bed and above catalyst bed respectively.

References

[1] Askins, J. W., Hinds, G. P. and Kunreuther, F.; *Chem. Eng. Prog.* 1951 **47** 401.
[2] Danckwerts, P. V.; *Chem. Eng. Sci.* 1953 **2** 1.
[3] Gilliland, E. R. and Mason, E. A.; *Ind. Eng. Chem.* 1949 **41** 1191, 1952 **44** 218.
[4] Kramers, H. and Alberda, G.; *Chem. Eng. Sci.* 1953 **2** 173.
[5] Schoenemann, K.; *Dechema-Monographien* 1952 **21** 203.
[6] Voice, E. W.; *J. Iron and Steel Inst.* 1949 **163** 312, 1950 **166** 315.

D4

INVESTIGATION OF AIR-FLOW IN A SPRAY-DRIER BY TRACER AND MODEL TECHNIQUES

By G. PLACE, M.A. (ASSOCIATE MEMBER)*, K. RIDGWAY, B.Sc. (ASSOCIATE MEMBER)† and Prof. P. V. DANCKWERTS, G.C., M.A., S.M. (MEMBER)‡

SUMMARY

This paper describes an investigation of the behaviour of the air flowing through a large spray-drier, making use of a tracer technique, and experiments in a 1/80 scale model of the drier, using water as the fluid.

Introduction

The drier is shown diagrammatically in Fig. 1. The slurry of detergent to be dried is sprayed in through a number of nozzles near the top. The air enters through a plenum chamber which opens into the drier through twelve ports situated at the top of the lower cone. These ports are directed downwards at 25° to the horizontal and tangentially at 17½° to the radial direction. However, by means of adjustable vanes in the ports it is possible to vary the angle of swirl in the horizontal plane. The so-called "swirl angle" is the angle of the deflector vanes relative to the radial direction. The combined cross-sectional area of the inlet ports is roughly 1/10 that of the cylindrical tower.

During the experiments the operating variable of air flow, inlet temperature, and slurry flow remained reasonably constant. The "swirl angle" varied from 25° to 33° (except for two runs at 22·5°).

The declared advantages[1] of imparting a swirl to the air are that it increases turbulence, and hence transfer-coefficients, between droplets and air; gives good mixing along the axis of flow, and thereby reduces the difference in air temperature between the top and bottom of the tower (an advantage when drying detergents); and tends to stabilise the flow-pattern. The bottom cone of the drier will also act as a cyclone separator for the disengagement of the powder.

Against these advantages must be set the likelihood that some undried droplets will be flung outwards by centrifugal force and stick to the upper part of the wall (this does in fact happen to some extent). There was also a suspicion that a considerable amount of the entering air might "channel" up the centre of the tower; if this air had substantially less than the mean residence-time it might leave unduly dry and hot without having made its fair contribution to the drying process, particularly since centrifugal force might be expected to decrease the concentration of droplets on the axis of the tower.

The object of the full-scale investigation was to find out more about the residence-time distribution and flow-pattern of the air in the drier by a tracer technique.

Principle of Method

Danckwerts[2] had previously suggested that it should be possible to find out a good deal about the flow-pattern of a fluid in a continuous-flow system by injecting a "pulse" of some tracer at the inlet, and subsequently measuring and recording the concentration of tracer as a function of time at varous points inside the system. In the present investigation it was more convenient to inject the tracer at a succession of points in the spray-drier; in each case the subsequent variation with time of the tracer-concentration in the air leaving the drier was recorded. Thence the average time taken by molecules to travel from a specified point to the exit can be deduced; or the time taken for any specified fraction (*e.g.* 50%) of the material starting from the neighbourhood of the point to pass out of the system can be calculated. This procedure should give considerable information about the flow-pattern of air through the drier, and in particular would be expected to show up regions of stagnation and of channelling or jet-like flow. It may be desirable to display the results in the form of contours of equal average exit-times; molecules starting from any point on such a contour-line take, on the average, the same length of time to reach the exit. A hypothetical example is shown in Fig. 2, which represents the cross-section of a cylindrical vessel. The contour lines are numbered in order of increasing average exit-times, and indicate channelling near the axis of the vessel, and stagnant regions near the corners.

However, the information obtained by tracer studies is not the same as would be given by a velocity-survey of the system. In turbulent pipe-flow, for instance, the velocity near the axis of the pipe is greater than that near the wall. Nevertheless, as shown by Taylor,[3] the exit-time for a pulse of fluid is the same whether it is injected on the axis or near the wall, because turbulence rapidly distributes the tracer across the pipe so that the mean velocity of the tracer molecules is equal to that of the fluid in the pipe, regardless of where the tracer is injected. In the case of a "perfectly-mixed vessel", of course, the concentration history at the exit is independent of the position of injection.

Further information may be obtained from the shape of the concentration-time curves and the degree of spread which they display. The spread is caused by mixing along the axis of flow or by transverse dispersal of the tracer into zones with varying axial velocities.

In addition to the internal exploration, tracer was injected into the air-stream entering the drier to get the over-all residence-time distribution.[4]

In the present investigation helium was used as a tracer because its concentration can be measured at low levels by means of a mass-spectrometer. It was assumed that helium molecules would behave in the same way as air molecules. This is probably justified, because the helium was so rapidly diluted with air in the turbulent conditions prevailing that buoyancy effects are likely to have been negligible. The molecular diffusivity of helium in air is considerably higher than the self-diffusivity of air, but molecular diffusion can play little part in the macroscopic aspects of the mixing effects observed; the mean residence-time of the air in the drier was of the order of 30 seconds, and in this time the

* Thomas Hedley & Co. Ltd., Whitley Road, Newcastle upon Tyne, 12.

† Department of Chemical Engineering, University College, Gower Street, London, W.C.

‡ Department of Chemical Engineering, Imperial College of Science and Technology, Prince Consort Road, London, S.W.7.

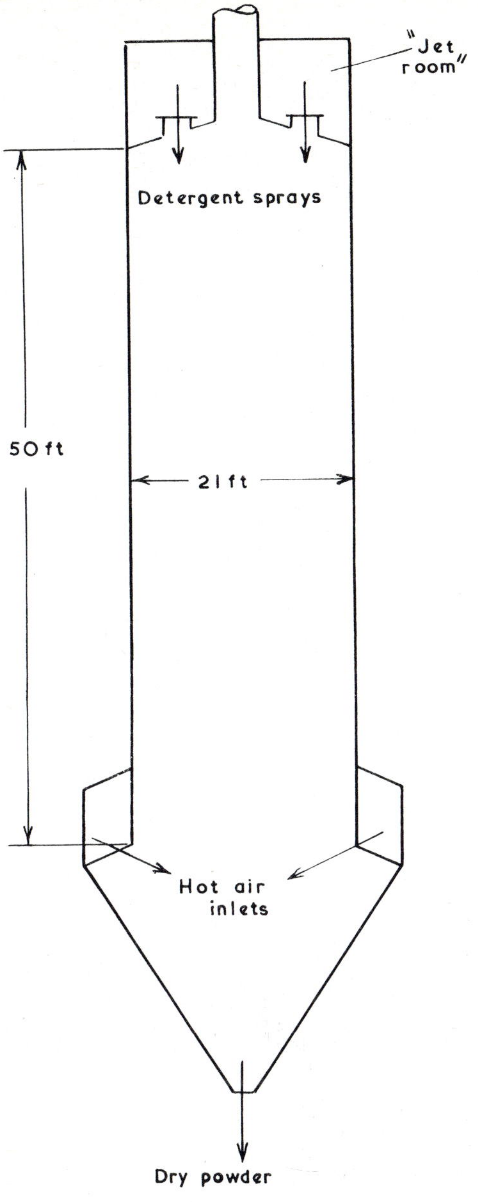

Fig. 1.—*The spray-drying tower*

root mean-square displacement of helium molecules diffusing in air ($\sqrt{2Dt}$) is only about 5 cm.

The method of injection was designed to concentrate the pulse into as short a time and space as possible, but the resulting initial cloud of helium was of finite and unknown dimensions. This is immaterial as long as the initial dimensions of the cloud are small compared to those of the drier; the larger the cloud, the less precisely can the flow-pattern be explored.

Each injection of helium was recorded on a chart of instrument deflection as a function of time. Since the sensitivity of

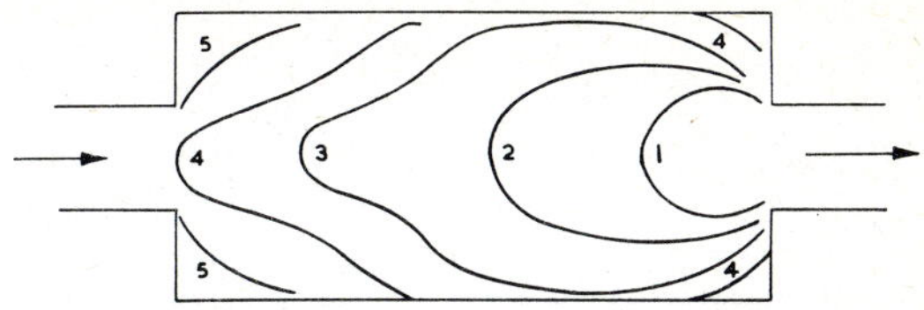

Fig. 2.—*Contours of equal average exit-times: hypothetical example, showing channelling along axis*

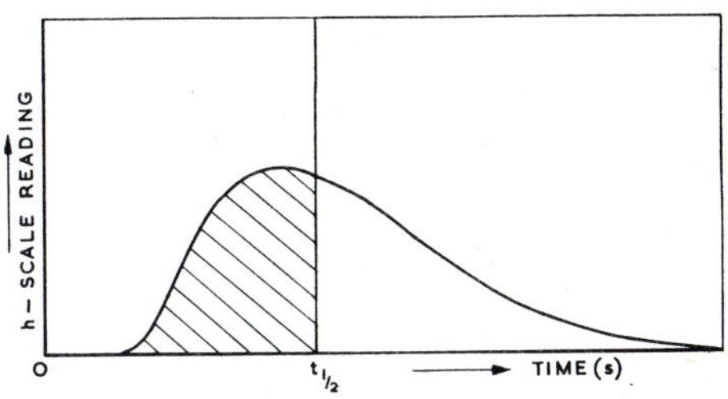

Fig. 3.—*50% exit time ($t_{\frac{1}{2}}$) divides area under curve in half*

the mass-spectrometer depends on the operating pressure in the ion-chamber (which varied from run to run, although constant in any one run) and since it tends to drift for other reasons, a laborious process of calibration and recalibration would have been necessary to convert the chart ordinates into absolute concentrations. In fact this is unnecessary for present purposes, so long as it can be assumed that the ordinate is proportional to the concentration of helium during any one run. This assumption seems quite justifiable on consideration of the physics of the processes occurring in the instrument. The height, h, of the ordinate at time t is then taken to be equal to Ac.* If c is expressed as a mole-ratio, and M is the molal flow rate of air, the molal rate of outflow of helium at time t is Mh/A. The total quantity of helium injected is therefore:

$$\frac{M}{A}\int_0^\infty h \,.\, \mathrm{d}t,$$

and the quantity which has left in time t after injection is

$$\frac{M}{A}\int_0^t h \,.\, \mathrm{d}t.$$

Thus the fraction of the helium injected which has left after time t is

$$\int_0^t h \,.\, \mathrm{d}t \Bigg/ \int_0^\infty h \,.\, \mathrm{d}t.$$

For instance, the time $t_{\frac{1}{2}}$ taken for half the helium to leave (the "50% time" or "median time") is that which divides the area under the curve into two equal parts (see Fig. 3; the shaded and unshaded areas are equal).

In some ways it would be logically more satisfactory to calculate the average time taken for the helium molecules to leave. In principle this can be done, since the average time is given by:[4]

* Symbols have the meanings assigned to them on p. 274.

$$\left(\int_0^\infty ht\,.\,\mathrm{d}t\right)\left(\int_0^\infty h\,.\,\mathrm{d}t.\right)$$

However, the evaluation of the top integral is very laborious if it has to be repeated for a large number of records; nor will it be very accurate, because of the heavy weight given to the "tail" of the curve, where the ordinate (concentration) is measured with low relative accuracy.

The curves obtained in these experiments were skew, with relatively sharp fronts and long tails. The average exit-time is thus greater than the median time, but in the cases where the comparison was made, the difference was found to be less than 15%.

Experimental

Helium injection

A rapid-acting lever-operated diaphragm valve allowed a burst of helium lasting ½ s or less to be injected into the tower from a storage vessel, thus fulfilling the necessary condition that the injected pulse must enter in a time short compared with the mean residence time. The amount of helium injected for each measurement was 0·5–2·3 ft^3 (S.T.P.). The lever-operated valve carried a pair of contacts which closed while the valve was opened and completed a circuit which gave a kink in the trace on the M.S. recorder and recorded the injection time.

The helium was fed into the tower through a vertical 1 inch diameter steel pipe projecting downwards through an opening in the top cone. The radial position of the injection point was varied by fitting horizontal sections of pipe of different lengths to the lower end of the vertical pipe. A baffle was fitted to the outlet to prevent the helium shooting out in a long jet.

Outlet gas sampling

The air sample at the outlet could be withdrawn from any desired position across the diameter of the outlet duct. A stream of air was drawn continuously through the spectrometer system, the total time-lag involved being negligible.

The spectrometer was a Metropolitan-Vickers model MS 6L; its external dimensions were approximately 3 ft × 3 ft × 5 ft high, and it was mounted near the offtake duct in the "jet-room" at the top of the drier. The sensitivity was approximately 1 inch deflection per 100 parts per million of helium at 10^{-4} mm operating pressure.

Results

Two injections were made with the helium-injection point at any given position on the lattice of Fig. 4, one with the sampling probe withdrawing outlet gas from a position 6 in. from the wall of the outlet duct, and the other with the sampling probe 6 in. from the centre of the outlet duct. There was in fact no systematic difference between the results from the centre and wall sampling points, and there were therefore treated as replicates. The exit times of the wall and centre samples for each station were averaged. Some specimen records are shown in Fig. 6.

Injections were also made at the entrance to the plenum chamber to determine the overall residence-time distribution of the air passing through the drier; 27 injections were made in the course of 2½ hours, and during this period the operating conditions did not change appreciably. The position of the sampling probe with respect to the centre of the duct was changed from run to run, but there was no correlation between the results and the sampling position. A summary of these results is given below:

	Mean	Range
Overall 5% time (s) . .	15·0	11·8–18·7
Overall 50% time (s) . .	25·9	22·0–30·1

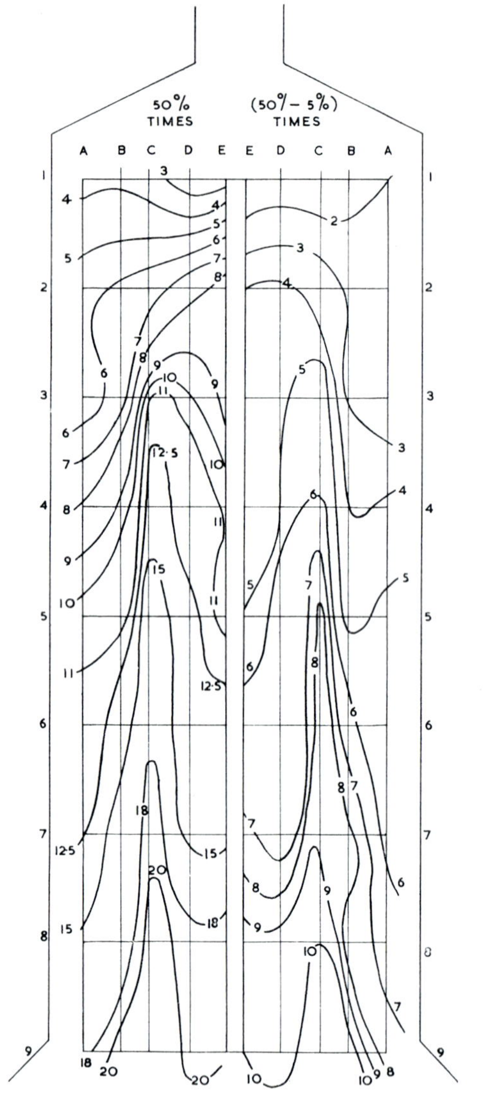

Fig. 4.—*Contours of equal 50% exit times and equal spreads (i.e. 50% time–5% time). Figures on contours are in seconds. Vertical distance between stations is 6 ft.*

The result of a typical run in which the helium was injected at the tower inlet is shown in the form of an F-diagram in Fig. 7. The mean residence-time was found in this case to be 28·5 s. The mean residence-time, calculated on the assumption that all the air in the tower was at the outlet temperature, was 28·6 s for all runs. In view of the scatter of the results, and the uncertainty of the average temperature in the drier, the measure of agreement for this particular run is not significant. The degree of back-mixing is moderate; the "hold-back"[4] is 0·15, corresponding to about 6 "mixing stages".

Discussion of Results of Tracer Experiments

One of the most striking features of the results was their wide scatter; this is shown, for instance, by the range of values

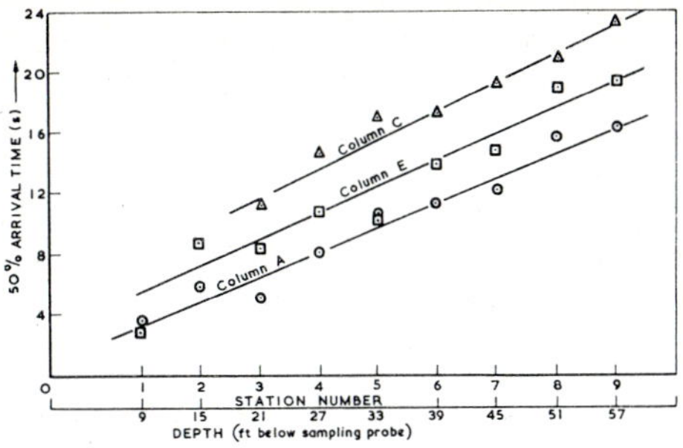

Fig. 5.—*50% exit time vs. height for injections on columns* A, C, E

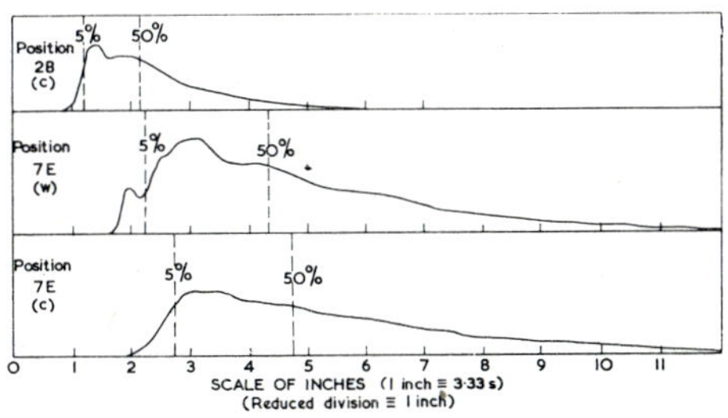

Fig. 6.—*Specimen records.* (C), (W) *mean that samples were taken near centre and wall of outlet duct respectively*

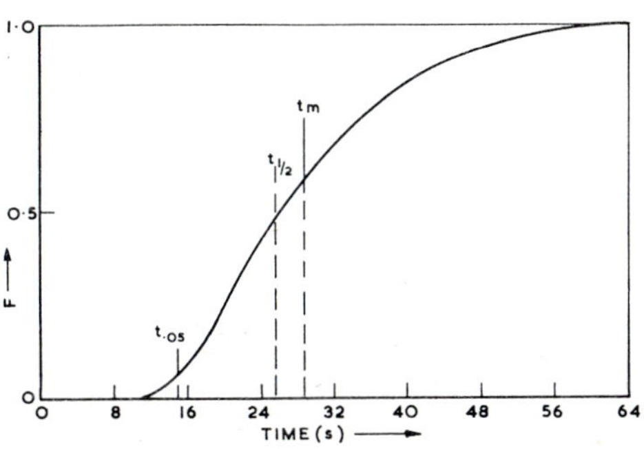

Fig. 7.—*F—diagram for run 210 (injection at inlet). F is the fraction of air which spends less than given time in passage.* $t_{1/2}$, $t_{.05}$ *are 5% and 50% exit times.* t_m *is mean residence-time*

for the overall 5% and 50% times measured under identical conditions. There were also substantial discrepancies between replicate internal injections. Instability of flow in the drier is also indicated by the irregular shapes of the curves for some individual runs. Oscillations in the slope, with a period of a few seconds, were observed in many runs (Fig. 6). The resulting fluctuations seriously limit the amount of information which can be obtained from a single injection. More precise information could, no doubt, be obtained by a frequency-response method using a sinusoidally-varying rate of injection, but the experiment would then become prohibitively elaborate.

Contours of equal 50% exit times and of equal "spreads" (*i.e.* 50% time − 5% time) are plotted in Fig. 4. Although the details are probably not significant, because of the variability in the individual results, their general shape is nevertheless clear. Some of these results are shown in Fig. 5 as plots of 50% time against height of injection for several radial positions. Two main points emerge from the results: firstly, 50% times for tracer injected at radial position C, between the bottom of the cylinder and a point between levels 2 and 3, are several seconds longer than for injections at the same level near the centre (position E) or the perimeter (position A); secondly, the mean vertical velocity of the tracer (deduced approximately from the slopes of the lines in Fig. 5) is not significantly different whether it is injected at positions A, B, or C. It appears that tracer injected at position C moves upwards in general with the same mean velocity, but suffers a delay or downward displacement at some stage. It will also be noticed that the spread is greatest at position C, where the delay is greatest.

The most plausible explanation for this seemed to be that there are three zones in the tower, flow being upward and having the same mean velocity in the inner and outer zones, while there is an intermediate zone where the flow is either slower or even downward; fluid injected into the intermediate zone (at about position C) spreads radially by turbulent diffusion into the centre or outer zones or both, where it is carried up to the exit with the mean velocity deduced from Fig. 5. Diffusion of the tracer out of the retarding zone will take time, and this will account for the enhanced spread between 5% and 50% exit times. Diffusion is essentially a process of exchange, so that when tracer is injected into the outer or inner zones, some of it must diffuse into the intermediate zone and be retarded; this would account for the long "tails" observed in the emergence of the tracer injected in the outer and inner zones. (The (95% − 50%) spread was generally about three times the (50% − 5%) spread.) This rather elaborate postulate was compared with the results of a model experiment, and with some published information on similar systems.

The tracer experiments give no indication of marked channelling up the centre of the drier. In view of the velocity measurements and model experiments discussed later, this is surprising. It is possible that tracer was never actually injected into the central core (model experiments suggest that this was only two or three feet in diameter). To inject tracer at the centre, a 1 in. pipe with a vertical section at least 18 ft long and a 12 ft horizontal limb was used. Only the upper end of the vertical section was rigidly fixed. Positive methods for locating the inner end of the pipe were not practicable, and it seems possible that the swirling air may have deflected the horizontal limb from the centre. On the other hand, the temperature profiles discussed later give no indication of channelling either.

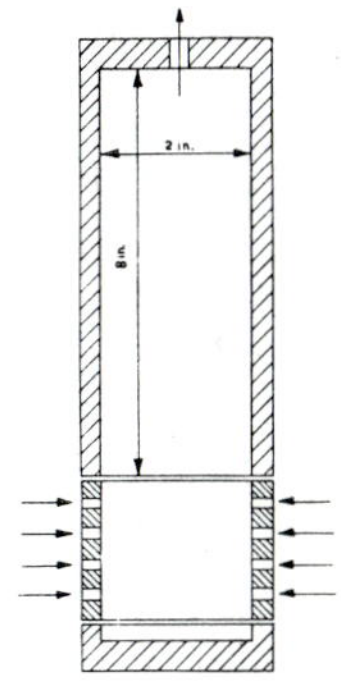

Fig. 8.—*Binnie's apparatus. Section* A *of cylinder rotates*

Previous Measurements

Chaloud, Martin, and Baker[1] have published some observations on a spray-drier similar to that on which our experiment was performed. The published information included profiles of vertical and tangential velocities at an unspecified level in the tower; only the shapes of the profiles are given, but no absolute velocities. There is also a plot of gas temperature against height in the tower (again without figures). The authors have been kind enough to supplement this information.

The striking features of the vertical velocity profile are[1] the central core having several times the mean vertical velocity, a minimum velocity at some intermediate zone, and an increase in velocity in the outer zone. No region of downward velocity is shown.

The tangential velocity profile shows the core rotating as a forced vortex (*i.e.* like a rigid body), while the tangential velocity outside the core is roughly inversely proportional to the square-root of the radius, except in the immediate neighbourhood of the wall. This behaviour has been observed in cyclone separators.[6, 7]

Four sets of temperature contours, communicated privately by Chaloud, show a number of curious features. Since inlet and outlet temperatures were different in each case, it is convenient to describe the temperature T at a point in the tower by a number θ, where

$$\theta = \frac{T - T_0}{T_i - T_0}$$

Thus θ varies from 1 at the inlet to 0 at the outlet (Fig. 9). The value of θ in all cases dropped abruptly at a point a foot or so inside the inlet to a value in the range $0{\cdot}65 - 0{\cdot}5$, showing that incoming air was rapidly mixed with cold air. This indicates a substantial degree of back-mixing, at any rate near the entry ports.

If the flow-pattern had been the same during each of the four surveys, the θ contours would have been identical. In fact, they differ enormously, both in shape and spacing. Although the inlet temperatures, outlet temperatures, and swirl angles varied somewhat from one survey to another, there seems to be no relationship between these variables and the temperature-patterns. We understand that the pattern changed greatly from one occasion to another under similar operating conditions. The result is a very substantial variation in the mean value of θ, or (for given inlet and outlet temperatures) in the mean temperature of the gas in the column (see Fig. 9). The lower the mean temperature, the higher the density and hold-up of gas in the column, and the higher the mean residence-time for a given flow-rate. Thus (as concluded from the tracer experiments) the residence-time may fluctuate widely.

Another point of interest is that there is no sign of a high-temperature core, although measurements were taken 6 in. from the axis. Temperatures on the axis were invariably lower than those at the wall. Channelling of air from the bottom of the tower up the axis (as suggested by the model experiments and velocity-traverse) would be expected to produce a hot core.

In each case the recorded temperature at the outlet was higher than that recorded in the upper part of the tower. This is no doubt due to heating of the air, just before it leaves the drier, by the spray of hot slurry.

Experiments with a Model

A transparent 1/80 scale model was used in which the fluid was water. Strict dynamical similarity could not be achieved with the flow-rates which were practicable, the Reynolds number being only about 1/20 of that in the plant. However, it was found that further reduction of flow by a factor of 5 did not affect the general nature of the flow-pattern, and Binnie's work[5] gave some reason to think that it would have been unaffected if the flow-rate had been indefinitely increased.

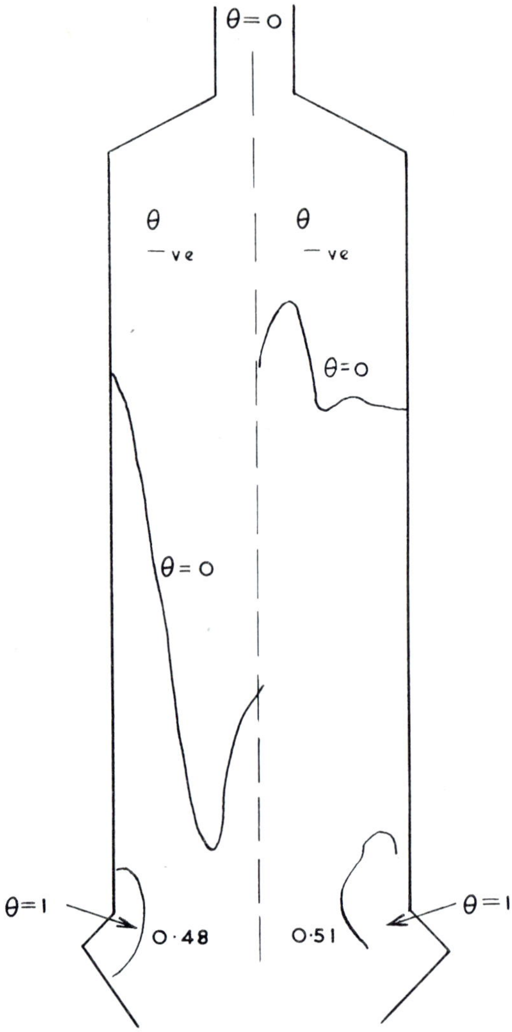

Fig. 9.—*Temperature profiles in two tests under similar conditions*

The water entered the model through twelve ports (of the correct cross-sectional area) inclined at 25° to the vertical and 17° to the radial direction. Most observations were made with deflector baffles fitted to the ports at 25° to the radial direction, to simulate the effect of the baffles in the plant.

Coloured solutions were injected into the model through steel capillary tubes. The following observations were made:

(*a*) Tracer injected close to the axis (within a distance corresponding to about 1 ft on the full-scale plant) moves vertically upwards at several times the mean vertical velocity without radial mixing. This core flows through the surrounding liquid almost as though it were surrounded by an impermeable tube (notice the clear cores in Figs. 11 and 12). Tracer injected into the core flows out along the axis of the exit-duct, surrounded by clear fluid.

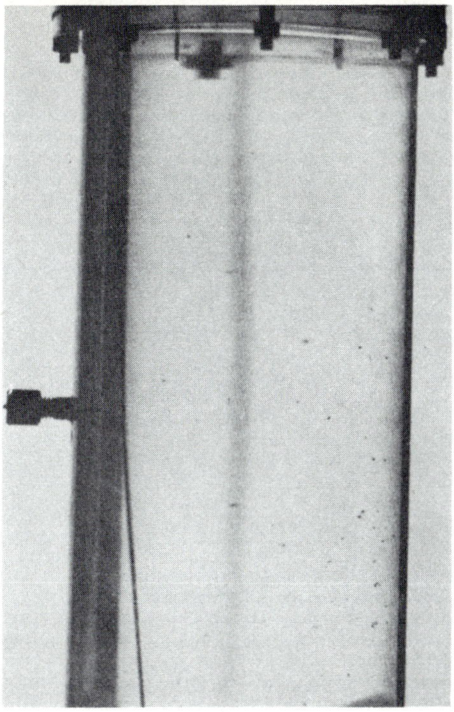

Fig. 10.—*Core formed by dye injected near surface of bottom cone*

(*b*) Tracer injected near the surface of the lower cone passes up the core (Fig. 10). The source of the fluid in the core is evidently a boundary-layer which moves downwards along the surface of the cone to the apex, where it turns upwards. This is a well-known type of secondary flow, which contributes to the effectiveness of cyclone separators.[7] Tracer injected near the top cone spirals inwards and passes through the outlet duct.

(*c*) Tracer injected just outside the core defined in Fig. 10 moves upwards less rapidly than the core; mixing takes place between fluid ascending in this zone and that further from the centre (Fig. 11).

(*d*) Tracer injected still further out ($\frac{1}{2}$ to $\frac{2}{3}$ of the radius; the limits of the zone depend on the flow-rate) moves downwards initially. Much of it is soon transferred by radial mixing into upward-moving zones, which are situated around the core and near the wall. The proportion of the tracer which flows up and out of the system by either route depends on the position of the injection point. In Fig. 12, for instance, most of the tracer seems to be diffusing into the upward-moving zone round the core, much less into the zone near the wall. Some of the tracer injected at the point shown in Fig. 12 eventually finds its way to the bottom of the tower; the photograph was taken before this had time to happen. The downward-moving zone extends almost the full height of the cylindrical part of the column.

(*e*) Tracer injected near the wall moves upwards, although at a speed considerably less than that of the fluid just outside the core. It seems to mix rapidly with the inner zones.

(*f*) If tracer is injected at the inlet to the model, the behaviour depends on the swirl-angle. When this is 17° (no deflectors) tracer passing up the core reaches the outlet well ahead of that moving upwards close to the wall. With deflectors set at 25°, however, the appearance of tracer in the core is delayed, but it subsequently catches up with that near the walls, so that they reach the outlet almost simultaneously. Thus an increase in the swirl-angle makes the overall residence-time distribution approximate more closely to piston-flow.

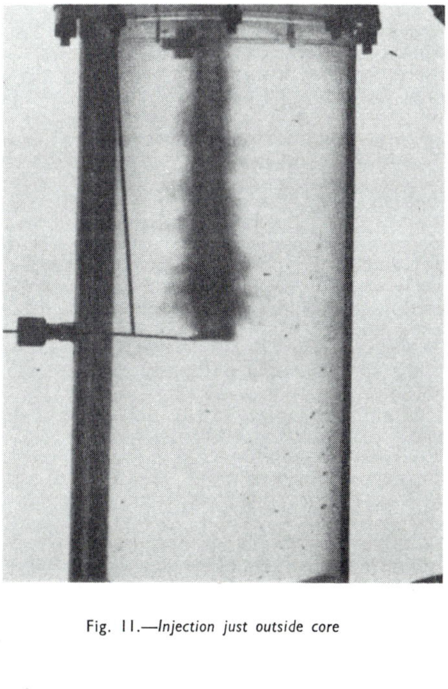

Fig. 11.—*Injection just outside core*

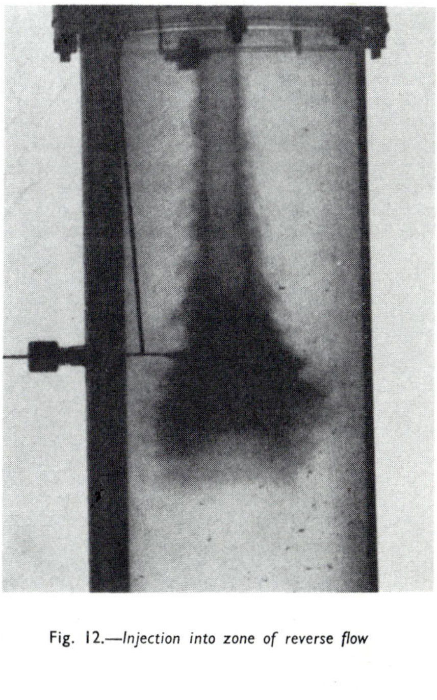

Fig. 12.—*Injection into zone of reverse flow*

(*g*) Considerable large-scale instability or surging is apparent in the outer zone (beyond half a radius). If tracer is injected continuously at such a position, the proportion moving upwards or downwards at a given moment varies substantially, with a time-scale of about one-third of the mean residence-time.

A system very similar to the spray drier model has been investigated by Binnie.[5] One of his experimental arrangements is illustrated diagrammatically in Fig. 8. Part of the cylinder wall rotates, and water enters through holes in the rotating part. The pattern of flow in this apparatus depended on the angular velocity and flow-rate, but in some circumstances the flow was upward near the centre and the walls and downward in an intermediate zone ("Regime III"), as in the present model.

It is difficult to make quantitative comparisons between Binnie's apparatus and the behaviour of the spray-drier model, because of the different ways in which the swirl is produced. However, Binnie's experiments indicate that if the flow is of the "Regime III" type for a particular value of flow-rate and swirl-velocity, it will continue to be of this type if both are increased in the same proportion. In the spray-drier model, using a fixed swirl-angle, an increase in flow-rate produces a proportionate increase in swirl-velocity; Binnie's results would lead us to expect, therefore, that the flow-pattern would continue to be of "Regime III" type at higher flow-rates than could be achieved with the model, and that the model therefore gives at any rate a qualitative guide to the behaviour of the full-scale plant.

Binnie comments that "viscous swirling motion in a tube is so complicated that little can be done to explain it in general terms".

Conclusions

Both the tracer and model experiments point to a somewhat unexpected result of the swirling flow in the drier, namely the zone of reversed flow. This is undoubtedly the main mechanism for the back-mixing observed in the drier; this mixing gives a more uniform temperature, which is said to be beneficial in the drying of detergents.[1]

The existence of an isolated core of air, with a high vertical velocity, channelling along the axis of the drier is indicated by the model experiments and velocity survey, but did not show in the tracer experiments or temperature survey. On balance it seems probable that this core exists; if so, it represents an undesirable short-circuiting of hot, dry air. The instability of flow observed in the model experiments, if representative of what occurs on the full scale, is certainly sufficient to account for the observed scatter in the tracer experiments. Another source of instability may be that the density of the air near the top of the drier is about 20% greater than that near the bottom. This effect was not simulated in the model experiments.

Little can be said about the effect of the falling particles on the flow-pattern. Visual observation suggests that the concentration of suspended particles is too low to have an appreciable effect.

The technique of internal injection of tracer, used here for the first time as far as is known, is capable of producing useful information. The ratio of the number of experiments required to the amount of information obtained is high in systems such as this, in which large-scale instabilities of flow occur. Model experiments carried out before the full-scale tracer experiments would be a useful guide in planning.

Acknowledgments

The help of the following is acknowledged:

The staff of the Instrument Division of Metropolitan Vickers Electric Co. Ltd.

Messrs. Leung, Huntley, Price, O'Connell, Roseman, and Honeyman, who, as vacation students, carried out the tracer experiments;

Mr. A. Reed, who carried out the model experiments, and Mr. D. Kristmanson who took the photographs;

In particular Messrs. Thomas Hedley & Co. Ltd. who supported this work and gave permission for its publication; also the staff of their factory at Grays, Essex, who gave active help in its execution.

Symbols Used

A = h/c.
c = concentration of tracer (mole ratio).
D = molecular diffusivity of helium in air.
h = height of ordinate on chart.
M = molal flow-rate of air.
t = time (s).
$t_{\frac{1}{2}}$ = median or 50% exit-time (s).
$t_{\cdot 05}$ = 5% exit time (s).
t_m = mean residence-time (s).
T_i = temperature of air at inlet.
T_0 = temperature of air at outlet.
T = temperature of air at point in drier.
$\theta = \dfrac{T - T_0}{T_i - T_0}$.

References

1 Chaloud, J. H., Martin, J. B., and Baker, J. S. *Chem. Engng Progr.*, 1957, **53**, 593.
2 Danckwerts, P. V. *Chemical Engineering Science*, 1958, **9**, 78.
3 Taylor, G. I. *Proc. Roy. Soc.*, 1954, **A223**, 446.
4 Danckwerts, P. V. *Chemical Engineering Science*, 1953, **2**, 1.
5 Binnie, A. M. *Quart. J. Mech.*, 1957, **10**, 276.
6 ter Linden, A. J. *Proc. Instn mech. Engrs, Lond.*, 1949, **160**, 233.
7 ter Linden, A. J. *Trans. Amer. Soc. mech. Engrs*, 1953, **75**, 433.

The manuscript of this paper was received on 1 *December*, 1958, *and the paper was presented at the Spring Meeting of the Institution in London on* 17 *March*, 1959.

D5

Flow-visualization by means of a time-reaction

By P. V. DANCKWERTS AND R. A. M. WILSON
Department of Chemical Engineering, University of Cambridge

(Received 29 October 1962 and in revised form 21 January 1963)

A continuous-flow system is fed with a liquid which turns blue a specified time θ after entering. A more or less stationary pattern is thus set up, those parts of the liquid which have spent times longer than θ in the system being blue while the rest is colourless. The value of θ can be changed at will, enabling details of the flow-pattern to be deduced. The interpretation of the finer points of the colour-pattern is made difficult by the effects of molecular interdiffusion of different parts of the liquid.

Denbigh, Dombrowski, Kisiel & Place (1962) have described the use of a time-reaction for flow-visualization. Two solutions are mixed, and the mixture turns blue quite suddenly a definite time after mixing. If the mixture is made up continuously and runs through a continuous-flow system a more or less stationary pattern is set up, only those parts of the liquid which at any instant have been in the system for more than a specified length of time being blue.

If a tracer such as a dye is injected uniformly into the fluid entering a continuous-flow system, only a momentary impression of the nature of the flow-pattern is obtained as the tracer spreads through the system. More can be discovered by injecting tracer at points in the interior of the system, but the injection tube itself disturbs the flow. On the other hand, if a time-reaction is used the pattern observed will be stationary, if the flow is steady. The boundary between blue and colourless liquid separates those parts of the system which are occupied by fluid 'older' and 'younger' than the reaction time θ, which can be varied at will. (The 'age' of the fluid here refers to the length of time for which it has been in the system.) The distribution of colour will reveal a good deal about the pattern of flow, regions of stagnation and recirculation, etc. If the flow is not steady, the fluctuations of the blue zones give a direct visual impression of the degree and nature of the unsteadiness, while a time-exposure photograph yields the time-average pattern.

Denbigh *et al.* used a reaction-mixture containing sulphite, iodate and starch. We have carried out experiments with a different reaction. In our case the mixture contained sodium persulphate, potassium iodide, sodium thiosulphate and starch. There is a slow reaction between persulphate and iodide ions, yielding iodine:

$$Na_2S_4O_8 + 2KI \rightarrow Na_2SO_4 + K_2SO_4 + I_2 \quad \text{(slow)}.$$

The iodine reacts instantaneously with the thiosulphate and is reduced back to iodide:

$$I_2 + Na_2S_2O_3 \rightarrow Na_2S_4O_6 + 2NaI \quad \text{(instantaneous)}.$$

When all the thiosulphate has been used up in this way, free iodine is liberated and the solution immediately turns blue because of the well-known starch–iodine reaction. The intensity of the blue colour does not depend on the amount of iodine present. The rate of production of iodine and destruction of thiosulphate is proportional to the concentrations of both persulphate and iodide. Provided a substantial excess of persulphate is present the rate of destruction of thiosulphate is virtually constant up to the moment of turning blue because the concentration of iodide remains unchanged. The time θ required for the mixture to turn blue is thus $\theta = C_0/kC_1C_2$, where C_0, C_1, C_2 are the initial concentrations of thiosulphate, persulphate and iodide. The rate-constant k has the value $1{\cdot}52 \times 10^{-4}$ l/g mole sec at 20° C, according to our measurements. In our experiments the rate of production of iodine after the moment of turning blue could also be regarded as constant, because the fractional extent of reaction of iodide was small on the time-scale of the experiments.

Three solutions were mixed at the entrance to the flow-system under study: (*a*) 0·5 M sodium persulphate (151 g/l); (*b*) 0·1 M potassium iodide (16·6 g/l) with 1 g starch per litre; (*c*) 1 M sodium thiosulphate (142 g/l). The first two solutions were metered in with rotameters, and in all the experiments described here the flow-rate of each was 400 cm³/min. The thiosulphate solution was added at a controlled rate (of the order of 1·5 cm³/min) by displacing it from a container by means of the oxygen and hydrogen generated by the electrolysis of 5 N H_2SO_4 (50 % wt.) between platinum electrodes with a 12 V supply. The rate of addition was controlled by adjusting the current, without changing the total rate of flow through the system appreciably. The time θ required (at 20° C) for the resulting mixture to turn blue was $63i$ sec, where i is the current in amperes.

The flow-system consisted of a cylindrical vessel of 200 cm³ capacity, about 5 cm diameter and 10 cm long. The mean residence-time of the liquid (i.e. volume divided by flow-rate) was 15 sec.

Figure 1 and figure 3, plate 1, illustrate an arrangement in which the liquid enters and leaves through radially directed tubes situated at opposite sides and ends of the vessel. Figures 3 (*a*), (*b*) and (*d*), plate 1, were taken with $\theta/\tau = 1{\cdot}0$ (τ being the mean residence-time). In these pictures, therefore, only fluid which has been in the system for a time longer than the mean residence-time is blue. Figures 3 (*a*) and (*b*) are instantaneous photographs, while figure 3 (*d*) is a time-exposure of duration 2τ. Figures 3 (*a*) and (*b*) show the degree of fluctuation in the spatial distribution of material of ages greater than the mean residence-time, while figure 3 (*d*) gives an idea of its time-average distribution. At first sight these pictures suggest that the flow-pattern is not far removed from 'piston-flow', in which all particles of fluid move along paths parallel to the axis and spend the same length of time in the system. In this case the time-reaction method would produce a piston-shaped blue region filling the outlet end of the cylinder and bounded by a plane perpendicular to its axis. However, in true piston-flow none of the liquid in the system would have an age greater than the mean residence-time. Figures 3 (*a*), (*b*) and (*d*) indicate that a substantial part of the system under discussion is occupied by liquid older than this, and as a corollary a substantial part of the liquid must spend times less than the mean residence-time in passage. Figure 3 (*c*), an instantaneous photograph with $\theta/\tau = 1{\cdot}25$, shows that an appreciable part of the liquid spends longer than $1{\cdot}25\tau$ in passage. Considerations concerning the spread of residence-times are important in the design of chemical reactors (for instance), where departures from piston flow may decrease output or affect the quality of the product.

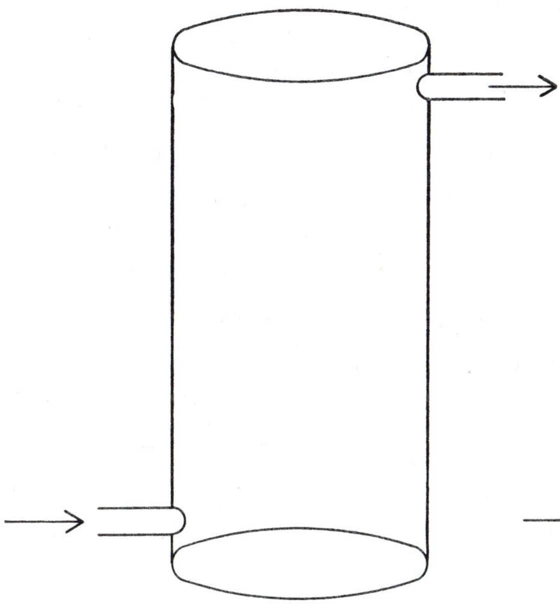

FIGURE 1. Inlet and outlet radially directed and at opposite sides and ends.

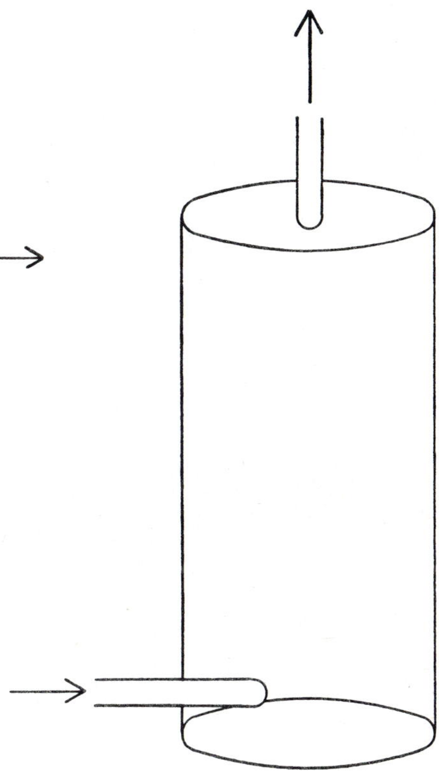

FIGURE 2. Tangential inlet and axial outlet.

Figure 2 and figure 4, plate 2, illustrate an arrangement in which the liquid is introduced tangentially at one end of the cylinder and leaves it via the centre of the other end. It is known that the flow in such systems is very complicated, with

(*a*)

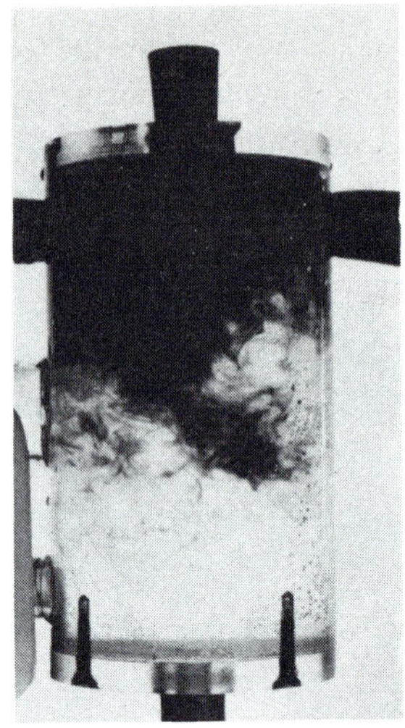

(*b*)

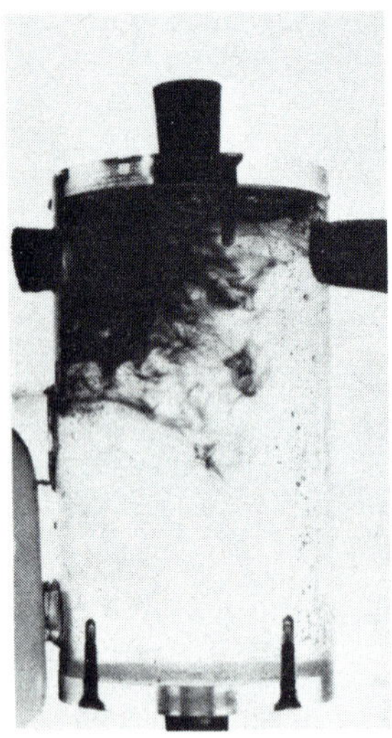

(*c*)

(*d*)

FIGURE 3. Inlet and outlet as shown in figure 1. (*a*) $\theta/\tau = 1{\cdot}0$, instantaneous; (*b*) $\theta/\tau = 1{\cdot}0$, instantaneous; (*c*) $\theta/\tau = 1{\cdot}25$, instantaneous; (*d*) $\theta/\tau = 1{\cdot}0$, time exposure.

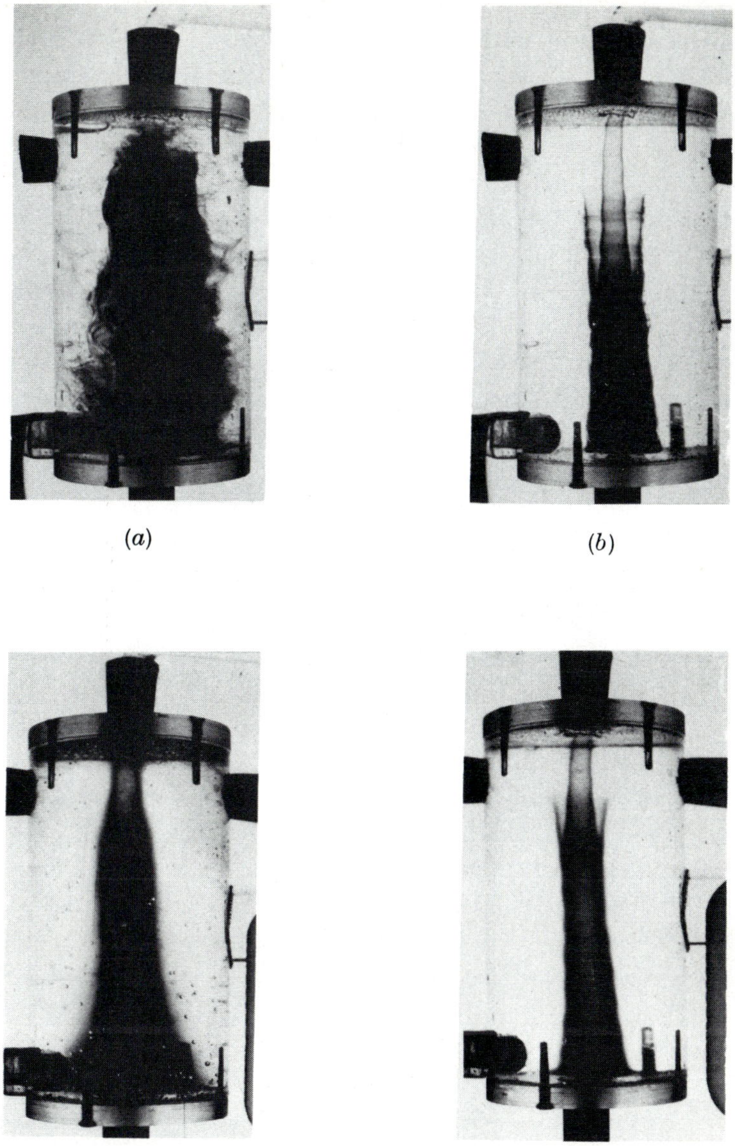

(*a*) (*b*) (*c*) (*d*)

FIGURE 4. Inlet and outlet as shown in figure 2. (*a*) $\theta/\tau = 1{\cdot}2$, instantaneous; (*b*) $\theta/\tau = 1{\cdot}6$, instantaneous; (*c*) $\theta/\tau = 1{\cdot}2$, time exposure; (*d*) $\theta/\tau = 1{\cdot}6$, time exposure.

regions of reverse flow (Binnie 1957; Bradley & Pulling 1959; Place, Ridgway & Danckwerts 1959). Figures 4 (*a*) and (*c*) are instantaneous and time-exposure photographs, respectively, with $\theta/\tau = 1{\cdot}2$. They show that there is a region near the axis containing liquid which has spent more than the average length of time in the system, and the comparison between figures 4 (*a*) and (*c*) indicates considerable unsteadiness of flow in the region midway between the walls and the axis. Figures 4 (*b*) and (*d*) are instantaneous and time-exposure photographs, respectively, with $\theta/\tau = 1{\cdot}6$, and show that there is an annular region containing liquid which has entered the system more than $1{\cdot}6\tau$ before. It is known (Place *et al.* 1959) that in a flow-system of this kind there is upward flow near the cylindrical walls and through a narrow 'tube' surrounding the axis, and reverse flow through an annular region between these; there must therefore be two zones where the mean velocity is zero. The reverse flow probably originates somewhere near the surface of the top end of the cylinder. This behaviour is illustrated by figures 4 (*a*) and (*b*); some of the blue liquid is stationary and some flowing backwards. The time-reaction method is particularly suitable for the study of complicated, relatively steady flow-patterns of this kind, which can be observed at leisure and in detail without disturbing the flow.

The crudest way of interpreting such observations, employed above, is to say that only those parts of the fluid which have been in the system for a time exceeding θ will be blue. Although this interpretation will serve for a qualitative assessment of the results, the true significance of the pattern observed is subtler than this. To the extent that molecular diffusion plays a role in determining the spatial distribution of molecules of different ages it is not strictly possible to think in terms of elements of fluid which maintain their identity during passage through the system. Each element, in the sense understood in hydrodynamics, contains a constantly changing population of molecules. It will then not in general be possible to ascribe an age to a macroscopic, chemically homogeneous region of the liquid, as it may contain an intimate mixture of molecules of different ages. In an isolated portion of solution the thiosulphate has an initial concentration C_0, and at time t after reaction starts its concentration is $C = C_0(1-t/\theta)$. At times greater than θ free iodine is being formed at the same equivalent rate as that at which thiosulphate was formerly disappearing. If we regard the concentration of iodine as a negative concentration of thiosulphate, it is also governed by the above relationship. If we mix two portions of liquid which have been reacting for different lengths of time (have different ages), one greater and one smaller than θ, the free iodine in one portion will react instantaneously with the free thiosulphate in the other. The resulting mixture will be blue if the average age of the mixture (weighted by volume) is greater than θ (i.e. if $C < 0$) and colourless if the average age is less than θ ($C > 0$). If, in the flow-system, there is diffusional mixing of molecules of the liquid which have been in the system for different lengths of time the fraction of the liquid which is blue is not necessarily the same as the fraction which has been in the system for a time greater than θ. A small amount of very old liquid can turn a large amount of young liquid blue. Since some of the liquid may remain in the system for a very long time, while none can have an age less than zero, the fraction of the liquid which is blue is likely in general to be larger than the fraction older than θ. On the other hand, in figure 4 (*b*) the down-flowing old blue liquid is decolourized before it leaves the system by mixing and reaction with an excess of fresh incoming fluid.

The parts of the system which are blue are in fact those regions where the local average age of the molecules is greater than θ. A deeper analysis of this statement would lead to a discussion of concepts such as the composition at a point, which are beyond the scope of this communication. Some of the concepts involved have been discussed by Zwietering (1959).

These considerations indicate that caution should be exercised in interpreting the finer points of the observations made with the aid of the time-reaction. The broad picture, however, can generally be taken at its face value.

REFERENCES

BINNIE, A. M. 1957 *Quart. J. Mech.* **10**, 276.

BRADLEY, D. & PULLING, D. J. 1959 *Trans. Inst. Chem. Engrs*, **7**, 34.

DENBIGH, K. G., DOMBROWSKI, N., KISIEL, A. J. & PLACE, E. R. 1962 *Chem. Engng Sci.* **17**, 573.

PLACE, G., RIDGWAY, K. & DANCKWERTS, P. V. 1959 *Trans. Inst. Chem. Engrs*, **37**, 268

ZWIETERING, Th. N. 1959 *Chem. Engng Sci.* **11**, 1.

D6

The Effect of Hold-Up and Mixing on a Stream of Fluctuating Composition

By P. V. Danckwerts, G.C., M.B.E., M.A., S.M. *and* E. S. Sellers,* M.A., M.Sc., A.M.I.Chem.E. . . . . *The authors discuss the relationship between the size of a gas holder and the reduction in the fluctuations in composition of the gas stream in passage through the holder*

THE following problem arose in connection with the manufacture of town gas, but is an example of a wider class of problem which is likely to be of interest to chemical engineers. When gas is manufactured, it is necessary not only to hold the mean calorific value at the required level, but also to avoid excessive fluctuations about the mean. The fluctuations may to some extent be "ironed out" by passing the gas into a holder of sufficient capacity, where it is mixed with the gas already in the holder. The final supply drawn from the holder will in general exhibit a smaller fluctuation in quality than the entering stream. The problem is to determine the relation between the size of the holder and the reduction in fluctuation.

If the fluctuations of the ingoing stream are a known function of the time, the problem is a simple one. We shall consider mainly the case in which the fluctuations are completely or partly random in nature, making a statistical treatment necessary. The method of solution can be applied to other systems in which a stream with fluctuating properties is subjected to hold-up and mixing, and also (as shown below) to systems in which some component of the stream undergoes a first-order reaction.

The Standard Deviation

The most convenient measure of the magnitude of the fluctuation is the standard deviation, which is the root-mean-square deviation from the mean :—

$$\sigma^2 = \overline{(x - \bar{x})^2} \quad . \; . \; . \; . \; . \; . \; . \; . \; (1)$$

where σ = standard deviation, x = value of variable at given time, $\bar{x}$ = mean value of x. The line above the expression on the right-hand side means that it is a time-average value. σ has the important property (unlike the mean deviation) that it will distinguish between the deviation of, say, one unit for ten hours and ten units for one hour, giving more weight to the latter.

Conditions of the Problem

The problem, therefore, is to relate the values of σ of the ingoing and outgoing streams to the size of the holder and the rate of flow. (As will be seen, however, more information is required about the fluctuations of the ingoing stream than merely the value of the standard deviation.) In order to render the problem determinate, the following assumptions will be made.

(*a*) The flow-rates of the ingoing and outgoing streams are equal and constant.

(*b*) The quantity x, when multiplied by a volume, constitutes an extensive quantity which is conserved—as will be the case if x denotes the calorific value of the gas-stream at any moment.

(*c*) The gas in the holder is completely mixed, so that its composition is uniform and identical with that of the outgoing stream. This condition is unlikely to be fulfilled in practice, and for this reason the solution obtained gives a lower limit to the size of the gas-holder required to effect a given reduction in σ. Any method by which the mixing of the gas in the holder can be improved will bring the outgoing value of σ closer to the value calculated on this assumption.

The Conservation Equation

The following nomenclature will be used :—

$x_i \;= \bar{x} + \delta_i$ = value of x in ingoing stream at time t.

$x_o \;= \bar{x} + \delta_o$ = value of x in outgoing stream at time t.

$\sigma_i^2 = \overline{(x_i - \bar{x})^2} = \overline{(\delta_i)^2}$

$\sigma_0^2 = \overline{(x_0 - \bar{x})^2} = \overline{(\delta_0)^2}$

v = volumetric flow-rate of both streams.

V = volume of gas-holder.

The conservation equation is :—

$$V \frac{dx_0}{dt} = v\,(x_i - x_0)$$

$$\therefore \quad V \frac{d\delta_0}{dt} = v\,(\delta_i - \delta_0) \quad . \; . \; . \; . \; . \; . \; (2)$$

The solution to this is :—

$$\delta_o = \frac{v}{V}\int_{\theta=0}^{t} e^{-v\theta/V} \,.\, \delta_{i\,(t-\theta)} \,.\, d\theta \quad . \; . \; . \; . \; . \; (3)$$

where $\delta_{i\,(t-\theta)}$ is the value of δ_i at time $(t - \theta)$. Since δ_i never exceeds some finite value, the values of $e^{-v\theta/V} \,.\, \delta_{i\,(t-\theta)}$ corresponding to the larger values of θ do not contribute appreciably to the integral. Provided, therefore, that t is large (*i.e.* the system has been running for some time), we can substitute ∞ for t as the upper limit of the integral and write :—

$$\delta_0 = \frac{v}{V}\int_{\theta=0}^{\infty} e^{-v\theta/V} \,.\, \delta_{i\,(t-\theta)} \,.\, d\theta \quad . \; . \; . \; . \; (4)$$

* The Department of Chemical Engineering, Cambridge

Now it is generally true, when $f(x)$ is a function of x, that :—

$$\left[\int_0^\infty f(x)\,dx\right]^2 = 2\int_0^\infty\int_0^\infty f(x)\,.\,f(x+r)\,dr\,dx \quad . \quad (5)$$

Hence from Eqn. (4) :—

$$(\delta_0)^2 = 2\frac{v^2}{V^2}\int_{\theta=0}^\infty\int_{r=0}^\infty \delta_{i\,(t-\theta)}\,.\,e^{-v\theta/V}\,.\,\delta_{i\,(t-\theta-r)}\,.\, e^{-\frac{v}{V}(\theta+r)}\,.\,dr\,d\theta \quad . \, . \, . \quad (6)$$

Averaging with respect to t:— $\sigma_0^2 = \overline{(\delta_0)^2} =$

$$2\frac{v^2}{V^2}\int_{\theta=0}^\infty\int_{r=0}^\infty e^{-2v\theta/V}\,.\,e^{-vr/V}\times \overline{\delta_{i\,(t-\theta)}\,.\,\delta_{i\,(t-\theta-r)}}\,.\,dr\,d\theta. \quad . \, . \, . \quad (7)$$

The Autocorrelation Coefficient

Now the quantity $R(r)$, defined by :—

$$R(r) = \frac{\overline{\delta_{i\,(t')}\,.\,\delta_{i\,(t'+r)}}}{\sigma_i^2} \quad . \, . \, . \, . \, . \, . \, . \quad (8)$$

(the average being taken over all values of t') is known as the coefficient of autocorrelation between values of δ_i separated by a time-interval r. Substituting in Eqn. (7) :—

$$\frac{\sigma_0^2}{\sigma_i^2} = 2\frac{v^2}{V^2}\int_0^\infty e^{-2v\theta/V}\,.\,d\theta\int_0^\infty e^{-vr/V}\,.\,R(r)\,.\,dr$$

$$= \frac{v}{V}\int_0^\infty e^{-vr/V}\,.\,R(r)\,.\,dr \quad . \, . \, . \, . \, . \, . \quad (9)$$

Hence, if $R(r)$ is determined as a function of r (for instance, from a representative section of a continuous record of values of x_i), the ratio of the standard deviations of the ingoing and outgoing streams can be computed for any given value of v/V. The results will, of course, only be of value if there is no significant long-term variation in the characteristics of the ingoing stream, so that the data concerning $R(r)$ remain applicable.

The calculation of $R(r)$ for various values of r is rather laborious (the possibility of dispensing with it is discussed later). Several machines have been designed for the purpose*, but if none is available the following method of computation is convenient.

(*a*) Tabulate the values of δ_i at uniform intervals of time, Δt. Let us call these values δ_1, δ_2, etc., numbered in the order in which they occur.

(*b*) Find the difference, D_1, of $(\delta_1 - \delta_2)$: the difference, D_2, of $(\delta_2 - \delta_3)$, and so on. Square each difference and add, giving ΣD^2.

(*c*) Find the sum, S_1, of $(\delta_1 + \delta_2)$; the sum, S_2, of $(\delta_2 + \delta_3)$, etc. Square each sum and add the squares, giving ΣS^2.

(*d*) Then, as can easily be shown,

$$R(r) = \frac{\Sigma S^2 - \Sigma D^2}{\Sigma S^2 + \Sigma D^2} \quad . \, . \, . \, . \, . \, . \quad (10)$$

r being in this case equal to Δt.

(*e*) Find the sums $(\delta_1 + \delta_3)$, $(\delta_2 + \delta_4)$, etc., and the differences $(\delta_1 - \delta_3)$, $(\delta_2 - \delta_4)$, etc. Repeat the above processes to find $R(r)$ for $r = 2.\Delta t$.

(*f*) Repeat for other values of r.

When $R(r)$ has been found for a sufficient range of r, Eqn. (9) can be integrated graphically or otherwise to give the ratio of the standard deviations for any assigned value of v/V. The range of r must be such that $R(r)\,.\,e^{-vr/V}$ becomes much less than unity for the largest value of r.

The Correlogram

The curve of $R(r)$ plotted against r is known as a correlogram, and its shape will depend on the processes giving rise to the fluctuations in the ingoing stream. A high value of $R(r)$ means that a high positive value of δ_i at any time t is likely to be followed by a high positive value at time $(t + r)$, and that large negative values will be similarly related. Hence, if there is any regular fluctuation of period λ, we shall expect to find exceptionally high values of $R(r)$ for $r = \lambda$, 2λ, etc. For values $\lambda/2$, $3\lambda/2$, etc., $R(r)$ will be small or negative. Two records of the calorific value of the gas entering a holder are shown in Fig. 1, together with their correlograms. Both records display a periodicity of two hours. The fact that the minima of the second correlogram lie above the zero-line is due to a long-scale drift in the recorded value of x_i ; this may have been caused by a drift in the zero of the recorder. Particular care should be taken to avoid such instrument-drift when obtaining records from which values of $R(r)$ are to be calculated.

Special Case I : Regular Periodic Fluctuations

It has already been mentioned that the solution of the problem is simple if δ_i is known as a function of the time. It becomes, in fact, merely a matter of integrating Eqn. (4) to find δ_0 ; once δ_0 is known as a function of time, its standard deviation is easily determined.

Suppose that the fluctuations in x_i are due to a regularly repeated cycle of operations, or to several such cycles superimposed. For instance, gas may be made in both vertical and horizontal retorts, each retort of each type being discharged a certain number of times in 24 hours. The combined gas from the two types of retort will display fluctuations in x which can probably be represented with sufficient accuracy by combining a small number of sine curves of different period (λ), amplitude (A) and phase (ε) :—

$$\delta_i = \sum_n A_n \sin\left[\frac{2\pi(t+\varepsilon_n)}{\lambda_n}\right] \quad . \, . \, . \, . \, . \, . \, . \, . \quad (11)$$

Under these circumstances it can be shown that :—

$$\sigma_i^2 = \overline{(\delta_i)^2} = \tfrac{1}{2}\,\Sigma\,A_n^2 \quad . \, . \, . \, . \, . \, . \quad (12)$$

Combining Eqns. (11) and (4), it is found that :—

$$\delta_0 = \Sigma\,\frac{A_n v/V}{v^2/V^2 + 4\pi^2/\lambda_n^2}\left\{\frac{v}{V}\sin\left[\frac{2\pi\,(t+\varepsilon_n)}{\lambda_n}\right] + \frac{2\pi}{\lambda_n}\cos\left[\frac{2\pi(t+\varepsilon_n)}{\lambda_n}\right]\right\} \quad . \, . \, . \, . \quad (13)$$

from which it can be shown that :—

$$\sigma_0^2 = \overline{(\delta_0)^2} = \tfrac{1}{2}\,\Sigma\,A_n^2\left[\frac{v^2/V^2}{v^2/V^2 + 4\pi^2/\lambda_n^2}\right] \, . \quad (14)$$

* Supplement to *Journal of the Royal Statistical Society*, 1946, **7**, 42, 62.

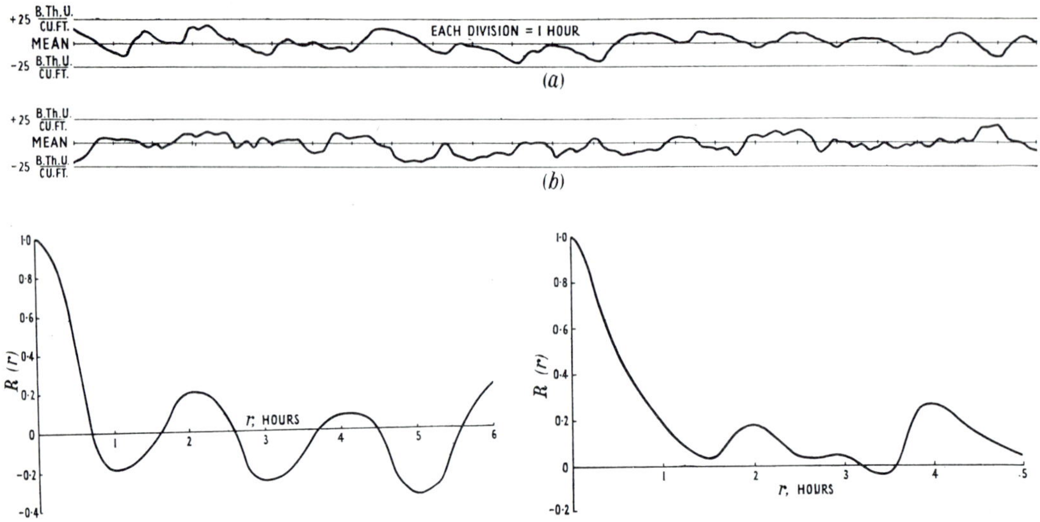

Fig. 1—Two records of the calorific value of gas entering a holder are shown in the top curves. The correlogram to curve (a) is shown below on the left, that for curve (b) on the right

In the case where δ_i can be represented by a single sine curve of period λ the result is simply :—

$$\frac{\sigma_0^2}{\sigma_i^2} = \frac{v^2/V^2}{v^2/V^2 + 4\pi^2/\lambda^2} \quad \ldots\ldots \quad (15)$$

Special Case II : Completely Random Fluctuations

In certain cases there may be no regular periodic component in the fluctuation of δ_i. For instance, the stream might be an effluent containing radioactive waste from a number of unrelated and irregularly timed laboratory experiments. In these circumstances the correlogram would look something like Fig. 2 (x_i being the degree of radio-activity of the effluent).

In such cases it may be possible to calculate σ_0^2/σ_i^2 without recourse to a correlogram, thus saving labour. The essential feature for the success of the method described is that $R(r)$ should be zero or very small for all values of r greater than some value ξ. This will not be the case if there is any periodicity in the fluctuations of the ingoing stream.

Reverting to Eqn. (9), it may be seen that the problem is solved if the functional relationship between $R(r)$ and r is known. It will be shown, however, that if the correlogram resembles that shown in Fig. 2, the calculated value of σ_0^2/σ_i^2 is not very sensitive to the exact form of the relationship. For instance, we might put :—

$$R\,(r) = e^{-\pi r^2/4P^2} \quad \ldots\ldots \quad (16)$$

or,

$$R\,(r) = e^{-r/P} \quad \ldots\ldots \quad (17)$$

These two functions are shown in Fig. 3, and it can be seen that they are of markedly different shape. In each case, $\int_0^\infty R(r)dr$ is equal to P.

Substituting Eqns. (16) and (17) in (9), we obtain respectively :—

Fig. 2—Correlogram for special case II; no regular periodic fluctuation

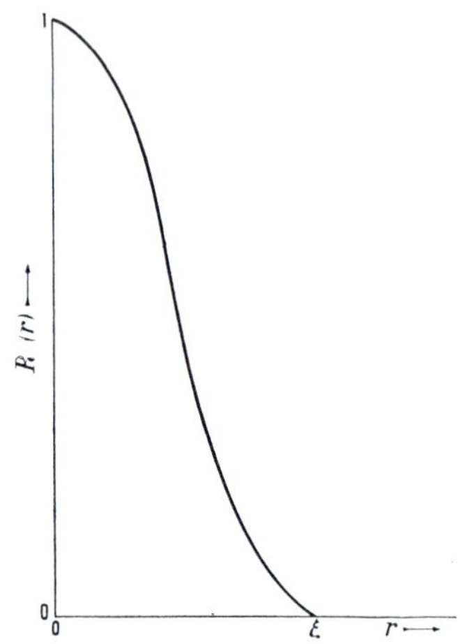

$$\frac{\sigma_0^2}{\sigma_i^2} = \frac{Pv}{V} \exp\left(\frac{P^2v^2}{\pi V^2}\right) erfc\left(\frac{Pv}{V\sqrt{\pi}}\right) \quad \ldots\ldots \quad (18)$$

(where $erfc(y) = 1 - \frac{2}{\sqrt{\pi}} \int_0^y e^{-y^2}dy$; values may be found in tables), and :—

$$\frac{\sigma_0^2}{\sigma_i^2} = \frac{Pv/V}{Pv/V + 1} \quad \ldots\ldots \quad (19)$$

These two functions are shown plotted against Pv/V in Fig. 4. Both tend to 1 as Pv/V becomes very large, and to Pv/V as the latter becomes very small. The difference between the two is nowhere more than 20 per cent. We may conclude that a knowledge of the actual form of the relationship between $R(r)$ and r is not essential. Provided it conforms generally to Fig. 2, a useful estimate of σ_0^2/σ_i^2 for a given v/V

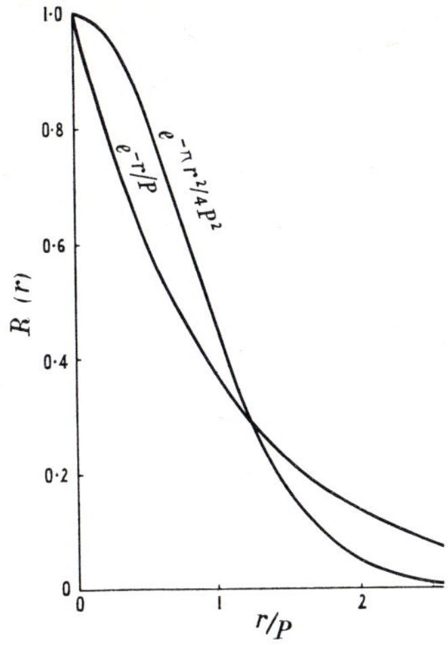

Fig. 3—Curves approximating to the correlogram for special case II

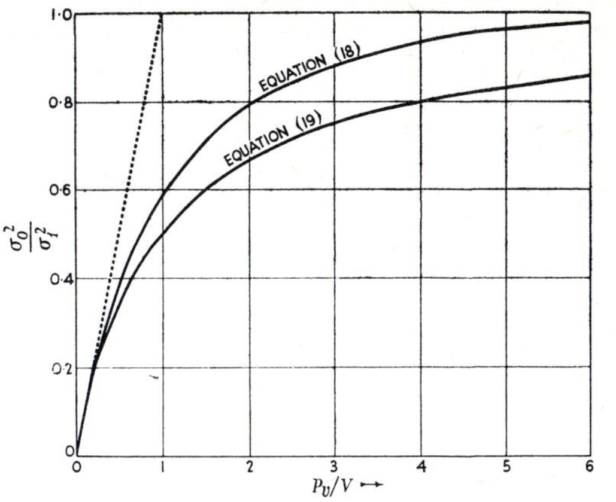

Fig. 4—Curves for the solution of equations (18) *and* (19)

can be obtained solely from a knowledge of P, the area under the curve, and the use of Fig. 4.

P may conveniently be determined in the following way from a continuous record of x_i covering a representative period. (The validity of the method will not be proved here). Choose a time-interval T, which must be large enough to contain a large number of troughs and peaks in the record, but should not be more than half the length of the record available. Compute the integral :—

$$J = \int_{t=\theta}^{T+\theta} \delta_i \, . \, dt \quad . \; . \; . \; . \; . \; . \; . \; . \quad (20)$$

for a number of values of θ lying within the period covered by the record. Compute $\overline{J^2}$, the mean square of the values of J. Then :—

$$P = \overline{J^2}/2T\sigma_i^2 \quad . \; . \; . \; . \; . \; . \quad (21)$$

Eqn. (21) is an approximation, which depends on T being much greater than ξ. It will be safe to accept the result if the value of P given by the equation is much less than the value of T.

The computation can be carried out as follows if an integraph is not available. The record must be in the form of a curve drawn on Cartesian co-ordinates. Draw in the line representing $x_i = \bar{x}$, and below it a parallel line, $x_i = h$, which does not cut the curve (Fig. 5). Cut out with a pair of scissors the area

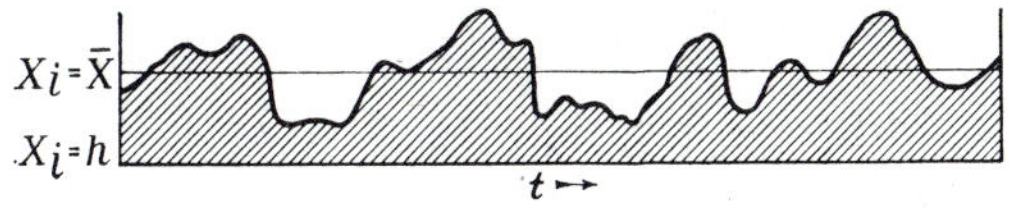

Fig. 5—Illustrating the method of estimating P for use in equations (18) *and* (19)

between $x_i = h$ and the curve (shown shaded). Cut this into a number of strips of equal width Δt; the strips must be numbered consecutively. Weigh each strip; let the weights be w_1, w_2, etc. If $T = n \, . \, \Delta t$, add the weights of strips 1 to n inclusive ($= W_1$); then the weights of strips 2 to $(n + 1)$ ($= W_2$), and so on. Then :—

$$\overline{J^2} = \overline{W^2}/W_0^2 - (hT)^2 \quad . \; . \; . \; . \; . \; . \quad (22)$$

where W_0 is the weight of a rectangle of paper measuring one unit of x_1 by one unit of t.

First-Order Reactions

Suppose that x_i represents the concentration in the ingoing stream of some substance A which is destroyed in the hold-up vessel by a first-order process. For instance, A might be a radioactive element and the stream be passed through a pond to allow its activity to diminish before it is allowed to run into a river. We then have :—

x_i = concentration of A in ingoing stream $= \bar{x}_i + \delta_i$
x_0 = concentration of A in outgoing stream $= \bar{x}_0 + \delta_0$ } $\bar{x}_i \neq \bar{x}_0$

k = velocity-constant of process leading to destruction of A.

Eqn. (1) must be replaced by :—

$$V \frac{dx_0}{dt} = v\,(x_i - x_0) - kVx_0 \quad . \; . \; . \; . \; . \; . \quad (23)$$

$$\therefore \quad x_0 = \frac{v}{V} \int_{\theta=0}^{\infty} e^{-\theta\,(v/V+k)} \, . \, x_{i\,(t-\theta)} \, . \, d\theta \quad . \; . \quad (24)$$

$$\therefore \quad \delta_0 = \frac{v}{V} \int_{\theta=0}^{\infty} e^{-\theta\,(v/V+k)} \, . \, \{x_{i\,(t-\theta)} - \bar{x}_i\} \, d\theta$$

$$= \frac{v}{V} \int_{\theta=0}^{\infty} e^{-\theta\,(v/V+k)} \, . \, \delta_{i\;(t-\theta)} \, . \, d\theta \quad . \; . \quad (25)$$

in place of eqn. (4). Hence :—

$$\frac{\sigma_0^2}{\sigma_i^2} = \frac{v}{V} \int_0^{\infty} e^{-r(v/V+k)} \, . \, R\,(r) \, . \, dr \quad . \; . \; . \quad (26)$$

in place of Eqn. (9). Eqn. (14) becomes :—

$$\sigma_0^2 = \tfrac{1}{2}\Sigma A_n^2 \left[\frac{v^2/V^2}{(v/V + k)^2 + 4\pi^2/\lambda_n^2}\right] \quad . \; . \; . \; . \quad (27)$$

and Eqns. (18) and (19) respectively :—

$$\sigma_0^2/\sigma_i^2 = \frac{Pv}{V} \, . \, exp\left[\frac{P^2}{\pi}\left(\frac{v}{V} + k\right)^2\right] erfc\left[\frac{P}{\sqrt{\pi}}\left(\frac{v}{V} + k\right)\right] \quad . \; . \; . \; . \; . \; . \quad (28)$$

$$\sigma_0^2/\sigma_i^2 = \frac{Pv/V}{Pv/V + Pk + 1} \quad \ldots\ldots\ldots\ldots \quad (29)$$

Also, averaging Eqn. (25) with respect to t, we have :—

$$\bar{x}_0 = \bar{x}_i \frac{v}{V} \int_{\theta=0}^{\infty} e^{-\theta(v/V+k)}\, d\theta$$

$$= \frac{\bar{x}_i}{1 + kV/v} \quad \ldots\ldots\ldots\ldots \quad (30)$$

The foregoing equations give the effect of the hold-up on both the mean value and the standard deviation of x. By diminishing the mean value, the destruction of the substance A also helps to diminish the standard deviation.

Acknowledgements

The authors are indebted to members of the staff of the Statistical Laboratory, Cambridge, for valuable advice and suggestions; to the management of Walsall gas works for proposing the problem; and to the management of the Cambridge gasworks for access to records.

SECTION E

Mixtures and Mixing

The papers in SECTION E are concerned with the purpose, definition and measurement of "goodness of mixing", concepts which had received very little attention in 1953. Paper E1 starts with the obvious question - what is the practical criterion for goodness of mixing? It obviously depends on the reason for carrying out the mixing process and the main interest of the paper is to focus attention on this purely pragmatic aspect of the subject.

Paper E2 essays a quantitative discussion which will lead to the definition and measurement of the properties of some non-uniform mixtures. Only "fine-grained"mixtures (in the sense of Paper E1) are considered - i.e. those in which the samples withdrawn to determine the uniformity of the mixture contain many independent particles. The concept of "concentration at a point" is introduced but has never been rigorously defined here or elsewhere as far as I know. In general, I would say that Paper E2 presents a valuable disciplinary background to discussions about mixing but that the practical tests proposed are difficult or impracticable and that those who have tried to carry them out have usually not been successful.

Paper E3, so far as I know, was the first suggestion that in the case of homogeneous chemical reactors not only the residence-time distribution but the scale of segregation of the mixture of the reactants might be important in determining the performance of the reactor. The concepts of segregation and molecular mixing (which are not uniquely determined by the residence-time distribution) are not fully spelled-out and indeed I do not think they have been since. Clearly, if no reaction occurs between two fluids which are being mixed, their degree of mixedness on the molecular scale is irrelevant as far as chemical reactions are concerned; whereas if A and B molecules react instantaneously another limiting situation is encountered. In general the reaction will be of finite speed and is related to the degree of molecular mixing which can be studied as described in Papers E4 and E5. A voluminous body of literature has been devoted to such topics in the last 22 years (starting with the masterly analysis by Th.N.Zwietering, Chem. Engg Sci. 1959 11 1).

Paper E5 is concerned with "molecular mixing" of a jet of water with its surroundings. Most other investigations of jets have been concerned only with the time-average concentration of jet fluid at a point, but told nothing about the extent of molecular mixing between the jet and the ambient fluid. The composition at a point might be pure A half the time and pure B half the time; the time-average composition at the point would be half A and half B but there would be no reaction between them at the point in question. In Paper E5 liquids A and B undergo an instantaneous reaction when mixed on the molecular scale and the time-average extent of reaction at a point is determined by means of an indicator. If the jet fluid is A, unreacted A can be detected further downstream than the point at which the time-average concentrations of A and B are chemically equivalent.

Similarly, in Paper E6 a hot-air jet flowed into an environment of cold air. The temperature at a given point fluctuates about its time-average value so that a temperature T can be detected for part of the time at points further downstream than that at which the average temperature is T. Heat conduction in a gas is effected in the main by the mixing of hot and cold molecules, so temperature makes a good tracer for molecular mixing.

Such considerations would appear to be very relevant to jet reactors such as flames, but Papers E5 and E6 appear to have attracted little attention - possibly because they were published in an inappropriate journal. (It should be noted that the crosses in Fig.12. Paper E6, are <u>not</u> experimental points).

THEORY OF MIXTURES AND MIXING

P. V. DANCKWERTS

The University, Cambridge

The fundamentals of chemical engineering are closely allied with classical physics. The chemical engineer, preoccupied with the working of practical plant, is often neglectful of the principles of his subject. This is the first of a series of articles intended to stimulate fundamental thought. In his review of mixing theory Mr Danckwerts suggests some possible future developments. The gaps which will be apparent in the discussion are a reflection of our present ignorance.

THE types of mixing process encountered in industry are of great diversity, alike in the characteristics of the materials to be mixed, the reasons for mixing them and the methods used to do so. This diversity is illustrated by the following examples: fuel gas and air (to promote rapid burning in a limited volume), blending of lubricating oils, chalk and clay slurries (cement manufacture), powdered chalk and clay (dry manufacture of cement), rubber and carbon black, powdered pigments, whiting and oil (putty), gravel and cement slurry, crushed lime with coal and coke (carbide), ammonium nitrate and T.N.T. (amatol), solution of pigment and water (rapid dilution and fine precipitate), hot gas and cold gas (rapid 'quenching' of reaction, production of aerosols).

Industrial mixing operations require capital outlay for machinery, and the time and power which they consume contribute to the operating costs of a process. Expense apart, the rapidity or intimacy with which materials are mixed may affect the success or efficiency of a chemical process. It is also of considerable economic importance to be able to specify the equipment required to carry out a given mixing task, without introducing wasteful safety factors.

While the practical problems of mixing will no doubt have to be solved largely by empirical means, the development of the theory, which has so far been almost completely neglected, should serve to define the subject itself more clearly, and suggest improvements in practice. The first requirement is for rational and quantitative methods of defining and measuring 'goodness of mixing', applicable as widely as possible and avoiding purely arbitrary methods. This would enable mixing tasks to be specified and the efficacy of mixing processes to be compared. A theory of the absolute energetic efficiencies of mixing processes is also required, since it is important to discover how much of the energy supplied to the mixer is essential and how much is dissipated fruitlessly.

The mechanics of mixing processes is the most difficult part of the subject to treat theoretically. Even when considering the case which seems simplest —the mixing of true fluids of similar properties— intractable problems of hydrodynamics are encountered. The mixing of non-Newtonian liquids, plastic substances or solid particles, or intermixing of these various types of material, raises even more formidable difficulties. Progress in this field is likely to depend less on analytical treatment than on well planned experiments, regulated by physical insight and the use of dimensional analysis.

Generally speaking, the more quickly a given mixing task is to be performed, the more energy must be used and *vice versa*. In practice most arrangements for mixing represent a more or less economical compromise between time and energy. However, convenience, constructional simplicity and ease of maintenance may be of overriding importance. When designing mixers for foodstuffs, for instance, ease of sterilization may be the primary consideration. Clearly, no theory of mixing processes can take such ulterior considerations into account. In this article only the mixing of stable, pre-formed particles will be considered. Mixing processes which are accompanied by size reduction or the transfer of material between phases raise problems which require separate treatment.

SCALE OF SCRUTINY

Unless the reasons for making up a mixture are known, it is impossible to decide whether it is well or badly mixed. Any mixture, if scrutinized closely enough, will show regions of segregation—that is, the composition will vary from point to point. The size of the regions of segregation which can be tolerated will vary from one case to another. The term 'scale of scrutiny' will be applied here to the minimum size of the regions of segregation in the mixture which would cause it to be regarded as imperfectly mixed for a specified purpose. Defined in this imprecise way, the scale of scrutiny can only be expressed as an order of magnitude (length, volume or area), but the concept is a useful one.

Suppose, for instance, that wheat of two qualities, *A* and *B*, is stored in a single bin. If it is desired merely to mix the wheat so that the proportion of *A* to *B* shows no trend with the level in the bin, then the scale of scrutiny is comparable with the depth of the bin. If, on the other hand, it is intended that a bushel of wheat taken from anywhere in the bin has the average proportion of *A* to *B*, within close limits, then the scale of scrutiny is a volume considerably smaller than a bushel.

Suppose that the efficiency of mixing of air and a fuel gas in a furnace or internal combustion engine is being investigated. The purpose of mixing is to promote encounters and reaction between oxygen and fuel molecules. The scale of scrutiny in this case is comparable with the mean free path of the molecules.

Again, consider the blending of pigments to give an apparently uniform colour. The scale of scrutiny depends on the resolving power of the eyes and the distance from which the mixture is to be viewed. For a distance of about one foot it will be somewhat larger than the grain of a photogravure screen ; for a distance of ten feet it will be a few millimetres square—comparable with the size of the spots of colour in a *pointilliste* painting.

It may be possible to define the scale of scrutiny more rigorously, and use it as a quantitative specification for a mixture. However, the matter is probably less simple than the foregoing discussion suggests, since in some cases ' goodness of mixing ' depends not only on the scale of the regions of segregation but on their intensity, as will be shown later.

COARSE-GRAINED MIXTURES

The components of any mixture consist of particles which are capable of relative movement. These particles may be large, like lumps of coal, or small, like molecules. If the scale of scrutiny is not large enough to embrace many particles, the mixture may be called ' coarse-grained '. Mixing is then essentially a process of ' randomization ' or shuffling of the positions of these particles. The result can be illustrated by considering a mixture of black and white marbles on a tray. Initially the marbles are unmixed (*Figure 1*a). When they are stirred up the random movements of the marbles eventually produce a mixture with a configuration such as that shown in *Figure 1*b. Dispersal is not perfect in this mixture ; local fluctuations in concentration can be seen, some black marbles being entirely surrounded by white ones and *vice versa*. Samples consisting of triangular blocks of ten marbles taken at random will show considerable variations in mean composition. Larger samples will show smaller variations and *vice versa*. On the other hand, the mixture must be considered ' fully randomized ' or completely mixed, since further stirring does not reduce the local fluctuations in composition. In a configuration such as that shown in *Figure 1*c dispersal is more uniform ; each marble is surrounded by four of the other colour and only two of its own, and any triangular block of ten will be half black and half white. However, this configuration cannot be regarded as a mixture ; in practice it will never be brought about by a random mixing process, and stirring will degrade it to a configuration similar to *Figure 1*b. Satisfactory mixing is thus only possible in principle if the scale of scrutiny is a volume great enough to include several particles of each component.

Fluctuations in coarse-grained mixtures are of practical importance when, for instance, it is desired to take a representative sample from a mixture of seeds, or make up a batch of tablets of identical composition from a granular mixture.

A precise statistical treatment of fluctuations is easily worked out for coarse-grained mixtures of a simple type like the black and white marbles already

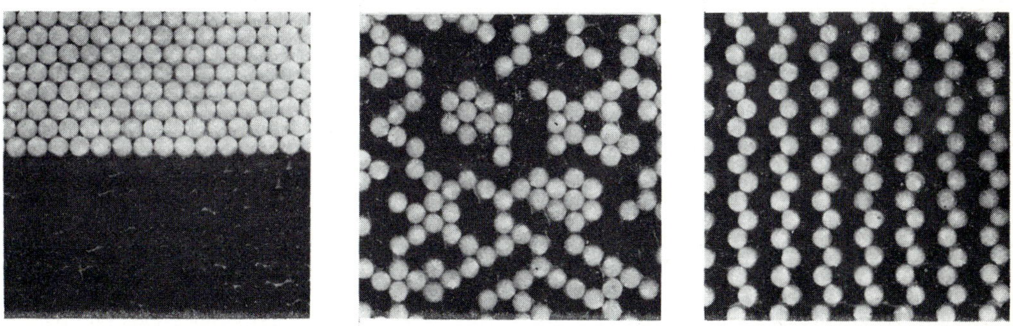

a b c

Figure 1. Idealized coarse-grained mixture: a *unmixed;* b *fully randomized;* c *non-random but fully dispersed*

discussed. In such a case it is possible to decide, by observing the variation between samples, whether or not the mixture from which they are drawn is fully randomized, and to express the results quantitatively as an idex of 'goodness of mixing'[1]. D. BUSLIK[2] has considered mixtures of particles which have sizes varying over a wide range, so that the fine particles tend to segregate in the voids between the large ones. G. HERDAN[3] describes a statistical procedure which can be used to test the uniformity of certain types of mixture, such as ammonium nitrate crystals in a matrix of T.N.T.

FINE-GRAINED MIXTURES

The scale of scrutiny may be great enough to embrace a very large number of particles of each component. Under these circumstances the mixture may be called 'fine-grained'. For instance, when the scale of scrutiny is determined by the resolving power of the eye, such mixtures as smoke in air or pigments in paint are fine-grained. *Figure 2* represents such a mixture. Mixtures of mutually soluble substances, or mixtures of gases, are fine-grained so long as the scale of scrutiny is much larger than the volume of the mixture divided by the number of molecules of any component.

When a fine-grained mixture is fully randomized it will be uniform in composition for practical purposes. Fluctuations, which are obtrusive in coarse-grained mixtures, become unimportant in fine-grained mixtures (unless attractive forces between the particles of one component tend to create clusters). For practical purposes, we can speak of the 'composition at a point' in the mixture; a 'point' in the sense used here means a region much smaller than the scale of scrutiny but much larger than the ultimate particles. In a fully randomized mixture the composition can be said to be the same at all points. In an imperfect mixture the composition may vary smoothly and continuously from point to point.

The author has considered elsewhere[4] the problem of defining and measuring the goodness of mixing of fine-grained mixtures when the scale of scrutiny is much less than the total volume of the mixture—for instance, when we are concerned with 'spottiness' in a mixture of two liquids rather than segregation between the top and bottom of the containing vessel. Two distinct quantities then seem to be required to describe the condition of an imperfect mixture with any precision—namely, the scale and the intensity of segregation.

The scale of segregation (linear or volumetric) is a measure of the size of the regions of segregation, or 'clumps' of the unmixed components, in an imperfect mixture. Since the clumps will usually be of irregular shape and diffuse outline, a special statistical method must be used to define and determine their scale. This is based on the degree of correlation between the compositions at neighbouring points and is similar to that used in the statistical theory of turbulence, which deals with a problem of the same sort.

Figure 2. Fine-grained mixture (scale of scrutiny visually determined)

The intensity of segregation is a measure of the departure of the composition from the mean value, averaged over all the points in the mixture. The intensity does not depend on the size of the clumps, but on the extent to which A has been diluted by B and *vice versa*. It has a value of unity when the composition at every point in a mixture of A and B is either pure A or pure B, and it will be zero when the composition is uniform. W. R. HAWTHORNE and his co-workers[5] have defined a quantity called 'unmixedness', which is closely related to the intensity of segregation, and devised an ingenious method of measuring it in air–fuel gas mixtures.

The criterion of goodness of mixing may sometimes be the scale of segregation, sometimes the intensity, but more frequently, in the author's opinion, some combination of the two. For instance, if two pigments are blended, the intensity can be high if the scale is small (*i.e.* streaks of the pure colours are permissible if they are too small to be distinguished by the naked eye). A somewhat larger scale may be offset by a lower intensity (large streaks will be indistinguishable if they are diluted by the surrounding colour). If the scale is very large, portions of the mixture drawn for use from different parts will not match; however, if the intensity is sufficiently low, the discrepancy will be negligible. If it is desired that a series of samples of a given size drawn from the mixture shall display not more than a given specified variation in composition, it is the *product* of the scale and the intensity which must be reduced to a certain value by mixing.

MINIMUM WORK REQUIRED FOR MIXING

Much of the current experimental work on mixing is implicitly concerned with the efficiency of using mechanical energy to perform a given task of mixing. Even if we ignore the vagueness with which the task is usually defined, such studies can only give comparative rather than absolute results unless they can be related to an *ideal* efficiency, analogous to the Carnot efficiency for heat engines. Only in this way can we discover whether the upper limit of efficiency is being approached and whether further refinements are worth while.

It is necessary to distinguish at least three types of mixture which differ fundamentally in their behaviour.

Positive mixtures (*e.g. gases or miscible liquids*)—Work must be done to separate the components of such mixtures. If the components are put in contact, they will mix spontaneously and with a decrease of free energy. It is thus possible in principle to extract work from the mixing process. Even if this is ruled out as impracticable, irreversible mixing can take place without the expenditure of any work, provided that unlimited time is available. If (as will usually be the case) it is desired to reduce the intensity of segregation to some specified level in a time much shorter than that required by unassisted molecular diffusion, it is necessary to reduce the scale of segregation by mechanical means, and hence increase the rate of decay of the intensity, as will be explained later. The reduction of the scale involves relative displacement of parts of the mixture, so that work must be done against viscous forces. It may be possible, by considering a systematic series of displacements, to evaluate the minimum amount of work required to reduce the scale and intensity of segregation to specified values in a specified time. The actual work done in mixing two fluids will inevitably be greater than the ideal. For instance, the production of turbulence, although usually the most convenient way of mixing mobile fluids, involves the dissipation of some work in movements which do not help to reduce the scale of segregation.

Negative mixtures (*such as suspensions of solids in liquids*)—These require work for their formation but, more important, the components separate unless work is continually expended on them. In some cases it is possible to calculate the power ideally required. For instance, the power needed to maintain a single isolated particle in suspension is the product of the (submerged) weight of the particle and its free-fall velocity (*Figure 3*). In an exceedingly dilute suspension (*i.e.* average distance between particles very much larger than particle diameter) containing N identical particles, the power required is N times this product. In denser suspensions the rate of fall of the particles is less, with a corresponding decrease in the ideal power requirement per particle. The principle can be extended to suspensions of particles with different rates of fall.

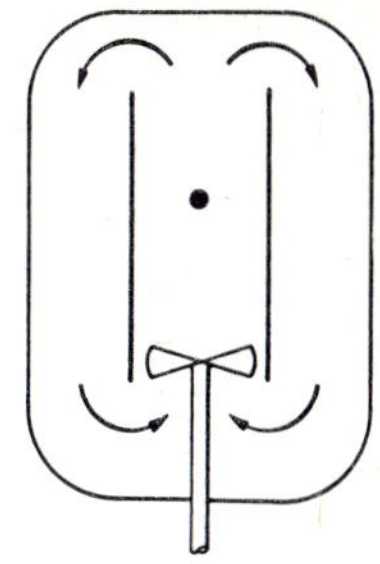

Figure 3. Illustrating the ideal power requirement for the suspension of a single particle

Neutral mixtures—Such mixtures are static. They show no tendency to segregate once mixed, but the components show no tendency to mix spontaneously (*e.g.* solid particles, plastic pastes, suspensions of solids in thixotropic liquids). Neutral behaviour may be destroyed by treatment such as heating or vibration. It may be possible to evaluate the ideal work required to form particular types of neutral mixture, but there seems to be no unifying principle involved.

SCALING UP FROM MODELS

It is possible in principle to use models to predict the behaviour of full scale mixing equipment, providing the requisite dimensional similarity can be established. Little rigorous work of this kind has been done, probably because there has been no clearly conceived definition of goodness of mixing or procedure for measuring it.

As an example of the use of dimensional analysis, consider the mixing of two fluids of similar properties in a vessel of specified shape and size, by any kind of paddle, impeller or circulating pump. The relevant quantities are the time t required to achieve a given volume scale v and intensity I of segregation in a mixture of density ρ and viscosity μ. The interdiffusion coefficient of the two fluids is D, the scale of the system is given by some characteristic linear dimension L, and the linear velocity at some specified point in the system (*e.g.* at the tip of a paddle blade) is u. Then

$$I=F_1\left[\frac{uL\rho}{\mu},\frac{ut}{L},\frac{u^2}{gL},\frac{\mu}{D\rho}\right]$$

$$\frac{v}{L^3}=F_2\left[\frac{uL\rho}{\mu},\frac{ut}{L},\frac{u^2}{gL}\right]$$

The first of the groups within the brackets is the Reynolds number, which determines the flow pattern in the tank, in so far as this depends on inertia and viscosity. The second is a scale factor relating the mixing time on the model to that in the full sized plant. The third is the Froude number, which is of importance when gravitational forces help to determine the flow pattern—for instance, when a deep depression forms on the surface of a swirling liquid or when a heavy liquid has to be dispersed into a light one[6]. The quantity $\mu/D\rho$ is the Schmidt number, and determines the rate at which the intensity of segregation decays by molecular diffusion.

It is possible by experiments on a scale model to find relationships between ut/L on the one hand and I or v/L^3 on the other, and provided the Reynolds, Froude and Schmidt numbers are the same for both, these relations will be equally valid for the full scale plant. Thus the performance of the mixer and the effect of alterations in the design or running speed, or in the properties of the mixture, can be determined from the model. It is not possible to keep both the Froude and Reynolds numbers the same without changing to a fluid of different properties (or altering the gravitational field). However, if either number is unimportant in determining the flow pattern it may be allowed to assume different values in the model and full scale plants. The experiment is then simplified, as the same fluids can be used in both cases, thus automatically making the Schmidt number the same. If the same fluids cannot be used in both cases, dimensional similarity may be very difficult to achieve. To ensure complete similarity when mixing liquids of markedly different properties, the ratios of the densities and viscosities of the two liquids used, and the relative variation of these and of the diffusivity with composition, must also be the same in the two systems.

H. FOSSET and L. E. PROSSER carried out experiments on models designed to simulate the mixing of tetraethyl lead fluid with gasoline in tanks by jet entrainment. They measured the time t required to achieve an arbitrarily defined degree of uniformity under various conditions. They showed the value of the Reynolds number to be unimportant over the range of conditions studied. Provided the Froude number (referring to the relative movement of the dense and light liquids) exceeded a certain critical magnitude, the value of the ratio ut/L was found to be roughly constant. The results in the models were in general agreement with those in full-sized tanks.

H. KRAMERS and his collaborators[7] have investigated mixing by stirrers in tanks. Over the range of variables covered, the Froude and Reynolds numbers were found to be unimportant, and ut/L was roughly constant for an arbitrarily defined degree of mixing.

J. H. RUSHTON[8] and others have carried out extensive model experiments on the power required to drive paddle and impeller agitators under various circumstances.

A similar analysis can be applied to other types of mixtures, such as solid particles or suspensions. Dimensional similarity may, however, be difficult or impossible to achieve in such cases, and also when such physical properties as stickiness, roughness, plasticity *etc* are involved. The results of model experiments should then be treated with great caution.

MECHANISMS OF MIXING

When liquids are mixed by impellers or paddles, there are regions of intense activity where the scale of segregation is being broken down, and other regions where the liquid is circulating quietly. It is important to design mixers so as to ensure that all elements of the material pass periodically through the active zones, none being trapped in dead spaces for long. The function of the processes taking place in the active zones is to reduce the scale of segregation as rapidly as possible by projecting jets of one liquid into another, by setting up turbulence which projects one material into another in eddies, by dragging viscous streamers of one material into another, by 'slicing' and relative displacement of parts of the liquid *etc*. When the components have been interspersed in clumps molecular diffusion can reduce the intensity of segregation by virtue of its tendency to obliterate differences of concentration between neighbouring regions. The smaller the scale of segregation, the more rapidly will the intensity of segregation decay as a result of molecular diffusion.

Any theory of the mixing which occurs in a moving fluid implies a theory of the motions of the parts of the fluid. Unfortunately the theory of turbulent motion, even in the simplest cases, presents notoriously difficult problems. S. CORRSIN[9] has considered the decay of temperature 'spottiness' in a region of decaying isotropic turbulence (the treatment applies equally to composition spottiness), and even under these idealized conditions the treatment is difficult and the results of limited value. This work is important, however, as being the first attempt at a fundamental approach to the subject. A recent paper by G. K. BATCHELOR[10], concerning the effect of turbulence on fluid lines and surfaces, may also be relevant.

It is probably true to say that when mixing is brought about by turbulence, energy is most efficiently applied in creating turbulence which (so far as possible) has a scale at any point equal to the local scale of segregation.

Mixing often takes place between a jet of fluid and a coaxial jet or surrounding static 'atmosphere'. Under certain conditions (*e.g.* in a candle flame) flow may be laminar, so that mixing occurs solely by molecular diffusion ; the process may then be subject to precise analysis. Generally, however, mixing is caused primarily by turbulence having characteristics which vary from one region to another. The complicated hydrodynamics of such systems cannot yet be analysed in quantitative terms. Experimental work on jet mixing has produced a good deal of information about the distribution of time-average concentrations in the region of intermixing[11]. However, only Hawthorne *et al*[5],

Figure 4. Mixing by jet entrainment (after A. M. Binnie[18]*)*

with their measurements of unmixedness, appear to have tried to measure any characteristics of the mixture other than its mean composition at various positions. *Figure 4* shows mixing of a jet with static liquid. The interchange of material in eddies produces clumps which then decay in intensity by molecular diffusion.

Materials having the consistency of dough or rubber are usually mixed in kneading machines of various types which act principally by folding, or cutting and superposition, accompanied by stretching, rolling out or some other process of deformation. *Figure 5* suggests diagrammatically the way in which the scale of segregation is reduced in such

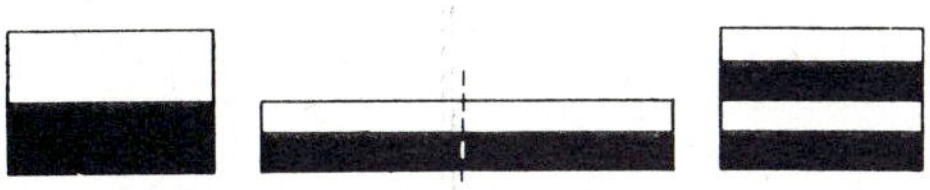

Figure 5. Illustrating the reduction of scale of segregation in a kneading machine by cycles of superposition followed by deformation

machines. In a well designed kneader every part of the material will pass through a cycle of such processes at fixed, or statistically distributed, intervals of time and it should be possible to work out the general form of the expression relating scale of segregation to mixing time. A. BROTHMAN and his colleagues[12] have proposed the view that mixing processes should be regarded as producing a progressive increase in interfacial area between the components. Although their treatment is less generally applicable than they suggest, it may be appropriate to the types of mixture produced by kneading, in which diffusion plays little or no part.

It should not be forgotten that many kneading processes (*e.g.* the incorporation of carbon black in rubber and the mixing of dough) are accompanied by complicated physicochemical processes which may bring about changes in the mechanical properties of the materials.

When solid particles are mixed, their shape, density and size distribution (and for very small particles, roughness, stickiness and electrical properties) are important. It is important to prevent material of one kind segregating in some part of the mixer. Such experiments as have been done have not led to a clear insight into the mechanics of the process, or to any very useful empirical relationships. J. VISMAN and D. W. VAN KREVELEN[13], and N. K. MAITRA and J. M. COULSON[14] have suggested methods by which the efficiencies of mixers may be measured and compared.

BACK-MIXING IN CONTINUOUS FLOW SYSTEMS

In all that has so far been said, it has been implicitly assumed that the mixing process involves two or more distinct materials. However, in many modern industrial processes, in which streams of material flow continuously through various pipes and vessels, mixing may occur between parts of the material that differ primarily in the length of time which they have spent in the system. When a fluid flows through a long pipe, it is frequently assumed that flow is piston-like—that is, molecules which enter the pipe at a given moment move with uniform velocity along parallel paths and leave at the same moment. There would then be no back-mixing. Actually, flow of this type can never occur with true fluids, but in long pipes it is more nearly attained with turbulent than with laminar flow. On the other hand, a stream of material may flow through a vessel of considerable capacity, in which it is subjected to effective agitation. Under these circumstances, an element of material entering at any instant may be uniformly dispersed throughout the vessel before an appreciable fraction of it has passed the outlet. Such a vessel is usually said to be ‘effectively stirred’ or ‘completely mixed’ (although the outflowing stream may in fact display segregation if the scale of scrutiny is made much less than the volume of the vessel). The completely mixed flow system represents an ideal which can be very closely approached under certain conditions.

However, there is a wide variety of types of continuous flow system in which the degree of back-mixing is somewhere between these two extremes—for instance, blast furnaces, chromatographic columns, catalyst beds (fixed and fluidized), oil pipe lines, tubular reactors, reservoirs, settling tanks, continuous fermenters, gasometers, heat exchangers, gas scrubbers and so on. The author has reviewed elsewhere[15] some ways in which back-mixing in such systems can be measured and allowed for in calculating the performance of the equipment.

A particularly interesting problem is offered by a continuous flow blender, that is a hold-up vessel (such as a gasometer) through which flows a continuous stream of material of fluctuating composition. The effect of the blender is to 'iron out' the fluctuations to an extent determined by the holding time, the degree of back-mixing and other factors. Completely mixed blenders have been discussed in detail elsewhere[16,17].

Figure 4 is reproduced by permission of A. M. Binnie and the Editor of Engineering.

REFERENCES

1 LACEY, P. M. C. *Trans. Instn chem. Engrs, Lond.* 21 (1943) 53

2 BUSLIK, D. *Bull. Amer. Soc. Test. Mat.* 165 (1950) 66

3 HERDAN, G. *Small Particle Statistics* Chapter 14 Elsevier: Amsterdam, 1953

4 DANCKWERTS, P. V. *Appl. sci. Res., Hague A* 3 (1952) 279

5 HAWTHORNE, W. R., WEDDELL, D. S. and HOTTEL, H. C. *Third Symposium on Combustion, Flame and Explosion Phenomena* p 266 Williams & Wilkins: Baltimore, 1949

6 FOSSET, H. and PROSSER, L. E. *Proc. Instn mech. Engrs, Lond.* 160 (1949) 224

7 KRAMERS, H., BAARS, G. M. and KNOLL, W. H. *Chem. Engng Sci.* 2 (1953) 35

8 RUSHTON, J. H. *Chem. Engng Progr.* 47 (1951) 485; *Industr. Engng Chem.* 44 (1952) 2931

9 CORRSIN, S. *J. aero. Sci.* 18 (1951) 417

10 BATCHELOR, G. K. *Proc. roy. Soc. A* 213 (1952) 349

11 FORSTALL, W. and SHAPIRO, A. H. *J. appl. Mech.* 17 (1950) 399
BARON, T. and ALEXANDER, L. G. *Chem. Engng Progr.* 47 (1951) 181

12 BROTHMAN, A., WOLLAN, G. N. and FELDMAN, S. M. *Chem. metall. Engng* 52 (1945) 102

13 VISMAN, J. and VAN KREVELEN, D. W. *Ingenieur* 63 (1951) 1

14 MAITRA, N. K. and COULSON, J. M. *J. imp. Coll. chem. Engng Soc.* 4 (1948) 135

15 DANCKWERTS, P. V. *Chem. Engng Sci.* 2 (1953) 1

16 — and SELLERS, E. S. *Ind. Chemist* 27 (1951) 395; *Coke & Gas* 14 (1952) 247

17 BEAUDRY, J. P. *Chem. Engng* 55 (1948) 112

18 BINNIE, A. M. *Engineering, Lond.* 153 (1942) 504

BIBLIOGRAPHY

PERRY, J. H. *Chemical Engineers' Handbook* Section 17 McGraw-Hill: New York, 1950

Conference on Mixing and Agitation in Liquid Media *Trans. Instn chem. Engrs, Lond.* 29 (1951) 277

E2

THE DEFINITION AND MEASUREMENT OF SOME CHARACTERISTICS OF MIXTURES

by P. V. DANCKWERTS

Chemical Engineering Department of the University of Cambridge

Summary

The systematic study of mixing processes requires a quantitative method of expressing "goodness of mixing", based on conveniently-made measurements. In this paper, mixtures of mutually soluble liquids, fine powders, or gases are considered. It is shown that the important features of such mixtures can be expressed by two statistically-defined quantities, the scale and the intensity of segregation, and methods of measuring these are suggested. The discussion also throws light on some of the factors which affect the efficiency of mixing processes.

Notation

a = Concentration (volume-fraction) of component A.
$\bar{a}$ = Mean concentration of A in mixture.
b = Concentration (volume-fraction) of component B.
$\bar{b}$ = Mean concentration of B in mixture.
c = $a - \bar{a}$.
d = Diameter (of circle, strip, sphere or pipe).
D = Diffusivity.
I = Intensity of segregation (defined by eqn. (14)).
J = Content of A in line sample (eqn. (5)).
k = Reaction-velocity constant.
K = Content of A in volume sample.
l = Length of section of pipe.
m = Concentration of reactant M in solution A (moles per unit volume).
n = Concentration of reactant N in solution B.
r = Distance apart of two points in mixture.
$R(r)$ = Correlation coefficient between points in mixture distance r apart.
S = Linear scale of segregation (defined by eqn. (3)).
t = Time.
T = See eqn. (12).
u = Velocity of flow.
v = Volume of sample.
V = Volume scale of segregation (defined by eqn. (4)).
x = Distance from one end of line-sample, or general space-coordinate.
X = Length of line-sample.
ξ = Value of r for which $R(r)$ falls to 0.
ϱ = Mean reaction rate per unit volume of mixture.
σ^2 = Variance of quantity denoted by subscript.

§ 1. *Introduction.* The study of mixing processes suffers at the present time from the lack of any quantitative method of expressing "goodness of mixing". To be of practical value, such a quantitative description must fulfil certain conditions: (*a*) It must be related as closely as possible to the properties of the mixture which we assess qualitatively when we judge it to be well or badly mixed; (*b*) It must be possible to make the required measurements conveniently; (*c*) The method of classification should be applicable without modification to as many different types of mixture as possible; (*d*) It should not depend on purely arbitrary tests, leading to quantities of doubtful physical significance.

The following notes suggest a system of describing mixtures which seems to go some way towards fulfilling the above requirements. The treatment is suitable chiefly for mixtures in which the ultimate particles (i.e. the smallest capable of independent movement in the mixture) are very small compared to the size of the portions which will normally be taken for use or for analysis. Under these conditions it is possible to speak of the "concentration at a point", and a series of samples taken from a "completely mixed" mixture will have the same composition. The discussion will refer mainly to a mixture of two mutually soluble liquids (which may be a solution and a pure solvent respectively), but many of the conclusions are applicable to gases, powders and pastes, and to multi-component systems. Mixtures of coarse particles, which are not extremely small compared to the samples taken, present a problem which must be approached in a different way [1]). It will also be assumed that the mixture is uniform in texture; that is to say, it cannot be divided into two parts of equal size in which the mean concentration or the scale or intensity of segregation (as defined later) differ significantly. This is the most important limitation on the practical value of the definitions and tests which will be proposed. Large-scale segregation caused, for instance, by sedimentation, or by dead space in a mixer, is of great practical importance, but its study cannot conveniently be combined with that of the small-scale characteristics, or texture, which are the subject of the present discussion. The subject of large-scale segregation in continuous-flow systems will be dealt with on another occasion.

Three different situations have been borne in mind: (*a*) we may be confronted with a mass of material already mixed, and desire to investigate its texture; (*b*) we may desire to determine the effectiveness of a continuous mixer, from which a stream of material is issuing; (*c*) we may desire to follow the progress of the mixing of a batch of material.

§ 2. *The mixing process.* If two mutually soluble liquids are mixed together two things happen. In the first place the liquids are broken up into "clumps", which are intermingled (this will happen even if the liquids are not mutually soluble). The shape of the clumps will depend on the mechanism of the mixing-process; they may, for instance be compact, or in the form of long streaks. Up to a point, at least, the average size of these clumps will continue to decrease as mixing is continued.

At the same time, molecular interdiffusion of the two liquids occurs across the boundaries of the clumps. This process is spontaneous and will continue even if the mechanical mixing is stopped. Diffusion will ultimately reduce any mixture of mutually soluble liquids to complete uniformity, but the process is slow unless the liquids have first been broken up into small clumps. Unless diffusion (or some analogous process) occurs, continuous stirring will not produce a homogeneous mixture; the composition at any point will be that of one of the pure components, and it will vary discontinuously from one region to another. In the case of powders, independent motion of the ultimate particles will produce an effect similar to molecular diffusion, but only so long as some kind of mechanical agitation is continued.

The breaking-up and the interdiffusion are, in the case of liquids, largely independent processes which produce distinguishable results. The former reduces the size of the clumps, while the latter tends to obliterate differences of concentration between neighbouring regions of the mixture. It therefore seems desirable to use two quantities to describe the degree of mixing — namely the *scale of segregation* and the *intensity of segregation*. Owing to the lack of geometrical regularity in the mixture, both quantities must be defined and determined by statistical methods.

§ 3. *The scale of segregation.* The measure of the scale of segregation adopted here is analogous to the "scale of turbulence" used in the statistical theory of turbulence [5]). Considering a mixture of two liquids, A and B, let their concentrations (volume-fractions) at any point be a and b respectively, and their mean concentrations in the mixture as a whole be $\bar{a}$ and $\bar{b}$. Then

$$a + b = 1, \quad \bar{a} + \bar{b} = 1. \tag{1}$$

Suppose the concentrations (a_1, a_2) are measured at two points in the mixture a distance r apart. The deviations of the two concentrations from the mean are multiplied together to give the product $(a_1 - \bar{a}).(a_2 - \bar{a})$. If a large number of such pairs are taken, the points in each pair being the same distance r apart, we can find the mean value, $\overline{(a_1 - \bar{a})\,(a_2 - \bar{a})}$, of the corresponding products of the deviations, and also the mean square deviation, $\overline{(a - \bar{a})^2}$, of all the concentrations from the mean. The quantity

$$R(r) = \frac{\overline{(a_1 - \bar{a})\,(a_2 - \bar{a})}}{\overline{(a - \bar{a})^2}} = \frac{\overline{(b_1 - \bar{b})\,(b_2 - \bar{b})}}{\overline{(b - \bar{b})^2}} \tag{2}$$

derived from these measurements is called the *coefficient of correlation* between values of a (or b) at points separated by a distance r. The denominator is usually called the *variance* of a (or b), and may be written σ_a^2 or σ_b^2 (the two are of course equal).

$R(r)$ may have any value from 1 to 0. (Negative values will only be encountered when there is some form of long-range segregation present, or some regularity in the pattern of the mixture. Such cases are not considered here, and attention is confined to systems for which $R(r) \geqslant 0$ for all values of r). $R(0)$, the value when $r = 0$, is identically equal to 1, and in general a value close to 1 means that a concentration much above the average at a given point is likely to be associated with an above-average concentration a distance r

away, or that below-average concentrations are similarly related. A value of 0 means that there is only a random relationship between the concentrations at points distant r apart. In an imperfect mixture, $R(r)$ will be greater than 0 for small values of r, because points close together will often be in the same clump. However, when r exceeds a certain value the relationship between the two concentrations will become a random one (provided there is no large-scale segregation or regular periodicity in the mixture) and $R(r)$ will fall to 0, whatever the proportions of A and B in the mixture. $R(r)$ will therefore vary with r somewhat in the manner of fig. 1, although the shape of the curve will differ from case to case.

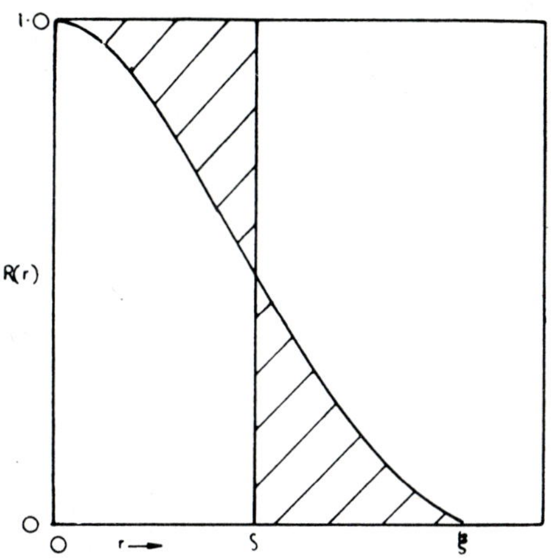

Fig. 1. Shaded areas are equal.

The curve is known as the *correlogram* of the mixture. (It is shown in Appendix C that the slope of the curve is always zero at $r = 0$ in systems in which diffusion occurs). Figs. 2 and 3 show correlograms for two two-dimensional models of mixtures. (The slopes of these curves are not zero at $r = 0$, because the "clumps" have sharp edges).

The value of r for which $R(r)$ falls to zero will be called ξ. (If $R(r)$ approaches zero asymptotically, ξ is assigned a value such that $\int_0^\xi r^2 R(r)\,\mathrm{d}r$ differs from $\int_0^\infty r^2 R(r)\,\mathrm{d}r$ by a fraction which is small enough to be ignored in evaluating V — see (4)).

There are two ways in which the correlogram of a mixture can be used to define a scale of segregation. The *linear scale*, S, is defined as the area under the correlogram:

$$S = \int_0^\infty R(r)\,\mathrm{d}r = \int_0^\xi R(r)\,\mathrm{d}r. \tag{3}$$

The *volume scale*, V, is defined as 2π times the area under the curve $r^2\,R(r)$ vs. r:

$$V = 2\pi \int_0^\infty r^2\,R(r)\,\mathrm{d}r = 2\pi \int_0^\xi r^2\,R(r)\,\mathrm{d}r. \tag{4}$$

The relationship between S and V depends on the shape of the clumps and hence of the correlogram. If the latter were linear, for example, V would be equal to $4\pi S^3/3$, i.e. to the volume of a sphere of radius S. In general, mixtures having equal S may have different

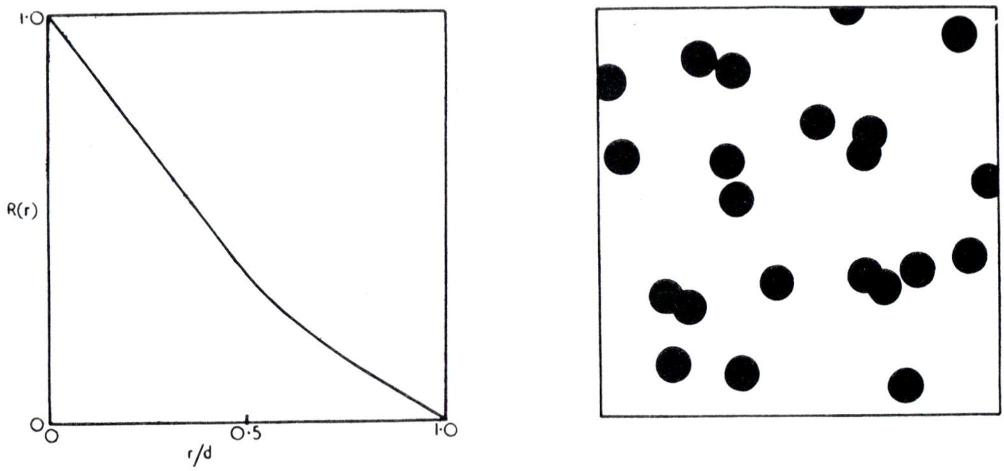

Fig. 2. Correlogram. Diameter of circles d, $S/d = 0.42$.

V, and the ratio of S^3 to V in a given system may change during the course of mixing. When the clumps are not extremely elongated (i.e. when the mixture is mottled rather than streaky) V will be of the same order of magnitude as $4S^3$. S and V will both normally diminish during the mixing process. The ease with which each can be measured is likely to determine whether S or V is used as a measure of the scale of segregation in a given situation.

The clumps in a mixture may vary widely in size and shape, and both will generally be indeterminate because the boundaries will be diffuse. It is not possible, therefore, to refer with any precision to an "average diameter" or an "average volume" of the clumps. S and V, on the other hand, are precisely-defined quantities and their magnitudes vary in the same sense as the size of the clumps. For this reason it is convenient to use them as unambiguous measures of the scale of segregation, even though their significance is not easily visualised.

Fig. 2 illustrates the point that if clumps of component A are scattered in an excess of B, S will indicate the size of the clumps of A. Fig. 3 shows the effect of elongation of the clumps on the shape of the correlogram and the value of S. The latter would increase indefinitely as the rectangles were lengthened. Provided the black elements are scattered at random, the shapes of the correlograms and the values of S do not depend on the ratio of black areas to white.

A random collection of spheres of diameter d, analogous to fig. 2, will give $S/d = 0{\cdot}38$, $V/S = 4{\cdot}7$.

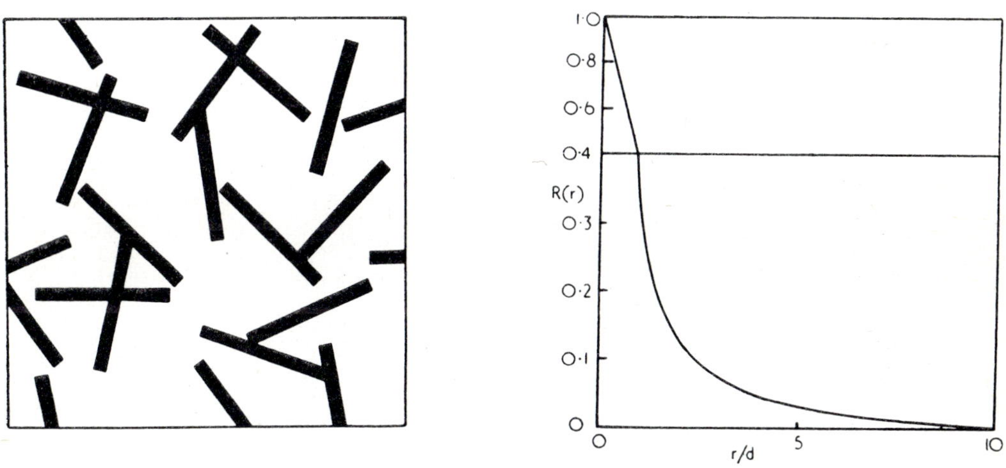

Fig. 3. Correlogram. Width of strips d, length of strips $10d$, $S/d = 1.1$.

§ 4. *Measurement of the scale of segregation.* There are a number of ways in which S and V can be determined. The method used in any given situation will depend on the circumstances.

(*a*) By measuring the concentration at a large number of points in the mixture, $R(r)$ can be determined as a function of r, and hence S and V found. Even if practicable, such a method is likely to be laborious.

(*b*) In a batch mixing-process, the concentrations at two fixed points a distance r apart might be measured continuously while the mixture moved past them (electrical or optical methods suggest themselves). $R(r)$ could be calculated from records of the readings from each point, or by a modification of one of the methods devised for measuring the scale of turbulence [3]). The value of $R(r)$ for a number of values of r must be found by employing a number of pairs of measuring points at various distance apart.

(*c*) A different and more convenient type of method can be used for finding S if a means can be found for measuring the total "content" of one of the components along a straight line joining two points. The content, J, of component A in a line-sample of length X is defined by

$$J = \int_0^X a \, dx \tag{5}$$

where a is the concentration of A at a point distance x from one end of the line. For instance, if both the components are transparent, one being coloured and one colourless, the content of the coloured material in a path of length X is proportional to the colour-density of this thickness of the mixture. In this case the content could conveniently be measured by setting up a light-source and photo-cell a distance X apart, both immersed in the mixture. The method seems particularly suitable for following a batch mixing-process.

Suppose that J is determined for a large number of such line-samples in different regions of the mixture. (In a batch-process this would probably be done by letting the mixture flow across a fixed line during the mixing process. The observations must then be confined to a period short enough so that the scale S does not alter appreciably). If the mixture is imperfect, the content J will fluctuate, and the scale of segregation, S, is related to the magnitude of the fluctuations by the expression (derived in Appendix A)

$$S = \frac{\sigma_J^2}{2X\,\sigma_a^2}, \tag{6}$$

where σ_J^2, σ_a^2 are the variances of J and a respectively and are defined by

$$\sigma_J^2 = \overline{(J - \bar{a}X)^2} \equiv \overline{J^2} - (\bar{a})^2\,X^2; \tag{7}$$

$$\sigma_a^2 = \overline{(a - \bar{a})^2} \equiv \overline{a^2} - (\bar{a})^2. \tag{8}$$

$\bar{a}$ is the mean concentration of a, which is assumed to be known. $\overline{J^2}$ and $\overline{a^2}$ are the mean square values of J and a respecively. The former can be determined in various ways from the output of the instrument which measures J. Hot-wire ammeters or other instruments capable of measuring root-mean-square voltages directly would probably provide the most convenient method.

$\overline{a^2}$ might be derived from the output of a simular instrument with a value of X very much smaller than S or by any method by which point-concentrations could be conveniently and continuously measured. Methods of determining $\overline{a^2}$ by following the progress of a chemical reaction are discussed in § 6.

Equation (6) is approximate, the error being of the order ξ/X. Assuming that ξ has the same order of magnitude as S, this implies that X must be considerably greater than S if this method of measuring the latter is to be used. On the other hand, a large value of X means a small value of the *fractional* fluctuations of J, with consequent difficulty in determining σ_J^2 accurately. The value of X most suitable in given circumstances must de betermined by trial and experience. If two substantially different values of X give the same value of S, it may be concluded that the latter contains no serious error.

This method is likely to be more convenient than (*b*), because it does not require independent measurement of $R(r)$ for a number of different values of r, and because the computation, automatic or otherwise, leading from the output of the measuring instruments to S will be simpler.

(*d*) In some cases it may be preferable to determine V, using a method based on the analysis of samples. Suppose a number of samples of volume v are taken from the mixture, and the content, K, of A found in each sample. The variance of K among the samples is related to V by the expression (derived in Appendix B)

$$V = \frac{\sigma_K^2}{2v\,\sigma_a^2}, \tag{9}$$

where

$$\sigma_K^2 = \overline{(K - v\bar{a})^2} \equiv \overline{K^2} - v^2(\bar{a})^2 \tag{10}$$

and σ_a^2 is defined by (8). $\overline{a^2}$ might be determined by taking a number of samples much smaller than V, or by one of the methods mentioned in (*c*) above. Equation (9) is an approximation; its use is justifiable only if $v \gg V$, and if the the samples are of a reasonably compact shape (e.g. cubes, spheres, short cylinders).

(*e*) When it is desired to measure the scale of segregation in a stream of material flowing continuously from a mixer through a pipe, method (*c*) above may be used if the diameter d of the pipe is much greater than S. Method (*d*) may also be used, subject to the same condition, if it is possible to measure continuously the content K of component A in a length l of the pipe. The volume v of liquid in this "sample" is $\pi d^2 l/4$, and hence, substituting in equation (9),

$$V = 2\sigma_K^2/\pi d^2 l \sigma_a^2, \tag{11}$$

(*f*) A case which is perhaps more likely to be encountered in practice is that in which a continuous stream of liquid leaves a mixer through a pipe whose diameter is very much less than S. The concentration will then be virtually uniform at all points at a given cross-section at a given moment, and the mixture will display segregation only in a longitudinal direction. Suppose a method is available for measuring continuously the concentration, a, of component A at a given cross-section; let the velocity of flow be u. Then the quantity

$$J = u\int_0^T a\,.\mathrm{d}t \tag{12}$$

is the content of A in a line-sample of length uT, the line being directed along the axis of the flow. The linear scale of segregation in this direction is thus seen from (6) to be

$$S = \sigma_J^2/2uT\sigma_a^2. \tag{13}$$

If a record of a is obtained as a function of the time t, the variance of J (as defined by (12)) can be found by integrating over a number of periods of length T (a convenient method of doing this has been described elsewhere [2]) and σ_a^2 can easily be computed from the same record. Hence S can be obtained from (13). The method described will not generally be applicable unless the velocity-profile across the pipe is nearly flat. In the case, for instance, of fully-developed laminar flow a given cross-section will not be uniform in composition, but will display ring-shaped zones of varying concentration due to the passage of successive clumps down the pipe; ostensible values of S found by the above method will then have little or no significance. On the other hand the method can be used if the fluid is in well-developed turbulent flow (Reynold's number $>$ 10,000), or if it displays the phenomenon of "Bingham flow" (e.g. extrusion of a plastic material), or even when it is in laminar flow, provided in the last case that the measurements are made close to the pipe-entry. It must be remembered that flow through a pipe, and subsequent discharge into a vessel, will themselves promote mixing, so that the scale of segregation measured at a given point in a pipe has only a local significance. Its value may nevertheless be a useful indication of the effectiveness of a mixer.

§ 5. *The intensity of segregation.* The intensity, I, of segregation is conveniently defined by the following mathematically identical expressions

$$I = \frac{\sigma_a^2}{\bar{a}.\bar{b}} = \frac{\sigma_b^2}{\bar{a}.\bar{b}} = \frac{\sigma_a^2}{\bar{a}(1-\bar{a})} = \frac{\sigma_b^2}{\bar{b}(1-\bar{b})}. \tag{14}$$

Thus defined, I has the value 1 when segregation is complete (i.e. when the concentration of A or B at every point is either 1 or 0), and the value 0 when the concentration is uniform. In general I reflects not the relative amounts of A and B nor the size of the clumps, but the extent to which the concentration in the clumps departs from the mean. If B is present in large excess, the value of I will depend primarily on the extent to which the clumps of A in the mixture have become diluted by B.

When two miscible liquids are mixed, the value of I is progressively reduced. However, as previously mentioned, this reduction is not caused directly by the mechanical process of mixing, but depends on molecular interdiffusion of the two components. In the absence of diffusion the mixture would remain "grainy", as when immiscible liquids are emulsified. It is shown in Appendix C that the fractional rate of decay of I is given by the relationship:

$$-\frac{1}{I}\cdot\frac{dI}{dt} = -6D\left[\frac{d^2R}{dr^2}\right]_{r=0} \tag{15}$$

Now $-[d^2R/dr^2]_{r=0}$ is a measure of the sharpness with which the correlogram bends over from its initially flat top. Circumstances which emphasize this change of slope therefore favour the rapid production of a homogeneous mixture. These are: (*a*) A small scale of segregation. If two mixtures are geometrically similar — that is, identical in pattern, but different in geometrical scale — the respective rates of decay of I will be proportional to D/S^2. This illustrates the importance of designing mixers so as to reduce the scale of segregation as quickly as possible. (*b*) Steep average concentration gradients at the boundaries of the clumps. These will generally be produced by the same processes which progressively decrease the scale of segregation — that is, by flattening or elongating the clumps, or by subdividing them by "slicing" action or by the dispersal of material in eddies.

If the purpose of mixing A and B is to enable them to react with one another, I represents an inverse measure of the effectiveness of the mixing. It is shown in Appendix D that if the local rate of reaction is proportional to the concentrations of both A and B, the total rate of reaction in the mixture will be greatest when $I = 0$. When $I \neq 0$, there is a linear relationship between I and the total reaction rate, which may be used as described in the next section to determine I.

§ 6. *Measurement of the intensity of segregation.* Any method of measuring $\overline{a^2}$ (such as those already mentioned) will enable I to be found from (8) and (14). Alternatively a method based on the measurement of the rate of a chemical reaction might prove more convenient when it is desired to follow the progress of mixing in a batch process. Its use would normally be limited to circumstances in which the liquids A and B could be chosen to suit the investigator. The reaction chosen must have a rate which is either proportional to the square of the concentration of one component, or to the products of the concentrations of the two components. The components A and B will normally be relatively dilute solutions of the actual reagents, so that changes in temperature and physical properties as a result of the reaction will be small. It is necessary to choose reaction systems such that the *fractional* rate of reaction is relatively slow, so that the mean concentration of the various reactants do not change appreciably during the period over which I is to be observed.

(*a*) Reaction between A and B. Solution A contains m moles per unit volume of reactant M; solution B, n of reactant N. The local rate of reaction between A and B at a point where the volume-fraction of A is a, is $[kmna(1 - a)]$ moles of M per unit volume per unit time, where k is the reaction-velocity constant. Then, as shown in Appendix D

$$I = 1 - \varrho/kmn\bar{a}(1 - \bar{a}), \tag{16}$$

where ϱ is the mean rate of reaction per unit volume of the mixture.

(*b*) Second-order reaction of A. Solution A contains m moles per unit volume of M, and is destroyed at a local rate of km^2a^2 moles of M per unit volume per unit time. Then

$$I = \frac{\varrho}{km^2\bar{a}(1 - \bar{a})} - \frac{\bar{a}}{(1 - \bar{a})}. \tag{17}$$

The choice of suitable reactions for the determination of I offers scope for considerable ingenuity. (Data on the kinetics of a large number of reactions in solution are given by Moelwyn-Hughes[4]). It must be possible either to measure the mean concentration of one of the reactants with great accuracy, or to measure the concentration of a product of the reaction (a lower standard of accuracy is permissible in this case) or to measure the reaction-rate directly. Reactions of which the progress can be followed by optical methods may prove useful, because of the ease and accuracy of the measurements. The following types of system are worth consideration.

(*a*) Polarimetric methods, in combination with mutarotation or racemisation reactions.

(*b*) Colorimetric methods, in combination with reactions giving coloured products.

(*c*) Photometric methods, in combination with chemiluminescent reactions. This method has the attraction that it is the reaction rate itself which is measured and not the integral extent of the reaction. For this reason it might be particularly suited to cases where I changed very rapidly.

An optical method of judging the extent of a reaction by sighting through the mixture (as in (*a*) or (*b*) above) will generally give information about the "content" of reactant or product in a "line-sample" of the mixture. The mean value of the content of the line-sample will change progressively as the reaction proceeds. However, superimposed on this steady change will be fluctuations due to segregation in the mixture, which may make it difficult to determine the rate of reaction. This difficulty can be overcome by taking a line-sample sufficiently large in relation to the scale of segregation to display only small fractional variations in content. Under some circumstances (depending on the rate of movement of the mixture, the rate of reaction and the rates of change of I and S) it may be possible to determine both the mean value and the variance of the content of the line-sample from readings taken over a period of time too short to allow the extent of the reaction or the degree of mixing to change appreciably. This would enable both I and S to be calculated as functions of the time of mixing from a single series of readings.

§ 7. *Conclusions.* A knowledge of the scale and intensity of segregation in a mixture will provide a good deal of information about its texture, in a quantitative form which allows different degrees of mixing to be compared. The two quantities are virtually independent and represent aspects of "goodness of mixing" which cannot be defined by a single quantity.

The determination of the scale and intensity, even by the simplest methods suggested, involves in effect making a considerable number of measurements and averaging them. This is inevitable, as the only useful information which can be given about the degree of mixing is of a statistical nature. It seems unlikely that any simpler programme of measurements than those suggested here would yield quantities which would be of any value in a systematic investigation of mixing processes. However, simplified procedures could probably be devised for specific industrial purposes. For instance, the continuous measurement of $\overline{J^2}$ or $\overline{K^2}$ alone might be used to monitor the texture of a stream of constant mean composition, without the additional determination of $\overline{a^2}$. An increase in $\overline{J^2}$ or $\overline{K^2}$ would indicate an increase in either the scale or intensity of segregation, and hence a general decrease in the efficiency of the mixing process.

It is suggested that the statistical treatment of mixtures advocated in this paper may be useful in developing a general theory of mixing processes. For instance, if turbulence is used to bring about mixing it seems likely that power is most economically applied in producing turbulence with a scale comparable to the scale of segregation.

Appendix A. *Determination of S from variance of content of line-samples.* Put $c = a - \bar{a}$. The content, J, of a line-sample of length X is then given by

$$J = \int_0^X a\,\mathrm{d}x = \bar{a}X + \int_0^X c\,.\mathrm{d}x, \qquad (18)$$

from which follows

$$(J - \bar{J})^2 = [\int_0^X c\,.\mathrm{d}x]^2 = 2\int_{r=0}^{X}\int_{x=0}^{X-r} c(x)\,.\,c(x+r)\,.\,\mathrm{d}x\,.\mathrm{d}r, \qquad (19)$$

$c(x)$, $c(x+r)$ being the values of c at points distant x, $(x+r)$ from one end of the line. So

$$\sigma_J^2 = \overline{(J-\bar{J})^2} = 2\int_{r=0}^{X}\int_{x=0}^{X-r} \overline{c(x).c(x+r)}.\mathrm{d}x.\mathrm{d}r. \qquad (20)$$

But $\overline{c(x).c(x+r)}$ is independent of x, and equal to $\sigma_a^2\,R(r)$ (see (2)).

Hence

$$\sigma_J^2 = 2\sigma_a^2 \int_{r=0}^{X} (X-r)\,R(r)\,\mathrm{d}r \qquad (21)$$

Now $R(r)$ falls to zero at some value $r=\xi$; if $X \gg \xi$, (21) becomes

$$\sigma_J^2 \approx 2\sigma_a^2\,X \int_0^{\infty} R(r)\,\mathrm{d}r = 2\sigma_a^2\,XS, \qquad (22)$$

which leads to (6).

The error is

$$2\sigma_a^2 \int_0^{X} r\,R(r)\,\mathrm{d}r = 2\sigma_a^2 \int_0^{\xi} r\,R(r)\,\mathrm{d}r < 2\sigma_a^2\,\xi S. \qquad (23)$$

Appendix B. *Determination of V from variance of content of volume samples.* The content, K, of the sample is given by

$$K = \iiint a\,\mathrm{d}x\,\mathrm{d}y\,\mathrm{d}z, \qquad (24)$$

where the integral is taken over the volume, v, of the sample. Putting $c = a - \bar{a}$:

$$K = \bar{a}v + \iiint c\,\mathrm{d}x\,\mathrm{d}y\,\mathrm{d}z; \qquad (25)$$

but $\bar{K} = \bar{a}v$, hence

$$K - \bar{K} = \iiint c\,\mathrm{d}x\,\mathrm{d}y\,\mathrm{d}z. \qquad (26)$$

For a large number, n, of samples of volume v:

$$n\sigma_K^2 = n\overline{(K-\bar{K})^2} = \iiint_{x,y,z}\iiint_{x',y',z'} \Sigma\,c(x,y,z).c(x',y',z').\mathrm{d}x\mathrm{d}y\mathrm{d}z.\mathrm{d}x'\mathrm{d}y'\mathrm{d}z', \qquad (27)$$

where $c(x, y, z)$, $c(x', y', z')$ are the values of c at the points (x, y, z), and (x', y', z') respectively, and both sets of integrals are taken over the whole volume. However, the value of the sum $\Sigma\,c(x, y, z).c(x', y', z')$ will be independent of x, y, z, x', y', z' and equal to $n\overline{c^2}\,R(r)$ for a given value of r, where $r = \sqrt{\{(x-x')^2 + (y-y')^2 + (z-z')^2\}}$, and will be zero for $r > \xi$. Integrating over x', y', z' about any point (x, y, z) gives therefore:

$$4\pi n\overline{c^2}\int_{r=0}^{\xi} r^2 R(r)\, dr . dx\, dy\, dz, \text{ or } 2nV\overline{c^2} . dx\, dy\, dz.$$

Equation (27) therefore becomes:

$$\sigma_K^2 \approx 2V\overline{c^2}\iiint dx\, dy\, dz = 2Vv\overline{c^2} \tag{28}$$

and since $\overline{c^2} = \sigma_a^2$, we have

$$V \approx \sigma_K^2 / 2v\sigma_a^2. \tag{29}$$

The integral in (28) has been taken over all the points (x, y, z) in the sample. Since we have previously integrated over a radius ξ about each of these points, it is clear that the expression given in (28) actually "overlaps" the surface of the sample, and should be corrected by an additional term analogous to that in equation (23). The fractional error introduced into (29) by ignoring this correction will be less than $\xi A/V$, where A is the area of the surface of the sample. Assuming that the value of ξ is of the same order of magnitude as $V^{\frac{1}{3}}$ and that the sample is, for example, cubical in shape, the error will be of the order $(V/v)^{\frac{1}{3}}$.The error will be increased if the shape of the sample is not reasonably compact.

Appendix C. *Effect of Diffusion. Rate of Decay of I.* Let $c = a - \bar{a}$. $c(p)$ is the value of c at some point p in the mixture. $\overline{c(p + r)}$ is the mean value of c at all points distant r from p. Then (2), the definition of $R(r)$, can be put in the form

$$\overline{c^2}\, R(r) = \overline{c(p) . \overline{c(p + r)}}, \tag{30}$$

the product $c(p). \overline{c(p + r)}$ being averaged over all points p in the mixture. Differentiating (30) with respect to r (at constant t) and putting $r = 0$ gives

$$\overline{c^2}\left[\frac{\partial R}{\partial r}\right]_{r=0} = \overline{c(p)\left[\frac{\partial \overline{c(p + r)}}{\partial r}\right]_{r=0}}. \tag{31}$$

Now $[\partial\, \overline{c(p + r)}/\partial r]_{r=0}$ is the average, at any instant, of the values of $\partial c/\partial r$ measured in all directions from the point p. Provided there are no discontinuities in $\partial c/\partial r$, this average will be zero, because the value of $\partial c/\partial r$ measured from the point in any direction will be equal and opposite in sign to that measured in the opposite direction. Now it is easily shown that in a system in which diffusion occurs, discontinuities in the concentration-gradient cannot persist for a finite length of time (unless the diffusivity D is a discontinuous function of c). For such systems, therefore

$$[\mathrm{d}R/\mathrm{d}r]_{r=0} = 0, \tag{32}$$

and the correlogram will have a flat top.

Consider a small spherical element of the mixture, centre p, radius δr. Equating the rate of inflow of component A by diffusion to the rate of rise of concentration, we have

$$\frac{4}{3}\pi(\delta r)^3 \frac{\partial c(p)}{\partial t} = 4\pi(\delta r)^2 D \left[\frac{\overline{\partial c(p+r)}}{\partial r}\right]_{r=\delta r}. \tag{33}$$

Using equations ((31) and (32) we have

$$\left[\frac{\overline{\partial c(p+r)}}{\partial r}\right]_{r=\delta r} = \delta r \left[\frac{\overline{\partial^2 c(p+r)}}{\partial r}\right]_{r=0} \tag{34}$$

Differentiating (30) twice with respect to r, and putting $r = 0$ we have

$$\overline{c^2}\left[\frac{\partial^2 R}{\partial r^2}\right]_{r=0} = \overline{c(p)\left[\frac{\overline{\partial^2 c(p+r)}}{\partial r^2}\right]_{r=0}}. \tag{35}$$

Substituting (34) and (35) in (33) we find

$$\overline{c(p)\frac{\partial c(p)}{\partial t}} = 3D\,\overline{c^2}\left[\frac{\partial R^2}{\partial r^2}\right]_{r=0}. \tag{36}$$

Whence

$$\frac{1}{\overline{c^2}} \cdot \frac{\mathrm{d}\overline{c^2}}{\mathrm{d}t} = 6D\left[\frac{\mathrm{d}^2 R}{\mathrm{d}r^2}\right]_{r=0} \tag{37}$$

and from the definition of I (equation (2)):

$$-\frac{1}{I}\frac{\mathrm{d}I}{\mathrm{d}t} = -6D\left[\frac{\mathrm{d}^2 R}{\mathrm{d}r^2}\right]_{r=0}. \tag{38}$$

Appendix D. *Determination of I from reaction-rate.*

(*a*) 2nd-order reaction between A and B. The local rate of reaction per unit volume is

$$k\,m\,n\,a\,(1 - a)$$

The total reaction rate in the system is therefore

$$\varrho v = kmn \int\int\int a(1 - a)\,\mathrm{d}x\,\mathrm{d}y\mathrm{d}z \tag{39}$$

the integral being taken over the whole volume v, and ϱ being the mean rate of reaction per unit volume. But

$$\frac{1}{v} \int\int\int a(1 - a)\,\mathrm{d}x\,\mathrm{d}y\,\mathrm{d}z = \bar{a} - \overline{a^2}, \tag{40}$$

from which

$$\bar{a} - \overline{a^2} = \varrho/kmn, \tag{41}$$

and using (7)

$$\bar{a} - (\bar{a})^2 - \sigma_a^2 = \varrho/kmn \tag{42}$$

whence

$$I = 1 - \varrho/kmn\bar{a}(1 - \bar{a}). \tag{43}$$

(*b*) 2nd-order reaction of *A*. The local rate per unit volume is km^2a^2. The total reaction-rate in the system is therefore

$$\varrho v = km^2 \int\int\int a^2\,dx\,dy\,dz \tag{44}$$

so

$$\varrho/km^2 = \overline{a^2}. \tag{45}$$

Using (7)

$$\sigma_a^2 = \overline{a^2} - (\bar{a})^2 \equiv \varrho/km^2 - (\bar{a})^2, \tag{46}$$

$$I = \varrho/km^2\bar{a}(1 - \bar{a}) - \bar{a}/(1 - \bar{a}). \tag{47}$$

REFERENCES

1) Buslik, D., A.S.T.M. Bull. **165** (1950) 66.
2) Danckwerts, P. V., and E. S. Sellers, Industrial Chemist **27** (1951) 395.
3) Fluid Motion Panel, Aeronautical Research Committee, "Modern Developments in Fluid Dynamics", Vol. I, p. 269, Oxford University Press (1938).
4) Moelwyn-Hughes, E. A., "Kinetics of Reaction in Solution", Oxford University Press (1947).
5) Taylor, G. I., Proc. Roy. Soc. A **151** (1935) 421.

E3

The effect of incomplete mixing on homogeneous reactions

Professor P. V. DANCKWERTS*

Résumé—La vitesse d'une réaction homogène dépend du nombre de fois que les molécules en réaction entrent en collision. Si le mélange de réaction est composé de deux courants entrant dans un réacteur à régime continu, ou bien de parties d'un seul courant qui diffèrent quant à leur temps de résidence dans le réacteur, la vitesse moyenne de réaction dépend en général du degré de mélange des molécules. Elle ne se laisse généralement pas prédire simplement à la base de la distribution des temps de résidence.

L'auteur discute la conception " mélange à l'échelle moléculaire ", qu'il définit quantitativement, et il montre comment il peut être mesuré. Quelques exemples d'importance pratique sont donnés à titre d'illustration.

Le problème d'un traitement généralisé des vitesses de réactions homogènes pour le cas de mélanges imparfaits est discuté d'une manière tentative.

Zusammenfassung—Die Geschwindigkeit einer homogenen Reaktion wird von der Zahl der Zusammenstösse zwischen den Molekeln der Reaktionsteilnehmer bedingt. Besteht das Reaktionsgemisch aus zwei Strömen, die in einen kontinuierlichen Reaktor eintreten, oder aus früheren und späteren Teilen eines einzigen Stromes, so wird die durchschnittliche Reaktionsgeschwindigkeit im allgemeinen von der Durchmischung der Molekeln bedingt. Sie lässt sich im allgemeinen nicht einfach aus der Verteilung der Verweilzeiten vorhersagen.

Der Begriff der " Durchmischung in molekularen Dimensionen " wird besprochen sowie quantitativ definiert, und es wird gezeigt, wie sie gemessen werden kann. Einige für die Praxis wichtige Beispiele werden gegeben.

Die Frage einer verallgemeinerten Behandlung der Geschwindigkeiten homogener Reaktionen bei unvollständiger Mischung wird erörtert.

Summary—The rate of a homogeneous reaction depends on rate of encounter between reactant molecules. When the reaction mixture consists of two reactant streams flowing into a continuous reactor, or of earlier and later parts of a single stream, the average rate of reaction is in general dependent on the degree of mixing on the molecular scale. It cannot generally be predicted simply from the distribution of residence times.

The author analyses the concept of " mixing on the molecular scale ", gives it a quantitative definition, and shows how it can be measured.

Some illustrations of practical interest are given.

The problem of a generalised treatment of imperfectly mixed homogeneous reactors is tentatively discussed.

ONE OF the problems involved in designing continuous-flow reactors from first principles is that of taking into account the flow-behavior of the stream of reactant. The problem is simple if complete mixing or perfect piston-flow can be taken to represent the facts with sufficient accuracy. In other cases it is possible by various methods [1] to establish experimentally the distribution of residence-times of the material—that is, the length of time which various proportions of it spend in the reactor. In the case of a homogeneous first-order or pseudo-first-order reaction this is sufficient to enable one to calculate the output of the reactor, since the chance of a molecule reacting then depends solely on the duration of its stay in the reactor. In the case of reactions with more complex kinetics (such as those normally encountered) the chance of a molecule reacting depends on the other molecules which it encounters during its stay, and a great deal more information would be required to predict the output of the reactor.

The problem of developing a general treatment of continuous-flow reactors for reactions other than first-order is an extremely difficult one. It would be desirable to have a design method whereby a knowledge of the kinetics of the reaction on the one

* Imperial College of Science and Technology, London.

hand and of the hydrodynamic and diffusional behavior of the reactant stream on the other could be combined as required to calculate the output of different types of reactors under different conditions. At the moment it is only possible to solve problems of this sort in special cases—for instance, the completely-mixed reactor in which the material is homogeneous on the molecular scale; or tubular reactors with longitudinal diffusion. At the moment there is no method whereby the history of the molecules flowing through an arbitrarily-chosen reactor can be determined and the results formulated in such a way that the output could be predicted for a reaction of known kinetics.

My purpose in this paper is merely to suggest some lines of approach to the general problem. These by no means converge to a complete solution of the type I have described.

Segregation and Mixing

Consider a " well-stirred " continuous tank reactor. In such a reactor a stream of fluid flows continuously through the tank, its mean residence time (i.e. quantity in tank divided by rate of throughput) being τ. An element of fluid entering the tank is uniformly dispersed through its volume in a time much less than τ. As is well known, the distribution of residence-times in the tank is then such that a fraction $[(1/\tau)e^{-t/\tau}\,dt]$ of the material passing through the tank spends in it a period of time lying between t and $(t+dt)$. Suppose the entering fluid is a mixture of two reactants, the reaction between them being second-order. If we wished to calculate the degree of reaction in the outflowing fluid we might assume one of two limiting circumstances:

(*a*) The incoming fluid is broken up into discrete fragments or streaks which are small compared to the tank and uniformly dispersed in it, but in which molecules entering together remain together indefinitely: one may describe this by saying that the fluid remains completely segregated.

(*b*) The inflowing material is dispersed on the *molecular scale* in a time much less than τ; the environment of any particular molecule does not tend to contain an excess of molecules which entered at the same time as itself and the mixture is chemically uniform.

Although the distribution of residence-times will be that of the "well-stirred reactor" in both cases, the average rate of reaction in the tank will be different. For instance, supposing the concentration of each reactant in the ingoing stream is C equivalents per unit volume and the velocity-constant is k, the average rates of reaction (equivalents per unit volume per unit time) will be respectively:

$$\text{(a)}\quad \frac{c}{\tau}\left[1-\frac{1}{\tau}\int_0^\infty \frac{e^{-t/\tau}}{1+kct}\,dt\right] = \frac{c}{\tau}\left[1+\frac{e^{1/kc\tau}}{kc\tau}\cdot Ei\left(-\frac{1}{kc\tau}\right)\right];$$

$$\text{(b)}\quad \frac{c}{\tau}\left[1-\frac{\sqrt{1+4kc\tau}-1}{2kc\tau}\right].$$

The rate of reaction is greater in case (*a*) for all values of $kC\tau$ [2].

Effect of Mixing on Molecular Scale

It is always possible to calculate the output of a continuous-flow homogeneous reactor on the assumption that molecules entering at the same time remain segregated, and that there is no intermixing on the molecular scale with molecules which enter earlier or later. All that is required is a knowledge of the batch kinetics and the residence-time distribution. It has been shown that for a second-order reaction and the " completely-mixed " residence-time distribution the effect of molecular mixing is to reduce the reaction-rate. A more general statement can be made. Suppose we take two portions of reaction-mixture which differ in their degree of reaction, and mix them on the molecular scale; the total rate of reaction after mixing will be less than before if the reaction is of an order greater than unity, greater than before if the order is less than unity and unchanged if the order is unity. This is proved in the Appendix for reaction-mixtures in which the rate of reaction is proportional to $(1-f)^n$, as would be the case for a stoichiometric mixture undergoing a reaction of order n. It is also proved for non-stoichiometric mixtures undergoing second-order reactions. It can be concluded that mixing on the molecular scale between portions of reactant mixture which have been in the reactor for different lengths of time will render the output of the reactors less than if the material were completely segregated when the order of reaction is greater than unity, and vice versa. (In industrial practice the situation is often a great deal more complicated, as the nature of the product, or the final yield of the desired product, may depend on the presence or absence of such mixing.)

In a stirred-tank reactor the degree of segregation of the contents is probably of greatest practical interest when two separate reactant streams enter the tank. If the two reactants remained completely segregated they would be separated by sharp interfaces, which would provide the only site for the reaction. Normally, the segregation is reduced by interdiffusion of the two reactants, whereby molecules of each type migrate into an environment containing molecules of the other type, and the rate of reaction will be increased by influences tending to decrease the segregation. The rate of disappearance of segregation, of course, depends on the diffusivity and on the area of the interface or the average size of the " clumps " of the two components in the tank [3].

It is often necessary to feed a reactant (A) in a concentrated form, to achieve the desired average concentration in the reactor, which will be lower than in the feed because of reaction and dilution with the other reactant. Segregation of reactant A means that there will be parts of the mixture in which the concentration of A is much higher than the average, and this may lead to undesirable side-reactions. This phenomenon is noticed in organic nitrations, where the feed of concentrated nitrating acid persists for a time in the form of streaks.

A number of studies have been made of the mixing of a turbulent jet with its surrounding atmosphere [4]. The mixing takes place primarily by eddy diffusivity, whereby " clumps " of the two fluids are intermingled along the margin of the jet. Molecular diffusion then tends to reduce the segregation of the " clumps " and to even-out local variations in composition. The actual degree of segregation may be of great importance in some applications. For instance, a solid may be precipitated by injecting one reactant solution into another; the rate of precipitation and hence the particle-size of the precipitate will be determined by the rate of reduction of segregation in the region of mixing. If it is desired to quench a gaseous reaction mixture by mixing it with cold gas, the rapidity of quenching and hence the reaction yield will depend on the rapidity with which the segregation is destroyed.

In the case of a burning jet of gas, if all or part of the oxygen required for combustion is derived from the air surrounding the jet, the length of the flame and the blow-off velocity will depend on the segregation of the two gases in the region of mixing. In the case of laminar flames (e.g. candle flames) the degree of segregation of the fuel jet is high, and burning is confined to a thin zone on the edge of the jet; if the jet is turbulent, clumps of air and fuel will intermingle, molecular diffusion will reduce the segregation of the components, and the volume in which the primary encounters between oxygen and fuel molecules can occur is much increased.

Methods of Measuring Segregation

Hawthorne, Weddell and Hottel [5] measured the degree of segregation in the margin of a turbulent flame by an ingenious method. Suppose a position is chosen on the margin of the flame where the *time average* composition is such that there is just sufficient air for complete combustion of the fuel. Since the mixture consists of partially-segregated clumps of fuel and air, the instantaneous composition will fluctuate, showing at one moment an excess of fuel and at the next an excess of air. If the gas at this position is sampled continuously through a cooled tube (to prevent reaction between successive increments of the sample) the sample will consist of a mixture of unreacted air and fuel in equivalent proportions together with products of combustion. If segregation of the clumps of gas were complete the sample would consist entirely of unreacted fuel and air, while if the segregation were zero the sample would be completely burnt. (There are of course some assumptions about the intrinsic speed of the reaction involved in this argument.) From the actual composition of the sample it is possible to calculate the mean deviation from the mean composition: if a " normal " distribution of deviations is assumed, the mean square deviation can be deduced.

It is not difficult to set up a satisfactory quantitative definition of the " degree of segregation " between the two components, A and B, of a binary mixture. The average volume fraction of A is $\bar{a}$, and of B is $(1-\bar{a})$, while the values at a given point and instant are a, $(1-a)$. The mean square deviation of the composition from the mean is $\overline{(\boldsymbol{a}-\bar{a})^2}$. The degree or intensity of segregation can now be defined as

$$I = \frac{\overline{(\bar{a}-a)^2}}{\bar{a}(1-\bar{a})}.$$

The averages may be taken over space (as over the contents of a steady-state continuous-flow reactor) or time (as in the flame-sampling method described above). I varies from zero when mixing is uniform on the molecular scale (a is the same at all points), to unity when segregation is complete ($a = 1$ or 0 at every point) [3].

It should be possible to measure the intensity of segregation in a binary mixture in the following way. The two components are solutions of substances which react together with known second-order kinetics. Suppose these are fed to a continuous reactor at volumetric rates in the proportion of $\bar{a}$ to $(1-\bar{a})$; the concentrations of the reactants in the two feed solutions are (A) and (B). If the two feeds were instantaneously mixed on the molecular scale, the initial rate, R_0, of reaction (per unit volume) would be $k(A)(B)\bar{a}(1-\bar{a})$. If the two fluids remained completely segregated the reaction rate would be zero. In any region of the reactor where diffusion has mixed the two fluids in a ratio of a to $(1-a)$, the concentrations of the two reactants will be $a(A)$ and $(1-a)(B)$, and the local rate of reaction per unit volume will be $k(A)(B)a(1-a)$. Thus the average rate R, of reaction in the reactor will be $k(A)(B)\overline{a(1-a)}$.

Now

$$\overline{a(1-a)} = \bar{a}+(\bar{a})^2-2a\bar{a}-\overline{(\bar{a}-a)^2} = \bar{a}(1-\bar{a})-\overline{(\bar{a}-a)^2}$$

so that

$$I = 1-\frac{R}{R_0}.$$

The reaction must be sufficiently slow so that there is no great depletion of the reactants during their passage through the system, and would not be even if they were mixed immediately on the molecular scale; to put it another way, the quantities $\ln 2/k(A)$ and $\ln 2/k(B)$ which are the half-lives of each species of reactant when highly diluted by the other, must be much higher than the residence-time of the major part of the material in the reactor.

A similar method of measuring I depends on feeding a reactant solution and a diluent in the proportions of $\bar{a}$ to $(1-\bar{a})$. The reactant solution contains two reactants which undergo a second-order reaction, their initial concentrations in the feed being (A) and (B).

Then $R_0 = k(A)(B)(\bar{a})^2$, and $R = k(A)(B)\overline{a^2}$.

Since

$$\overline{(\bar{a}-a)^2} = \overline{(\bar{a})^2-2a\bar{a}+a^2} = \overline{a^2}-(\bar{a})^2,$$

$$I = \frac{\bar{a}}{1-\bar{a}}\left(\frac{R}{R_0}-1\right).$$

It is possible to define and measure in a rather similar way the segregation of a single stream of fluid passing through a system. Here we consider mixing on the molecular scale between elements of fluid which have been in the system for different lengths of time or have different " ages ".

Previously, when discussing the mixing of fluids of different compositions, I have used the concept of " concentration at a point ". In the present case I wish to refer to the " age of the fluid at a point ". In neither case are geometrical points referred to, of course, but rather averages taken over volumes small compared to that of the system, but large enough to contain many molecules, and centred on geometrical points. The logical basis of the arguments developed here depends on a specification of the magnitude of this volume. This specification is developed later; meanwhile I shall take it for granted. The " age ", t, of the fluid at a given point and instant is the average time which has already been spent in the system by the molecules contained in the small volume specified. The average value of t for all points in the system is τ, the mean residence-time. The degree of segregation may then conveniently be defined as

$$J = \frac{\overline{(\tau-t)^2}}{\overline{\theta^2}}.$$

The quantity $\overline{(\tau-t)^2}$ is of course the mean square deviation of the ages of all points in the system from the mean age, τ. The quantity $\overline{\theta^2}$ is defined as follows. Let $F(t)$ be a distribution function such that a fraction $F(t)\,dt$ of the molecules in the system at any instant have spent in it times between t and $(t+dt)$. Then

$$\overline{\theta^2} = \int_0^\infty (\tau-t)^2 F(t)\,dt.$$

The value of $\overline{\theta^2}$ can be evaluated if the residence-time distribution for the system is known. For a " well-stirred " system it has the value of τ^2. When segregation is complete, each molecule is surrounded by other molecules of the same age; thus the fraction of " points " in the system having age t is the same as the fraction $F(t)$ of molecules in the system having age t. Under these circumstances $\overline{\theta^2} = \overline{(\tau-t)^2}$, and $J = 1$, while if molecular mixing is complete $t = \tau$ for every point in the system, and $J = 0$.

The value of J for the contents of a continuous flow system might be measured as follows. The inflowing stream of fluid contains reactants A and

B which undergo second-order reaction with velocity constant k. The concentrations of the two reactants in the entering stream are both proportional to the time which has elapsed since an arbitrary starting-point $t' = 0$, so that these concentrations at time t' are $(A)t'$ and $(B)t'$ respectively. As before we assume that the reaction is so slow that all but a small fraction of reactant molecules entering the system leave without reacting. If at time t' we take a sample of material from the system it will contain a proportion of molecules which entered before $t' = 0$; but as t' increases this proportion will decrease, no matter what part of the system the sample is taken from. It will be supposed that t' is so large that the system has been thoroughly purged of material present at $t' = 0$.

Now a molecule which at time t' has been in the system for a length of time t must have entered the system at time $(t'-t)$. (In the well-purged system $t' > t$ for virtually all molecules in the system.) Thus the average concentrations of A and B in the molecules constituting a " point " of age t will be $(A)(t'-t)$ and $(B)(t'-t)$ respectively, and the rate of reaction (per unit volume) at this point will be $k(A)(B)(t'-t)^2$. Thus the average rate of reaction, R (per unit volume), in the system as a whole at time t' is $k(A)(B)\overline{(t'-t)^2}$ or $k(A)(B)(\overline{t^2}-2\bar{t}t'+(t')^2)$. From this it can be calculated that

$$\frac{dR}{d\ln t'} = t'\frac{dR}{dt'} = k(A)(B)2t'(t'-\bar{t})$$

and it follows that:

$$\overline{t^2} = (t')^2 + \frac{1}{k(A)(B)}\left(R - \frac{dR}{d\ln t'}\right)$$

Since $\overline{(\tau-t)^2} = \overline{t^2}-\tau^2$, J can be evaluated by measuring R as a function of t'. The expression on the right-hand side of the above equation should of course be independent of t' if all the stipulated conditions are fulfilled.

Definition of a " Point "

As already mentioned, the phrases " concentration at a point " and " age of fluid at a point " imply the concentration or age averaged over a region small compared to the whole system but large enough to contain many molecules. In specifying the size of this region it must be borne in mind that if it is too small a spurious appearance of segregation would result from the statistical fluctuations of composition from one point to another, even in a system completely mixed on the molecular scale. On the other hand, if the region specified is too large genuine segregation will be concealed. The size of the region specified must be compatible with the methods of measuring segregation by means of chemical reactions, as described above.

In a mixture which is not homogeneous, the regions of segregation will form streaks or " clumps ". These will generally be irregular and indefinite in outline; nevertheless, by using a technique similar to that used in the statistical treatment of turbulence, it is possible to define a " scale of segregation " which is a measure of the size of the regions of inhomogeneity [3]. The specification for a " point " in the sense used here is a region large enough to contain many molecules but much smaller than the scale of segregation. If the radius of a " point " is r, the composition will then be negligibly different at points separated by distances comparable with r; and r can be varied over a wide range without affecting the validity of the specification.

If the scale of segregation is so small as to constitute a volume containing only a few molecules the approach used above is inappropriate, and indeed the whole concept of " concentration " breaks down.

Fast Reactions

The methods of measuring segregation suggested above make use of reactions in which the fractional conversion is negligible, the reaction being used merely as an indicator of molecular mixing. Such reactions may be described as extremely slow, in the sense that the half-life of the reaction-mixture, if uniformly mixed, would be very much greater than the mean residence-time in the reactor. Thus the analysis given in this paper constitutes only the solution to a " degenerate " case of the problem of second-order reactions. However, the study of segregation is perhaps likely to be most useful in practice when applied to extremely fast reactions, such as combustion or precipitation. Under comparable hydrodynamic and diffusional conditions a very much smaller proportion of the volume of the fast reaction system will contain a molecular mixture of the several reactants; for instance, in the case of a flame a mixture of unburnt fuel and air is present only in the very thin flame-front, whereas the major part of the system contains either fuel or air, but not both. In spite of this wide difference in the configuration of fast and slow reaction-mixtures, I think that segregation studies using slow reactions are likely to prove helpful in the design of fast reaction systems.

Models

Dimensional analysis shows that the intensity of segregation, as defined above, is determined in continuous-flow systems of given geometry only by the Reynolds and Schmidt numbers (unless there is a free liquid surface, in which case the Froude number must also be considered). Since the Schmidt number (ν/D_m) is a property of the fluid, there should be no difficulty in using scale models with the same fluid as on the full scale. It would be convenient to use liquid models of gaseous reaction systems, because of the wider choice of reactions in the liquid phase; unfortunately, however, *Sc* for liquids is in general much higher than for gases and it is unlikely that anything more than qualitative indications could be obtained from liquid models of gaseous systems.

Received 20 December 1956.

Appendix

(a) *Second-Order Reactions*

A mixture of reactants A and B contains these initially in the concentrations (A) and (B), where $(B) = (A)\,(1+x)$. Thus when the fraction of A reacted is f, the concentrations are $(A)\,(1-f)$ and $(A)(1-f+x)$ respectively. Consider two portions of the reaction mixture having different degrees of reaction, f_1 and f_2; the volumes of the two portions are in the ratio a to $(1-a)$. The rate of reaction averaged over both portions is then

$$R_1 = ka(A)^2(1-f_1)(1-f_1+x)+ +k(1-a)(A)^2(1-f_2)(1-f_2+x).$$

If however the two portions are mixed, the resulting concentrations of A and B are $(A)[a(1-f_1)+(1-a)(1-f_2)]$ and $(A)[a(1-f_1+x)+(1-a)(1-f_2+x)]$ respectively. The rate of reaction in the mixture will then be

$$R_2 = k(A)^2[a(1-f_1)+(1-a)(1-f_2)] \times \times [a(1-f_1+x)+(1-a)(1-f_2+x)].$$

It is easily shown that

$$R_2 = R_1 - a(1-a)(f_1-f_2)^2,$$

and since a is less than unity, R_2 is less than R_1.

(b) *Reactions of nth Order*

A mixture of reactants in stoichiometric proportions undergoes an nth order reaction. Thus if the rate at the beginning of the reaction is R_0, the rate when a fraction f of each of the reactants has disappeared will be $R_0(1-f)^n$. As before, two portions with fractional reaction f_1 and f_2 are taken in the volumetric proportions a to $(1-a)$. The rates before and after mixing, R_1 and R_2, are given by

$$R_1/R_0 = a(1-f_1)^n+(1-a)(1-f_2)^n,$$

$$R_2/R_0 = [a(1-f_1)+(1-a)(1-f_2)]^n.$$

If a and f_1 are held constant, and the ratio

$$\frac{1-f_2}{1-f_1} = \gamma$$

is allowed to vary, both R_1 and R_2 may be considered functions of γ. It is found that

$$\frac{dR_1}{dR_2} = \frac{dR_1}{d\gamma}\Big/\frac{dR_2}{d\gamma} = \left[\frac{\gamma}{a+(1-a)\gamma}\right]^{n-1}$$

which is always > 0, since $\gamma > 0$, $a < 1$.

When $\gamma = 1$, $R_1 = R_2$ and $dR_1/dR_2 = 1$. Also

$$\frac{d^2R_1}{dR_2^2} = \frac{d}{d\gamma}\left(\frac{dR_1}{dR_2}\right)\Big/\frac{dR_2}{d\gamma} = \frac{n-1}{n} \cdot \frac{a\gamma^{n-2}}{(1-f_1)^n(1-a)[a+(1-a)\gamma]^{2n-1}}$$

which is > 0 if $n > 1$, and is < 0 if $n < 1$. Also, when $\gamma = 0$, we have

$$R_1/R_2 = a^{1-n}$$

which is > 1 if $n > 1$, and is < 1 if $n < 1$. These various conditions can only be satisfied, as can be seen from the figure, if for all values of a and γ, $R_1 > R_2$ when $n > 1$, $R_1 < R_2$, when $n < 1$, $R_1 = R_2$ when $n = 1$.

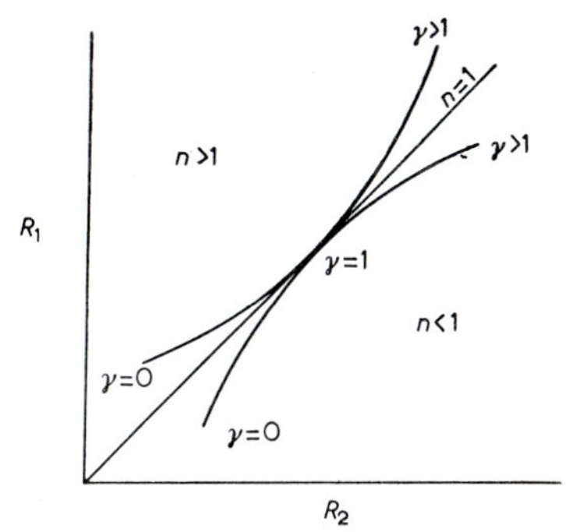

Fig. 1. Curves of R_1 vs. R_2 for arbitrary value of f, and a, with $n > 1$, $n = 1$ and $n < 1$.

Notation

a	volume fraction
(A)	concentration of reactant A (equivs. per unit vol.)
(B)	concentration of reactant B (equivs. per unit vol.)

C concentration (equivs. per unit vol.)
D_m molecular diffusivity

$$Ei(-x) = \int_x^{\infty} \frac{e^x}{x} dx \text{ (tabulated)}$$

f fraction of reactant reacted
I intensity of segregation in binary mixture
J intensity of segregation of successive elements of fluid
k reaction-velocity constant
n order of reaction
r radius of " point " in mixture
R rate of reaction (equivs. per unit time per unit vol.)
t length of time for which material has been in system
t' time since start of experiment
x as defined in text
γ as defined in text
ν kinematic viscosity
τ average residence-time

A bar over a quantity represents an average taken over time or space.

References

[1] Danckwerts, P. V.; *Chem. Eng. Sci.* 1953 **2** 1; *Ind. Chemist* 1954 (March) 102.

[2] Gilliland, E. R., Mason, E. A. and Oliver, R. C.; *Ind. Eng. Chem.* 1953 **45** 1177.

[3] Danckwerts, P. V.; *Appl. Sci. Res.* A 1952 **3** 279.

[4] Forstall, W. and Shapiro, A. H.; *J. Appl. Mech.* 1950 **17** 399; Baron, T. and Alexander, L. G.; *Chem Eng. Prog.* 1951 **47** 181.

[5] Hawthorne, W. R., Weddell, D. S. and Hottel, H. C.; *Third Symposium on Combustion, Flame and Explosion Phenomena*, p. 266. Williams & Wilkins, Baltimore, 1949.

E4

Measurement of Molecular homogeneity in a mixture

(*Received* 14 *June* 1957)

I have suggested elsewhere [1] how the degree of mixing of two fluids on the molecular scale (or its converse, the degree of segregation) could be determined by observing the rate of a second-order reaction in the mixture. This method gives essentially the mean square concentration of component A in a non-homogeneous mixture of A and B.

The method to be described here appears to be more powerful, in that it will give the actual distribution of deviations from the mean concentration in the mixture. It is also experimentally simpler.

The mixture consists of two aqueous solutions, A and B. A contains m molal NaOH, while B contains mx molal HCl. In any portion of the mixture in which A and B are mixed on the molecular scale in proportions a, $(1 - a)$ the local composition will be

$$\text{alkaline if } a > x(1-a), \text{ or } a > \frac{x}{1+x};$$

$$\text{acid if } a < x(1-a), \text{ or } a < \frac{x}{1+x}.$$

Phenolphthalein or some other suitable indicator is added to one or the other or both of the solutions. Thus three cases may be considered :

(1) *Phenolphthalein in concentration n added to solution A.* The solution will be red in the alkaline regions ; in an alkaline region in which the proportion of A is a, the concentration of phenophthalein in the red form will be na. Thus if the fraction of the volume of the mixture in which the porportion of A lies between a and $(a + da)$ is $f(a)\, da$, the average concentration of phenolphthalein in the red form will be

$$n_R = n \int_{x/1+x}^{1} a f(a)\, da \tag{1}$$

$$\therefore \frac{1}{n}\frac{dn_R}{dx} = -\frac{x}{1+x} \cdot f\left(\frac{x}{1+x}\right) \cdot \frac{d}{dx}\left(\frac{x}{1+x}\right). \tag{2}$$

$$\therefore \frac{(1+x)^3}{nx}\frac{dn_R}{dx} = -f\left(\frac{x}{1+x}\right). \tag{3}$$

Thus by varying x, the ratio of the strengths of the two solutions, and measuring the corresponding colour-density of the mixture, it is possible in principle to determine the fraction of the volume of the mixture which contains A and B mixed on the molecular scale in any given proportion. In a mixture with a high intensity of segregation, consisting mostly of undiluted A and B, $f(a)$ will have high values at $a = 0$ and $a = 1$, and will be low at intermediate values of a. When the segregation is small and mixing on the molecular scale nearly complete, $f(a)$ will be small at $a = 0$ and $a = 1$, and will pass through a maximum at a value of a corresponding to the volume fraction of A in the mixture as a whole. The " intensity of segregation " as previously defined [1] can be calculated if $f(a)$ is known.

(2) *Phenolphthalein in concentration n added to solution B.* By a similar argument,

$$n_R = n \int_{x/1+x}^{1} (1 - a) \cdot f(a) \cdot da, \tag{4}$$

and

$$\frac{(1 + x)^3}{n} \frac{dn_R}{dx} = -f\left(\frac{x}{1 + x}\right). \tag{5}$$

In this case also (a) can be determined by varying x. However, the colour is developed only in the regions in which molecular mixing has taken place, so that it may be easier to make the requisite colour measurements with sufficient accuracy in case (b) than in case (a). Case (b) will also provide a better visual demonstration of the progress of molecular mixing, particularly when $x = 0$ – that is, when B contains phenophthalein but no acid – since the whole region of molecular intermixing will then be coloured.

(3) *Both solutions contain phenophthalein in concentration n.* Then

$$n_R = n \int_{x/1+x}^{1} f(a) \cdot da, \tag{6}$$

and

$$\frac{(1 + x)^2}{n} \frac{dn_R}{dx} = -f\left(\frac{x}{1 + x}\right). \tag{7}$$

In this case there is a sharp boundary separating regions in which $a > \frac{x}{1 + x}$ and the concentration of red phenolphthalein is uniformly equal to n, from regions in which $a < \frac{x}{1 + x}$ and the solution is colourless. The arrangement is thus particularly suitable for the study of macroscopic mixing processes, since it emphasises the outlines of the blobs and streaks of the components in an inhomogeneous mixture.

It is interesting to consider the application of the various versions of the method to a specific case, such as the mixing of a central jet into a coaxial stream in a pipe. In case (a), if solution A were the central stream, it would form a solid red jet which, for a given flow-ratio, would either dwindle and vanish downstream or spread to fill the whole tube, depending on the value of x. If solution B were the central stream, red and white would be reversed. In case (b) the central jet would be colourless and surrounded by a red shell, which would either vanish or fill the whole tube downstream, depending on the value of x. Case (c) will be similar to case (a), except that the boundary of the central jet will be more sharply defined; if conditions are such that the central jet disappears downstream, it will offer a close analogy to a flame. In all cases the time-average concentration of colour at any point can be determined by taking a time-exposure photograph of the system. The mathematical problem of determining the radial concentration distribution from such a photograph will be discussed elsewhere.

There are three assumptions involved in the arguments leading to equations (3), (5) and (7) which are probably justified in many circumstances, although they must be borne in mind. These are that the full colour-change of phenolphthalein occurs over a very small increment of a; that acid, alkali, phenolphthalein and water can be regarded as being affected to the same degree by processes leading to molecular mixing; and that the colour-change of phenophthalein can be regarded as an instantaneous reaction.

P. V. DANCKWERTS

REFERENCE

[1] DANCKWERTS P. V. The effect of incomplete mixing on homogeneous reactions. Paper presented at the first European Symposium on chemical engineering, *Chemical Reaction Engineering*. Amsterdam, May, 1957.

E5

Studies in turbulent mixing—I

Dilution of a jet

D. KRISTMANSON and P. V. DANCKWERTS
The Department of Chemical Engineering, Pembroke Street, Cambridge.

(*Received* 17 *March* 1961)

Abstract—(a) A photographic method has been used to determine the concentration-distribution of jet-fluid when a circular turbulent jet of liquid mixes with an unlimited liquid environment. The time-average concentration $\bar{a}$ at a point (x, r) is given by

$$\bar{a} = 4{\cdot}78 \frac{d}{x} \exp - \left(\frac{r}{0{\cdot}135\, x}\right)^2 .$$

(b) The process of dilution on the molecular scale of jet liquid by ambient liquid has been studied by using an alkaline jet in an acid environment, both containing an indicator. A plume of blue indicator results ; outside the plume all jet fluid has been diluted below some specified concentration, m. It is found that all plumes are the same shape, having a width r at a distance x from the orifice given approximately by

$$\frac{r}{x} = 0{\cdot}32 \sqrt{\left(\log_{10} 6{\cdot}41 \frac{d}{xm}\right)} .$$

The point on the axis at which all jet fluid has been diluted to a concentration m or less is 1·34 times as far from the orifice as the point at which $\bar{a} = m$.

E6

Studies in turbulent mixing—II

A hot-air jet

R. A. M. WILSON† and P. V. DANCKWERTS

Department of Chemical Engineering, Pembroke Street, Cambridge

(*Received* 30 *April* 1964)

Abstract—Time-average temperatures and temperature fluctuations have been measured in a turbulent jet of hot air. Heat serves here as a tracer for the mixing of molecules. The results are presented in terms of the RMS temperature (or concentration) fluctuations at different points in the jet, and also in terms of the fraction of time for which the temperature (or concentration of jet fluid) at a specified point exceeds a specified value. The intimacy of mixing at comparable points in jets of water, air and smoke is compared; the effect of the magnitude of the molecular diffusivity on the intimacy of mixing is appreciable. If the concentration-measuring device has a low spatial resolving power, the effect of a high diffusivity is simulated.

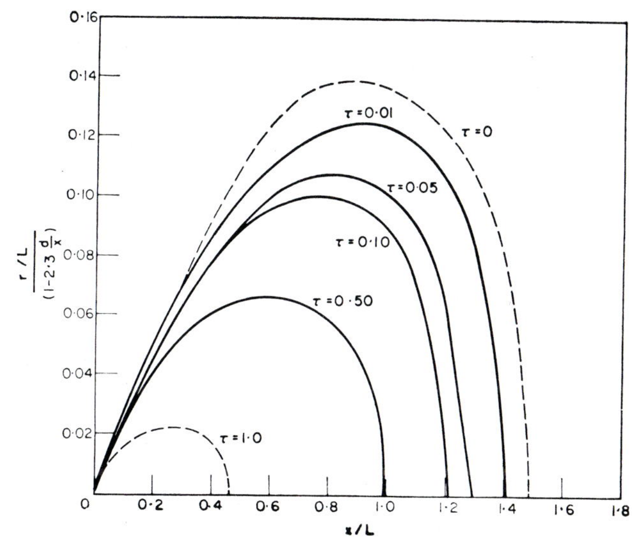

FIG. 9. τ is the fraction of the time for which the local temperature exceeds the time-average temperature on the axis at a distance L from the nozzle (valid for $L/d \gg 40$, $\bar{\theta}_0 = 200°C$).

d diameter of nozzle
r radial distance from axis of jet
x distance downstream from nozzle

SECTION F

Miscellaneous

The 3 papers included in this section are represented by their abstracts, for reasons of space. They have nothing in common with each other or with the rest of my work.

Paper F1 describes a "poor man's Rotameter" which makes use of a cylindrical tube made of Perspex (= U.S. Lucite) rather than an expensive tapered glass tube. It might be useful in countries which wish to conserve foreign currency. The figure shows how the meter is made. The Rotameter-type bob can be hollowed-out and weighted with varying amounts of fine lead shot, to measure different ranges of flow. The device gives rise to a substantial pressure-drop in the line. The fluid mechanics are interesting; at a certain critical flow-rate the bob (if not restrained by cross-wires) is blown right out of the perforated tube, however long it may be.

Paper F2 (Parts I and II) deal with the formation of large granules out of damp powder rolled in a drum. Granules of the sizes dealt with here (1/3 to 1 in.) might be required, for instance when processing iron ore. It was found that the granules increase in size as rolling progresses because large granules crush small ones and the material so liberated is incorporated into the remaining granules. The shape of the size-distribution curve, plotted with suitable dimensionless coordinates, remains unchanged during this process. I must say that the experimental techniques (involving coloured sand) and the mathematical analysis are extremely elegant, but I doubt whether the paper has proved of great value to industry.

Paper 3 describes some experiments relating to mass-transfer coefficients in packings comprised of crossed wooden slats. Such packings have been used in cooling towers and absorbers and give rise to relatively low pressure drops. The coefficients were measured by observing the rate of disappearance of films of naphthalene. The coefficients varied from point to point on the surface and were related to the pattern of air flow, as rendered visible by smoke filaments. The naphthalene-evaporation technique is very elegant and revealing, and could be applied to other types of packing. The experiments were conducted in a carefully-designed wind-tunnel but this is not necessary when observing a single element in an array of packing, as the flow pattern and degree of turbulence in the air stream impinging on the element is then determined by the upstream elements. The techniques and the dimensional analysis used are worthy of further application.

F1

Perforated-tube flowmeter

P. V. DANCKWERTS and A. K. SIKDER

Department of Chemical Engineering, Imperial College, London S.W.7

(*Received* 28 *March* 1960)

Abstract—A cheap and simply-constructed visual flow-meter is described. It consists of a Rotameter-type bob fitting fairly closely inside a perforated tube. Most of the fluid flows out of the holes below the bob ; the height of the bob indicates the flow-rate. A simple theoretical treatment is given. Following experiments with air and water, formulae are given which can be used to specify the leading dimensions of a meter for a given duty.

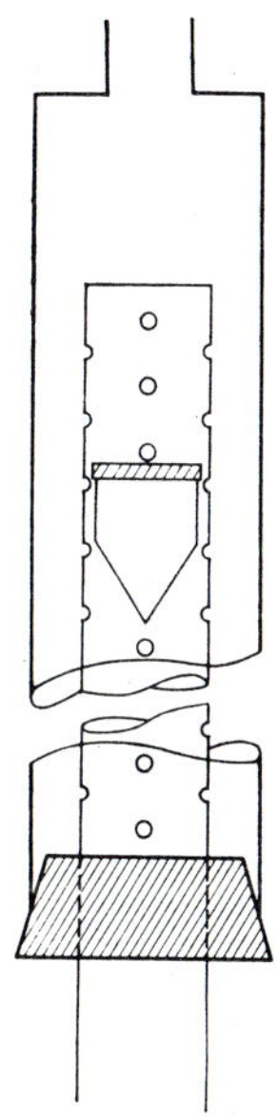

FIG. 1. Flow meter.

GRANULE FORMATION BY THE AGGLOMERATION OF DAMP POWDERS

PART I: THE MECHANISM OF GRANULE GROWTH

By C. E. CAPES, Ph.D.,* and Professor P. V. DANCKWERTS, G.C., M.B.E., M.A., S.M. (MEMBER)†

SYNOPSIS

The mechanism by which granules (up to about 1 in. dia.) are formed in a tumbling mass of moist particles has been investigated.

Uniformly sized sands (different batches having mean particle diameters ranging from 44 μ to 165 μ) were granulated, usually with water as binding liquid. It was found that granules, once formed, continue to grow at a significant rate only when liquid contents are used which are equal to between about 90% and 110% of the amount required to fill the voids between the sand particles when they are highly compacted. The mechanism of granule growth was investigated by methods involving two sands of contrasting colours. At the beginning of the process the "crumbs" of sand originally present in the drum stick together to form small granules. Subsequent growth occurs by the crushing of the smallest granules in the load, the fragments then sticking to the larger surviving granules. The larger the granule the greater the rate at which its diameter increases in this way.

Sands of various particle sizes were granulated with water and with alcohol–water solutions. The results show that when closely sized sands are granulated the maximum particle size which allows satisfactory granulation is reduced when liquids of lower surface tension are used as binders.

GRANULE FORMATION BY THE AGGLOMERATION OF DAMP POWDERS

PART II: THE DISTRIBUTION OF GRANULE SIZES

By C. E. CAPES, Ph.D.* and Professor P. V. DANCKWERTS, G.C., M.B.E., M.A., S.M. (MEMBER)†

SYNOPSIS

Damp sand of uniform particle size was granulated by tumbling in a drum. After the early stages of growth successive granule-size distributions for a given charge become coincident when plotted in a certain dimensionless form. This "self-preserving" behaviour is shown to be compatible with the mechanism of granule growth described in Part I. Simple postulates about the mode of growth lead to a mathematical expression, containing one parameter, which represents the observed size distribution fairly closely.

MASS TRANSFER FROM A GRID PACKING TO AN AIR STREAM

By Prof. P. V. DANCKWERTS, G.C., M.B.E., M.A., S.M. (MEMBER)*† and C. ANOLICK B.Sc. (GRADUATE)*

SUMMARY

A technique has been developed for investigating the detailed mechanism of gas-side mass transfer from packings. The method may also prove convenient for determining the performance of different types of packing in the laboratory.

An element of the packing is coated with a thin film of naphthalene of known uniform thickness. The disappearance of this film is followed as it evaporates in an air-stream. This allows local mass-transfer coefficients to be determined.

The method has been applied to grid packings made from rectangular slats and light has been thrown on the details of the transfer process. The behaviour and effectiveness of different shapes and arrangements of slats have been compared.

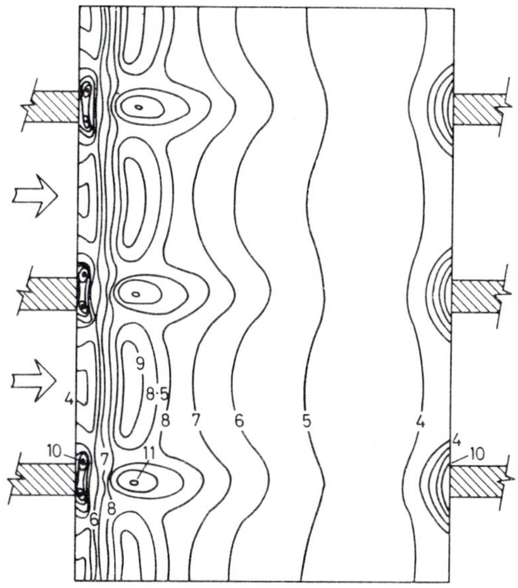

Velocity = 11·0 ft/s
Pitch = $1\frac{1}{2}$ in.
Slats = 3 in. × $\frac{1}{4}$ in.

Fig. 13.—*Contours of equal mass transfer* ($k/u \times 10^3$)

k = mass transfer coefficient
u = velocity (both ft/s)

Subject Index

(References are mainly to keywords in titles and abstracts; page numbers refer to title pages)

Index of Co-Authors